ANTISENSE RNA AND DNA

MODERN CELL BIOLOGY

SERIES EDITOR

Joe B. Harford
Cell Biology and Metabolism Branch
National Institute of Child Health and Human Development
National Institutes of Health
Bethesda, Maryland 20892

Volume 11—Antisense RNA and DNA
James A.H. Murray, Editor

ANTISENSE RNA AND DNA

James A. H. Murray, Editor

Institute of Biotechnology
University of Cambridge
Cambridge, England

 **WILEY-LISS**

A JOHN WILEY & SONS, INC., PUBLICATION
New York • Chichester • Brisbane • Toronto • Singapore

Address all Inquiries to the Publisher
Wiley-Liss, Inc., 605 Third Avenue, New York, NY 10158-0012

Library of Congress Cataloging-in-Publication Data

Antisense RNA and DNA / edited by James A.H. Murray.
    p.   c.m. — (Modern cell biology ; v. 11)
    Includes bibliographical references and index.
    ISBN 0-471-56130-4
    1. Antisense nucleic acids. I. Murray, James A.H. II. Series.
    [DNLM: 1. DNA, Antisense. 2. Gene Expression Regulation. 3. RNA,
Antisense. W1 MO124T v. 11 / QU 58 A6331]
QH573.M63 vol. 11
[QP623.5.A58]
574.87 s—dc20
[574.87'328]
DNLM/DLC
for Library of Congress
                                                        92-4911
                                                       CIP

For Jenny and Oswyn

Contents

CASE STUDIES

Control of RNA Viruses

c-*myc* Expression in HL-60 Cells

RIBOZYMES

Contributors

Sudhir Agrawal, Worcester Foundation for Experimental Biology, Shrewsbury MA 01545 [305]

Brenda L. Bass, Department of Biochemistry, University of Utah, Salt Lake City UT 84132 [159]

Eduardo R. Bejarano, Centre for Biotechnology, Imperial College of Science, Technology and Medicine, London SW7 2AZ, United Kingdom [137]

Arturo Bevilacqua, Departments of Human Genetics, Pediatrics, and Communicable Diseases, University of Michigan School of Medicine, Ann Arbor, MI 48109-0618; present address: Department of Psychology, University ''La Sapienza'' of Rome, 00185 Rome, Italy [87]

Lawrence Bogorad, The Biological Laboratories, Harvard University, Cambridge MA 02138 [121]

C. Thomas Caskey, Howard Hughes Medical Institute and Institute for Molecular Genetics, Baylor College of Medicine, Houston TX 77030 [97]

Alan Colman, School of Biochemistry, University of Birmingham, Birmingham B15 2TT, United Kingdom [213]

Nigel Crockett, Institute of Biotechnology, University of Cambridge, Cambridge CB2 1QT, United Kingdom [1]

Anthony G. Day, Centre for Biotechnology, Imperial College of Science, Technology and Medicine, London SW7 2AZ, United Kingdom; present address: Chemical Laboratory, University of Cambridge, Cambridge CB2 1EW, United Kingdom [137]

Geneviève Degols, Laboratoire de Biochimie des Proteines, UA CNRS 1191, Université Montpellier II, Sciences et Techniques de Languedoc, 34095 Montpellier Cedex 5, France [255]

Robert P. Erickson, Departments of Human Genetics, Pediatrics, and Communicable Diseases, University of Michigan School of Medicine, Ann Arbor MI 48109-0618; present address: Departments of Pediatrics and Molecular and Cellular Biology, University of Arizona Health Sciences Center, Tucson, AZ 85724 [87]

Tim Hunt, Department of Biochemistry, University of Cambridge, Cambridge CB2 1QW, United Kingdom; present address: Imperial Cancer Research Fund, Clare Hall Laboratories, South Mimms, Potters Bar, Hertfordshire, EN6 3LD, United Kingdom [195]

Marcelo Jacobs-Lorena, Department of Genetics, Case-Western Reserve University School of Medicine, Cleveland OH 44106-4901 [77]

Gerald F. Joyce, Departments of Chemistry and Molecular Biology, Scripps Research Institute, La Jolla CA 92037 [353]

Motoya Katsuki, Department of DNA Biology, Tokai University School of Medicine, Bohseidai, Isehara, Kanagawa 259-11, Japan [109]

The numbers in brackets are the opening page numbers of the contributors' articles.

Minoru Kimura, Department of DNA
Biology, Tokai University School of Medicine,
Bohseidai, Isehara, Kanagawa 259-11, Japan;
present address: Department of Molecular and
Cellular Biology, Medical Institute of
Bioregulation, Kyushu University, Kyushu,
Japan [109]

Makoto Koizumi, Faculty of Pharmaceutical
Sciences, Hokkaido University, Sapporo 060,
Japan [373]

Bernard Lebleu, Laboratoire de Biochimie
des Proteines, UA CNRS 1191, Université
Montpellier II, Sciences et Techniques de
Languedoc, 34095 Montpellier Cedex 5,
France [255]

Josef M.E. Leiter, Kreutzstrasse 19, 8156
Otterfing, Germany [305]

Jean-Paul Leonetti, Laboratoire de Biochimie
des Proteins, UA CNRS 1191, Université
Montpellier II, Sciences et Techniques de
Languedoc, 34095 Montpellier Cedex 5,
France [255]

Lee Leserman, Centre d'Immunologie,
INSERM-CNRS de Marseille, Luminy, France
[255]

Brian Levy, Departments of Human Genetics,
Pediatrics, and Communicable Diseases,
University of Michigan School of Medicine,
Ann Arbor MI 48109-0618 [87]

Conrad P. Lichtenstein, Centre for
Biotechnology, Imperial College of Science,
Technology and Medicine, London SW7 2AZ,
United Kingdom [137]

Patrick Machy, Centre d'Immunologie,
INSERM-CNRS de Marseille, Luminy, France
[255]

Makoto Matsukura, Department of Child
Development, Kumamoto University Medical
School, 1-1-1 Honjo, Kumamoto City 860,
Japan, and Department of Pediatrics, Ashikita
Institute for Handicapped Children, Ashikita
869-54, Kumamoto Prefecture, Japan
[285]

Paul S. Miller, Department of Biochemistry,
School of Hygiene and Public Health, The
Johns Hopkins University, Baltimore MD
21205 [241]

Jeremy Minshull, Department of Biochemistry,
University of Cambridge, Cambridge CB2
1QW, United Kingdom; present address:
Department of Physiology, University of
California School of Medicine, San Francisco
CA 94143-0444 [195]

M. Indrees Munir, Howard Hughes Medical
Institute, Baylor College of Medicine, Houston
TX 77030 [97]

James A.H. Murray, Institute of
Biotechnology, University of Cambridge,
Cambridge CB2 1QT, United Kingdom [1]

Eiko Ohtsuka, Faculty of Pharmaceutical
Sciences, Hokkaido University, Sapporo 060,
Japan [373]

Rekha Patel, Department of Genetics,
Case-Western Reserve University School of
Medicine, Cleveland, OH 44106-4901; present
address: Department of Molecular Biology,
Cleveland Clinic Foundation, Cleveland OH
44195-5001 [77]

Steven R. Rodermel, Department of Botany,
Iowa State University, Ames IA 50011 [121]

Belinda J.F. Rossiter, Institute for Molecular
Genetics, Baylor College of Medicine,
Houston TX 77030 [97]

Masahiro Sato, Department of DNA Biology,
Tokai University School of Medicine,
Bohseidai, Isehara, Kanagawa 259-11, Japan
[109]

Christopher M. Thomas, School of
Biological Sciences, University of Birmingham,
Birmingham B15 2TT, United Kingdom [51]

David M. Tidd, University of Liverpool
Cancer Research Campaign Department of
Radiation Oncology, Clatterbridge Hospital,
Bebington, Wirral, Merseyside L63 4JY,
United Kingdom; present address: Department
of Biochemistry, University of Liverpool,
Liverpool L69 3BX, United Kingdom [227]

Richard Y. To, Fred Hutchinson Cancer Research Center, Seattle WA 98104 **[267]**

Jean-Jacques Toulmé, Laboratoire de Biophysique Moléculaire, INSERM CJF 90-13, Université de Bordeaux II, F-33076 Bordeaux, France **[175]**

Eric Wickstrom, Department of Chemistry, Department of Biochemistry and Molecular Biology, and Department of Surgery, University of South Florida, Tampa FL 33620 **[317]**

Kazushige Yokoyama, Gene Bank and Frontier Research Program, RIKEN (The Institute of Physical and Chemical Research), 3-1-1 Koyadai, Tsukuba, Ibaraki 305, Japan **[335]**

Preface

Antisense RNA and DNA explores the use of antisense and catalytic nucleic acids for regulating gene expression. Antisense nucleic acids are single-stranded RNAs or DNAs that are complementary to the sequence of their target genes. Their specificity results from base-pairing interactions between the target and the antisense sequences. Catalytic nucleic acids or ribozymes similarly derive specificity from base-pairing interactions but, in addition, contain within the nucleic acid sequence a catalytic activity capable of cleaving the target. Antisense regulation has now been shown to be successful in a large number of systems, and in this book a collection of reviews has been assembled that are representative of the entire field of antisense regulation using RNA (covering both natural and artificial regulatory systems), DNA oligonucleotides (including the many modifications that have been tested), and ribozymes. The emphasis is directed toward the practical, with the discussion of results both positive and negative aiming to distill the experience from a selection of leading laboratories that can assist other researchers in the design of successful experiments. The breadth of coverage is such that, hopefully, it will enable some of the strong underlying principles that link antisense control in different systems to emerge, in addition to providing a review of this exciting field that will be of use both to existing researchers and to those contemplating antisense techniques for the first time.

The use of complementary or antisense nucleic acids to regulate gene activity has developed from nothing into a major area of research activity in the space of a little over ten years. Its attraction is in the promise of being able to control any gene or virus for which sequence information is available. It requires either the stable expression of antisense RNA or ribozymes in transgenic organisms or cells, the microinjection of RNA, or the application of small antisense oligonucleotides to cells and animals. The technology for the delivery of such molecules to target cells within the body is now developing rapidly, opening up real possibilities for their use for therapeutic purposes in humans. Indeed, interest from the biotechnology and pharmaceutical sectors has been one of the major forces pushing forward the development of antisense strategies, in the hope of "magic bullet" type designer drugs against cancers and otherwise intransigent viral infections such as AIDS. This is reflected in a number of chapters in the second part of this book, and in particular in two collections of case studies that offer an insight into the relative merits of different approaches to the same problem. To biotechnologists the interest lies in the possibility of manipulating gene activity in a highly specific manner and in the ability to control virus susceptibility in transgenic organisms. In these cases the use of stably inherited constructs expressing antisense RNA is more appropriate, and these are discussed in the first section of the book. Throughout the volume it should be apparent that in basic research antisense regulation is a powerful tool for investigating gene function, since it is dominant in action and therefore applicable to almost any organism. Gene activity can be reduced or abolished without the need to remove the endogenous alleles.

Despite their potential power, antisense techniques are both simple and elegant in concept, requiring only a stretch of RNA or DNA that is complementary to the target, usually messenger RNA. This complementarity leads to the formation of duplex hybrids within the cell, and the target is rendered inactive. In practice, the application of antisense techniques has been less straightforward than was initially hoped, partly due to the wide range of different strategies that can be used and the many different stages during the life of an mRNA that can be affected. Today, however, the understanding of how and why antisense nucleic acids work is growing, making possible sensible choices in the selection of antisense agents and the much wider application of the approach.

When first conceived, it was the aim of this book to provide as wide a coverage of the whole antisense field as possible in order to emphasize the important similarities and parallels that exist between different systems. This was in the belief that by bringing together results from many separate fields, it might be possible to draw a picture that is more than the sum of its individual parts, and thus to learn lessons of wider applicability. In doing this, the second aim, which is shared with the new Series Editor for the *Modern Cell Biology* series, Joe Harford, was not simply to produce a catalog of past successes in antisense research that could have been assembled from journal articles, but to attempt a practical flavor—not a "recipe" book of which there are many other excellent examples, but rather one with a leaning toward the discussion of experimental approaches, of what is and is not possible, and of what has and has not been useful. If a measure of success has been achieved in this, it will be reflected in whether the book finds its way out of the library and into the laboratory as a source of ideas and approaches.

I am, of course, deeply indebted to all the contributors to this book, without whose hard work and patience during its preparation nothing would have happened. To them goes any credit for its utility and interest, for it is in their laboratories that a good deal of the work has been carried out. In particular I would like to thank Chris Thomas, author of the review on bacterial antisense regulation, and Jean-Jacques Toulmé, who wrote the review of antisense DNA regulation by oligonucleotides, for the additional effort required to produce such wide surveys. I would also like to thank Chris Lowe both for his support in getting established in Cambridge and for the initial idea of tackling the subject of antisense RNA and DNA, Catherine Boulter for invaluable support, advice, assistance, and proof-reading, Nigel Crockett for his great help in co-authoring the first chapter of the book, and indeed all the members of my lab who have tolerated and perhaps enjoyed my absences, without which the editorial work could not have been completed. Finally, I would like to thank Peter Brown and Brian Crawford of Wiley-Liss for their encouragement in the earlier stages, and William Curtis, John Hanley, and other members of the Wiley-Liss editorial and production team who have worked hard on this book.

James A.H. Murray
Cambridge
January 1992

Antisense RNA and DNA: 1—49
© 1992 Wiley-Liss, Inc.

Antisense Techniques: An Overview

James A.H. Murray and Nigel Crockett

I. INTRODUCTION

It is indeed a paradox that it is possible to isolate virtually any gene from any desired organism and to establish its DNA sequence and expression pattern in great detail, yet there exist very few techniques that allow us to approach its in vivo function. The amount of DNA sequence data is accumulating rapidly, making the problem of determining the biochemical and developmental roles of the open reading frames identified ever more acute. Antisense methods are an important tool in this process, since, analogous to the use of inhibitors for studies of proteins, they provide a method of preventing or reducing gene function by targeting the genetic material or its expression. Their application requires only a knowledge of the DNA or RNA sequence of the target gene.

The realization of the genetic code by gene expression requires an information flow from gene to protein. This depends at each stage on specific base pairing between complementary nucleic acids to ensure the accuracy of transmission and interpretation of the information, and herein lies the basis for antisense methods. Our understanding of the molecular details of these processes has developed from the model of the complementary nature of the two strands of the double helix proposed by Watson and Crick to an appreciation that both single-stranded DNA and RNA can form homo- and heterodimers of DNA:DNA, RNA:DNA, and RNA:RNA hybrids and that such complementary interactions are of intrinsic importance to transcription, RNA processing, and translation.

The fundamental importance of specific base pairing to the function of nucleic acids offers the possibility of interfering with the expression of target genes in a highly selective manner by using a complementary or *antisense* sequence. In general terms, the antisense sequence will hybridize to its target and block expression by one of several possible means. It may act by preventing other complementary interactions from occurring that are necessary for expression or occluding a process that requires a single-stranded substrate, by stimulating the action of nucleases specific for the double-stranded regions of DNA:RNA or RNA:RNA hybrids, or in certain cases by inactivating the target directly by intercalation or cleavage. It should therefore be apparent that antisense strategies can in principle be used for any gene for which sequence information is available and in any system into which antisense nucleic acid can be introduced by external application, by microinjection, or by nuclear expression from a construct containing an artificial antisense gene.

Thus the ultimate objective of the antisense approach is to use a sequence complementary to the target gene to block its expression and thereby create a mutant cell line or organism in which the level of a single chosen protein (or, in certain cases, the replication of a virus) is selectively reduced or abolished. The aim may be to produce a phenotype to investigate the function of the gene in developmental or cellular processes, to reduce expression of a gene causing an undesirable phenotype in transgenic plants or animals, or to interfere with oncogenic transformation or viral infection. The antisense approach is clearly particularly attractive for those eukaryotic organisms that do not have well-defined or amenable genetic systems because of long generation time, genome complexity, or difficulty in isolating recessive mutations in diploids. In contrast, antisense techniques enable a "phenocopy" of the gene mutation to be produced using only the cloned sequences. However, as will become clear, even in classic genetic organisms such as *Drosophila* antisense methodology can be a convenient short-cut to answering certain types of question.

Other "reverse genetic" approaches besides antisense are also available for assisting in the assignation of function of a cloned gene. The most clear-cut and effective way to extinguish the function of a gene is to delete or disrupt its coding sequences in a heritable way by homologous recombination. However, this approach is laborious, not least because both alleles of a gene must be removed and it is at present restricted to a few organisms [yeast, Winston et al., 1983; *Dictyostelium*, De-Lozanne and Spudich, 1987; plants, Paszkowski et al., 1988; Lee et al., 1990; mice; Schwartzberg et al., 1989; and human cells, Song et al., 1987]. Ectopic expression, in which a gene is introduced back into cells or an organism under a heterologous promoter to cause alteration in the temporal or spatial pattern of its expression, and overexpression are alternative approaches that share with antisense a possible dominant phenotype, exhibited despite the presence of wild-type copies of the gene. There are, however, many gene products that for a number of reasons can show no phenotype when their level is increased. A way around this problem involves the fact that it is sometimes feasible to overexpress a mutant form of a gene whose product acts as part of a complex so that it competes with the wild-type protein and disrupts the normal activity of the complex to produce a dominant negative phenotype.

In comparison, antisense is broadly available and applicable and has the important advantage that a range of phenotypes can be produced corresponding to various levels of expression. The site of inactivation and its developmental effect can be manipulated by the choice of promoter for antisense genes or by the timing of external application or microinjection. This is particularly an advantage for genes with an essential cellular or developmental function or when it is desirable to examine the role of a gene in a particular subset of cells in an organism. A further feature of the antisense approach is that it is possible to manipulate its specificity by selecting either unique regions of the target gene or regions where it shares homology to other related genes and thereby investigate the role of functions encoded by gene families. It is, however, important to realize that antisense strategies have not been as universally straightforward or as easy to apply as was initially hoped, nor has the interpretation of results always been unambiguous, and this has perhaps led to their premature dismissal in certain instances. Many chapters in this book, and the millions of dollars currently being invested in the biotechnological application of antisense technology [Klausner, 1990], bear witness to the claim that antisense approaches can be successful. Now that the initial euphoria and subsequent disenchantment has given way to a greater pragmatism, it is clear that there is an important future for antisense approaches not only as a research tool but also in biotechnology for manipulating the characteristics of transgenic animals and plants and in pharmacology as therapeutic agents [Uhlmann and Peyman, 1990]. It is the purpose of this book to draw together the practical lessons from a wide range of systems, with the successes and failures, to provide not only an overview of the field but also a source of relevant practical information on which to build future experiments.

In this overview, we follow this discussion of the general principles that lie behind the antisense dogma with an outline of the classes of antisense agents available and how they are applied. We then go on to consider in more detail evidence for the mechanisms of antisense regulation, since one of the least satisfactory and satisfying aspects of antisense is the lack of consistent and generally applicable information on its precise mode of action, which makes a rational design of antisense strategies difficult. This is partly due to the variety of different agents used, partly to a dearth of systematic data relating to questions of mechanism, but is perhaps chiefly a reflection of the large number of variables within the multistep process from gene to protein. Any one of these may or may not be a target, depending on mRNA structure, accessibility, transport, and other factors that can differ from gene to gene, cell type to cell type, and organism to organism.

After attempting to draw some conclusions on mechanism, we switch attention exclusively to antisense RNA and discuss practical considerations of its use and possible reasons for failure. The chapter concludes with a brief survey of the use of antisense RNA in various eukaryotic systems, concentrating on those not covered in pages 77–174.

II. ANTISENSE TOOLS

The materials with which the antisense approach can be applied are broadly divided into antisense RNA, antisense DNA oligomers, and catalytic RNAs or ribozymes. In this section we briefly review the features of each class of reagent, and, since the choice of the type of nucleotide sequence to be used as the antisense reagent is inevitably a compromise, we concentrate on the characteristics that are most important in the consideration of how each approach can be used. Figure 1 summarizes the various methods available for the introduction of antisense reagents into cells. We also point the reader to other chapters in the book where various points are covered in much greater detail. In addition, a considerable number of reviews have appeared that cover aspects of antisense regulation, and these are summarized in Table I.

A. Enabling Technologies

In addition to methods for establishing the DNA sequence of genes and for introducing vectors carrying new genetic information back into cells and organisms, two important developments were required for the application of certain antisense techniques. The first of these was the discovery that purified RNA polymerases encoded by bacteriophages such as Sp6, T7, and T3 can be used in vitro to produce large quantities of specific RNA by cloning the desired DNA sequence downstream of the appropriate promoter [Melton et al., 1984; Davanloo et al., 1984]. Since these RNA polymerases will transcribe on a linear molecule, the template can be cut with a convenient restriction endonuclease to define the 3' end of the RNA to be synthesized, and run-off transcription will then produce RNAs of a defined length. The ability to synthesize RNA efficiently in vitro is important not only for microinjection of antisense RNAs but also for producing radiolabeled strand-specific probes that can be used to detect independently the presence of sense and antisense transcripts in cells by Northern blotting or nuclease mapping techniques.

The second advance, of importance for strategies using oligonucleotides, is the increased efficiency and availability of automated oligonucleotide synthesizers. Together with improvements in synthetic chemistry this now allows a much greater range of analogs to be produced by manipulation of the procedures. Readers are referred to other sources for details of oligonucleotide synthesis [Gait, 1984; Uhlmann and Peyman, 1990].

B. Antisense RNA

In the early 1980s, the realization that natural antisense RNA is involved in the regulation of plasmid copy number in bacteria was followed by the discovery of a considerable number of prokaryotic examples in which antisense RNA is implicated in regulation of plasmid and phage systems and in the expression of certain chromosomal genes. This then led to the development of artificial antisense RNA regulatory systems, both in *Escherichia coli* and in eukaryotes. Antisense RNA in prokaryotes is reviewed in detail by Thomas (this volume).

1. Nuclear expression of RNA by engineered antisense genes. The principle of regulation by antisense RNA is extremely simple in that RNA that is complementary to the target mRNA is introduced into cells, resulting in specific RNA:RNA duplexes being formed

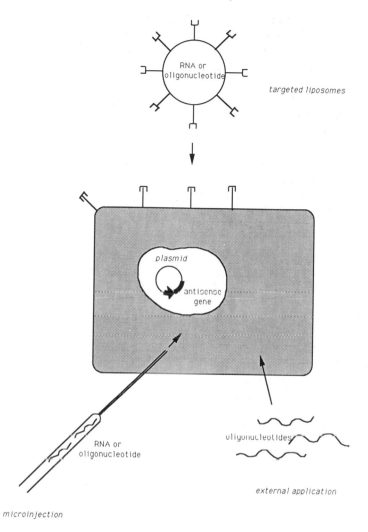

Fig. 1. *Approaches that can be used to introduce antisense nucleic acids into cells are expression from antisense gene constructs introduced in a transient or stable system; microinjection of RNA or oligonucleotides; liposomes carrying RNA or oligonucleotides and targeted with antibodies; and addition of oligonucleotides to cell culture medium.*

by base pairing between the antisense substrate and the target mRNA.

This can be achieved in vivo by the introduction and expression of an antisense gene sequence, that is, one in which part or all of the normal gene sequences are placed under a promoter in inverted orientation so that the "wrong" (complementary) strand is transcribed into a noncoding antisense RNA that then hybridizes with the target mRNA and interferes with its expression. Such an antisense expres-

sion vector can be constructed by standard procedures and introduced into cells by any of the normal means of transferring DNA, such as transfection, electroporation, microinjection, or, in the case of viral vectors, by infection. The type of transformation and choice of vector will determine whether expression is transient or stable, while the promoter used for the antisense gene will influence the level, timing, inducibility, or tissue specificity of the antisense inhibition.

TABLE I. Published Reviews on Antisense RNA and DNA, Catalytic RNA, and Related Areas

Authors	Antisense topic	Notes
Books		
Wickstrom [1991]	Prospects for antisense therapy	Regulation of viruses and oncogenesis
Cohen [1989a]	Antisense oligonucleotides	
Melton [1988]	Antisense RNA and DNA	Papers presented at the 1987 Banbury Meeting, Cold Spring Harbor
Review articles		
General		
Colman [1990]	Antisense RNA and DNA	Review of applications in cell and developmental biology
Hélène and Toulmé [1990]	Antisense RNA and DNA	Useful general review
Holt and Lechner [1990]	Antisense RNA and DNA	Short discussion of practical uses
Weintraub [1990]	Antisense RNA and DNA	*Scientific American* review
Walder [1988]	Antisense RNA and DNA	Short commentary on prospects
Van der Krol et al. [1988b]	Antisense RNA and DNA	Useful general review
Antisense RNA		
Mol et al. [1990]	Antisense RNA in plants	Short review
Takayama and Inouye [1990]	Antisense RNA	Detailed and useful review
Hiatt et al. [1989]	Antisense RNA in plants	Application in tomato of antisense PG
Izant [1989]	Antisense RNA	Applications in cytoskeleton
Knect [1989]	Antisense RNA	Application to cytoskeleton studies in *Dictyostelium*
Inouye [1988]	Antisense RNA in prokaryotes	Discusses design of artificial antisense
Simons [1988]	Natural antisense RNA	Brief review of bacterial systems
Simons and Kleckner [1988]	Antisense RNA in prokaryotes	Naturally occurring systems and mechanisms
Van der Krol et al. [1988a]	Antisense RNA in plants	Brief review
Green et al. [1986]	Antisense RNA	Review of prokaryotic and eukaryotic gene regulation
Weintraub et al. [1985]	Antisense RNA	Short discussion of early results
Antisense DNA		
Miller [1991]	Antisense methylphosphonates	Review
Dolnick [1990]	Antisense oligonucleotides	Short review of pharmacological applications
Sherman [1990]	Antiviral oligonucleotides	Short discussion of relevant considerations
Sigman and Chen [1990]	Chemical nucleases	Chemical cleavage agents and site-specific nucleases
Uhlmann and Peyman [1990]	Antisense oligonucleotides	Detailed and comprehensive review of all aspects
Zon [1990]	Antiviral oligonucleotides	Short discussion of anti-HIV potential
Cohen [1989]	Antisense oligonucleotides	Pharmaceutical possibilities
Stein and Cohen [1989]	Antisense oligonucleotides	Review of potential role in cancer therapy
Loose-Mitchell [1988]	Antisense oligonucleotides	Brief discussion of pharmaceutical possibilities
Marcus-Sekura [1988]	Antisense oligonucleotides	Practical approaches
Stein and Cohen [1988]	Antisense oligonucleotides	General review
Toulmé and Hélène [1988]	Antisense oligonucleotides	Short review

TABLE I. Published Reviews on Antisense RNA and DNA, Catalytic RNA, and Related Areas (Continued)

Authors	Antisense topic	Notes
Ribozymes and catalytic RNA		
Cech [1990]	Self-splicing of group I introns	Full review of structure, reactions, and chemistry; especially *Tetrahymena* intron
Perriman and Gerlach [1990]	Ribozyme technology	Brief survey of recent work on all ribozyme types
Rossi and Sarver [1990]	Ribozymes as antiviral agents	Review of *trans*-acting ribozymes with potential intracellular uses
Symons [1989b]	Self-cleaving pathogenic RNAs	Viruses and virusoids; hammerhead structures
Michel et al. [1989]	Group II catalytic introns	Conserved sequence and structure
Cech [1988]	Group I introns	Conserved sequence and structure
Cech [1987]	Chemistry of RNA enzymes	Mechanistic and enzymatic considerations
Cech and Bass [1986]	Catalytic RNA	General review
Waugh and Pace [1985]	Catalytic RNA	*Tetrahymena* intron and RNase P

Such artificial antisense genes have been demonstrated to regulate the expression of both endogenous and introduced target genes and have therefore found broad application in an impressive range of prokaryotic and eukaryotic systems, as discussed in Section V and elsewhere in this book, showing how biochemical and developmental functions can be targeted to answer questions about their biological role.

2. Microinjection of in vitro transcribed RNA. An alternative approach to antisense genes makes use of purified phage polymerase to produce RNA with subsequent microinjection of the in vitro transcribed antisense transcripts. Clearly some systems such as *Drosophila* (see Patel and Jacobs-Lorena, this volume), the mouse embryo (Levy et al., this volume), and *Xenopus* (see chapters by Bass and by Colman, this volume) are more amenable to this approach than others, because of the difficulty of injecting or accessing certain cells, and it is normally more laborious than using antisense genes. Application in multicellular organisms is normally difficult, being restricted in the case of *Drosophila*, for example, to the period of embryogenesis before cellularization occurs. However, unlike the antisense gene approach, microinjection is not restricted by the availability of suitable promoters or vectors, and it may be the only approach for cells that are transcriptionally inactive. Indeed, a major advantage is that a much larger excess of antisense RNA over target can be injected than can normally be expressed from endogenous antisense genes, although this is compensated for somewhat because of the relatively lower efficiency of injected antisense RNA in inhibiting gene expression. This is the consequence of a more subtle difference between the two approaches, in that RNA is normally injected into the cytoplasm and thus any potential points of antisense regulation within the nucleus are missed. It can therefore be concluded that duplex formation by injected antisense RNA must act by preventing translation or by provoking attack by double-stranded specific RNases. We return to this point in Section III.

The most obvious difference between injected RNA and nuclear antisense expression is that the effect of injection will of necessity be transient, as the RNAs are degraded or diluted during growth. This can, however, be used to advantage in certain situations, since by injecting the antisense RNA, and therefore ablating the gene function, at a specific time, it may be possible to show that the gene has a key role within a particular window of development.

3. Delivery of antisense RNA by liposome fusion. A recent approach that may be useful for both in vitro and in vivo delivery of antisense reagents, including RNA, is the use of liposomes. Encapsulation into liposomes protects

RNA (and DNA) against degradation by nucleases. Liposomes may be targeted to certain cells by coupled antibodies that ensure specific interaction with cells that carry corresponding surface antigens into which they are taken up by receptor-mediated endocytosis. This is discussed further in the chapters by Degols et al. (Section II.D) and by To (Section III.B.3), this volume.

C. Antisense DNA: Inhibition by Oligodeoxynucleotides and Their Derivatives

Antisense oligodeoxynucleotides are short sequences of single-stranded DNA, usually less than 30 nucleotides (nt) in length, synthesized by chemical means in vitro, and are complementary to a specific intracellular target, normally mRNA. More recently, oligonucleotides designed to form triple helices with double-stranded DNA, or selected to interact with proteins, have been developed, and these are mentioned below. The general use of oligonucleotides and their analogs to modulate gene expression is referred to as *antisense DNA* regulation in this book, as distinct from the use of antisense genes for nuclear expression of *antisense RNA*. The whole subject of gene regulation by antisense DNA is reviewed in detail in this volume by Toulmé, and specific examples are given in the chapters following that (see also Table I), so here we will discuss only general principles.

Because antisense oligonucleotides are produced by chemical synthesis, they must be applied either externally to cells or microinjected (see Fig. 1), and as a consequence their effects will normally be transient. However, they offer the important advantage that the full power of synthetic organic chemistry can be used to design oligomers with particular modified linkages or terminal groups, which results in improved properties. It is on such modified oligonucleotides that most hope is pinned for developing antisense therapeutics for human antiviral or anticancer treatment, and such work is the subject of several chapters in this book.

1. Unmodified phosphodiester oligodeoxynucleotides: "n-oligos." It was with short DNA oligonucleotides rather than with RNA

that the first demonstrations of an antisense effect were made, initially using in vitro translation systems (see Toulmé, this volume, for references), and subsequently demonstrating direct inhibition of Rous sarcoma virus (RSV) replication in cultured cell lines by synthetic oligonucleotides added to the culture medium [Zamecnik and Stephenson, 1978]. Subsequent experiments have shown that inhibition of various viruses (see chapters by To and by Agrawal and Leiter) and the expression of genes (see particularly Wickstrom's chapter) can be achieved, clearly indicating that unmodified oligonucleotides (''n-oligos'') are internalized by cells and can interact with intracellular targets. However, it is a common observation that n-oligos are rapidly attacked and degraded by nucleases present in the serum component of culture medium, which can only be partly overcome by heat inactivation (see chapters by Toulmé (Section III.B); Tidd; Matsukura; and Wickstrom) [Holt et al., 1988]. Moreover, after uptake, oligonucleotides are subject to continuing degradation within cells, though at a greatly reduced rate compared with their attack in the culture medium.

It was in fact surprising that n-oligos can have an effect simply by adding them to the cell culture medium, since it was not expected that such highly charge molecules could cross the plasma membrane. It is now generally believed that their uptake occurs by receptor-mediated endocytosis (see discussion by Tidd [Section IV], this volume), but it is clearly not a very efficient process. Wickstrom (this volume) calculates that only 1%–2% of added oligonucleotides were internalized by cells after 4 hours, but gel electrophoresis showed that after uptake oligonucleotides were relatively stable, reducing to 25% of their original level after 24 hours. In contrast, degradation of oligonucleotides in the culture medium was complete after 8 hours. It is primarily in search of solutions to these problems of poor uptake and nucleolytic attack that derivatives of oligonucleotides that would be more stable and lipid soluble were sought.

2. Oligonucleotide analogs with backbone modifications. The dual aims of improving both cell uptake and resistance to nucleases were tackled by synthesizing derivatives with substitutions in place of one of the nonbridging oxygens in the internucleotide bond of the backbone. In particular, methylphosphonates, where $-CH_3$ replaces the $-O$ (discussed by Miller, this volume), and phosphorothioate (where $-S$ replaces $-O$, abbreviated to "S-oligos") have been explored (see Matsukura, this volume), both of which show greater extra- and intracellular longevity because of increased nuclease resistance. These and other modified oligos are shown in Figure 2. The elimination of the negative charge in the internucleotide phosphate bridge in methylphosphonates considerably enhances cellular uptake, probably by improving lipid solubility, and it appears that such oligonucleotides enter cells slowly until the intracellular and extracellular concentrations are approximately equal, presumably by simple diffusion through the membrane. However, in contrast, phosphorothioate oligonucleotides do not penetrate cells efficiently, requiring their use at relatively high concentrations.

Of particular relevance in a consideration of the use of analogs is their mechanism of antisense action. This is discussed in greater detail in Section III, but it is necessary to appreciate here that unmodified n-oligos mediate their effects because they stimulate the cleavage of their target RNA by RNase H, which cuts the RNA component of RNA:DNA duplexes. Phosphorothioate oligonucleotide:RNA duplexes are also substrates for RNase H, whereas methylphosphonate oligonucleotide:RNA duplexes are not, and must therefore act as physical blocks that prevent splicing or translation machinery from accessing the RNA. A promising approach, which is discussed by Tidd (this volume), uses "sandwich" oligonucleotides of a central core of nucleotide units linked by normal phosphodiester bonds, protected from 5' and 3' exonucleolytic attack by terminal nucleotides linked by methylphosphonate bonds.

A second type of backbone modification is exhibited by the α-oligomers, which have an unnatural glycosidic configuration (Fig. 2) that results in nuclease resistance and in an increase in the temperature of dissociation (T_m) of the RNA:α-DNA duplex and therefore an improvement in the stability of the hybrid. Although not a substrate for RNase H, they can inhibit translation when targeted at the cap site of an mRNA [Boiziau et al., 1991].

3. Terminally modified oligonucleotides. The alternative route to improving oligonucleotide efficiency is by attaching terminal groups. This may again be aimed at improving their membrane penetration, for which 3'-conjugated groups such as polylysine and cholesterol have been used (see Degols et al., this volume), or they may be designed to increase oligonucleotide reactivity. This latter type of modification is reviewed by Toulmé (this volume) and includes acridine rings (which increase the affinity toward RNA), alkylating and cross-linking moieties (psoralen), and metal complexes that can cleave the RNA target (Fig. 2). Acridine has the additional advantages of improving cell uptake and of allowing this to be followed by the green fluorescence it produces. An example of its effective use in an oligonucleotide targeted against the common 5' end of mRNAs for variant surface glycoproteins in the unicellular flagellate parasite *Trypanosoma brucei* is given by Verspieren et al. [1987].

It will be clear from this brief discussion that there are many backbone and terminal modifications to oligonucleotides that might improve their performance. Cohen [1989b] points out that 15 classes of compound can be synthesized simply by substituting each of the four groups attached to the phosphorus in the internucleotide bond with $O-$, CH_3, or $S-$ in various combinations. Many of these have not been synthesized, and fewer have been tried with additional terminal groups.

Although there are signs that a more systematic approach toward combining the advantages of various types of analog is emerging, such as the methylphosphonate−phosphodiester chimeras (see Tidd, this volume), it is true that no clear generalizations have emerged on the use of anti-mRNA oligonucleotides to date. This is not to imply that notable successes in gene

Y	B	X	R
— O⁻ ß-nucleotide phosphodiester (naturally occurring)	A	4'-N-(aminoethyl)amino-methyl-4,5',8-trimethylpsoralen (Kulka et al., 1989)	
— O⁻ α–nucleotide phosphodiester (Morvan et al., 1990)	C		—H deoxyribonucleotide
— S⁻ nucleoside phosphorothioate (chapters 13 and 17)	T	ethylenediamine tetraacetic acid (Boidot-Forget et al.,1988)	
— CH₃ nucleoside methylphosphonate (chapter 14)	G	5'-p(N-2-chloroethyl-N-methylamino) benzylamide (Vlassov et al., 1988)	—OCH₃ 2'-O-methylribonucleotide (Shibahara et al., 1989)
— O — CH₂CH₃ alkyl phosphotriester (Marcus-Sekura et al., 1987)	U		
— N—CH₂CH₂CH₂CH₃ H nucleoside butylarnidate (Jäger et al., 1988)		N-6-thiophenyl-2-methoxy acridine (Stein et al., 1988)	

regulation have not been observed, since this is clearly the case in many chapters in this book and elsewhere, but that shortcomings in the experimental systems often do not allow the modes of action, specificity, and toxicity to be thoroughly studied. It is in the rabbit reticulocyte cell-free system and in *Xenopus* (see Minshull and Hunt, this volume; Colman, this volume) that most systematic analysis is probably possible, and the reader is particularly pointed to the discussions by Minshull and Hunt (this volume) on specificity and interpretation of results.

It is clear that the in vivo use of oligonucleotides as potential therapeutic agents will require further improvements in cell uptake to reduce nonspecific toxicity effects and the cost per dose [Geiser, 1990]. The possibility of using liposomes as a delivery system, as mentioned above for RNA [see Degols et al., this volume; Pidgeon et al., 1990], offers great potential in vivo advantages of protection from nucleases as well as targeting to specific cells by antibodies. However, a further problem here is the short half-life of conventional liposomes in serum caused by uptake into the reticuloendothelial system (liver and spleen), although this could perhaps be alleviated by newer generations of liposomes such as those described by Gabizon and Papahadjopoulos [1988].

Despite the manifest problems, there is increasing interest in the pharmacological potential of oligonucleotides [Zon, 1990] (see references in Table I), accompanied by considerable commercial investment [Klausner, 1990].

Fig. 2. *Generic structure of antisense DNA oligomer* (**top**) *and an indication of the range of analogs that have been tested for antisense regulatory activity (**bottom**). B, base (adenine, cytosine, thymine, guanine, or uracil) through which hydrogen bonding to the target occurs; Y, different substituents on the phosphorous atom in the nucleotide phosphodiester linkage in place of the naturally occurring oxygen; X, different 5' terminal groups that can in principle be used with any internucleoside linkage; R, normally hydrogen, but the use of a 2'-O-methylribonucleoside has also been reported. A full review has been published by Uhlmann and Peyman [1990].*

Because costs are falling [Geiser, 1990], antisense treatments may soon be tried against life-threatening viral infections such as AIDS.

4. Triple helix-forming oligonucleotides. There has been recent excitement [Riordan and Martin, 1991; Charles, 1991] about the possibility of interfering directly with gene expression or viral replication by targeting DNA with oligonucleotides that bind in a sequence-specific manner in the major groove of a double-stranded target to form a triple helix or triplex structure (Fig. 3a). Such "anti-gene" oligonucleotides make use of the fact that recognition sites still remain in the major and minor grooves of the Watson-Crick double-stranded DNA structure. In particular, it is possible for thymine to form hydrogen bonds with adenine, while the adenine is simultaneously involved in Watson-Crick bonding with thymine (Fig. 3b). Similarly, protonated cytosine can bind to the guanine of a G.C base pair. These patterns of hydrogen bonding are referred to as *Hoogsteen base pairing*. As a consequence, a pyrimidine (cytosine and thymine) oligonucleotide can form a triple helix structure within the major groove of a purine (adenine and guanine) stretch of a double-stranded DNA molecule (Fig. 3a), with specificity provided by Hoogsteen base pairing. Such structures can form on supercoiled and relaxed DNA targets [Maher et al., 1989]. However, the binding is pH dependent, the triple helical structures with C or T on the Hoogsteen strand being stable in acid conditions but, because of the requirement for the cytosine to be protonated to bind to a G.C base pair, dissociating with increasing pH. Povsic and Dervan [1989] have shown that triple helix formation can be stabilized and extended to physiological pH by substituting the analog 5-bromouracil and/or 5-methylcytosine for thymine and cytosine, respectively. An example of the type of triple helix-forming oligonucleotide that can be used is shown in Figure 3c.

The potential advantage of the triple helix over the classic antisense approach for regulating genes lies in the enormously reduced number of molecules per cell that must be inactivated to inhibit expression if the DNA of the

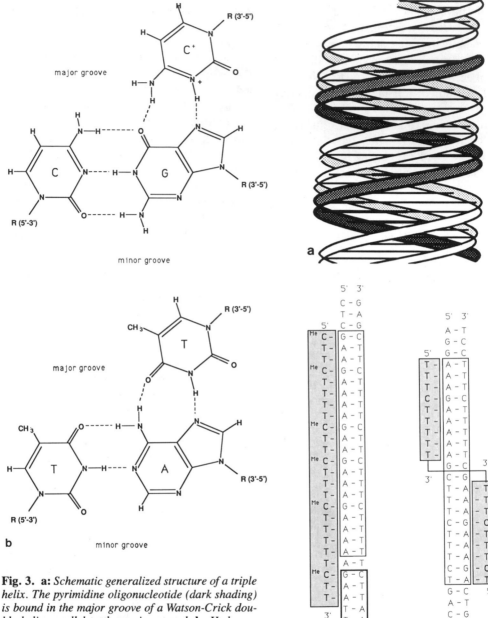

Fig. 3. a: *Schematic generalized structure of a triple helix. The pyrimidine oligonucleotide (dark shading) is bound in the major groove of a Watson-Crick double helix parallel to the purine strand.* **b:** *Hydrogen bonding interactions involved in Watson-Crick base pairing (across) and in Hoogsteen triple helix formation (upwards). Hoogsteen hydrogen bonding of thymine to Watson-Crick A.T and protonated cytosine to G.C base pairs is shown.* **c:** *Example of triple helix-forming oligonucleotide. This sequence was used by Strobel and Dervan [1991] to block EcoRI methylase at a specific site in the yeast genome. Subsequent dissociation of the oligonucleotide and digestion with EcoRI resulted in cleavage only at the protected site, all other sites in the genome having been methylated. In this case oligonucleotides with C and T, ^{Me}C and*

T, or ^{Me}C and ^{Br}U were tested and were effective up to pH 7.4, 7.6, and 7.8, respectively. **d:** *''Switchback'' or alternate-strand triple helix–forming oligonucleotide capable of binding parallel to both purine strands of a DNA duplex in the opposite orientation, as described by Horne and Dervan [1990].*

gene is targeted rather than the mRNA, since many copies of an mRNA are produced from a single equivalent of DNA. The steady-state level varies in lymphoid cells, for example, from about 10 per cell for rare mRNAs, to 200 copies for nonabundant, and up to 30,000 copies for each of the members of the "abundant class" [Schröder et al., 1989], and constant synthesis occurs. In contrast, most genes and proviruses are present at low number, often only one to two per cell, and DNA replication occurs only before cell division, which is far less frequent than the rate at which genes are transcribed into new mRNA molecules. In particular, the possibility of triple helix-forming oligonucleotides linked to cross-linking, alkylating, or cleavage reagents [Povsic and Dervan, 1989; Strobel and Dervan, 1990; Perrouault et al., 1990] offers the prospect of long-term inactivation of target genes or proviruses at low therapeutic doses, since such agents may act to stabilize the triple helical binding and/or inactivate the DNA.

Although it has been shown in vitro that triple helix-forming agents can block enzymatic access to DNA by restriction enzymes or methylases [François et al., 1989; Maher et al., 1989; Hanvey et al., 1990; Strobel and Dervan, 1991] and by a eukaryotic transcription factor [Maher et al., 1989], there is only one example to date in which it has been speculated that regulation of gene expression under physiologic conditions may be attributable to triple helix formation [Cooney et al., 1988]. In this case, regulation of c-*myc* expression by a mixed-sequence oligonucleotide was observed, and a triplex structure was proposed but not proven, although supported by indirect evidence including the effectiveness of the oligonucleotide in the 100 nM range. The oligonucleotide sequence was purine rich, so Hoogsteen base pairing could not have been involved, and the molecular basis for the recognition has not yet been established. It does, however, raise the possibility that an alternative pattern of recognition of double-stranded DNA by purine oligonucleotides could exist, in addition to the established pyrimidine oligonucleotide Hoogsteen motif.

At present, however, a major limitation of the triple helix approach is the available code, since it is restricted to the recognition of homopurine tracts by pyrimidine oligonucleotides. The runs of 15–18 purines that are required for a target site occur only rarely within genes that would be a potentially interesting target for regulation by triple helix formation, and, for the yeast experiments described above, the target was introduced by prior transformation. A partial way around this problem is the use of alternate-strand multimeric crossover oligonucleotides that produce so-called switchback triple helix formation, which consists essentially of two short (8- to 9-mer) pyrimidine oligonucleotides joined at their 3' ends by a 3'–3' linkage. Such oligonucleotides are therefore capable of simultaneous binding to a purine segment on one strand and a second purine segment running in the opposite direction on the other strand. This was demonstrated by Horne and Dervan [1990] with the oligonucleotide illustrated in Figure 3d. This alternate-strand purine binding greatly increases the potential number of sequences that can be read to include those of the type $5'$-(purine)$_m$-(pyrimidine)$_n$-$3'$ and $5'$-(pyrimidine)$_m$-(purine)$_n$-$3'$, where m and n need be no greater than 8–9 bp.

However, it would clearly be desirable to extend the triple helical recognition code to a general solution in which all four base pairs of Watson-Crick double strands could be recognized at 37°C and physiological pH. Griffin and Dervan [1989] have shown that the code can formally be extended to the recognition of T–A base pairs by G within the environment of an otherwise pyrimidine oligonucleotide but that limitations on sequence composition are likely. The existence of other third-strand recognition interactions leading to G.G–C and A.A–T triplets has been explored by Letai et al. [1988], by binding to homopolymer agarose affinity columns, but the utility of these results to the recognition of mixed sequences has not been shown. Possibly more fruitful approaches may be the design of non-natural analogs to complete the triplex code or the incorporation of abasic residues that would allow certain Watson-Crick bases to remain unread.

Further possibilities that remain to be explored are whether triple helix formation can also act to *activate* transcription by altering the local parameters of DNA conformation appropriately and the possibility of using analogs to improve the affinity or efficacy of binding. Hélène and coworkers [reviewed by Hélène and Toulmé, 1990] have shown that α-anomers of oligonucleotides can also form triple helical structures with normal double-stranded DNA, in which the phosphodiester backbone may have the opposite orientation to that of a β-oligomer. Despite current interest, much work is required to demonstrate the action of triple helix formation in cells in culture and in vivo and to define the parameters required for its action, including the response to DNA repair and replication, before it can be seriously considered as a therapeutic tool.

5. Oligonucleotides selected against protein targets. A recent approach [Riordan and Martin, 1991] has been the use of in vitro selection techniques to generate compounds with high affinity for biological target molecules in a manner analogous to that described by Joyce (this volume) for optimization of ribozyme function. In general terms it involves the use of the polymerase chain reaction (PCR) or other enzymatic techniques to amplify those oligonucleotides from a random pool that demonstrate affinity for the target molecule. The starting material is a pool of up to 10^{13} different oligonucleotide species, each with a different sequence. These are incubated with the target molecule or passed through an affinity column of the target, and bound oligonucleotides are separated from the unbound. The bound oligonucleotides with greater affinity for the target are then eluted and amplified and then subjected to further rounds of binding, elution, and amplification until the selection procedure results in an enriched pool that contains only those oligonucleotides with the highest binding affinity. In this way oligonucleotides are selected that have, by chance, the correct three-dimensional structure to bind to the target molecule under the selection conditions. It is then possible to test the ability of the selected oligonucleotide

to inhibit the biological activity of the target. Such oligonucleotides, dubbed "aptamers," have been generated to small organic molecules [Ellington and Szostak, 1990] as well as to large proteins, such as transcription factors [Blackwell and Weintraub, 1990; Blackwell et al., 1990] and T4 DNA polymerase [Tuerk and Gold, 1990].

D. Ribozymes

An extension of the antisense approach is to confer catalytic activity on the antisense RNA molecule such that cleavage of the target mRNA substrate occurs, resulting in its inactivation without affecting the antisense molecule. This is then free to dissociate and attack other substrate molecules, thus acting catalytically. In contrast, "conventional" antisense RNAs are required in stoichiometric amounts, since they must either remain hybridized to block mRNA function or target the duplex for degradation, in which case they are destroyed along with their substrate. The potential attraction of catalytic RNA systems for gene regulation or antiviral strategies is therefore that a lower concentration of antisense RNA molecules may be necessary for effective regulation.

RNA enzymes, or *ribozymes*, a term first used by Kim and Cech [1987], have been described in a number of naturally occurring systems and are responsible for cleavage and ligation of specific phosphodiester bonds within RNA molecules. Here we briefly review the main classes, concentrating on the progress that has been made on engineering them into catalytic antisense RNAs, i.e., enzymatic cleavage agents whose specificity is provided by antisense RNA sequences linked to the catalytic centre of the ribozyme. Further reviews are listed in Table I.

1. Group I introns: The *Tetrahymena* ribozyme. The first ribozyme activity to be discovered was that responsible for the autocatalytic removal of the intron from the precursor of the 26S ribosomal RNA of the ciliate *Tetrahymena thermophila*. In this and other introns (termed *group I*) excision of the intron and ligation of the exons to form the mature RNA occur as a result of specific folding of intron sequences to produce the enzymatically active unit [for review, see Cech, 1988]. The reaction requires either Mg^{2+} or Mn^{2+} as a divalent cation and either free guanosine or an internal G residue. The first step of self-splicing is the specific binding of free guanosine, which following cleavage of the 5' intron–exon junction becomes covalently attached to the 5' end of the intron. The newly exposed end of the 5' exon is then joined to the 3' exon, with concomitant excision of the 414 nucleotide intron (see Joyce, Fig. 1, this volume). In the case of the *Tetrahymena* ribozyme, specificity of the initial cleavage reaction is determined by binding of the substrate sequence of six conserved residues, CUCUCU, at the 3' end of the upstream exon to a complementary "internal guide sequence" situated near the 5' end of the enzyme. The site of cleavage is immediately downstream of the CUCUCU sequence.

Subsequent work on adapting the *Tetrahymena* ribozyme has been stimulated by the demonstration that the reaction can also proceed when the target and catalytic sequences are separated on two different RNA molecules, resulting in ribozyme-mediated cleavage of a single-stranded RNA substrate. The minimal recognition site requirement for the wild-type ribozyme has been shown to be CUCU [Woodson and Cech, 1989], whereas Doudna et al. [1989] and Murphy and Cech [1989] have shown that by altering the guide sequence and conditions almost any substrate can be recognized. This has led to the first commercially available ribozyme product launched by U.S. Biochemicals (Cleveland, OH), which cleaves RNA targets after the sequence CUCU. However, the use of engineered *Tetrahymena* or other group I–derived ribozymes to bring about specific cleavage reactions on novel target sites in vivo has not been demonstrated, and their principle use may be for in vitro RNA studies.

Recently an elegant approach using rounds of selection and amplification has been used to select ribozymes with increased catalytic activity against DNA substrates [Robertson and Joyce, 1990; North, 1990], and Joyce (this volume) describes the use of this system for optimization of ribozyme function, using group 1 catalytic RNA.

2. "Hammerhead" ribozymes. The "hammerhead" ribozyme motif has been identified in a number of plant viroid and satellite RNAs, consists of three stems and a catalytic center of 13 conserved nucleotides, and is responsible for self-catalyzed cleavage of multimers of RNA during the viroid replication cycle [see Symons, 1989b, for review]. In naturally occurring hammerhead ribozymes the reaction mediated is intramolecular, since the target and catalytic strands are part of the same molecule and cleavage occurs after the triplet GUC (GUA in the lucerne transient streak virus). However, Haseloff and Gerlach [1988] have demonstrated that with ribozymes derived from the encapsulated satellite RNA of tobacco ringspot virus (+ strand sTRSV) the reaction can also proceed efficiently *in trans* and could be adapted to cleave at defined sites in an mRNA sequence after GUC triplets by attaching the catalytic region to the appropriate antisense guide sequences. Such catalytic RNAs can be as short as 19 nt, as demonstrated by Uhlenbeck [1987]. As discussed in detail by Koizumi and Ohtsuka (this volume), it is now possible to generalize the cleavage reaction such that any RNA can be cleaved after the sequence NUX (N = A,G,C,U; X = A,C,U), with specificity again determined by the surrounding guide sequence, as shown in Figure 4. In this way Koizumi and coworkers have been able to design hammerhead ribozymes that can distinguish between RNAs that differ only in a single nucleotide [Koizumi and Ohtsuka, this volume; Koizumi et al., 1989].

Ribozymes based on the hammerhead model have generated considerable interest both for the commercial potential they offer for controlling viruses and gene expression [e.g., O'Neill, 1989] and as possible antiviral therapeutic agents, particularly against HIV [see Rossi and Sarver, 1990; Rossi et al., 1990]. They have also been used by a number of investigators in attempts to bring about specific RNA cleavage in cell extracts and in vivo. In general, however, the results have been somewhat disappoint-

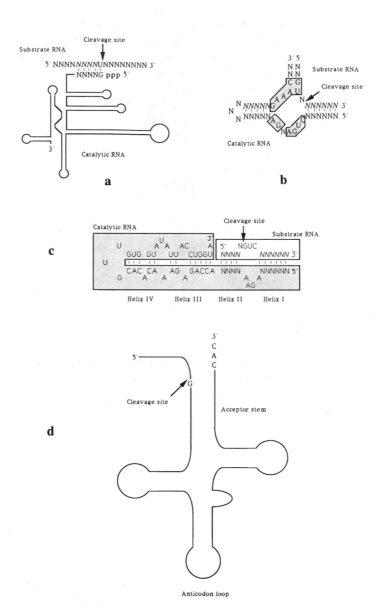

Fig. 4. a: Tetrahymena *ribozyme as adapted to carry out* trans-*cleavage reactions [Zaug et al., 1986] and generalized by Murphy and Cech [1989]. The secondary structure is shown only schematically. N and N are any complementary nucleotides.* **b:** *Generalized "hammerhead"* trans-*acting ribozyme as described by Uhlenbeck [1987], Mei et al. [1989], and Koizumi and Ohtsuka (this volume).* **c:** *"Hairpin"* trans-*acting ribozyme as described by Hampel et al. [1990]. There are no specific sequence requirements in helices I and II.* **d:** *Schematic diagram of* E. coli tRNA^{Phe} *precursor molecule showing cleavage site by RNase P and the known sequence requirements [McClain et al., 1987; Forster and Altman, 1990].*

ing, suggesting that as yet ribozymes cannot be regarded as generally useful laboratory reagents for inhibiting gene expression. The general structure of ribozymes used in these experiments is as illustrated in Figure 4, with a catalytic domain linked on each side to "tails" that have the antisense sequence to the target and therefore act as guide sequences to provide the specificity. These are normally in the range of 8 to 12 nt long and are indicated here by the notation $a–b$ guide sequence for a 5' antisense guide sequence of a nucleotides and a 3' guide sequence of b nucleotides.

Cameron and Jennings [1988] tested the same three ribozymes (all with 8–8 guide sequences) against the mRNA for chloramphenicol acetyltransferase (CAT), which were used by Haseloff and Gerlach [1988] in vitro, for their ability to block CAT expression in monkey (COS1) cells. The ribozyme genes were cloned under the SV40 early promoter into a replicating plasmid vector, as part of the 3' noncoding region of a firefly luciferase carrier gene, because the short RNA carrying only the ribozyme sequence was found to be unstable. At a high molar excess of approximately 1,000-fold, one of the ribozymes repressed CAT expression by about 60%, whereas a noncatalytic antisense stretch of the same length had no effect. However, no products of the predicted ribozyme cleavage could be detected.

The efficiency of antisense RNA, DNA, and ribozymes against the U7 SnRNA to inhibit in vitro splicing of histone pre-mRNA processing was tested by Cotten et al. [1989]. It was found that antisense RNA to the U7 sequence was effective at a sixfold excess, whereas full-length (65-mer) or 18-mer single-stranded DNA was required at 60- and 600-fold excess, respectively. The ribozyme (with 12–11 guide sequences) did not preserve the efficiency of the antisense RNA, being required at a 1,000-fold excess for complete inhibition of processing, perhaps because its antisense guide sequences were short compared with those of the antisense RNA.

Cotten and Birnstiel [1989] also tested the anti-U7 ribozyme in vivo, but in this case incorporated the ribozyme sequence into a transcrip-

tion unit, placed between the A and B box internal promoter sequences of a *Xenopus* tRNA[Met] gene, which was therefore transcribed by RNA polymerase III when injected into *Xenopus* oocytes. Previous studies have also shown the utility of Pol III–directed transcripts for the expression of antisense RNA [Jennings and Molloy, 1987]. However, although high rates of ribozyme-tRNA could be expressed, resulting in up to 10^{10} ribozyme-tRNA molecules per injected oocyte, it was not as stable as native tRNA. Most of the ribozyme-tRNA also remained in an unprocessed pre-tRNA form and did not enter the cytoplasm efficiently where the target U7 SnRNA was located, although sufficient ribozyme-tRNA did appear to enter the cytoplasm to destroy the U7 SnRNA. However, calculations showed that once again a large excess of the ribozyme over target (500-to 1,000-fold) within the cytoplasmic compartment, consistent with the results discussed above. Cleavage products were again not detected, raising the possibility that the observed reduction in U7 SnRNA was due to antisense effects rather than to ribozyme-mediated destruction.

Saxena and Ackerman [1990] were, however, able to demonstrate correct cleavage with a hammerhead ribozyme (with 8–8 guide sequences) both of a model template corresponding to the α-sarcin domain of 28S ribosomal RNA and of the endogenous 28S RNA in *Xenopus* oocytes. The cleavage of the model substrate occurred equally well in the nucleus and cytoplasm. Although the ribozyme blocked protein synthesis efficiently, it is unclear whether this was due to cleavage of 28S or simply to hybridization of the ribozyme to its target sequence without cleavage, since a control oligonucleotide was equally effective.

Chuat and Galibert [1989] have investigated the activity of hammerhead ribozymes expressed in *Escherichia coli* against *lac*Z mRNA. In the strain used, part of the mature β-galactosidase protein (the β-peptide) is expressed from a defective *lac*Z gene on an F' episome and part from the α-peptide gene encoded on the M13 mp8 vector used. An oligonucleotide with the sequence coding for the ribozyme (with 9–8 guide sequences) was inserted into the polylinker region of the *lacZ* α-peptide gene in the M13 vector mp8 in such a way that the reading frame was not disrupted. In one orientation transcription gave rise to a ribozyme designed to cleave a sequence located approximately 270 nt downstream on the same transcript. The activity of this ribozyme would therefore have resulted in an untranslatable α-peptide mRNA. In the other orientation there would be no active ribozyme produced. The effectiveness was tested by examining the ability of cells infected with this construct or its control to form the blue plaques on indicator plates that demonstrate mature β-galactosidase activity. This intramolecular reaction, with the ribozyme and target carried on the same transcript molecule, was found to be effective and white plaques resulted, indicating destruction of α-peptide mRNA, although no cleavage products were demonstrated. The ability of a ribozyme to be active *in trans* was also tested by targeting a part of the β-peptide mRNA with a ribozyme in the same position. Although this did not result in white plaques, it could not be ruled out that some cleavage occurred, which, however, left sufficient intact mRNA for color development.

A plasmid designed for in vivo release of ribozymes or other RNAs has been described by Taira et al. [1990], in which transcript carries two ribozymes. The first is targeted against another gene, but, to remove extraneous downstream sequences that could be present from inefficient or variable transcription termination and that could prevent correct folding, a second ribozyme is encoded downstream on the same RNA that is designed to cleave immediately 3′ to the first ribozyme. In this way a precise 3′ end to the first ribozyme (or, in principle, any RNA) could be produced in vivo, although only in vitro results are presented. A similar approach has also been reported by Dzianott and Bujarski [1989].

Sarver et al. [1990a,b] and Rossi et al. [1990] produced transgenic HeLa cells that expressed a ribozyme against the *gag* gene of human immunodeficiency virus HIV-1 and that also expressed the CD4 receptor, allowing their infection with HIV-1. During acute infection a reduction in p24 antigen of 20–40-fold was observed, although it was not possible to demonstrate conclusively that this was caused by ribozyme-mediated destruction of RNA in vivo. However, nuclear extracts from cells expressing the ribozyme exhibited correct cleavage of in vitro transcribed substrate. It is unclear from the results presented how large was the excess of ribozyme over target.

3. "Hairpin" Ribozymes. A second catalytic motif that has thus far only been found in the 359 nt minus strand of satellite RNA of tobacco ringspot virus ([−]s TRSV) is the so-called hairpin ribozyme (see Fig. 4) [Haseloff and Gerlach, 1989]. The model of (−)s TRSV function suggests that the minimum sequence consists of a 50 nt catalytic domain and a 14 nt substrate domain organized into four helices. Helices III and IV form part of the catalytic domain, whereas I and II form between the target and catalytic center. However, intermolecular interactions can be substituted in helices I and II to produce a *trans*-acting ribozyme in which the catalyst and target are separate molecules [Feldstein et al., 1989, 1990]. Hampel et al. [1990] showed that new target specificities can be created by altering sequences of the helices I and II components of the ribozyme strand.

4. RNase P. The processing of the larger precursors of pre-tRNA into mature transfer RNA involves a series of enzymatic reactions, including the cleavage by RNase P to generate the 5′ terminus of the tRNA. In *E. coli*, and probably in other organisms as well, RNase P consists of a catalytic RNA (M1 RNA) and a protein subunit (C5 protein) [Pace et al., 1987]. Although both protein and RNA components are required for in vivo activity of RNase P, in vitro the M1 RNA alone is capable of cleavage of pre-tRNA to yield authentic 5′ termini. Since both the specificity and the K_m of this reaction appear to be the same with or without the protein component, it can be concluded that the RNA alone is responsible for the enzymatic reaction [Guerrier-Takada et al., 1983]. Cleavage is believed to occur by a mechanism similar to that of the group I introns, since products with 5' phosphate and 3' hydroxyl groups are produced, as by the *Tetrahymena* ribozyme.

RNase P is an interesting candidate for engineering as a catalytic RNA because it is currently the only known naturally occurring example of a *trans*-acting ribozyme and because its activity appears to be enhanced, and perhaps modulated, in vivo by a protein component. McClain et al. [1987] have shown that substrate specificity can be altered, and, although enzyme–substrate interaction appears to involve "external guide sequences" (EGS) at the 3' end of the acceptor stem of the pre-tRNA and a distal conserved CCA triplet, recent work suggests that with appropriate changes to the EGS it may be possible to target a wide range of RNA molecules [Forster and Altman, 1990].

5. Conclusions. From the above discussion it will be clear that, although ribozymes can be efficient tools for in vitro cleavage of RNA, considerable further work is required before they can be regarded as suitable for routine use. Particular problems appear to be caused by the instability of ribozymes, because of the susceptibility of RNA to nuclease attack, and by the low catalytic rate and consequent requirement for a large excess of ribozyme over target sequences. On the positive side, however, there does not appear to be evidence for a deleterious effect on the growth rate of cells expressing ribozyme-containing sequences [Sarver et al., 1990a; Cameron and Jennings, 1989].

III. MECHANISMS OF ANTISENSE INHIBITION

We have already referred to the general lack of evidence for the molecular basis of the mechanism of action of antisense agents. In this section we consider the available evidence that points toward the probable mechanisms operating in vivo with antisense RNA and DNA. Ribozyme action is covered in the previous section.

Figure 5 illustrates the numerous points at which antisense inhibition may act in the stages from the production of the mRNA to its translation into functionally active protein. These may, however, be simplified into two basic alternative explanations that can account for antisense effects, and we examine the apparent relative importance of these in particular examples. They are 1) the extent to which antisense: sense hybrid formation directly blocks mRNA function, as against 2) the importance of the destruction or inactivation of the target mRNA induced by interaction with the complementary antisense sequence.

A. Antisense RNA

The literature on attempts, both successful and unsuccessful, to regulate genes by antisense RNA is now quite large, but only in a minority of cases is much information pertinent to the molecular mechanisms of the inhibition included. It is generally assumed that, in the presence of antisense RNA, mRNA:antisense hybrids are produced, with the result in successful cases that a substantial reduction in detectable levels of the target protein is observed. A common, though not universal, result is that antisense transcripts cause a reduction in steady-state sense mRNA levels, perhaps because of increased turnover, or specific duplex attack by double-stranded RNases similar to *E. coli* RNase III [Robertson, 1990], or perhaps related to the interferon-induced double-stranded ribonuclease reported by Meegan and Marcus [1989]. This is supported by the failure in many cases to detect sense:antisense hybrids in cells. An excess, often a large excess, of antisense over sense transcripts is normally necessary, and there is generally a correlation between the amount of antisense RNA and the degree of inhibition.

Beyond these observations the data that exist on the presence of duplex RNA, the localization of the duplex, the compartment of inhibition (nucleus or cytoplasm), or the compartment in which degradation occurs are often contradictory between different systems. These are all questions that must be addressed for a full understanding of antisense RNA mechanism.

Perhaps the most detailed work is that which has been carried out by Cornelissen [1989] and Cornelissen and Vandewiele [1989] in transgenic

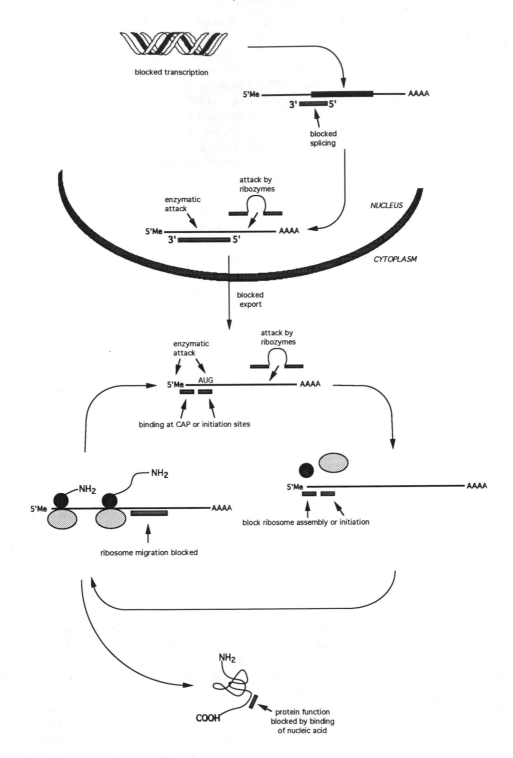

tobacco plants transformed with a target gene conferring resistance to the herbicide bialophos (*bar* gene) and then transformed again with an antisense construct. These results are discussed further by Rodermel and Bogorad (Section V.A, this volume) and thus are only briefly mentioned here as a means of illustration. In this careful study both protein and RNA levels were determined, and it was established that antisense RNA expression results in regulation at two levels—reduced *bar* (sense) transcript level per cell and reduced translation of *bar* mRNA—so that less protein is produced per surviving transcript. This is not due to a block in the transport of the target RNA to the cytoplasm, since the presence of antisense RNA did not affect the relative distribution of *bar* mRNA between the nucleus and cytoplasm, although the absolute amount was reduced. This part of the inhibition was therefore due to a direct effect on translation. However, little or no stable RNA duplex could be seen, indicating that interference must result from unstable interactions.

It could also be concluded that these controls on transcript level and translation are independent and occur in different cellular compartments. Evidence for this conclusion is that the cytoplasmic half-life of *bar* mRNA was unaffected by the presence of antisense RNA, so that there is no evidence for degradation of the *bar* mRNA transcript in the cytoplasm. The reduction in *bar* mRNA must therefore occur in the nucleus and would be consistent with nuclease degradation of the duplex structure, but does not formally exclude a direct effect

on transcription or processing of the nascent transcript that led to a blockage of further transcript accumulation. It also suggests that antisense as well as sense transcripts are attacked, since they are equivalent components of a duplex structure. This is supported by the observation by Cornelissen [1989] that transformation to introduce further antisense genes resulted in a fourfold decrease in sense *bar* mRNA, but only a slight increase in steady-state anti-*bar* RNA. In general, therefore, the reduction of sense mRNA produced is not a function of the steady-state level of antisense RNA, but may rather be controlled by the number of antisense transcripts synthesized. Moreover, plants transformed with only the antisense vector were always found to have higher levels of antisense RNA than if sense mRNA was also present, showing that antisense (hence noncoding) RNA is not unstable per se (Cornelissen, personal communication). We can therefore conclude that both sense and antisense RNA are destabilized by duplex formation, as would be expected if rather rapid degradation of duplex structures is occurring.

1. The Cornelissen model. It seems clear that antisense regulation involves hybridization between RNA molecules in vivo and is therefore influenced by many factors, including the secondary structure of the particular RNAs and their availability for hybridization due to association with proteins or nucleoprotein complexes. Therefore the ability of RNAs to interact and cause antisense inhibition will differ in each case. It is also plausible that double-stranded RNases may be more active in certain cell types or that particular hybrids may resist attack. As a consequence, it is likely that the actual mechanism and site of antisense RNA inhibition will show differences between systems or even between different genes or constructs, since there will be a different balance of mechanisms that contribute to the overall inhibitory effect. However, we would like to propose that the model developed by Cornelissen [1989] may be more widely applicable to antisense RNA control, and in the next section, after explaining the model further, we examine the extent to which

---

Fig. 5. *Potential points of regulation by antisense nucleic acids during gene expression from DNA to protein. The mechanism of action of antisense is either to block processes that require physical access to the sense sequence or to promote cleavage of the sense strand by the action of the antisense itself (e.g., ribozymes) or by stimulating intracellular degrading enzymes. The sense target is represented by a straight line and the antisense by a hatched line. At each step regulation can in principle occur by either antisense RNA or oligonucleotides, except for the blocking of transcription by triple helix formation and the blocking of protein function, which have only been described for oligonucleotides.*

it holds true for other published data. The main tenets are these:

1. The primary site of interaction and regulation is in the nucleus. This could be due to antisense transcript interfering with synthesis and/or processing of the target or blocking its export from the nucleus, but it is more likely that the antisense transcript triggers nuclease degradation by base paring to free or nascent transcripts. Sense and antisense transcripts are not distinguished and are degraded equally. In addition to a possible role for RNase III–like double-stranded RNases, it is possible that the double-stranded RNA unwinding/modifying activity discussed by Bass (this volume) could be responsible for some or all of the antisense effect. This activity is widespread if not ubiquitous in animal phyla (it has not been tested in plants), is normally located exclusively in the nucleus, and results in the modification of adenosine to inosine residues in RNA duplexes. The consequence is a gradual unwinding of the duplex because I–U base pairs are less stable than A–U base pairs. However, this also results in the effective degradation of both RNA strands, since the sequence is rendered not only undetectable by hybridization techniques but also untranslatable. It remains to be seen what contribution the unwinding/modifying activity makes to antisense regulation, but it is possibly of considerable importance, perhaps explaining, for example, by its absence in yeast, the difficulties in observing antisense regulation.

2. Transcripts appear in the cytoplasm if they have escaped the nuclear inhibitory interaction. There is little or no degradation of RNA:RNA duplexes in the cytoplasm, but if both sense and antisense transcripts accumulate the efficiency of translation of the sense mRNA is reduced. The reduction in translation does not require the formation of stable duplexes, but it is likely that inhibition results from transient sense:antisense hybrids that have to be unwound by the translation machinery, thus reducing the rate of translation initiation, elongation, or termination. This proposal is consistent with observations that translating ribosomes are capable of unwinding duplex regions (such as hairpin structures), since the model requires only that the rate of progress be slowed, as has been suggested by Wolin and Walter [1988]. It is possible that the inhibition of translation is more effective when the antisense RNA covers the AUG initiation codon and adjacent region, in analogy to the inhibitory effects of oligonucleotides that do not promote RNase H cleavage (see below), the effects of cDNA fragments on translation in vitro [Liebhaber et al., 1984; Shakin and Liebhaber, 1986; Lawson et al., 1986], and the inability of ribosomes to recognize upstream AUGs that are buried in strong secondary structure [van Duijn et al., 1988]. However, the question of which region of a gene is best targeted in eukaryotes has not been rigorously examined (see also discussion below).

2. Causes of nuclear inhibition

a. Duplex degradation. The majority of published examples of antisense regulation using nuclear expression of antisense RNA are not inconsistent with the model that it is in the nucleus that most of the inhibitory effect is observed. It explains several observations, including the inefficiency with which plant viruses that replicate in the cytoplasm are regulated by antisense RNA (Bejarano et al., this volume), and the relatively large amounts of RNA that must be injected into the cytoplasm to obtain regulation in *Xenopus* or *Drosophila* (Patel and Jacobs-Lorena). It is a very common observation that a specific decrease in the detectable level of the target RNA is induced by the presence of antisense RNA [Crowley et al., 1985; Khokha et al., 1989]. The reduction can vary from severalfold [Rothstein et al., 1987; Smith et al., 1988; Sheehy et al., 1988; Rodermel et al., 1988; Kasid et al., 1989] to the extent that the target mRNA is no longer detectable [Knecht and Loomis, 1987; Giebelhaus et al., 1988; Levi and Ozato, 1988], and in some cases it has been shown to be proportional to the amount of antisense RNA expression [McGarry and Lindquist, 1986; Chang and Stolzfus, 1985], although such a straightforward relationship is not universally true [Sheehy et al., 1988]. It is likely also that the abundance of antisense RNA is similarly reduced over the level it would reach if the sense RNA were not present. Few experiments are available to verify this directly, because it requires that the target is an exogenous gene introduced by transformation. It has been noted by several workers that antisense RNA can have a significant effect even if little or none is detectable by techniques that measure steady-state RNA levels [Crowley et al., 1985; Knecht and Loomis, 1987; Nishikura and Murray, 1987; Hamilton et al., 1990; Smith et al., 1990b; Bird et al., 1991]. This may be because of inherent instability, but equally possible that duplex formation with its target leads to rapid destruction of the antisense RNA. The results of Robert et al. [1990] using antisense RNA against a heat-inducible β-glucuronidase gene in transgenic tobacco are also suggestive that increased sense RNA levels cause a reduction in antisense RNA levels.

A further example of when it is possible to examine the response of antisense RNA to sense RNA is the work of Sheehy et al. [1988] discussed by Rodermel and Bogorad (Section V.A, this volume) where the target polygalacturonase gene (PG) is expressed only in ripe tomato fruit, and in this case a small reduction in antisense RNA level does occur at the time of sense mRNA induction during fruit development. This is also a case where the steady-state level of antisense RNA is much lower than the normal level of the sense mRNA after induction. Nevertheless, inhibition is effective, causing a 70–90% reduction in PG enzyme levels. The reason is only revealed by nuclear run-off experiments, which show that the *rate* of synthesis of the antisense RNA is much higher than the rate at which sense mRNA is transcribed, which therefore relies on its much greater stability to accumulate. It is probable that effective antisense regulation is observed, because when sense mRNA is switched on during fruit ripening, it encounters an excess level of antisense RNA and is therefore unable to accumulate. Crowley et al. [1985] made a similar observation in *Dictyostelium* (see Section V.D), since they were

unable to detect antisense RNA despite observing effective antisense ablation of discoidin gene expression. A concomitant reduction in sense RNA was also observed. Run-on transcription experiments in isolated nuclei demonstrated that both genes are transcribed, ruling out an effect of antisense on RNA transcription.

From these examples, one may conclude that it is likely that degradation of duplex RNA by destruction or modification is often an important mechanism in antisense RNA regulation. Furthermore, it is clear that hybridization to cytoplasmic RNA (which is the RNA normally extracted and used for Northern blots) is a poor indicator of the relative rate of transcription of sense and antisense genes, or even of the steady-state ratio of sense:antisense RNA in the nucleus, where most antisense inhibition occurs, and either of these is probably a better indicator for antisense inhibition. Cytoplasmic levels probably only indicate RNA that has escaped either degradation or entrapment in the nucleus.

b. Duplexes and blocking of export. In only two examples have a stable sense:antisense duplex been detected when an antisense gene is being expressed, and in both cases it was confirmed to be in the nuclear fraction [Kim and Wold, 1985; Yokoyama and Imamoto, 1987] (see Section IV, below). In the case of Kim and Wold [1985], antisense thymidine kinase (TK) RNA was expressed as part of a chimeric transcript consisting of the coding region of dihydrofolate reductase (DHFR) fused to the antisense TK RNA. Since the level of antisense RNA is the same as that of the coding DHFR transcript in this system, it was possible to produce elevated levels of antisense RNA expression by selecting for increasing resistance to the drug methotrexate. The target TK gene was previously introduced by transfection. Kim and Wold [1985] found that a high level of antisense TK transcript was required to produce inhibition. A steady-state level of $\approx 5{-}10 \times 10^3$ antisense TK molecules per cell (a 300-fold excess over TK mRNA transcripts) produced a reduction of 90% in TK enzyme levels. About 40% of the antisense RNA was found in the nucleus and 60% in the cytoplasm. Sense:antisense duplexes were detected in the nucleus by virtue of their resistance to the single-stranded RNases A and T1. It was calculated that about 50 molecules of duplex were present per cell, corresponding to more than 50% of the sense TK mRNA but only 1% of the antisense transcripts. Unusually, however, in this example there is no evidence of increased degradation of RNA in the duplex, since the level of sense mRNA is identical in control cells and in the cells with a high level of antisense RNA. Inhibition resulted from an altered distribution of sense TK transcripts in the cell, since in control cells most are in the cytoplasm, whereas in the antisense-expressing cells more than 95% is confined to the nucleus. It seems likely therefore that in the absence of duplex degradation in these particular cells, a very high level of antisense RNA is required to drive the equilibrium of hybridization toward duplex formation, as is also needed with the injection of RNA into the cytoplasm, as discussed below. It should also be noted that in the absence of the DHFR selection for very high antisense RNA levels, it is probable that inhibition would not have been observed, and TK inhibition would have joined the list of examples of antisense experiments in which no regulation could be achieved [e.g., Salmons et al., 1986; Gunning et al., 1987; Kerr et al., 1988].

The reason for the lack of observed duplex degradation in this case is unknown. It is possible that this particular cell line does not contain the requisite enzyme activities, although in other reported experiments using mouse cells [NIH3T3, Nishikura and Murray, 1987; Edwards et al., 1988] there is a reduction in target mRNA caused by the antisense. Possibly it is due to a particular feature of the TK mRNA itself or of the duplex.

From this result we can conclude that duplexes though normally, are not necessarily, degraded, but that if formed they are not exported from the nucleus. It would thus seem that two mechanisms for nuclear antisense inhibition exist, the first in which there is degradation of duplexes (the majority of reported cases) and a second in which stable duplexes

form that are then unable to be exported from the nucleus. This only seems to occur when a very high ratio of antisense:sense RNA is produced, as in this case by selection for increased drug resistance.

Despite this extended discussion of the experimental results of Kim and Wold [1985], it should be repeated that duplexes, if detected at all, are normally at a very low level (e.g., low levels of cytoplasmic hybrids [Salmons et al., 1986]), and cases in which no degradation of the target mRNA is reported are rare. One such example is that of γ-actin [Gunning et al., 1987], in which antisense RNA did not alter target levels; neither did it cause any reduction in protein expressed. It is possible that there was a failure of the antisense to access the target or that an insufficiently high excess of antisense RNA was achieved to drive duplex formation and block nuclear export in this case. Two further cases of successful antisense inhibition that do not appear to involve nuclear sites of regulation are discussed in Section III.A.3 below [Ch'ng et al., 1989; Sullenger et al., 1990].

c. Transcription. We have already discussed evidence from nuclear run-on experiments that discounts interference with transcription as a mechanism that can account for the observed reduction of sense target in the presence of antisense RNA. Other results, including nuclear run-off transcription [Sheehy et al., 1988], also demonstrate no effect on transcription of the sense gene in the presence of antisense RNA. However, the experiments reported by Yokoyama and Imamoto [1987] show clear evidence that expression of antisense against the endogenous c-*myc* gene in the human promyelotic leukemia cell line HL-60 is accompanied by a reduction in the rate of transcription of the c-*myc* gene and monocytic differentiation of the cells. Unfortunately the interpretation is complicated by the fact that monocyte differentiation is itself associated with downregulation of c-*myc* expression (see Wickstrom, this volume). It is therefore possible that the reduction in c-*myc* transcription is not directly caused by the presence of antisense RNA, but is a consequence of such cells already having started to differentiate. In other words, antisense RNA may act at another level (e.g., duplex degradation, translation block) to reduce c-*myc* levels sufficiently for differentiation to be triggered, which then as a secondary effect leads to a reduction in c-*myc* transcription. The results presented by Yokoyama and Imamoto [1987] indicate that antisense also clearly acts at levels other than transcription (see also Yokoyama, Table III, this volume), which could provoke the initial differentiation event.

The more recent results presented by Yokoyama (this volume) would also support this conclusion, since a 74 kD factor is described that is induced by the antisense c-*myc* RNA and reduces endogenous c-*myc* transcription. In elegant experiments in which the 74 kD protein is introduced by liposome fusion, it was also capable of reducing c-*myc* expression and triggering differentiation into macrophages. It remains to be seen if this factor is induced by the duplex of antisense RNA per se, or in response to falling c-*myc* levels, and whether it is related to the occurrence of a natural antisense c-*myc* transcript (Table II).

d. Splicing. Although Munroe [1988] has demonstrated that antisense RNA can inhibit splicing in vitro (see also Section V.B. below) there is little evidence that interference with RNA processing contributes to antisense RNA regulation in vivo. In one of the few examples, Stout and Caskey [1990] demonstrated that effective inhibition of hypoxanthine guanine phosphoribosyltransferase (HPRT) can occur when intron-specific antisense RNA complementary to sequences adjacent to splice donor or acceptor sites of the first intron of the mouse HPRT gene are used. However, intron sequences further from the first exon were not effective, suggesting that inhibition of splicing may occur in this case.

However, when Izant and Sardelli [1988a,b] microinjected antisense sequences designed to inhibit splicing of a β-globin transcript into *Xenopus* oocyte nuclei, there was no effect on splicing caused by the antisense RNA, whereas antisense oligonucleotides caused rapid RNase H–mediated cleavage of the target (see Section III.B, below).

Other evidence pointing to the possible role of the inhibition of splicing is more indirect: Chang and Stolzfus [1985] established a quail cell line containing a Rous sarcoma virus (RSV) mutant deleted in the *env* gene that could be rescued by transfection of a plasmid carrying *env* so that infectious virus particles were produced. Cotransfection of an antisense *env*-expressing plasmid blocked this virus production, irrespective of the presence of an intron in the sense *env* gene, indicating that the antisense effect did not appear to involve inhibition of splicing. Similarly Ch'ng et al. [1989] were not able to demonstrate an effect on splicing.

3. Cytoplasmic inhibition. The model of antisense action that we have developed here predicts that inhibition that occurs in the cytoplasm is likely to be relatively inefficient compared with antisense regulation in the nucleus, since it is due not to destruction of duplex RNA but to a reduction in the translation rate caused by a need to unwind the duplex. Therefore the presence of antisense should not lead to an increased rate of degradation of either sense or antisense RNA.

In agreement with the first of these premises, there are relatively few examples of antisense RNA regulation where it is clear that the site of inhibition is cytoplasmic, though the microinjection of antisense RNAs into *Xenopus* oocytes, *Drosophila*, mammalian cells, or embryos is clearly a case in point. Data on the fate of the injected RNA are only available for *Xenopus*, to our knowledge (see Section V.F), and Melton [1985], Harland and Weintraub [1985], and Wormington [1986] report that duplexes are not rapidly degraded in *Xenopus* oocyte cytoplasm, since both sense and antisense transcripts were reported to have half-lives of 8–12 hours [Harland and Weintraub, 1985] or >48 hours [Melton, 1985; Wormington, 1986], provided that the in vitro transcribed RNAs were capped during transcription with diguanosyl 5' triphosphate (GpppG). Uncapped RNAs are known to be rapidly degraded, with $t_{1/2} < 1$ hour [Harland and Weintraub, 1985]. Moreover, the level of sense RNA is unaltered in both the control oocytes and those injected with antisense RNA, indicating that inhibition of translation is the most likely mechanism. As expected, therefore, a rather large excess of antisense RNA must be injected. In contrast, when plasmids are injected into fertilized *Xenopus* eggs so that antisense RNA is expressed in the nucleus after the midblastula transition, effective antisense inhibition is observed, with a dramatic reduction in the level of the target mRNA [Giebelhaus et al., 1988]. This illustrates clearly the difference in the mechanisms involved in antisense inhibition in the nuclear and cytoplasmic compartments.

An interesting exception to this is to be found in the results of Strickland et al. [1988], who investigated the effect of antisense RNA on the activation of dormant maternal mRNA transcripts for tissue plasminogen activator (t-PA) in maturing mouse oocytes. Primary oocytes contain untranslated t-PA mRNA transcripts, which during meiotic maturation to secondary oocytes are polyadenylated, translated, and degraded. The effect of antisense RNAs against 5', central, and 3' regions of the t-PA mRNA was tested by microinjection of a 400-fold excess of in vitro transcribed RNA into the cytoplasm of primary oocytes, which were then allowed to mature. Inhibition of t-PA accumulation was observed for all three antisense RNAs, but was particularly effective for the antisense targeted at the 3' end of the gene (the 3' antisense), which only required a four-fold excess to give regulation. The 5' and middle antisense RNAs did not affect the untranslated t-PA mRNAs in primary oocytes, but caused destruction of the target as it became accessible on maturation. In contrast, the 3' antisense resulted in the cleavage of the t-PA mRNA in the primary oocytes, resulting in a stable 5' and an unstable 3' fragment. Further experiments showed that injection of RNA complementary to the final 103 nt of the 3' untranslated region caused cleavage, and thereby prevented the normal polyadenylation, translation, and destabilization of the mRNA. From these results it would appear that this 3' region is accessible in primary oocytes and that it is

involved in activation of this dormant maternal mRNA.

Interestingly, destruction of the duplex RNA was observed with the 5' and middle antisense RNAs. This may be due to the presence of a nuclease in the cytoplasm of maturing oocytes, or alternatively, mouse oocytes may parallel *Xenopus* in the localization of the double-stranded RNA unwinding/modifying activity, since it is known that the activity is released from the nucleus into the cytoplasm during *Xenopus* meiotic maturation (see Bass, this volume).

The use of liposomes to target oligonucleotides or RNAs to cells has already been mentioned, and this is another case where delivery is presumably into the cytoplasm, although as with microinjection partitioning of RNAs into the nucleus cannot be excluded. Renneisen et al. [1990] discuss the use of antibody-targeted liposomes carrying antisense RNA against HIV *env* gene to inhibit viral replication and were able to reduce *tat* expression by 90% and abolish gp160 expression. As might be expected, translation inhibition appeared to be responsible, since there was no dramatic change in mRNA levels.

There is a small number of examples of endogenously expressed antisense RNA that appears to exert its effect in the cytoplasm. Ch'ng et al. [1989] also report results that a small region at the 3' end of a transcript can be an extremely efficient target. They describe inhibition of creatine kinase B (CK-B) in human U937 cells, using a retrovirus vector to produce stable cell lines expressing antisense RNA as part of the retroviral transcript, and show that targeting the 3' end of the CK-B mRNA results in the blocking of expression with only one to two antisense molecules per sense transcript. However, the relative transcription rates are not reported, so it is possible that the antisense transcript is synthesized and degraded at a higher rate.

In these experiments, the antisense RNA had no effect on the abundance of the target, its processing or export, or on the polysome profile. The antisense effect required the last 17 codons and the 3' noncoding regions, without which no inhibition was observed. It is not clear whether a particular feature of the CK-B mRNA causes ribosomes to be stopped, or whether a more general effect of antisense inhibition is involved, in which ribosomes pausing at the translation stop codon [Wolin and Walter, 1988] are unable to unwind the untranslated region beyond. Such accentuated pausing could then result in a block of translation or in premature translation termination. However, it should be noted that contrary results have been obtained in *Xenopus* and in other systems [e.g., Sumikawa and Miledi, 1988], so the general effectiveness of the 3' region is still unclear.

Sullenger et al. [1990] showed that an antisense transcript contained in a chimeric tRNA gene and inserted into the 3' long terminal repeat (LTR) of a retroviral vector resulted in extremely high levels of antisense RNA accumulation, equivalent to 15%–25% of the polyadenylated RNA present in the cell. Antisense RNA to the *gag* gene was able to inhibit replication of Moloney murine leukemia virus by up to 97% in these cells, despite the very high level of infecting viral transcripts that are normally present (up to 1% of polyadenylated RNA). These authors suggest that inhibition is at the level of mRNA translation, since the abundance of viral RNA is not affected by the presence of antisense message.

It is unclear why no nuclear-mediated effects are seen in these last two examples quoted. We can only offer the observation that a full-length retroviral transcript is involved as either target [Sullenger et al., 1990] or antisense RNA [Ch'ng et al., 1989].

4. Homologous cosuppression. A phenomenon has recently been described that may have relevance to the interpretation and design of some antisense RNA experiments and to the general use of transgenes in plants [reviewed by Jorgensen, 1990]. The observation is that introduction of extra copies of a gene in a sense orientation can lead to suppression of expression of the introduced transgene and of both alleles of an endogenous gene. This effect, which has been termed *cosuppression*, has been observed with endogenous genes both when a

further copy is introduced by transformation and when a transformed plant carrying a heterologous transgene is retransformed with the same gene. Napoli et al. [1990] and van der Krol et al. [1990] attempted to overexpress the chalcone synthase (CHS) gene in petunia petals, but found that, rather than the expected increase in pigmentation, a partial or total block in CHS was observed in up to 50% of the transformed plants. In such white flowered plants, developmental induction of CHS occurred, but the level was reduced 50-fold. Moreover, the phenotype was variable both in somatic growth (within the same plant "revertant" wild-type sectors with normal CHS levels arose) and in germinal transmission (although the effect cosegregated with the transgene, the stability and extent of the phenotype was variable).

Cosuppression does not require the complete gene sequence to be introduced, since Smith et al. [1990a] found that downregulation of PG expression occurred when extra copies of a truncated (sense) PG gene were introduced. Elkind et al. [1990] also showed that suppression of the phenylalanine ammonia lyase (PAL) gene could be suppressed in tobacco by an introduced gene from a heterologous species (bean). Moreover, several related genes in the PAL family were all affected. Kawchuck et al. [1991] have also reported sense and antisense inhibition of potato leafroll luteovirus replication (see Bejarano et al., Section III.c, this volume).

Various mechanisms have been proposed to account for cosuppression [Jorgensen, 1990], including methylation, the formation of RNA: DNA triple helix structures that could block transcription, and other as-yet undescribed molecular phenomena, while parallels have also been drawn with *trans*-interaction in fungi. However, Grierson et al. [1991] and Mol et al. [1991] have explored the possibility that cosuppression actually involves antisense RNA, through promoters that transcribe the "wrong" strand of the introduced sense DNA. Such promoters could be present either within or near the T-DNA borders used to transfer all these genes into plants (see description in Bejarano et al., this volume). Frankham [1988] has proposed a similar hypothesis to account for some of the position-effect variation observed in many transgenic organisms. For the cosuppression examples discussed, possible explanations would be read-through from the transcription of the resistance gene used to select for these constructs in the opposite orientation to the transgene. Although antisense RNA has not been detected in RNase protection experiments [van der Krol et al., 1990], as we have already seen this is also the case with many antisense experiments because of rapid duplex degradation. Further work, including nuclear run-on, is required to establish if fortuitous antisense RNA is produced and to ascertain the molecular mechanism for cosuppression. In any case it may require rethinking of the structure of vectors for transgenic plants and possibly other organisms, and care will be needed in the interpretation of some experiments, since, for example, in a bona fide antisense construct adventitious sense transcription of the same gene would reduce the antisense effect.

B. Antisense DNA

The mechanism of antisense DNA inhibition is explored further in the second section of this book. In vitro experiments, of the type discussed by Minshull and Hunt (this volume) have demonstrated that the hybrid arrest of translation is due to cleavage of the RNA component in RNA:DNA hybrids by RNase H. It is believed that a similar mechanism operates in vivo, since RNase H is widespread in eukaryotic cells [Wagner and Nishikura, 1988], reflecting its role in DNA replication. There are, however, few examples where the involvement of RNase H has been demonstrated unambiguously in vivo, although Izant and Sardelli [1988b] show cleavage of RNA in Xenopus at the site targeted by the oligonucleotide. Tschudi and Ullu [1990] found that treatment of permeabilized trypanosome cells with E. coli RNase H and oligonucleotides to snRNAs results in cleavage and loss of trans-spicing. Further indirect evidence comes from the use of analogs, since while phosphothioate oligonucleotides also direct RNase H cleavage in vitro, methylphos-

phonates and α-oligomers do not and mediate their effects by steric interference and are therefore generally required at higher concentrations.

For oligonucleotides that do not mediate RNase H cleavage, there is evidence for both the inhibition of splicing and the initiation of translation (see Miller, Section III.A, this volume). Kulka et al. [1989] show that inhibiting splicing of herpes simplex virus with a methylphosphonate oligomer resulted in a 98% reduction in virus titer at a concentration of 100 μM. Boiziau et al. [1991] show that α-oligomers complementary to either the cap region or the region containing the AUG translation initiation codon can block translation of β-globin in rabbit reticulocyte or wheat germ lysate systems, with the anti-cap oligomer 5–10 times more efficient than the anti-AUG. Oligomers to the coding region had no effect, consistent with the idea that translating ribosomes can unwind regions of duplex structure. When the oligomers were tested by injection into Xenopus oocytes, only the anti-cap oligomer was effective. In these studies a methylphosphonate 17-mer targeted to the cap showed no effect at concentrations of up to 50 μM.

The efficiency of RNase H–independent inhibition can be improved by introducing terminal groups (see Fig. 2), which promote intercalation, cross-linking, and nonenzymatic cleavage of the RNA. Such modifications are discussed by Toulmé (this volume), and by To (Section III.A.2, this volume).

IV. GENERAL CONSIDERATIONS IN THE USE OF ANTISENSE RNA

The questions of "which target?" and "which antisense?" for antisense DNA oligonucleotides form the central themes of Toulmé's review (this volume), so we will consider only points relevant to the use of antisense RNA, drawing on one of the most comprehensive sets of data on artificial regulation of a gene by antisense RNA. This work by Daugherty et al. [1989], and a preceding study by Pestka et al. [1984], examined the effect of antisense RNAs corresponding to different regions of the *lac*Z gene on the induction response of the chromosomal *lac*Z gene of *E. coli*.

Despite the important differences between prokaryotes and eukaryotes in the mechanisms of transcription and translation, which are coupled processes in bacteria, we believe that these results point to some general principles relevant for the design of antisense experiments in eukaryotes. The *lac*Z gene (encoding the enzyme β-galactosidase) is repressed in normal growth, but is induced by lactose or gratuitous inducers such as isopropyl β-D-thiogalactoside (IPTG). Antisense RNAs, expressed from the thermoinducible λ P_L promoter, were able to inhibit this induction and therefore block *lac*Z expression. A number of different constructs were tested by Daugherty et al. [1989] whose results are summarized in Figure 6.

A. Extent of Inhibition Achievable

As appears to be the case in eukaryotes, the extent of inhibition observed in *E. coli* was proportional to the amount of antisense RNA produced, with a large excess of antisense over target mRNA required for maximum inhibition. No detectable β-galactosidase was produced when 180-fold more antisense than sense *lac*Z mRNA was achieved, and with a 10–20-fold excess inhibition was about 60%. As discussed in Section III.A.2 above, duplex degradation is an important mechanism in antisense regulation in eukaryotes, and in prokaryotes the involvement of the double-stranded specific RNase III has been demonstrated in the natural regulation of the λ CII gene by OOP RNA (see Thomas, Section V, this volume) and suggested to explain the destabilization of *lpp* mRNA by artificial antisense [Coleman et al., 1984], but no evidence is produced in this study for specific degradation of the target RNA. However, it seems likely that in prokaryotes much of the antisense effect with engineered constructs operates by inhibiting translation, either by blocking ribosome binding or by slowing ribosomal progress when regions of RNA:RNA duplex are reached. The use of antisense RNA in bacteria may therefore be comparable to microinjection of antisense RNA into the

eukaryotic cytoplasm, since in this case the major site of the inhibition in the nucleus is by-passes and a large excess of antisense is similarly required.

In general one may conclude that a large excess of antisense RNA is likely to improve the chance of a successful outcome. Daugherty et al. [1989] investigated approaches for maximizing the abundance of antisense RNA, both increasing its synthesis by inserting a transcription terminator downstream of the antisense gene and improving stability by including an AUG initiation codon at the 5' end of the antisense RNA (see Fig. 6). It is clear that in eukaryotes it is also desirable to consider measures that will contribute to the abundance of the antisense transcripts and their stability, since, although there are several reports of unstable antisense RNAs giving rise to antisense inhibition (see Section III.A.2, above) presumably by effective interactions in the nucleus, additional translational interference can occur when antisense RNAs survive to reach the cytoplasm [Cornelissen, 1989]. Various techniques have been used to increase expression levels of antisense transcripts, such as the use of strong promoters [Gunning et al., 1987] and linking the antisense transcript to a selectable marker [Kim and Wold, 1985; Edwards et al., 1988; Delauney et al., 1988] (see also Bejarano et al., this volume), secondary rounds of transformation with the antisense vector (Yokoyama, this volume], the use of polymerase III promoters [Jennings and Molloy, 1987] and in particular the use of genes in which the antisense transcript forms part of a chimeric tRNA gene [Sullenger et al., 1990; for a similar approach with ribozymes, see Cotten and Birnstiel, 1989]. It has also been proposed that stem–loop structures, such as those that occur in the natural *mic*F antisense RNA (Fig. 7), may assist in protecting antisense RNA from degradation [Coleman et al., 1984; Inouye, 1988]. Other features known to enhance mRNA stability in eukaryotes could also be considered [Atwater et al., 1990].

It is also important to give consideration to the type of gene to be targeted. Rarely is it possible to ablate expression completely, and in

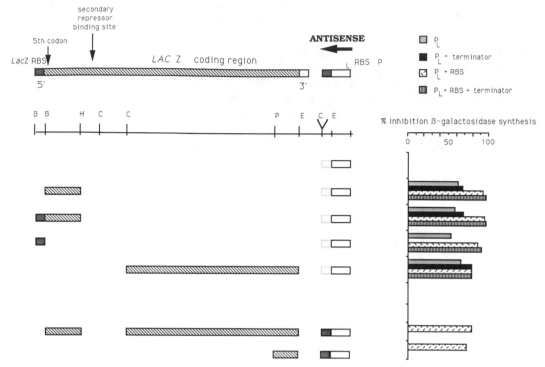

Fig. 6. *Illustration of the plasmid constructions utilized in the production of antisense lacZ RNA by Daugherty et al. [1989]. The letters B, C, E, H, and P represent the restriction sites BamHI, ClaI, EcoRI, HinfI, and PvuII, respectively. The antisense constructs contained various regions of the lacZ gene, which is shown at the top 5'–3'. The fragments indicated in the lower part of the figure by filled bars were cloned behind the λ P_L promoter to produce the antisense RNA. The histogram to the right details the percentage inhibition of β-galactosidase synthesis afforded by each of the constructs. The first four constructs were tested with the λ P_L promoter alone and also with a functional P_L ribosome binding site (RBS; which increases RNA halflife) or a terminator (which increases its abundance), and different shadings in the histogram illustrate their relative effectiveness. The lower two constructs were only tested with an RBS, and without a terminator. None of the constructs includes the secondary repressor binding site, which otherwise titrates the lac repressor.*

certain cases enzyme levels can be reduced to as little as 1% of wild type and still produce a normal phenotype [Gelbart et al., 1976]. Thus antisense RNA regulation is most useful when the phenotype produced is likely to be proportional to gene expression so that a partial reduction in levels of the target protein is likely to be informative, particularly when it is desired to correlate expression level with the degree of phenotypic severity. In cases when reduction in the target gene may be lethal, an inducible promoter can be used [Holt et al., 1986; McGarry and Lindquist, 1986] (see also Munir et al., this volume), although even with a con-

stitutive promoter position effects are likely to produce a range of transformants expressing different levels of antisense RNA.

B. Where to Target?

In the study of Daugherty et al. [1989], greatest inhibition was observed when the ribosome binding site (RBS) of the target message was included in the antisense RNA, although inhibition was still apparent when only the carboxyterminus of the gene was targeted. Very efficient inhibition has been reported in bacteria with constructs lacking the target mRNA RBS, e.g., covering codons 5–119 of the mRNA [Daugherty et al.,

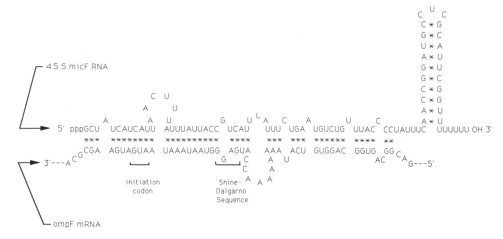

Fig. 7. *Proposed secondary structure for the naturally occurring antisense micF RNA and its interaction with ompF mRNA [Andersen et al., 1987]. micF hybridizes to 79% of nucleotides in the overlap region with its target.*

1989] or codons 4–45 [Coleman et al., 1984], but this may reflect the fact that the ribosome covers about 40 nt of the mRNA, so duplex formation within this region may inhibit binding even if it does not include the RBS.

In eukaryotes, knowing which part of the mRNA to target is somewhat more difficult, probably because of the greater number of stages at which regulation can occur. We suggest that it is likely that, when inhibition occurs within the nucleus, there is no consistent difference whichever part of the RNA is targeted, and success will depend on its relative availability as determined by secondary structure in different regions and on steric blocking caused by bound proteins. Kim and Wold [1985], who found that inhibition was caused by a failure to export duplex RNA from the nucleus, saw no difference between a construct that contained sequences homologous only to the 3′ or to both the 5′ and 3′ regions of the target. However, for antisense RNA that escapes the nucleus and causes inhibition of translation, it is probable that the 5′ end of the mRNA is more likely to be an effective target, since inhibition of initiation seems more effective than inhibition of elongation. This view is partially supported by experiments involving injection of antisense

RNA into the cytoplasm of *Xenopus* oocytes, in which RNAs that did not hybridize to the 5′ end of the target message were either ineffective [Melton, 1985] or less efficient [Harland and Weintraub, 1985; Sumikawa and Miledi, 1988] than others that covered the initiation region, although Wormington [1986] found a 3′ transcript to be equally as effective as a full-length sequence. Whether these observations can be explained by differences in secondary structure or whether computer programs that model mRNA folding [e.g., Konings et al., 1987] can be used to predict good regions for targeting with antisense remains untested.

When nuclear expression of antisense RNA has been used, success has been reported with complete or 5′-targeted antisense RNAs [Holt et al., 1986; Izant and Weintraub, 1985; Trevor et al., 1987; Nishikura and Murray 1987; McGarry and Lindguist, 1986; Stout and Caskey, 1987; Amini et al., 1986], and with 3′-targeted antisense [Chang and Stolzfus, 1987; Kim and Wold, 1985; Giebelhaus et al., 1988; Delauney et al., 1988; Ch'ng et al., 1989]. However, although the 3′ may be equally effective [Chang and Stolzfus, 1987], there are few clear examples in which it is *more* effective than the 5′ end [Ch'ng et al., 1989, discussed in Section III.A.3,is possibly an exceptional case]. In contrast, Kasid et al. [1989] observed that a full-length transcript was more effective against *raf* oncogene expression in human carcinoma cells than a transcript complementary only to the 3' part of the coding region.

It is therefore difficult to predict the best target for antisense RNA because of the large number of variables involved and the lack of systematic studies in eukaryotic systems. Similarly, there is little information available on the optimal length for antisense transcripts. Daugherty et al. [1989] found that maximum inhibition of β-galactosidase activity was achieved with a fragment that had 381 nt complementary to the RBS and 5' end of the *lacZ* gene and that up to 94% inhibition was observed with a sequence complementary to only 33 nt overlapping the RBS. Similarly, Hirashima et al. [1989] reported that repression of SP phage RNA was most effective with a short antisense RNA of 30 nt (again overlapping the RBS) and that additional downstream sequence had little effect. These observations that relatively short antisense RNAs can be effective in bacteria are consistent with the short lengths of most naturally occurring antisense RNAs (see, e.g., *micF/ompF* interaction, Fig. 7), although we should note that natural systems are designed to provide *sufficient* regulation for correct biologic responses rather than the *maximal* regulation that may be demanded by the researcher. There may, however, be cases when a longer antisense transcript can give a greater level of inhibition, since this would appear to be the case when there is no complementarity to the mRNA RBS in the antisense RNA [Daugherty et al., 1989; Ellison et al., 1985].

However, there are arguments that suggest that relatively short anti-mRNAs should be more effective. First, at least in bacteria it appears that a short mRNA is produced at a faster rate and is not degraded significantly more rapidly [Daugherty et al., 1989]. Second, the rate of hybridization of a larger antisense transcript to its target will be slower than that of a shorter molecule [Wetmur and Davidson, 1968].

Most antisense experiments have been carried out with relatively long transcripts, though it is difficult to conclude that these are more effective. Wormington [1986] injected a 140 nt RNA into *Xenopus* oocytes and found it less effective than longer antisense RNAs, as was a 420 nt RNA to the 3' of a target gene reported by Sumikawa and Miledi [1988]. For intracellularly expressed RNAs, Holt et al. [1988] showed efficient regulation with a 196 nt antisense RNA to the 5' end of c-*fos* introduced into mouse 3T3 cells under a mouse mammary tumor virus promoter, while Jennings and Molloy [1987] targeted the 5' end of the SV40 large T antigen mRNA in COS cells with a 145 nt antisense RNA and reduced T antigen–dependent plasmid replication by 50%–95%. In plants Delauney et al. [1988] found a 225 nt targeted to the 5' end of the mRNA less effective than an antisense encompassing the whole RNA, whereas Kim and Wold [1985] found a full-length antisense less efficient than shorter transcripts. However, none of these experiments can be regarded as conclusive, and a pragmatic approach to the choice and length of antisense would if possible be to test initially a full-length (or near, but including 5' nontranslated sequences), a 5', and possibly a 3' fragment for their potency. It is sensible to check that no strong secondary structure exists in the antisense RNA, which would lead to intramolecular hybridization competing with the antisense–sense interaction [Rhodes and James, 1990].

A final consideration in the choice of antisense RNA is the degree of homology required with the target. The interaction of the naturally occurring *E. coli* rgulation between antisense *micF* and *ompF* mRNA (Fig. 7) indicates that a perfect match is not necessary, nor need all regions of homology between antisense and target be continuous [Daugherty et al., 1989; Kim and Wold, 1985]. In eukaryotes, too, regulation does not demand complete homology between antisense RNA and target, since the mouse c-*fos* gene can be regulated by antisense RNA to the human gene with which it shows >80% homology [Holt et al., 1988], and all three closely related *Dictyostelium* discoidin-1 genes could be regulated by a single antisense construct [Crowley et al., 1985; Ahern et al., 1988]. However, McGarry and Lindquist [1986] showed that RNA to *Drosophila* hsp26 did not affect the other three small heat-shock proteins, although there is 77% homology in a stretch of 330 nt between hsp23, hsp26, and hsp28 [Ingolia and Craig, 1982]. It would appear that, provided that sequences are more than 80% homologous, antisense regulation can successfully be used.

V. OVERVIEW OF ANTISENSE RNA REGULATION OF GENES IN EUKARYOTES

In this section we review some examples of antisense RNA regulation in various eukaryotes in which regulation has been attempted, either by indicating the relevant chapters in this book or by briefly summarizing below.

A. Naturally Occurring Antisense RNA

Naturally occurring antisense RNA has been extensively implicated in regulation of replication and expression in prokaryotes, and this is reviewed by Thomas (this volume). A considerable number of antisense transcripts have also been discovered in eukaryotes, which are listed in Table II [see also Takayama and Inouye, 1990, for review], but there is little evidence that these are normally involved in gene regulation, although the existence of two mutations whose phenotypes appear to be due to antisense RNA produced by rearranged genes has been reported. The first of these is the mouse *mld* mutation, which results in hypomyelination and is discussed by Katsuki et al. The second is the possible involvement of antisense RNA in the semidominant *niv*-525 allele in the plant *Antirrhinum majus* [Coen and Carpenter, 1988]. Such effects may suggest that antisense may have a more significant role in eukaryotic gene regulation than has henceforth been discovered.

B. In Vitro Systems

Antisense RNA has not been extensively used with in vitro systems. Munroe [1988] demonstrated that in vitro splicing can be inhibited in a HeLa cell extract and that an activity exists that promotes rapid annealing of complementary RNAs in the extract by antisense RNA. Splicing was found to be sensitive both to sequences annealing across the slice junction and to sequences complementary to the exon downstream of the 3' splice site. There is not much evidence that antisense RNA can inhibit splicing in vivo (see Section III.A.2, above), so it must be uncertain how significant these results are to the mechanism of antisense RNA action in whole cells. It is interesting to consider the possible existence of a hybridization-promoting activity in vivo.

Nicole and Tanguay [1987] examined the specificity with which antisense RNA against *Drosophila* hsp23 blocks translation in a rabbit reticulocyte lysate. Antisense hsp23 transcripts were preannealed with mRNA extracted from heat-shocked cells under various conditions of stringency and then translated in vitro. The antisense hsp23 sequence contained the whole coding sequence and 5' and 3' flanking regions. There is considerable homology (up to 77% over a stretch of 330 bp) between the genes for hsp23, hsp26, and hsp28 within the coding regions, but the 5' leader and initiation region are variable. Despite the homology, inhibition of translation was specific for hsp23 under both low and high stringency conditions, indicating that it is the 5' leader and initiation regions that are important for blocking of in vitro transcription. This is consistent with the ability of elongating ribosomes to destabilize also RNA:DNA duplexes once translation has initiated, while they are unable to initiate if the 5' nontranslated region and AUG initiation codon are covered [Liebhaber et al., 1984; Shakin and Liebhaber, 1986].

An interesting use of antisense RNA to analyze the phenomenon of unmasking of maternal mRNAs is described by Standart et al. [1990]. The mRNA for ribonucleotide reductase and cyclin A are masked in clam oocyte extracts and are not translated when the extract is mixed with rabbit reticulocyte lysate. However, they can be unmasked by phenol extraction or gel filtration in high salt. By using

TABLE II. Naturally Occurring Antisense Transcripts

Species	Target gene	Target complementarity	Function	References
Plants				
Tomato leaf tissue	7S RNA (integral part of signal recognition particle [SRP]	299 nt in length, 36–53 nt complementarity with viroid species	Possible primary cellular target for viroid species PSTV, CSV, CEV, TASV, and TPMV	Haas et al. [1988], Symons [1989a]
Antirrhinum majus	*Nivea* locus, chalcone synthase gene (CSH)	*niv-525*, semidominant allele carrying inverted duplication of 207 bp promoter region of *niv*, allowing antisense RNA formation	Possible hybridization of antisense RNA to first 40 bp of *niv* mRNA leader	Coen and Carpenter [1988]
Barley	α-Amylase	RNA complementary to almost full-length mRNA of both isozymes of α-amylase	Unknown function	Rogers [1988]
Cucumber	Cucumber mosaic virus (CMV) RNA 3 and 4	CMV satellite-RNA contains two noncontiguous regions of homology to CMV of 15 bp and 18 bp	Proposed to bind to a 33 bp sequence in CMV. Possible regulation of viral coat protein synthesis	Rezaian and Symons [1986]
Mammalian				
Mouse	L27' ribosomal protein mRNA	Two antisense transcripts; 1.0 kb, 1.8 kb. Complementary to region +8 – more than +346 of L27' mRNA	Unknown function. Possible posttranscriptional regulation of L27' mRNA	Belhumeur et al. [1988]
Mouse	Herpes simplex virus type 1	LAT; transcript associated with latent infection (2.3 kb). 360 bp complementarity with an ICP-0 transcript	Possible suppressor of expression of an ICP-0 transcript	Wagner et al. [1988]
Mouse	Myelin basic protein gene (MBP). *mld*, mutant allele with gene-2 intact and exons 3–7 inverted in gene-1	Antisense RNA complementary to exons 3 and 7 of gene-1	Possible posttranscriptional regulation	Okano et al. [1988], Tosic et al. [1990] (see also Katsuki, this volume)

Mouse	DNA "expression sequences"	1.2 kb and 3.0 kb mRNAs transcribed from opposite strands of same DNA molecule. 133 bp complementarity between 3′ termini of two RNA species	Possible posttranscriptional regulation	Williams and Fried [1986]
Mouse	DHFR gene	Antisense transcripts, 180–240 nt in length, initiated from opposite strand of 5′ flanking region of DHFR gene. All are complementary, with first 10 nt of major DHFR mRNA	Unknown function	Farnham et al. [1985]
Mouse	Herpes virus α-gene mRNA	Antisense transcripts to the whole of α-protein ICP-0, abundant in latently infected neurons	Possible role in HSV pathogenesis	Stevens et al. [1987]
Mouse	c-myc	Antisense RNAs initiate in first intron of c-myc	Pausing of c-myc transcription might be caused by antisense RNA binding at exon 1–intron 1 boundary in c-myc mRNA	Nepveu and Marcu [1986], Piechaczyk et al. [1988]
Human	c-myc	Transcription of both strands upstream of exon 1 in HL60 cells	Antisense transcripts appeared unstable. Function unknown	Bentley and Groudine [1986]
Rat	Mitochondrial DNA. D-loop–containing region houses origin of replication on one strand and promoters for both strands	DNA transcribed from both strands	Antisense RNA might modulate the processes of replication and transcription	Sbisà et al. [1988]

TABLE II. Naturally Occurring Antisense Transcripts (Continued)

Species	Target gene	Target complementarity	Function	References
Rat	GnRH (gonadotropin-releasing hormone)	GnRH gene transcribed from both strands (GnRH and SH). Antisense transcript present in heart, GnRH expressed in CNS	Unknown function of SH mRNA	Adelman et al. [1987]
Yeast				
Yeast	2 μm circle plasmid *RAF* gene	Full length of RAF gene transcribed on opposite strand into ~600 nt and ~1950 nt transcripts	Function unknown, but antisense transcripts are regulated according to plasmic copy number	Jayaram et al. [1985], Murray et al. [1987]
Yeast	*HAP3* (regulates *CYC1* expression and is homologous to mammalian CP1a transcription factor)	3 kb antisense transcript; sense transcript is 570 nt	Function unknown	Hahn et al. [1988]
Yeast	*RAD10*	Overlapping antisense transcription unit in 3′ region	Antisense transcript also found in equivalent human gene ERCC-1	Van Duin et al. [1989]
Yeast	Actin	Antisense transcript starts in intron and goes through 5′ region and upstream sequences	Unknown function	Thompson-Jager and Domdey [1990]
Others				
Chicken	Myosin heavy chain (MHC-mRNA)	tcRNA 102 (isoforms A and B). 34 nt at 3′ end of both these RNAs complementary with 5′ end of MHC-mRNA	Proposed that tcRNA 102 is the natural antisense RNA for specific muscle mRNAs	Heywood [1986]
Drosophila	Ddc (Dopa decarboxylase) gene	Adjacent gene to Ddc transcribed on opposite strand	88 bp complementarity between 3′ termini of Ddc mRNA and antisense RNA. Possible RNA duplex	Spencer et al. [1986]

antisense RNAs to different regions of the mRNA, Standart et al. [1990] were able competitively to unmask the mRNAs for these two proteins in high salt conditions and to map the 3′ region involved in masking. The antisense message is assumed to hybridize to the target, rendering it unrecognizable to the masking proteins. Translation was not inhibited by the duplex because of ribosomal unwinding, as described above; however, the use of a 5′ antisense RNA did block translation, so the additional involvement of the 5′ end in the masking phenomenon could not be excluded.

C. Yeast

Antisense RNA has not been used extensively in yeast (*Saccharomyces cerevisiae*). This is in part due to the availability of alternative tools for the deletion of genes by homologous recombination and their control by fusion to regulatable promoters, but is reinforced by a prevailing belief that antisense techniques do not work in yeast, a scepticism also supported by reports in the literature [Oeller and Johnston, 1985; Law and Devenish, 1988]. However, there has been a report of engineered antisense RNA regulation [Xiao and Rank, 1988], indicating that there may be no intrinsic reason why antisense techniques cannot be applied and suggesting that the several known cases of natural antisense transcripts that have been detected could have a role in regulation.

Xiao and Rank [1988] describe the only case of successful application of antisense RNA technology in yeast, which was to regulate the *ILV2* gene, causing a reduction in enzyme activity and a growth requirement for isoleucine and valine. The *ILV2* gene encodes acetolactate synthase (ALS), an enzyme common to the biosynthesis of isoleucine and valine. This enzyme is inhibited by sulfometuron methyl, and mutations of *ILV2* that confer resistance as the result of a single amino acid change have been isolated and are referred to as *SMR1*. Two fragments of the cloned *SMR1* allele were used in antisense constructs. Both included the 5′ nontranslated region of the gene, the longer fragment being 2 kb, which extended from −133

(relative to translation start) and included more than 90% of the coding region, whereas the shorter fragment of 400 bp covered the region from −189 and included 30% of the coding region. These fragments were cloned in an antisense orientation between the highly inducible galactose promoter (P_{GAL10}) and a transcription terminator and transformed into yeast on a high copy number plasmid. Induction of antisense RNA by galactose led to a reduction in ALS enzyme activity of around 45%, which was enhanced to around 60% when a yeast strain that lacked the normal inducer of *ILV2* expression and thus had lower endogenous levels of the target RNA was used. This reduction in ALS levels produced a temporary growth requirement for isoleucine and valine, which was, however, overcome after a longer period of growth. The 2.0 kb antisense construct was somewhat more efficient than the shorter transcript, which reflected a greater abundance of the longer RNA due to a higher copy number for this plasmid.

Although Xiao and Rank [1988] do not state the relative abundance of antisense RNA to sense message, from the Northern hybridization data presented there is clearly a massive excess, as would be expected from the very strong galactose promoter on a high copy number plasmid. Despite this, the extent of inhibition is relatively modest, so that although one may conclude that antisense inhibition *can* occur in yeast, it does not seem likely to become an important tool. As we have speculated above, perhaps the apparent absence of the double-stranded RNA undwinding/modifying activity from yeast (Bass et al., this volume) may help to explain the ineffectiveness of antisense RNA. In the light of these results it is not clear how much significance should be attached to reports of naturally occurring antisense transcripts in yeast, although several examples have been described (see Table II).

D. *Dictyostelium*

The cellular slime mould *Dictyostelium* normally grows as single cells, but can be induced by starvation to undergo a developmental program that leads to the aggregation of around

10^5 cells to form a multicellular organism that produces a fruiting body. This makes it a useful system for studying processes of development in an amenable model system. Although the technique of homologous recombination is available in *Dictyostelium* [De Lozanne and Spudish, 1987], antisense RNA has still proved to be a useful tool in studying cellular and developmental processes [see Knecht, 1989, for review].

Crowley et al. [1985] used an antisense construct to disrupt expression of the discoidin gene, which encodes a developmentally regulated lectin that serves a function similar to that of fibronectin in higher animals. Transformants showed a 90% reduction in steady-state discoidin mRNA and protein levels from all three endogenous discoidin genes. Such cells were unable to stream on a plastic surface, a phenocopy of the *discoidin*⁻ mutation. In these experiments very low levels of both sense and antisense RNA were detected, although nuclear run-on with isolated nuclei demonstrated that both sense and antisense genes are actively transcribed and that the rate of transcription of the sense gene was not affected by the presence of the antisense gene. Thus the absence of hybridized sense or antisense RNA suggests that degradation of both occurs in the nucleus.

Knecht and Loomis [1987] investigated the function of myosin heavy chain (*mhc*A) protein by antisense RNA inactivation. Stable transformants had levels of MHCA protein reduced by 250-fold and grew slowly, generating multinucleate progeny, which suggests that cytokinesis was impaired. The cells were unable to form multicellular aggregates and undergo morphogenesis. By changing the food source from liquid media to bacteria, these investigators were able to increase the level of endogenous *mhc*A expression. The extent of antisense inhibition was therefore reduced, and the increased levels of MHCA protein allowed cells to develop normally. In this case, too, the level of sense transcript was strongly reduced when antisense RNA

was present, but, in contrast to Crowley et al. [1985], antisense RNA could be detected by Northern analysis.

The role of the cell surface cyclic adenosine monophosphate (cAMP) receptor in development was investigated in work described by Klein et al. [1988], a good example of the direct testing of gene function by antisense RNA. The cAMP receptor was cloned by antibody screening of λ gt11 libraries, and the cDNA was cloned in a vector that expressed the gene in sense and antisense orientations. Expression from the sense construct caused undifferentiated cells to bind cAMP specifically, which wild-type cells do not normally do, whereas the antisense construct blocked cAMP production in starved cells. Such starved cells failed to enter the aggregation stage of development, suggesting an essential role for the cAMP receptor protein in the developmental program.

Attempts to regulate the *ras* gene of *Dictyoselium* [Ahern et al., 1988] were not successful, because no antisense *ras* transformants could be obtained, probably because of the essential role of *ras* for growth. Although a regulated promoter was used, a low level of anti-*ras* expression may have resulted in the failure to observe transformants. The control sense construct did not have any effect on growth. The same authors also tested the ability of antisense RNA to block expression of developmentally regulated genes of unknown function and were successful with both genes tested. In these cases no steady-state antisense RNA was observed, suggesting again a rapid degradation of sense:antisense duplexes.

It therefore seems that antisense inhibition works particularly well in *Dictyostelium*. It has been suggested that the small genome size and the low complexity may favor the kinetics of duplex formation, but the failures in yeast discussed above suggest that this cannot be a complete explanation. It may be significant that nuclear destruction of duplex hybrids appears to be very rapid in *Dictyostelium*, which may correlate with the effective antisense regulation.

E. *Drosophila*

The use of antisense RNA with *Drosophila* is comprehensively reviewed by Patel and Jacobs-Lorena (this volume) and is not covered further here.

F. *Xenopus*

Early experiments were carried out in *Xenopus* oocytes to demonstrate the principle that antisense regulation could block translation. Melton [1985] and Harland and Weintraub [1985] injected antisense RNA, which blocked translation of subsequently injected or coinjected sense transcripts. Similar results were also obtained by Sumikawa and Miledi [1988]. Wormington [1986] showed that endogenous transcripts could also be targeted by blocking expression of ribosomal protein L1. However, the interest in the *Xenopus* system lies in the developmental program initiated after fertilization utilizing maternally encoded mRNAs. When such mRNAs were targeted in fertilized eggs, inhibition was not observed [Bass and Weintraub, 1987; Rebagliati and Melton, 1987]. This led to the discovery of the RNA duplex unwinding/modifying activity (discussed in detail by Bass, this volume). This is, however, unable to account for the failure of these experiments, since although the duplexes are unwound the RNA is concomitantly modified, rendering it untranslatable. It therefore seems probable that these maternal RNAs may be in a form inaccessible to hybridization. On a similar note, one may consider both that Strickland et al. [1988] found that only the 3′ end of the t-PA message is accessible in mouse oocytes (Section III.A.3) and the evidence for specific proteins bound to maternal mRNA in clams discovered by Standard et al. [1990] (Section V.B).

The use of RNA expressed from introduced antisense genes has not been widespread with *Xenopus*, but was successfully demonstrated by Giebelhaus et al. [1988]. Since no stable transgenic system is available, antisense genes must be introduced by injection of plasmids,

and this limits the time period over which regulation can be achieved. Nevertheless the plasmids persisted through to the tadpole stage and after the injected genes were switched on at midblastula transition, and throughout the rest of development no protein 4.1 mRNA was detected in the embryos expressing antisense 4.1 RNA. This resulted in effective antisense regulation of membrane skeleton protein 4.1 and demonstrated its involvement in eye development as necessary for the normal interdigitation of the photoreceptor outer segments with the pigment epithelium layer in the retina. The further injection of a plasmid that expressed a truncated sense transcript rescued the antisense effect and permitted normal embryogenesis, proving the specific involvement of the antisense regulation of protein 4.1 in the abnormalities observed.

This success of injecting antisense genes into *Xenopus* suggests that this approach may be valuable for investigating the function of other genes during development. Additional information on antisense regulation in *Xenopus* is given by Bass, (this volume) and by Colman (this volume).

G. Mammalian Cells and Mice

The use of antisense RNA in mammalian cells and mice is covered in various aspects by Munir et al.; Katsuki et al.; To; Agrawal and Leitec; and Yokoyama (this volume). We have also used several specific examples in our discussion in Section III, and in Table III we have listed examples available in the literature. The reader is also referred to Takayama and Inouye [1990] for a further review with relevant coverage.

H. Plants

Material relevant to the use of antisense RNA in plants can be found in Section III of this chapter. It is also discussed by Rodermel and Bogorad (this volume) and by Bejarano et al. (this volume). Readers are also referred to the review by Mol [1990] and others listed in Table I.

TABLE III. Regulation of Mammalian and Avian Genes by Artificial Antisense RNA*

Target gene	Cell type	Target complementarity	Inhibition**	References
Human				
c-myc	HL-60	Exon 1	90%[b]	Yokoyama and Imamoto [1987], Yokoyama (this volume)
β-Actin	Fibroblast	5' and 3' untranslated regions	None observed	Gunning et al. [1987]
(2'–5') oligoadenylate synthetase	Osteosarcoma cells	3' Termini	Subsequently retracted	De Benedetti et al. [1987]
Creatine kinase	U937	3' Coding and noncoding	40%[a]	Ch'ng et al. [1989]
c-raf-1	SQ-20B	cDNA	>90%[c]	Kasid et al. [1989]
HIV-1 env	H9 T	Exon 2 of tat gene (using liposome fusion)	90%–100%[b]	Renneisen et al. [1990]
HIV	Jurkat cells	tat, rev, and vpu slice and initiation regions	70%[e]	Rhodes and James [1990]
T-cell leukemia virus type 1	Primary T lymph-ocytes	5' splice site, part of tax gene	Cell proliferation	Von Rüden and Gilboa [1989]
Mouse				
Endo B-cytokeratin	F9 derivative	5' Terminus	50%[b]	Trevor et al. [1987]
myc	F9	Exon 1; exons 2–3	Differentiation	Griep and Westphal [1988]
c-fos	F9	Exon 1 and 5' flanking	Block of c-fos induction	Levi and Ozato [1988], Levi et al. [1988]
c-fos	F9	Exon 1 and part of exon 2	Protein levels	Edwards et al. [1988]
c-fos	3T3	5' Terminus	Phenotypic	Holt et al. [1986]
c-fos	3T3	1.7 kb of 5' end	Phenotypic	Mercola et al. [1987]
c-fos	3T3	5' Noncoding + exon 1	95%[c]	Nishikura and Murray [1987]
H-ras	3T3	Exon 1	Not observed	Salmons et al. [1986]
MoMLV (Moloney murine leukemia virus)	3T3	gag gene pol gene	94%[a] 32%[a]	Sullenger et al. [1990]

			Induction of oncogenicity	
Tissue inhibitor of metalloproteinases	3T3	Full length		Khokha et al. [1989]
HSV-TK	L	5' and 3' Terminus	80%–90%[a]	Kim and Wold [1985]
HSV-TK	L	5' Noncoding	100%[a]	Izant and Weintraub [1984, 1985]
CAT	L	Entire cDNA	88%[b]	Izant and Weintraub [1985]
β-Actin	L	cDNA	Phenotypic	Izant and Weintraub [1985]
HPRT	Fibroblasts	Various, including intron sequences	Up to 100%[a]	Stout et al. [1987], Stout and Caskey [1990], Munir et al. (this volume)
Tissue plasminogen activator (t-PA)	Primary oocytes	3' noncoding (by microinjection)	90%[a]	Strickland et al. [1988]
Myelin basic protein	Transgenic mice	cDNA	Phenotypic	Katsuki et al. [1988], Katsuki et al. (this volume)
HPRT	Transgenic mice	Various	None observed	Munir et al. [1990], Rossiter et al. [1989], Munir et al. (this volume)
Monkey				
CAT	CV1, COS	Full length	No effect	Kerr et al. [1988]
SV40 large T-ag	COS	5' Terminus	50%–95%[d]	Jennings and Molloy [1987]
Rat				
c-*src* (pp60)	FR3T3	Entire cDNA	Phenotypic	Amini et al. [1986]
Quail				
RSV *env*	R(-)Q	*env* coding region 5' and 3' noncoding	80%[e]	Chang and Stolzfus [1985, 1987]
neo[f]	QT35	Full length	Replication inhibited	To et al. [1986]. To (this volume)

*Antisense RNA was expressed from introduced antisense gene constructs unless microinjection or liposome-mediated transfer of in vitro synthesized RNA is stated.

**Level of inhibition expressed as a reduction (in percentage) of [a]enzyme activity, [b]protein steady-state level, [c]RNA steady-state level, [d]T-antigen–dependent plasmid replication, and [e]virus replication.

VI. REFERENCES

Adelman JP, Bond CT, Douglass J, Herbert E (1987): Two mammalian genes transcribed from opposite strands of the same DNA locus. Science 235: 1514–1517.

Ahern KG, Rubino S, Hori R, Firtel RA (1988): Use of antisense mutagenesis to examine gene function during development of *Dictyostelium discoideum*. In Melton DA (ed): Antisense RNA and DNA. Current Communications in Molecular Biology. Cold Spring Harbor, New York: Cold Spring Harbor Laboratory, pp. 85–91.

Amini S, DeSeau V, Reddy S, Shalloway D, Bolen JB (1986): Regulation of pp60 c-*src* synthesis by inducible RNA complementary to c-*src* mRNA in polyomavirus-transformed rat cells. Mol Cell Biol 6:2305–2316.

Andersen J, Delihas N, Ikenaka K, Green PJ, Pines O, Ilercil O, Inouye M (1987): The isolation and characterization of RNA coded by the *micF* gene in *Escherichia coli*. Nucleic Acids Res 15:2089–2101.

Atwater JA, Wisdom R, Verma IM (1990): Regulated mRNA stability. Annu Rev Genet 24:519–541.

Bass BL, Weintraub H (1987): A developmentally regulated activity that unwinds RNA duplexes. Cell 48:607–613.

Belhumeur P, Lussier M, Skup D (1988): Expression of naturally occurring RNA molecules complementary to the murine L27' ribosomal protein mRNA. Gene 72:277–285.

Bentley DL, Groudine M (1986): A block to elongation is largely responsible for decreased transcription of c-*myc* in differentiated HL-60 cells. Nature 321: 702–706.

Bird CR, Ray JA, Fletcher JD, Boniwell JM, Bird AS, Teulieres C, Blain I, Bramley PM, Schuch W (1991): Using antisense RNA to study gene function: inhibition of carotenoid biosynthesis in transgenic tomatoes. BioTechnology 9:635–639.

Blackwell TK, Kretzner L, Blackwood EM, Eisenman RN, Weintraub H (1990): Sequence-specific DNA binding by the c-Myc protein. Science 250:1149–1151.

Blackwell TK, Weintraub H (1990): Differences and similarities in DNA-binding preferences of MyoD and E2A protein complexes revealed by binding site selection. Science 250:1104–1110.

Boidot-Forget M, Chassignol M, Takasugi M, Thuong NT, Helene C (1988): Site-specific cleavage of single-stranded and double-stranded DNA sequences by oligodeoxyribonucleotides covalently linked to an intercalating agent and an EDTA-Fe chelate. Gene 72:361–371.

Boiziau C, Kurfurst R, Cazenave C, Roig V, Thuong NT, Toulmé J-J (1991): Inhibition of translation initiation by antisense oligonucleotides via an RNase-H independent mechanism. Nucleic Acids Res 19: 1113–1119.

Cameron FH, Jennings PA (1989): Specific gene suppression by engineered ribozymes in monkey cells. Proc Natl Acad Sci USA 86:9139–9143.

Cech TR (1987): The chemistry of self-splicing RNA and RNA enzymes. Science 236:1532–1539.

Cech TR (1988): Conserved sequences and structures of group I introns: Building an active site for RNA catalysis—A review. Gene 73:259–271.

Cech TR (1990): Self-splicing of group I introns. Annu Rev Biochem 59:543–568.

Cech TR, Bass BL (1986): Biological catalysis by RNA. Annu Rev Biochem 55:599–629.

Chang LJ, Stoltzfus CM (1985): Gene expression from both intronless and intron-containing Rous sarcoma virus clones is specifically inhibited by anti-sense RNA. Mol Cell Biol 5:2341–2348.

Chang LJ, Stoltzfus CM (1987): Inhibition of Rous sarcoma virus replication by antisense RNA. J Virol 61:921–924.

Charles D (1991): A triple helix to cripple viruses. New Scientist 13 April 1991:19.

Ch'ng JLC, Mulligan RC, Schimmel P, Holmes EW (1989): Antisense RNA complementary to 3' coding and noncoding sequences of creatine kinase is a potent inhibitor of translation in vivo. Proc Natl Acad Sci USA 86:10006–10010.

Chuat JC, Galibert F (1989): Can ribozymes be used to regulate procaryote gene expression? Biochem Biophys Res Commun 162:1025–1029.

Coen ES, Carpenter R (1988): A semi-dominant allele, *niv-525*, acts in trans to inhibit expression of its wild-type homologue in *Antirrhinum majus*. EMBO J 7:877–883.

Cohen JS (ed) (1989a): Oligodeoxyribonucleotides: Antisense Inhibitors of Gene Expression. London: Macmillan Press.

Cohen JS (1989b): Designing antisense oligonucleotides as pharmaceutical agents. TIPS 10:435–437.

Coleman J, Green PJ, Inouye M (1984): The use of RNAs complementary to specific mRNAs to regulate expression of individual bacterial genes. Cell 37:429–436.

Colman A (1990): Antisense strategies in cell and developmental biology. J Cell Sci 97:399–409.

Cooney M, Czernuszewicz G, Postel EH, Flint SJ, Hogan ME (1988): Site-specific oligonucleotide binding represses transcription of the human c-*myc* gene in vitro. Science 241:456–459.

Cornelissen M (1989): Nuclear and cytoplasmic sites for anti-sense control. Nucleic Acids Res 17:7203–7209.

Cornelissen M, Vandewiele M (1989): Both RNA level and translation efficiency are reduced by anti-sense RNA in transgenic tobacco. Nucleic Acids Res 17:833–843.

Cotten M, Birnstiel ML (1989): Ribozyme mediated destruction of RNA in vivo. EMBO J 8:3861–3866.

Cotten M, Schaffner G, Birnstiel ML (1989): Ribozyme, antisense RNA, and antisense DNA inhibition of U7 small nuclear ribonucleoprotein-mediated histone pre-mRNA processing in vitro. Mol Cell Biol 9:4479–4487.

Crowley TE, Nellen W, Gomer RH, Firtel RA (1985): Phenocopy of discoidin I-minus mutants by antisense transformation in *Dictyostelium*. Cell 43:633–641.

Daugherty BL, Hotta K, Kumar C, Ahn YH, Zhu JD, Pestka S (1989): Antisense RNA: Effect of ribosome binding sites, target location, size, and concentration on the translation of specific mRNA molecules. Gene Anal Tech 6:1–16.

Davanloo P, Rosenberg AH, Dunn JJ, Studier FW (1984): Cloning and expression of the gene for bacteriophage T7 RNA polymerase. Proc Natl Acad Sci USA 81:2035–2039.

De Benedetti A, Pytel BA, Baglioni C (1987): Loss of (2'-5')oligoadenylate synthetase activity by production of antisense RNA results in lack of protection by interferon from viral infections. Proc Natl Acad Sci USA 84:658–662 [retracted in Proc Natl Acad Sci USA 1987 84:6740].

De-Lozanne A, Spudich JA (1987): Disruption of the *Dictyostelium* myosin heavy chain gene by homologous recombination. Science 236:1086–1091.

Delauney AJ, Tabaeizadeh Z, Verma DPS (1988): A stable bifunctional antisense transcript inhibiting gene expression in transgenic plants. Proc Natl Acad Sci USA 85:4300–4304.

Dolnick BJ (1990): Antisense agents in pharmacology. Biochem Pharmacol 40:671–675.

Doudna JA, Cormack BP, Szostak JW (1989): RNA structure, not sequence, determines the 5' splice-site specificity of a group I intron. Proc Natl Acad Sci USA 86:7402–7406.

Dzianott AM, Bujarski JJ (1989): Derivation of an infectious viral RNA by autolytic cleavage of in vitro transcribed viral cDNAs. Proc Natl Acad Sci USA 86:4823–4827.

Edwards SA, Rundell AY, Adamson ED (1988): Expression of c-*fos* antisense RNA inhibits the differentiation of F9 cells to parietal endoderm. Dev Biol 129:91–102.

Elkind Y, Edwards R, Mavandad M, Hedrick SA, Ribak O, Dixon RA, Lamb CJ (1990): Abnormal plant development and down-regulation of phenylpropanoid biosynthesis in transgenic tobacco containing a heterologous phenylalanine ammonia-lyase gene. Proc Natl Acad Sci USA 87:9057–9061.

Ellington AD, Szostak JW (1990): In vitro selection of RNA molecules that bind specific ligands. Nature 346:818–822.

Ellison MJ, Kelleher RJ, Rich A (1985): Thermal regulation of β-galactosidase synthesis using anti-sense RNA directed against the coding portion of the mRNA. J Biol Chem 260:9085–9087.

Farnham PJ, Abrams JM, Schimke RT (1985): Opposite-strand RNAs from the 5' flanking region of the mouse

dihydrofolate reductase gene. Proc Natl Acad Sci USA 82:3978–3982.

Feldstein PA, Buzayan JM, Bruening G (1989): Two sequences participating in the autolytic processing of satellite tobacco ringspot virus complementary RNA. Gene 82:53–61.

Feldstein PA, Buzayan JM, van-Tol H, deBear J, Gough GR, Gilham PT, Bruening G (1990): Specific association between an endoribonucleolytic sequence from a satellite RNA and a substrate analogue containing a 2'-5' phosphodiester. Proc Natl Acad Sci USA 87:2623–2627.

Forster AC, Altman S (1990): External guide sequences for *E. coli* RNase P, and a new function for the protein subunit. 1990 Cold Spring Harbor RNA Processing Meeting, p. 92.

Frankham R (1988): Molecular hypotheses for position-effect variegation: Antisense transcription and promoter occlusion. J Theor Biol 135:85–107.

François JC, Saison-Behmoaras T, Thuong NT, Helene C (1989): Inhibition of restriction endonuclease cleavage via triple helix formation by homopyrimidine oligonucleotides. Biochemistry 28:9617–9619.

Gabizon A, Papahadjopoulos D (1988): Liposome formulations with prolonged circulation time in blood and enhanced uptake by tumors. Proc Natl Acad Sci USA 85:6949–6953.

Gait MJ (ed) (1984): Oligonucleotide Synthesis: A Practical Approach. Oxford: Oxford University Press.

Geiser T (1990): Large-scale economic synthesis of antisense phosphorothioate analogues of DNA for preclinical investigations. Ann NY Acad Sci 616:173–183.

Gelbart W, McCarron M, Chovnick A (1976): Extension of the limits of the XDH structural element in *Drosophila melanogaster*. Genetics 84:211–232.

Giebelhaus DH, Eib DW, Moon RT (1988): Antisense RNA inhibits expression of membrane skeleton protein 4.1 during embryonic development of *Xenopus*. Cell 53:601–615.

Green PJ, Pines O, Inouye M (1986): The role of antisense RNA in gene regulation. Annu Rev Biochem 55:569–597.

Griep AE, Westphal H (1988): Antisense Myc sequences induce differentiation of F9 cells. Proc Natl Acad Sci USA 85:6806–6810.

Grierson D, Fray RG, Hamilton AJ, Smith CJS, Watson CF (1991): Does co-suppression of sense genes in transgenic plants involve antisense RNA? Trends Biotechnol 9:122–123.

Griffin LC, Dervan PB (1989): Recognition of thymine adenine base pairs by guanine in a pyrimidine triple helix motif. Science 245:967–971.

Guerrier-Takada C, Gardiner K, Marsh T, Pace N, Altman S (1983): The RNA moiety of ribonuclease P is the catalytic subunit of the enzyme. Cell 35:849–857.

Gunning P, Leavitt J, Muscat G, Ng SY, Kedes L (1987):

A human β-actin expression vector system directs high-level accumulation of antisense transcripts. Proc Natl Acad Sci USA 84:4831–4835.

Haas B, Klanner A, Ramm K, Sanger HL (1988): The 7S RNA from tomato leaf tissue resembles a signal recognition particle RNA and exhibits a remarkable sequence complementarity to viroids. EMBO J 7: 4063–4074.

Hahn S, Pinkham J, Wei R, Miller R, Guarente L (1988): The HAP3 regulatory locus of Saccharomyces cerevisiae encodes divergent overlapping transcripts. Mol Cell Biol 8:655–663.

Hamilton AJ, Lycett GW, Grierson D (1990): Antisense gene that inhibits synthesis of the hormone ethylene in transgenic plants. Nature 346:284–287.

Hampel A, Tritz R, Hicks M, Cruz P (1990): ''Hairpin'' catalytic RNA model: Evidence for helices and sequence requirement for substrate RNA. Nucleic Acids Res 18:299–304.

Hanvey JC, Shimizu M, Wells RD (1990): Site-specific inhibition of EcoRI restriction/modification enzymes by a DNA triple helix. Nucleic Acids Res 18: 157–161.

Harland R, Weintraub H (1985): Translation of mRNA injected into Xenopus oocytes is specifically inhibited by antisense RNA. J Cell Biol 101:1094–1099.

Haseloff J, Gerlach WL (1988): Simple RNA enzymes with new and highly specific endoribonuclease activities. Nature 334:585–591.

Haseloff J, Gerlach WL (1989): Sequences required for self-catalysed cleavage of the satellite RNA of tobacco ringspot virus. Gene 82:43–52.

Hélène C, Toulmé J-J (1990): Specific regulation of gene expression by antisense, sense and antigene nucleic acids. Biochim Biophys Acta Gene Struct Expression 1049:99–125.

Heywood SM (1986): tcRNA as a naturally occurring antisense RNA in eukaryotes. Nucleic Acids Res 14:6771–6772 [published erratum appears in Nucleic Acids Res 1987 15:384].

Hiatt WR, Kramer M, Sheehy RE (1989): The application of antisense RNA technology to plants. In Setlow JK (ed): Genetic Engineering, Vol 11. New York: Plenum, pp. 49–63.

Hirashima A, Sawaki S, Mizuno T, Houba-Herin N, Inouye M (1989): Artificial immune system against viral infection involving antisense RNA targeted to the 5′-terminal noncoding region of coliphage SP RNA. J Biochem Tokyo 106:163–166.

Holt J, Lechner R (1990): Sequence-specific anti-RNA reagents for cell physiology. Comments (US Biochemical Corp) 16, No. 4:1–21.

Holt JT, Gopal TV, Moulton AD, Nienhuis AW (1986): Inducible production of c-fos antisense RNA inhibits 3T3 cell proliferation. Proc Natl Acad Sci USA 83:4794–4798.

Holt JT, Redner RL, Nienhuis AW (1988): An oligomer complementary to c-myc mRNA inhibits proliferation of HL-60 cells and induces differentiation. Mol Cell Biol 8:963–973.

Horne DA, Dervan PB (1990): Recognition of mixed-sequence duplex DNA by alternate-strand triple-helix formation. J Am Chem Soc 112:2435–2437.

Ingolia TD, Craig EA (1982): Four small Drosophila heat shock proteins are related to each other and to mammalian α-crystallin. Proc Natl Acad Sci USA 79: 2360–2364.

Inouye M (1988): Antisense RNA: Its functions and applications in gene regulation—a review. Gene 72:25–34.

Izant JG (1989): Antisense ''pseudogenetics.'' Cell Motil Cytoskeleton 14:81–91.

Izant JG, Sardelli AD (1988a): Antisense inhibition of RNA splicing. In Melton DA (ed): Antisense RNA and DNA. Cold Spring Harbor, New York. Cold Spring Harbor Laboratory, pp. 141–144.

Izant JG, Sardelli AD (1988b): Anti-sense suppression of RNA maturation. J Cell Biol 107:102a.

Izant JG, Weintraub H (1984): Inhibition of thymidine kinase gene expression by anti-sense RNA: A molecular approach to genetic analysis. Cell 36:1007–1015.

Izant JG, Weintraub H (1985): Constitutive and conditional suppession of exogenous and endogenous genes by anti-sense RNA. Science 229:345–352.

Jager A, Levy MJ, Hecht SM (1988): Oligonucleotide N-alkylphosphoramidates: synthesis and binding to polynucleotides. Biochemistry 27:7237–7246.

Jayaram M, Sutton A, Broach JR (1985): Properties of REP3: A cis-acting locus required for stable propagation of the Saccharomyces cerevisiae 2μ circle plasmid. Mol Cell Biol 5:2466–2475.

Jennings PA, Molloy PL (1987): Inhibition of SV40 replicon function by engineered antisense RNA transcribed by RNA polymerase III. EMBO J 6:3043–3047.

Jorgensen R (1990): Altered gene expression in plants due to trans interactions between homologous genes. Trends Biotechnol 8:340–344.

Kasid U, Pfeifer A, Brennan T, Beckett M, Weichselbaum RR, Dritschilo A, Mark GE (1989): Effect of antisense c-raf-1 on tumorigenicity and radiation sensitivity of a human squamous carcinoma. Science 243: 1354–1356.

Katsuki M, Sato M, Kimura M, Yokoyama M, Kobayashi K, Nomura T (1988): Conversion of normal behavior to shiverer by myelin basic protein antisense cDNA in transgenic mice. Science 241:593–595.

Kawchuk LM, Martin RR, McPherson J (1991): Sense and antisense RNA-mediated resistance to potato leafroll virus in Russet Burbank potato plants. Mol Plant Microbe Interact 4:247–253.

Kerr SM, Stark GR, Kerr IM (1988): Excess antisense RNA from infectious recombinant SV40 fails to inhibit expression of a transfected, interferon-inducible gene. Eur J Biochem 175:65–73.

Khokha R, Waterhouse P, Yagel S, Lala PK, Overall

CM, Norton G, Denhardt DT (1989): Antisense RNA-induced reduction in murine TIMP levels confers oncogenicity on Swiss 3T3 cells. Science 243: 947–950.

Kim SH, Cech TR (1987): Three-dimensional model of the active site of the self-splicing rRNA precursor of *Tetrahymena*. Proc Natl Acad Sci USA 84:8788–8792.

Kim SK, Wold BJ (1985): Stable reduction of thymidine kinase activity in cells expressing high levels of antisense RNA. Cell 42:129–138.

Klausner A (1990): Antisense start-ups surveyed. Bio-Technology 8:303–304.

Klein PS, Sun TJ, Saxe CL, Kimmel AR, Johnson RL, Devreotes PN (1988): A chemoattractant receptor controls development in *Dictyostelium discoideum*. Science 241:1467–1472.

Knecht D (1989): Application of antisense RNA to the study of the cytoskeleton: background, principles, and a summary of results obtained with myosin heavy chain. Cell Motil Cytoskeleton 14:92–102.

Knecht DA, Loomis WF (1987): Antisense RNA inactivation of myosin heavy chain gene expression in *Dictyostelium discoideum*. Science 236:1081–1086.

Koizumi M, Hayase Y, Iwai S, Kamiya H, Inoue H, Ohtsuka E (1989): Design of RNA enzymes distinguishing a single base mutation in RNA. Nucleic Acids Res 17:7059–7071.

Konings DAM, van Duijn LP, Voorma HO, Hogegweg P (1987): Minimal energy foldings of eukaryotic mRNAs form a separate leader domain. J Theor Biol 127:63–78.

Kulka M, Smith CC, Aurelian L, Fishelevich R, Meade K, Miller P, Ts'o PO (1989): Site specificity of the inhibitory effects of oligo(nucleoside methylphosphonate)s complementary to the acceptor splice junction of herpes simplex virus type 1 immediate early mRNA 4. Proc Natl Acad Sci USA 86:6868–6872.

Law RH, Devenish RJ (1988): Expression in yeast of antisense RNA to ADE1 mRNA. Biochem Int 17: 673–679.

Lawson TG, Ray BK, Dodds JT, Grifo JA, Abramson RD, Merrick WC, Betsch DF, Weith HL, Thach RE (1986): Influence of 5′ proximal secondary structure on the translational efficiency of eukaryotic mRNAs and on their interaction with initiation factors. J Biol Chem 261:13979–13989.

Lee KY, Lund P, Lowe K, Dunsmuir P (1990): Homologous recombination in plant cells after *Agrobacterium*-mediated transformation. Plant Cell 2:415–425.

Letai AG, Palladino MA, Fromm E, Rizzo V, Fresco JR (1988): Specificity in formation of triple-stranded nucleic acid helical complexes: Studies with agarose-linked polyribonucleotide affinity columns. Biochemistry 27:9108–9112.

Levi BZ, Kasik JW, Ozato K (1988): c-*fos* antisense RNA blocks expression of c-*fos* gene in F9 embryonal carcinoma cells. Cell Differ Dev 25(Suppl):95–101.

Levi BZ, Ozato K (1988): Constitutive expression of c-*fos* antisense RNA blocks c-*fos* gene induction by interferon and by phorbol ester and reduces c-*myc* expression in F9 embryonal carcinoma cells. Genes Dev 2:544–566.

Liebhaber SA, Cash FE, Shakin SH (1984): Translationally associated helix-destabilizing activity in rabbit reticulocyte lysate. J Biol Chem 259:15597–15602.

Loose-Mitchell DS (1988): Antisense nucleic acids as a potential class of pharmaceutical agents. Trends Pharmacol Sci 9:45–47.

Maher LJ, Wold B, Dervan PB (1989): Inhibition of DNA binding proteins by oligonucleotide-directed triple helix formation. Science 245:725–730.

Marcus-Sekura CJ (1988): Techniques for using antisense oligodeoxyribonucleotides to study gene expression. Anal Biochem 172:289–295.

Marcus-Sekura CJ, Woerner AM, Shinozuka K, Zon G, Quinnan GVJ (1987): Comparative inhibition of chloramphenicol acetyltransferase gene expression by antisense oligonucleotide analogues having alkyl phosphotriester, methylphosphonate and phosphorothioate linkages. Nucleic Acids Res 15:5749–5763.

McClain WH, Guerrier-Takada C, Altman S (1987): Model substrates for an RNA enzyme. Science 238:527–530.

McGarry TJ, Lindquist S (1986): Inhibition of heat shock protein synthesis by heat-inducible antisense RNA. Proc Natl Acad Sci USA 83:399–403.

Meegan JM, Marcus PI (1989): Double-stranded ribonuclease coinduced with interferon. Science 244: 1089–1091.

Mei HY, Kaaret TW, Bruice TC (1989): A computational approach to the mechanism of self-cleavage of hammerhead RNA. Proc Natl Acad Sci USA 86:9727–9731.

Melton DA (1985): Injected anti-sense RNAs specifically block messenger RNA translation in vivo. Proc Natl Acad Sci USA 82:144–148.

Melton DA (ed) (1988): Antisense RNA and DNA. Cold Spring Harbor, New York: Cold Spring Harbor Laboratory.

Melton DA, Krieg PA, Rebagliati MR, Maniatis T, Zinn K, Green MR (1984): Efficient in vitro synthesis of biologically active RNA and RNA hybridization probes from plasmids containing a bacteriophage SP6 promoter. Nucleic Acids Res 12:7035–7056.

Mercola D, Rundell A, Westwick J, Edwards SA (1987): Antisense RNA to the c-fos gene: Restoration of density-dependent growth arrest in a transformed cell line. Biochem Biophys Res Commun 147:288–294.

Michel F, Umesono K, Ozeki H (1989): Comparative and functional anatomy of group II catalytic introns—A review. Gene 82:5–30.

Miller PS (1991): Oligonucleoside methylphosphonates as antisense reagents. BioTechnology 9:358–362.

Mol J, van Blockland K, Kooter J (1991): More about co-suppression. Trends Biotechnol 9:182–183.

Mol JN, van-der-Krol AR, van-Tunen AJ, van-Blokland R, de-Lange P, Stuitje AR (1990): Regulation of plant gene expression by antisense RNA. FEBS Lett 268: 427–430.

Morvan F, Rayner B, Imbach J-L (1990): α-Oligodeoxy-nucleotides (α-DNA): A new chimeric nucleic acid analog. In Setlow JK (ed): Genetic Engineering, Vol 12. New York: Plenum, pp. 37–52.

Munir MI, Rossiter BJ, Caskey CT (1990): Antisense RNA production in transgenic mice. Somat Cell Mol Genet 16:383–394.

Munroe SH (1988): Antisense RNA inhibits splicing of pre-mRNA in vitro. EMBO J 7:2523–2532.

Murphy FL, Cech TR (1989): Alteration of substrate specificity for the endoribonucleolytic cleavage of RNA by the *Tetrahymena* ribozyme. Proc Natl Acad Sci USA 86:9218–9222.

Murray JAH, Scarpa M, Rossi N, Cesareni G (1987): Antagonistic controls regulate copy number of the yeast 2 μ plasmid. EMBO J 6:4205–4212.

Napoli C, Lemieux C, Jorgensen R (1990): Introduction of a chimeric chalcone synthase gene into petunia results in reversible co-suppression of homologous genes in trans. Plant Cell 2:279–289.

Nepveu A, Marcu KB (1986): Intragenic pausing and antisense transcription within the murine c-*myc* locus. EMBO J 5:2859–2865.

Nicole LM, Tanguay RM (1987): On the specificity of antisense RNA to arrest in vitro translation of mRNA coding for *Drosophila* hsp23. Biosci Rep 7:239–246.

Nishikura K, Murray JM (1987): Antisense RNA of protooncogene c-*fos* blocks renewed growth of quiescent 3T3 cells. Mol Cell Biol 7:639–649.

North G (1990): Ribozymes: Expanding the RNA repertoire. Nature 345:576–578.

Oeller P, Johnston M (1985): Unobstructed transcription from opposing yeast promoters. Abstracts of 1985 Cold Spring Harbor Molecular Biology of Yeast Meeting, p. 355.

Okano H, Aruga J, Nakagawa T, Shiota C, Mikoshiba K (1991): Myelin basic protein gene and the function of antisense RNA in its repression in myelin-deficient mutant mouse. J Neurochem 56:560–567.

Okano H, Ikenaka K, Mikoshiba K (1988): Recombination within the upstream gene of duplicated myelin basic protein genes of myelin deficient shi^mld mouse results in the production of antisense RNA. EMBO J 7:3407–3412.

O'Neill G (1989): Genetic ''shears'' cut their way to market. New Scientist 29 July, '89:33.

Pace NR, Reich C, James BD, Olsen GJ, Pace B, Waugh DS (1987): Structure and catalytic function in ribonuclease P. Cold Spring Harbor Symp Quant Biol 52:239–248.

Paszkowski J, Baur M, Bogucki A, Potrykus I (1988): Gene targeting in plants. EMBO J 7:4021–4026.

Perriman RJ, Gerlach WL (1990): Manipulating gene expression with ribozyme technology. Curr Opinion Biotechnol 1:86–91.

Perrouault L, Asseline U, Rivalle C, Thuong NT, Bisagni E, Giovannangeli C, Le-Doan T, Helene C (1990): Sequence-specific artificial photo-induced endonucleases based on triple helix-forming oligonucleotides. Nature 344:358–360.

Pestka S, Daugherty BL, Jung V, Hotta K, Pestka RK (1984): Anti-mRNA: Specific inhibition of translation of single mRNA molecules. Proc Natl Acad Sci USA 81:7525–7528.

Pidgeon C, Weith HL, Darbishire-Weith D, Cushman M, Byrn SR, Chen J-K, Stowell JG, Ray K, Carlson D (1990): Synthesis and liposome encapsulation of antisense oligonucleotide-intercalator conjugates. Ann NY Acad Sci 616:593–596.

Piechaczyk M, Blanchard JM, Bonnieu A, Fort P, Mechti N, Rech J, Cuny M, Marty L, Ferre F, Lebleu B, et al. (1988): Role of RNA structures in c-*myc* and c-*fos* gene regulations. Gene 72:287–295.

Povsic TJ, Dervan PB (1989): Triple helix formation by oligonucleotides on DNA extended to the physiological pH range. J Am Chem Soc 111:3059–3061.

Rebagliati MR, Melton DA (1987): Antisense RNA injections in fertilized frog eggs reveal an RNA duplex unwinding activity. Cell 48:599–605.

Renneisen K, Leserman L, Matthes E, Schroder HC, Muller WE (1990): Inhibition of expression of human immunodeficiency virus-1 in vitro by antibody-targeted liposomes containing antisense RNA to the *env* region. J Biol Chem 265:16337–16342.

Rezaian MA, Symons RH (1986): Anti-sense regions in satellite RNA of cucumber mosaic virus form stable complexes with the viral coat protein gene. Nucleic Acids Res 14:3229–3239.

Rhodes A, James W (1990): Inhibition of human immunodeficiency virus replication in cell culture by endogenously synthesized antisense RNA. J Genet Virol 71:1965–1974.

Riordan ML, Martin JC (1991): Oligonucleotide-based therapeutics. Nature 350:442–443.

Robert LS, Donaldson PA, Ladaique C, Altosaar I, Arnison PG, Fabijanski SF (1990): Antisense RNA inhibition of β-glucuronidase gene expression in transgenic tobacco can be transiently overcome using a heat-inducible β-glucuronidase gene construct. BioTechnology 8:459–464.

Robertson DL, Joyce GF (1990): Selection in vitro of an RNA enzyme that specifically cleaves single-stranded DNA. Nature 344:467–468.

Robertson HD (1990): *Escherichia coli* ribonuclease III. Methods Enzymol 181:189–202.

Rodermel SR, Abbott MS, Bogorad L (1988): Nuclear–organelle interactions: Nuclear antisense gene inhibits ribulose bisphosphate carboxylase enzyme levels in transformed tobacco plants. Cell 55:673–681.

Rogers JC (1988): RNA complementary to α-amylase mRNA in barley. Plant Mol Biol 125:138.

Rossi JJ, Cantin EM, Zaia JA, Ladne PA, Chen J, Stephens DA, Sarver N, Chang PS (1990): Ribozymes as therapies for AIDS. Ann NY Acad Sci 616:184–200.

Rossi JJ, Sarver N (1990): RNA enzymes (ribozymes) as antiviral therapeutic agents. Trends Biotechnol 8:179–183.

Rossiter BJF, Munir MI, Stout JT, Caskey CT (1989): Repression of HPRT activity by antisense RNA in mammalian cells and transgenic mice. In Brakel CL (ed): Discoveries in Antisense Nucleic Acids. The Woodlands, TX: Portfolio Publishing Company, pp. 57–69.

Rothstein SJ, DiMaio J, Strand M, Rice D (1987): Stable and heritable inhibition of the expression of nopaline synthase in tobacco expressing antisense RNA. Proc Natl Acad Sci USA 84:8439–8443.

Salmons B, Groner B, Friis R, Muellener D, Jaggi R (1986): Expression of anti-sense mRNA in H-*ras* transfected NIH/3T3 cells does not suppress the transformed phenotype. Gene 45:215–220.

Sarver N, Cantin EM, Chang PS, Zaia JA, Ladne PA, Stephens DA, Rossi JJ (1990a): Ribozymes as potential anti-HIV-1 therapeutic agents. Science 247:1222–1225.

Sarver N, Hampel A, Cantin EM, Zaia JA, Chang PS, Johnston MI, McGowan J, Rossi JJ (1990b): Self-cleaving RNAs (ribozymes) as new modalities for anti-HIV therapy. Ann NY Acad Sci 616:606–609.

Saxena SK, Ackerman EJ (1990): Ribozymes correctly cleave a model substrate and endogenous RNA in vivo. J Biol Chem 265:17106–17109.

Sbisà E, Nardelli M, Saccone C (1988): Symmetric transcription of the replication origin of rat mitochondrial DNA. Gene 72:309–310.

Schröder HC, Wenger R, Gerner H, Reuter P, Kuchino Y, Sladic D, Muller WE (1989): Suppression of the modulatory effects of the antileukemic and antihuman immunodeficiency virus compound avarol on gene expression by tryptophan. Cancer Res 49:2069–2076.

Schwartzberg PL, Goff SP, Robertson EJ (1989): Germline transmission of a c-*abl* mutation produced by targeted gene disruption in ES cells. Science 246:799–803.

Shakin SH, Liebhaber SA (1986): Destabilization of messenger RNA/complementary DNA duplexes by the elongating 80 S ribosome. J Biol Chem 261:16018–16025.

Sheehy RE, Kramer M, Hiatt WR (1988): Reduction of polygalacturonase activity in tomato fruit by antisense RNA. Proc Natl Acad Sci USA 85:8805–8809.

Sherman MI (1990): Antisense and antiviral therapy. Ann NY Acad Sci 616:201–204.

Shibahara S, Mukai S, Morisawa H, Nakashima H, Kobayashi S, Yamamoto N (1989): Inhibition of human immunodeficiency virus (HIV-1) replication by synthetic oligo-RNA derivatives. Nucleic Acids Res 17:239–252.

Sigman DS, Chen CH (1990): Chemical nucleases: New reagents in molecular biology. Annu Rev Biochem 59:207–236.

Simons RW (1988): Naturally occurring antisense RNA control—A brief review. Gene 72:35–44.

Simons RW, Kleckner N (1988): Biological regulation by antisense RNA in prokaryotes. Annu Rev Genet 22:567–600.

Smith CJS, Watson CF, Ray J, Bird CR, Morris PC, Schuch W, Grierson D (1988): Antisense RNA inhibition of polygalacturonase gene expression in transgenic tomatoes. Nature 334:724–726.

Smith CJS, Watson CF, Bird CR, Ray J, Schuch W, Grierson D (1990a): Expression of a truncated tomato polygalacturonase gene inhibits expression of the endogenous gene in transgenic plants. Mol Gen Genet 224:477–481.

Smith CJS, Watson CF, Morris PC, Bird CR, Seymour GB, Gray JE, Arnold C, Tucker GA, Schuch W, Harding S, Grierson D (1990b): Inheritance and effect on ripening of antisense polygalacturonase genes in transgenic tomatoes. Plant Mol Biol 14:369–379.

Song KY, Schwartz F, Maeda N, Smithies O, Kucherlapati R (1987): Accurate modification of a chromosomal plasmid by homologous recombination in human cells. Proc Natl Acad Sci USA 84:6820–6824.

Spencer CA, Gietz RD, Hodgetts RB (1986): Overlapping transcription units in the dopa decarboxylase region of *Drosophila*. Nature 322:279–281.

Standart N, Dale M, Stewart E, Hunt T (1990): Maternal mRNA from clam oocytes can be specifically unmasked in vitro by antisense RNA complementary to the 3′-untranslated region. Genes Dev 4:2157–2168.

Stein CA, Cohen JS (1988): Oligodeoxynucleotides as inhibitors of gene expression: A review. Cancer Res 48:2659–2668.

Stein CA, Cohen JS (1989): Antisense compounds: Potential role in cancer therapy. In DeVita VT, Hellman S, Rosenberg SA (eds): Important Advances in Oncology. Philadelphia: JB Lippincott, pp. 79–97.

Stein CA, Mori K, Loke SL, Subasinghe C, Shinozuka K, Cohen JS, Neckers LM (1988): Phosphorothioate and normal oligodeoxyribonucleotides with 5′-linked acridine: Characterization and preliminary kinetics of cellular uptake. Gene 72:333–341.

Stevens JG, Wagner EK, Devi-Rao GB, Cook ML, Feldman LT (1987): RNA complementary to a herpesvirus α gene mRNA is prominent in latently infected neurons. Science 235:1056–1059.

Stout JT, Caskey CT (1987): Antisense RNA inhibition of endogenous genes. Methods Enzymol 151:519–530.

Stout JT, Caskey CT (1990): Antisense RNA inhibition of HPRT synthesis. Somat Cell Mol Genet 16:369–382.

Strickland S, Huarte J, Belin D, Vassalli A, Rickles RJ, Vassalli JD (1988): Antisense RNA directed against the 3' noncoding region prevents dormant mRNA activation in mouse oocytes. Science 241: 680–684.

Strobel SA, Dervan PB (1990): Site-specific cleavage of a yeast chromosome by oligonucleotide-directed triple-helix formation. Science 249:73–75.

Strobel SA, Dervan PB (1991): Single-site enzymatic cleavage of yeast genomic DNA mediated by triple helix formation. Nature 350:172–174.

Sullenger BA, Lee TC, Smith CA, Ungers GE, Gilboa E (1990): Expression of chimeric tRNA-driven antisense transcripts renders NIH 3T3 cells highly resistant to Moloney murine leukemia virus replication. Mol Cell Biol 10:6512–6523.

Sumikawa K, Miledi R (1988): Repression of nicotinic acetylcholine receptor expression by antisense RNAs and an oligonucleotide. Proc Natl Acad Sci USA 85:1302–1306.

Symons B (1989a): Viroids. Pathogenesis by antisense [news]. Nature 338:542–543.

Symons RH (1989b): Self-cleavage of RNA in the replication of small pathogens of plants and animals. Trends Biochem Sci 14:445–450.

Taira K, Oda M, Shinshi H, Maeda H, Furukawa K (1990): Construction of a novel artificial-ribozyme-releasing plasmid. Protein Eng 3:733–737.

Takayama KM, Inouye M (1990): Antisense RNA. Crit Rev Biochem Mol Biol 25:155–184.

Thompson-Jager S, Domdey H (1990): The intron of the yeast actin gene contains the promoter for an antisense RNA. Curr Genet 17:269–273.

To RY, Booth SC, Neiman PE (1986): Inhibition of retroviral replication by antisense RNA. Mol Cell Biol 6:4758–4762.

Tosic M, Roach A, de-Rivaz JC, Dolivo M, Matthieu JM (1990): Post-transcriptional events are responsible for low expression of myelin basic protein in myelin deficient mice: Role of natural antisense RNA. EMBO J 9:401–406.

Toulmé JJ, Hélène C (1988): Antimessenger oligodeoxyribonucleotides: An alternative to antisense RNA for artificial regulation of gene expression—A review. Gene 72:51–58.

Trevor K, Linney E, Oshima RG (1987): Suppression of endo B cytokeratin by its antisense RNA inhibits the normal coexpression of endo A cytokeratin. Proc Natl Acad Sci USA 84:1040–1044.

Tschudi C, Ullu E (1990): Destruction of U2, U4, or U6 small nuclear RNA blocks trans splicing in trypanosome cells. Cell 61:459–466.

Tuerk C, Gold L (1990): Systematic evolution of ligands by exponential enrichment: RNA ligands to bacteriophage T4 DNA polymerase. Science 249:505–510.

Uhlenbeck OC (1987): A small catalytic oligoribonucleotide. Nature 328:596–600.

Uhlmann E, Peyman A (1990): Antisense oligonucleotides: A new therapeutic principle. Chem Rev 90: 543–584.

Van der Krol AR, Mol JN, Stuitje AR (1988a): Antisense genes in plants: An overview. Gene 72:45–50.

Van der Krol AR, Mol JNM, Stuitje AR (1988b): Modulation of eukaryotic gene expression by complementary RNA or DNA sequences. BioTechniques 6:958–976.

Van der Krol AR, Mur LA, Beld M, Mol JNM, Stuitje AR (1990): Flavonoid genes in petunia: Addition of a limited number of gene copies may lead to a suppression of gene expression. Plant Cell 2:291–299.

van Duijn LP, Holsappel S, Kasperaitis M, Bunschoten H, Konings DAM, Voorma H (1988): Secondary structure and expression in vivo and in vitro of messenger RNAs into which upstream AUG codons have been inserted. Eur J Biochem 172:59–66.

van Duin M, van-Den-Tol J, Hoeijmakers JH, Bootsma D, Rupp IP, Reynolds P, Prakash L, Prakash S (1989): Conserved pattern of antisense overlapping transcription in the homologous human ERCC-1 and yeast RAD10 DNA repair gene regions. Mol Cell Biol 9:1794–1798.

Verspieren P, Cornelissen AWCA, Thuong NT, Hélène C, Toulmé JJ (1987): An acridine-linked oligodeoxynucleotide targeted to the common 5' end of trypanosome mRNAs kills cultured parasites. Gene 61:307–315.

Vlassov VV, Gaidamakov SA, Zarytova VF, Knorre DG, Levina AS, Nikonova AA, Podust LM, Fedorova OS (1988): Sequence-specific chemical modification of double-stranded DNA with alkylating oligodeoxyribonucleotide derivatives. Gene 72: 313–322.

von Rüden T, Gilboa E (1989): Inhibition of human T-cell leukemia virus type I replication in primary human T cells that express antisense RNA. J Virol 63:677–682.

Wagner EK, Devi-Rao G, Feldman LT, Dobson AT, Zhang YF, Flanagan WM, Stevens JG (1988): Physical characterization of the herpes simplex virus latency-associated transcript in neurons. J Virol 62:1194–1202.

Wagner RW, Nishikura K (1988): Cell cycle expression of RNA duplex unwindase activity in mammalian cells. Mol Cell Biol 8:770–777.

Walder J (1988): Antisense DNA and RNA: Progress and prospects. Genes Dev 2:502–504.

Waugh DS, Pace NR (1985): Catalysis by RNA. BioEssays 4:56–61.

Weintraub H (1990): Antisense RNA and DNA. Sci Am 262:40–46.

Weintraub H, Izant JG, Harland RM (1985): Anti-sense RNA as a molecular tool for genetic analysis. Trends Genet 1:22–25.

Wetmur JG, Davidson N (1968): Kinetics of renaturation of DNA. J Mol Biol 31:349–370.

Wickstrom E (ed) (1991): Prospects for Antisense Nucleic

Acid Therapy of Cancer and AIDS. New York: Wiley-Liss, Inc.

Williams T, Fried M (1986): A mouse locus at which transcription from both DNA strands produces mRNAs complementary at their 3′ ends. Nature 322:275–279.

Winston F, Chumley F, Fink GR (1983): Eviction and transplacement of mutant genes in yeast. Methods Enzymol 101:211–228.

Wolin SL, Walter P (1988): Ribosome pausing and stacking during translation of a eukaryotic mRNA. EMBO J 7:3559–3569.

Woodson SA, Cech TR (1989): Reverse self-splicing of the tetrahymena group I intron: Implication for the directionality of splicing and for intron transposition. Cell 57:335–345.

Wormington WM (1986): Stable repression of ribosomal protein L1 synthesis in *Xenopus* oocytes by microin-jection of antisense RNA. Proc Natl Acad Sci USA 83:8639–8643.

Xiao W, Rank GH (1988): Generation of an ilv brady-trophic phenocopy in yeast by antisense RNA. Curr Genet 13:283–289.

Yokoyama K, Imamoto F (1987): Transcriptional control of the endogenous MYC protooncogene by antisense RNA. Proc Natl Acad Sci USA 84:7363–7367.

Zamecnik PC, Stephenson ML (1978): Inhibition of Rous sarcoma virus replication and cell transformation by a specific oligodeoxynucleotide. Proc Natl Acad Sci USA 75:280–284.

Zaug AJ, Been MD, Cech TR (1986): The *Tetrahymena* ribozyme acts like an RNA restriction endonuclease. Nature 324:429–433 [published erratum appears in Nature 1987, 325:646].

Zon G (1990): Innovations in the use of antisense oligo-nucleotides. Ann NY Acad Sci 616:161–172.

ABOUT THE AUTHORS

JAMES A.H. MURRAY is a University Lecturer in Biotechnology at the University of Cambridge. While studying for his B.A. in Natural Sciences at Cambridge, he specialized in genetics. After graduating in 1983, he went to the European Molecular Biology Laboratory in Heidelberg, Germany, where he studied for his Ph.D. under Dr. Gianni Cesareni in the Gene Structure Division. His thesis work concentrated on developing plasmid expression systems for yeast and on the analysis of the maintenance mechanisms of the 2μ circle plasmid of yeast. In 1988 he took up a faculty position at the Institute of Biotechnology in the University of Cambridge, where he lectures on gene expression and genetic engineering in yeast. Dr. Murray's research group continues studies both on yeast plasmids and on techniques to manipulate the growth and genomes of higher eukaryotic organisms. His research papers have appeared in *EMBO Journal*, the *Journal of Molecular Biology*, and *Nucleic Acids Research*. Dr. Murray is a member of the editorial board of *Methods in Molecular and Cellular Biology*. He has also edited *Transgenesis—Applications of Gene Transfer*, published by John Wiley & Sons in 1992.

NIGEL CROCKETT is a postdoctoral research associate at the Institute of Biotechnology, Cambridge University. He received his B.Sc. in Chemistry in 1986 from the University of Manchester Institute of Science and Technology where he was awarded the Heslop Prize for chemistry. In Manchester he carried out research into anti-tumor platinum complexes with Professor Noel McAuliffe and was a member of a team working on the production of fuel grade oils from biomass. In 1989 he received his Ph.D. from Cambridge University for his work on Vitamin-B12 biosynthesis while working with Dr. Chris Abell, cloning essential genes on the biosynthetic pathway of vitamin-B12 and overexpressing their gene products for experiments investigating the enzyme active sites. He is currently working with Dr. James A.H. Murray on the regulation of gene expression in *Saccharomyces cerevisiae*, with particular emphasis on the natural amplification mechanism of the 2μ plasmid and on systems employed for the controllable expression of heterologous proteins in yeast.

Antisense RNA and DNA: 51–76
© 1992 Wiley-Liss, Inc.

Regulation of Gene Expression and Function by Antisense RNA in Bacteria

Christopher M. Thomas

I. INTRODUCTION

The role of RNA as a *trans*-acting regulatory molecule in bacteria has been recognized for a little over a decade, and the nature of this regulatory activity has been reviewed a number of times recently [e.g., Green et al., 1986; Inouye and Delihas, 1988; Polisky, 1988; Simons and Kleckner, 1988; Inouye, 1988; Simons, 1988; Takayama and Inouye, 1990]. This role for RNA was discovered initially through studies on replication and its control in bacterial plasmids; therefore these will be the focus of the first section of this chapter. However, regulatory RNA molecules have subsequently been discovered in many other bacterial systems, and these will form the basis of the second group of sections, which include examples from chromosomal gene regulation as well as phage and transposons. Finally, artificial RNA regulators have been developed for the controlled expression of cloned genes in bacteria, and this will be covered in the last main section.

Although the majority of this book focuses on the use of antisense RNA and DNA to regulate eukaryotic gene expression, these bacterial examples are important for a number of reasons. First, they illustrate the ways in which regulatory RNA can achieve its effects. Second, they reveal the properties of RNA molecules that are important for RNA–RNA interactions and that may be important for the design of artificial regulatory elements. Third, the potential role of accessory proteins in modulating RNA–RNA interactions is demonstrated. As well as highlighting them during the description of specific examples, these general points are discussed more extensively at the end of the chapter.

A. Analysis of RNA Structure

In building models to describe the mechanism of antisense RNA, a knowledge of the secondary and tertiary structures of the RNA molecules involved is vital. The likely stem and loop structures that an RNA molecule will adopt can be predicted using rules developed a number of years ago [Tinoco et al., 1973] and incorporated into a variety of computer programs such as STEMLOOP, which is part of the widely used Wisconsin (GCG) package [Devereux et al., 1984]. Further refinements of such methods have also been published [e.g., Jacobson et al., 1984; Konings and Hogeweg, 1989]. Evidence for the physical existence of the predicted structures has been obtained in a number of ways.

Genetic evidence based on the effect of primary mutations and second site suppressors of those mutations can provide evidence of stem structures. A mutation in a region proposed to encode both an essential transcript and its antisense repressor may affect repressor activity even though it reduces neither complementarity nor the strength of interaction between the two RNAs, e.g., a G to C or a C to G transversion. The observed effect may be due to the disruption of a stem structure. In this case the effect should be reversed if the opposite base in the stem structure is changed by a second transversion so as to be complementary once again. An example of such study is given by Case et al. [1989].

Biochemical evidence for the existence of such structures is commonly obtained in two ways. One approach is to use ribonucleases with specificity for double- or single-stranded RNA. Partial digestion of end-labeled RNA is carried out and the products analyzed by denaturing polyacrylamide gel electrophoresis and autoradiography. The size of the radioactive products observed locates regions of specific structure. Commonly used ribonucleases are RNase T1, which prefers G in single-stranded RNA (ssRNA); RNase A, which prefers C and T in ssRNA; RNase T2, which cleaves any base in ssRNA; and RNase V_1, which cleaves preferentially in double-stranded RNA (dsRNA) helical regions. A good example of this approach to determining RNA structure is given by Wagner and Nordström [1986].

A second approach is to use short synthetic oligodeoxyribonucleotides as probes for RNA structure. If the RNA complementary to the probe is in a single-stranded segment, then a short RNA–DNA duplex should be formed, and this will be cleaved during subsequent incubation with RNase H, producing a novel band or set of bands of characteristic length. However, if the target for the oligonucleotide is already sequestered in a double-stranded segment, then it will not form an RNA–DNA hybrid with the probe and no RNase H cleavage will occur. An example of this method has been published by Öhman and Wagner [1989].

The same methods can be used to follow the interaction between antisense RNA and its target. This is well illustrated in a recent study on the interaction of RNA I-RNA II from ColE1 [Tomizawa, 1990a]. Here the interaction between the two RNAs involves association between the single-stranded loops of the two RNAs. They thus become incorporated into dsRNA and are no longer susceptible to single-stranded–specific RNase T2, but are susceptible to RNase V_1. The time course and concentration dependence of this process can be followed by the appearance of the bands characteristic of the complex.

B. Determining the Action of Antisense RNA

An excellent short review of approaches to investigating antisense RNA function is given by Wagner et al. [1988]. Much early work on regulation by antisense RNA simply used transcriptional and translational probe genes such as *lac*Z fused to the expression signals of interest. Typically, antisense RNA has no effect on the transcription from the promoter region but does inhibit expression of a translational probe. However, there are many points after initiation of transcription at which gene expression can be modulated. These include premature transcriptional termination (attenuation), decreased mRNA functional half-life, and direct inhibition of translational initiation or elongation. To define the point of inhibition it is necessary to analyze the effects on mRNA size and abundance by Northern blot analysis. A direct inhibition of translation is most likely if the properties of mRNA are not affected. However, even in IS*10*, where the primary effect of RNA–OUT is on initiation of translation, subsequent cleavage by the double-stranded–specific ribonuclease RNase III occurs, causing irreversible inactivation of the mRNA [Case et al., 1989; see Robertson, 1990, for review of RNase III]. Distinguishing between the primary and secondary consequences of the antisense RNA requires either a genetic approach [e.g., to determine the effect of inactivating the gene for RNase III) or a biochemical approach (e.g., to develop an in vitro system in which to determine the effect of RNA–OUT on translation in the presence and absence of RNase III). In the case of the chromosomally encoded *mic*F RNA, Andersen et al. [1989] concluded that the primary effect

was on translation because boosting the level of *mic*F RNA increased the level of translation inhibition much more dramatically than it affected mRNA stability. At present therefore there is no single way to determine by which mechanism an antisense RNA species is having its effect. The following sections illustrate the variety of mechanisms found in well-studied examples. These naturally occurring examples are listed in Table I. References to the individual systems are given in the text, where they are described in more detail.

II. RNA REGULATORS OF PLASMID REPLICATION

Bacterial plasmids are extrachromosomal elements that replicate autonomously within the cytoplasm of their bacterial host. Particular plasmid species are found at characteristic numbers of copies per host chromosome or per cell. While plasmids generally use at least some of the host's DNA replication machinery, all plasmids thus far studied in detail have been found to encode their own copy number control mechanisms. Studies on the replication control system of ColE1 and related plasmids were the first to reveal RNA as a *trans*-acting repressor molecule [Tomizawa et al., 1981; Lacatena and Cesareni, 1981]. Subsequently other groups of plasmids have also been found to utilize RNA molecules as key elements in the control of their replication. RNA has one major advantage over protein as a system for sensing and regulating plasmid copy number, namely, most RNA species except tRNA and rRNA and certain stable mRNAs have half-lives of a few minutes, and therefore the concentration of such a repressor RNA, if transcribed constitutively from a plasmid DNA template, will respond rapidly to changes in plasmid copy number. This will lead to rapid inhibition or stimulation of plasmid replication as plasmid copy number rises or falls. The plasmid systems in which such regulatory RNAs have been found are described below.

TABLE I. Examples of Naturally Occurring Antisense RNA Regulators*

Target gene			Antisense RNA		
Name	Function	System	Name	Size (nt)	Target
Interference with transcription (transcriptional attenuation)					
repC	Replication	pT181	RNA I	87	L-RNA**
traJ	Transfer	F	FinP	105	fisO (L-RNA)
crp	cAMP receptor	E. coli	tic	92	L-RNA
Direct interference with translation (sequestering of ribosome-binding site)					
ant	Antirepression	P22	sar	71	L-RNA
ftsZ	Cell division	E. coli	DicF	53	L-RNA
mok	Plasmic stability	R1	Sok	67	L-RNA
ompF	Outer membrane	E. coli	micF	174	L-RNA
Q	Antitermination	λ	P_{aQ} RNA	220	L-RNA
sulA	Division inhibition	E. coli	isf	353	L-RNA
tnp	Transposition	Tn10	RNA_{OUT}	69	RNA_{IN}
Indirect interference with translation (RNase cleavage?)					
repA	Replication	R1	CopA	89	CopT
cII/O	Control/replication	λ	cII/O	77	cII/O intergenic region
Interference with other RNA functions (tertiary structure change?)					
RNA II	Replication	ColE1	RNA I	108	RNA II 5' end

*References containing details of these systems can be found in the text.
**L-RNA, leader RNA.

A. ColE1-Like Plasmids

ColE1 is a 6.4 kb (kilobase pair) plasmid from *Escherichia coli* that encodes the colicin E1 protein as well as immunity to the lethal effect of this protein. It is mobilizable by the conjugative apparatus of a variety of self-transmissible plasmids. ColE1 is maintained at about 15 copies per chromosome during exponential growth, a copy number high enough to ensure stable inheritance by random partitioning to daughter cells as long as the multimers that arise by homologous recombination between monomers are broken down [Summers and Sherratt, 1984]. No positively acting plasmid-encoded proteins are required for replication of ColE1.

Initiation of replication is absolutely dependent on the association of a preprimer RNA (RNA II) with the origin of replication, *ori*V [Itoh and Tomizawa, 1980; Dasgupta et al., 1987; Masukata et al., 1987] (Fig. 1). Initiation of leading strand synthesis normally occurs as a result of elongation by DNA polymerase

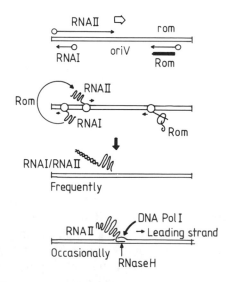

Fig. 1. *Summary of the replication control system of ColE1.*

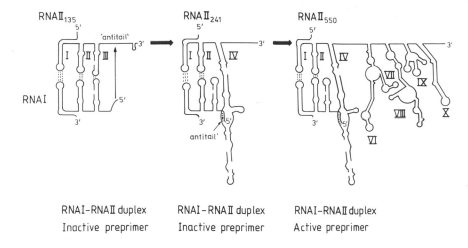

Fig. 2. *Alternative structures of the ColE1 RNA II and their interaction with RNA I.*

I from a primer produced by RNase H processing of the 555 base RNA II [Itoh and Tomizawa, 1980]. In the absence of RNase H there are at least two alternative routes to initiation of replication, but in all cases it depends on association of RNA II with the *ori*V region. RNA II can adopt at least two different secondary structures, only one of which can associate with *ori*V.

1. Control by RNA I. Control over this process is exerted by a 108 base antisense RNA (RNA I) that is transcribed from the opposite strand to RNA II [Tomizawa et al., 1981; Lacatena and Cesareni, 1981] and interacts with RNA II to form a dsRNA duplex [Tomizawa, 1984]. Sequestration into this form prevents RNA II from adopting the secondary structure that associates with *ori*V and can therefore be productively cleaved by RNase H to produce active primer [Masukata and Tomizawa, 1986]. However, the influence of RNA I on RNA II structure and function must be quite finely balanced, because a single point mutation can create an RNA transcript that at high temperature is not inactivated by RNA I binding [Fitzwater et al., 1988].

RNA I adopts a structure consisting of three hairpin loops (Fig. 2), and efficient interaction between RNA I and its target (RNA II) appears to depend on this secondary structure [Wong and Polisky, 1985]. RNA II can adopt a similar structure with complementary loops, but the existence of this form is only transient, loop III being replaced by loop IV when RNA II has grown to greater than 185 nucleotides in length (Fig. 2). Although full-length RNA II can still associate with RNA I, when it is longer than 360 nucleotides binding of RNA I no longer inhibits it from interacting with *ori*V and being processed to an active primer [Tomizawa, 1986]. Therefore the rate of interaction between RNA I and RNA II is critical for inhibition.

One set of data suggests that, once the conformational change produced by the appearance of the loops has occurred, RNA II becomes much less sensitive to RNA I [Wong and Polisky, 1985], formation of loop IV causing a sixfold reduction in the rate of interaction and further elongation of RNA II causing an even greater fall. Recent studies do not support these observations, suggesting that, if anything, the RNA II$_{241}$ species examined in both studies is more reactive than the shorter species, which adopts the same conformation as RNA I [Tomizawa, 1986, 1990a].

Interaction between RNA I and RNA II proceeds via unstable intermediates identified by the ability of related but different RNA Is to inhibit pairing, although they themselves cannot progress further than the initial interactions. There follows a step in which the single-stranded loops of RNA I and RNA II interact and form double-stranded sections that are resistant to RNase V_1. Conversion to full-length dsRNA duplex depends on the 5' ssRNA tail of RNA I, which interacts with its complementary region, which is partially single stranded in loop IV [Tomizawa, 1990a]. This region has been referred to as the "anti-tail," and its accessibility is essential for inhibition by RNA I. Point mutations that change the folding pattern so that the anti-tail does not go through this accessible stage cause an increase in copy number [Polisky et al., 1990]. This region of interaction with the tail of RNA I then spreads to create a duplex along the whole length of RNA I. The formation of loop IV can therefore be seen to facilitate the initiation of pairing between RNA I and RNA II [Tomizawa, 1986, 1990a]. Analysis of the interaction of RNA I with RNA II transcripts of various lengths has shown that a single-stranded tail is the essential component after the initial "kissing." If the RNA II is truncated to less than 111 nucleotides and no longer has a single-stranded tail (just three loops) it reacts very slowly, but the rate of reaction increases again with further truncation, which destroys a hairpin and creates a new ssRNA tail [Tomizawa, 1986]. This indicates that a stem–loop structure is susceptible to invasion by a complementary ssRNA tail.

Evidence that the inhibition by RNA I occurs as described above also comes from studies on mutations affecting the activity of RNA I. Mutations that reduce neither the complementarity of RNAI and RNA II nor the strength of the interaction between the two RNAs but disrupt the formation of the stems of the hairpin loops reduce the repression brought about by RNA I. Mutations in the loop part of these structures also reduce repression if they affect the complementarity or the strength of the interaction [Tomizawa, 1984; Lacatena and Cesareni,

1983; Tamm and Polisky, 1983]. Saturation mutagenesis using a phage λ–ColE1 hybrid provided good evidence that five bases in loop II of RNA I are the most critical for its activity, with two bases in loop I providing a secondary interaction (Fig. 3) [Lacatena and Cesareni, 1983].

The concentration of RNA I is proportional to plasmid copy number, and therefore plasmid replication is progressively inhibited as copy number rises. During growth with glycerol as sole carbon source, the RNA I promoter fires once every 30 seconds, the RNA II promoter fires once every 3 minutes, and it has been estimated that at steady state only 1 in every 15 RNA II molecules is processed productively [Lin-Chao and Bremer, 1987]. A fall in copy number is rapidly followed by a fall in RNA I concentration, since it has a half-life of 55 seconds.

2. Modulation of repression by Rom protein. The interaction between RNA I and RNA II is catalyzed by a plasmid-encoded protein [Tomizawa and Som, 1984], the product of the gene called *rom* or *rop* [Cesareni et al., 1982], which is a dimer with two identical subunits of 63 amino acids [Banner et al., 1987]. The production of active Rom is part of the plasmid replication control system and further reduces plasmid copy number about twofold. The Rom protein has been crystallized, and its structure has been determined by X-ray crystallography to a resolution of 1.7Å [Banner et al., 1987]. Each Rom monomer consists of two α-helices. Two monomers fit together head to tail so that the four α-helices run parallel to each other (Fig. 4). Extensive site-directed mutagenesis followed by functional analysis of mutant proteins suggests that the Rom structure is very stable, not being easily disrupted by single amino acid substitutions or by small insertions or deletions [Castagnoli et al., 1989]. Rom has a net negative charge, and replacement of basic side chains by nonbasic ones, thus further increasing the net acidity of Rom, does not have a major effect on activity. Therefore it seems unlikely that Rom

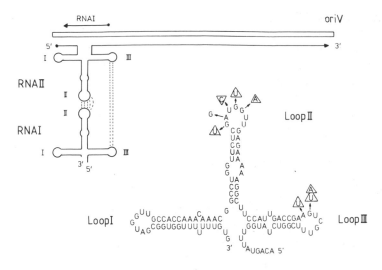

Fig. 3. *Summary of mutations affecting loop II and loop III in RNA I of ColE1.*

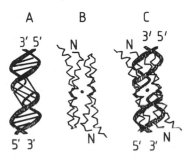

Fig. 4. *Proposed interaction of Rom with RNA I–RNA II paired loops.*

acts simply by reducing the electrostatic repulsion between the RNA I and RNA II. All except one of the few substitutions that have been shown to cause a major decrease in Rom activity are located at the ends of the cylinders formed by the paired α-helices.

Rom binds to an RNA I–RNA II complex with low affinity and protects the stems of both RNA I and RNA II from nuclease digestion (Fig. 4) [Helmer-Citterich et al., 1988; Eguchi and Tomizawa, 1990]. It recognizes the general stem–loop structure, rather than being highly sequence specific. It decreases the dissociation constant of the RNA I–RNA II complex and converts an unstable intermediate to a more stable one, which then follows a faster route of isomerization to form eventually the dsRNA duplex [Tomizawa, 1990b]. It not only speeds up the interaction of perfectly complementary RNA I and RNA II molecules but also suppresses the effect of point mutations that reduce the complementarity [Dooley and Polisky, 1987]. The degree of sequence specificity that Rom requires is not known. It is conceivable that protein engineering may be able to broaden or change the specificity of Rom so that it can be exploited in other contexts as well.

3. Modulation of repression by host genes. In addition to Rom, it has recently become clear that host factors can affect the inhibitory activity of RNA I. Host *pcn* mutations that result in a reduction in copy number of ColE1-like plasmids are easily isolated after growth of bacteria carrying recombinant plasmids carrying heterologous genes (e.g., interferon) whose expression is deleterious to the bacterial host. The chromosomal gene affected in *pcn*B mutants has been mapped, cloned, and sequenced [Lopilato et al., 1986; Liu and Parkinson, 1989]. Database comparisons have indicated that PcnB has homology to the likely tRNA-binding domain of a bacterial enzyme, tRNA nucleotidyltransferase, which adds CCA to tRNA, and this has led to the suggestion [Masters et al., 1989] that the effect of PcnB is due to its sequestration of RNA I, which resembles tRNA structurally, into a less active form. Inactivation of *pcn*B therefore leads to an increase in RNA I potency.

ColE1 therefore illustrates two general points:

1. Interaction of mRNA with antisense RNA can alter its potential secondary structure, but this sensitivity to functional inhibition may only be transient.

2. Inhibition by antisense RNA can be modulated either by a protein that catalyzes the formation of a dsRNA duplex or by a protein that apparently sequesters the inhibitory RNA.

B. IncFII Plasmids

In contrast to ColE1, IncFII plasmids of *E. coli* (R1, R100, R6-5, NR1) are large (100 kb, approximately), low copy number (one to two copies per chromosome), self-transmissible plasmids [for a review, see Womble and Rownd, 1988]. Replication depends on host proteins as well as on the product of the *rep*A gene (Fig. 5). A sequence, CIS, located between *rep*A and *ori*V induces RepA to act preferentially in *cis* [Masai and Arai, 1988]. Regulation of the amount of RepA controls plasmid replication. While initiation of transcription of the *rep*A gene is repressed by the interaction of repressor protein CopB with an operator overlapping one of the two *rep*A promoters [Riise and Molin, 1986], it is still transcribed constitutively at a low level from the *cop*B promoter (Fig. 5). A second repressor, CopA, a 91 nucleotide antisense RNA, represses *rep*A expression by forming a duplex with the *rep*A mRNA [Light and Molin, 1983; Womble et al., 1984]. As with ColE1, the CopA RNA adopts hairpin loop structures, the larger of which appears to be necessary for interaction with its target, as discussed in the next paragraph.

The sequence in the *rep*A mRNA that is complementary to CopA is designated CopT (T for target). The structures of CopA and CopT have

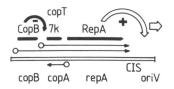

Fig. 5. *Summary of the IncFII replication control system.*

been probed with sequence- and structure-specific RNases [Wagner and Nordström, 1986; Öhman and Wagner, 1989]. Their structures are shown in Figure 6. Stem–loop II appears to be the most critical structure for inhibition, since mutations that cause changes in copy number map in this region [Wagner et al., 1988]. The majority of the mutations are found in the single-stranded loop and result in the change of a G–C pair between CopA and CopT to an A–U pair that is of lower thermodynamic stability. That such a small change can have a major effect suggests that the initial CopA–CopT interaction or "kissing" involves only a few base pairs, i.e., a subset of those bases in the single-stranded loop. In addition, the efficiency with which this "kissing" occurs may be critically influenced by the structure of the nucleotides in the single-stranded loop, since mutations that contract it to 4 nucleotides or increase it to 12 nucleotides drastically affect the interaction [Wagner et al., 1988].

The kinetics of the CopA–CopT interaction have been represented in a number of ways. In the approach of Persson et al. [1988], the interaction was split into two stages: a reversible binding characterized by forward and backward rate constants k_{f1} and k_{b1}; and an irreversible conversion to full-length dsRNA duplex characterized by the rate constant k_{f2}. The initial interaction was analyzed by taking either radioactive CopA or radioactive CopT and studying its conversion into the dsRNA complex as a function of concentration and of time in the presence of an excess of the nonradioactive component of the reaction. Under these circumstances the reaction proceeds with pseudo-first-

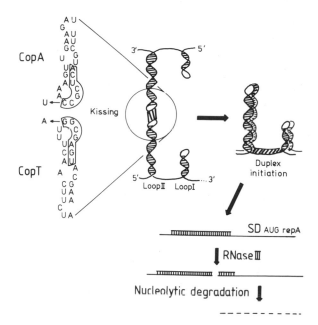

Fig. 6. *Structure of CopA and CopT of R1.*

order kinetics. The rate constant for binding was estimated to be approximately $1 \times 10^6 \, M^{-1}$ s^{-1} at 37°C. When the equivalent rate constant was estimated for interaction between CopA and CopT from copy number mutants, the size of the reduction in this rate constant was proportional to the change in copy number [Nordström et al., 1988; Persson et al., 1988]. For example, Nordström et al. [1988] found that for a mutant with a three-to fourfold higher copy number the rate of CopA–CopT interactions was reduced four- to fivefold. Therefore the effectiveness of repression can be estimated from the rate of interaction.

However, the above analysis does not allow the strength of the CopA–CopT interaction to be determined. The dissociation rates for the CopA–CopT complex were not determined because the initial reaction never reaches equilibrium. A way to avoid this problem is to use reaction substrates that do not progress further than the initial CopA–CopT complex. This can be achieved through the use of CopA derivatives that lack the single-stranded tails at the base of the loop II stem, since these tails are required for the irreversible conversion to dsRNA [Persson et al., 1990b]. Although such molecules cannot progress to the full-length RNA duplex, they do interfere efficiently with the interaction between normal CopA and CopT. By determining the inhibition constants it is possible to determine the equilibrium constants and therefore binding strength [Persson et al., 1990a]. Such estimates of CopA–CopT binding strength suggest that seven or eight nucleotides eventually interact in the initial CopA–CopT complex. Sensitivity to RNases suggests that it is the left side of the CopA loop II (Fig. 6) that enters the intermolecular double-stranded structure, the right side becoming more accessible to single-strand-specific nucleases (K. Nordström, personal communication).

Following the formation of this initial complex, the formation of the dsRNA duplex along the whole length of the CopA is initiated by the ssRNA at the 5′ side of the loop II stem, although when this is deleted the ssRNA on the other side can replace it [Persson et al., 1990b]. Studies with mutant CopA molecules suggest that it is the length of ssRNA (14 nucleotides or more) rather than the 3′ or 5′ position that is important. Although the initial CopA–CopT interaction does not progress to full-length pairing, it may align the loop II stems to bring the ssRNA close together.

Although CopA has been shown to inhibit the expression of translational probe genes fused to the R1 *rep*A, the CopA–CopT complex does not directly convert the translational signals of the *rep*A gene into an inaccessible form. It has been suggested that the CopA–CopT complex favors one of two alternative *rep*A mRNA structures in which the *rep*A translational signals are inaccessible [Dong et al., 1987], but in vitro analysis of *rep*A mRNA structure in the presence and absence of the CopA–CopT complex suggests that association with CopA does not influence the *rep*A mRNA secondary structure in this region [Öhman and Wagner, 1989]. An alternative explanation is provided by the observation that the CopA–CopT complex is cleaved by RNase III both in vivo and in vitro and that much less mRNA downstream of this cleavage site accumulates [Blomberg et al., 1990]. This suggests that CopA stimulates processing of the *rep*A mRNA, leading to limiting levels of active mRNA. This may still not be the complete story, because in vivo the CopA derivatives lacking the single-stranded tails at the base of the hairpin and thus unable to progress to the dsRNA CopA–CopT complex are still able to repress *rep*A expression (G. Wagner, personal communication).

The CopA system has additional complicating factors. First, the leader RNA prior to *rep*A codes for two additional polypeptides, of 7 and 3 kD, and translation of these small genes may affect the susceptibility to CopA and thus cause plasmid copy number to respond to the level of ribosomes in the cell, i.e., to growth rate [Wagner et al., 1987]. Second, there is a transcriptional pause site prior to the start of the *rep*A coding region, and it has been postulated that this may create a delay that might allow mRNA to adopt a CopA-sensitive structure that is not favored in the full-length transcript [Dong

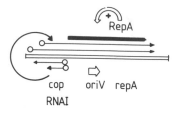

Fig. 7. *Summary of the replication control system of pT181.*

et al., 1987]. The *rep*A mRNA may therefore go through a window of susceptibility to repression, although the in vitro experiments referred to above [Öhman and Wagner, 1989] suggest that this should not be the case.

The IncFII plasmids illustrate a number of ways in which antisense RNA can function and the factors that can or may influence the inhibitory effect of the antisense RNA. In addition, plasmids that exhibit "runaway replication" at high temperature have been developed by replacing the *rep*A promoter with a thermoinducible λ P_L promoter [Larsen et al., 1984]. The complete escape from regulation when P_L is switched on is due to transcription from this promoter inhibiting CopA production. Thus the concentration of mRNA is increased and translational inhibition is reduced at the same time.

C. pT181 and Related Plasmids

While both of the above examples occur in gram-negative bacteria, there is increasing knowledge of similar systems in plasmids of gram-positive bacteria [Novick, 1989]. Replication control has been studied in the most detail for plasmids pT181 and pC194 from the gram-positive bacterial species *Staphylococcus aureus* [Carleton et al., 1984; Alonso and Tailor, 1987]. In pT181, two overlapping antisense RNA species are made complementary to the *rep*A gene [Kumar and Novick, 1985] (Fig. 7). While the original implication was that these species repress the translation of the *rep* gene mRNA, it now appears more likely that they cause atten-

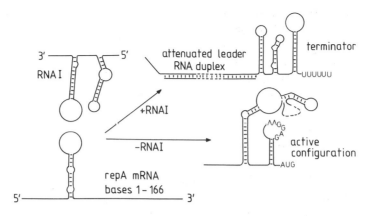

Fig. 8. *Attenuation of rep transcription by antisense RNA in plasmid pT181.*

uation, i.e., premature termination of transcription [Novick et al., 1989; Highlander and Novick, 1990]. This attenuation is brought about as a result of the antisense RNA–mRNA duplex influencing the choice of possible secondary structures and favoring the adoption of the one that includes a transcriptional terminator (Fig. 8). It is interesting to note that the ssRNA loop that is proposed to mediate the interaction between repressor and target in this system is 16 nucleotides in length, considerably longer than the apparent preferred loop size in the case of CopA–CopT interaction for plasmid R1.

D. R6K

Plasmid R6K differs from the examples described above in that its replication region contains three potential origins, α, β, and γ [Crosa, 1980], of which only α and β appear to be significantly active in vivo. The γ origin is repressed by a silencer function, which seems to consist of a countertranscript directed by a promoter region adjacent to the core origin activation sequences [Patel and Bastia, 1986]. This effect is not simply due to transcription from the silencer promoter causing interference as a result of RNA polymerase molecules moving in the opposite direction to the RNA polymerase responsible for activating the origin, since inhibition is still observed even when the

antisense RNA is synthesized *in trans*. Nor does it appear to be due to displacement of the replication initiator protein π by formation of R loops with the origin DNA, since π protein also has affinity for origin-specific RNA–DNA hybrids. Therefore it seems most likely that this silencer RNA acts by association with activator RNAs to form an inactive dsRNA molecule [Patel and Bastia, 1987]. Since the silencer promoter is weak, it seems likely that the silencer RNA is very stable. This is similar to the situation with RNA OUT of Tn*10* (see below) and is more consistent with its role as a silencer than as a means of sensing copy number.

E. Control of Plasmid Transfer–Replication

Conjugative transfer of the F-like sex factors of *E. coli* is controlled by two genes, *fin*P and *fin*O [Ippen-Ihler and Minkley, 1986]. Transfer of F itself is derepressed because *fin*O is inactivated by the insertion of IS*3* [Cheah and Skurray, 1986]. The Fin[+] (fertility inhibition) phenotype of other F-like plasmids is due to the fact that the *fin*O defect in F can be complemented by these plasmids. However, the *fin*P gene is plasmid specific and cannot be complemented by plasmids other than F. FinP is a 105 nucleotide RNA [Dempsey, 1987] and has

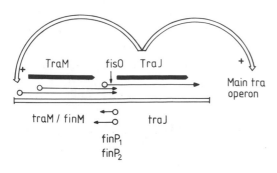

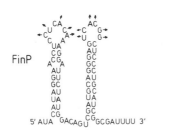

Fig. 9. *Control of tra gene expression by antisense RNA in IncF plasmids.*

the predicted structure shown in Figure 9 [Finlay et al., 1986]. It is complementary to the *tra*J mRNA, TraJ being a positively acting protein essential to activate *tra* operon transcription. Interaction between FinP and *fis*O (the target for FinP) is proposed to lead to transcriptional attenuation, and indeed a shorter RNA species (235 nucleotides in length) is made in the presence of FinP [Dempsey, 1987]. Comparison of FinP from a variety of plasmids with different inhibition specificities revealed major differences in the predicted loop regions [Finlay et al., 1986], consistent with these loops being critical for the inhibitory interaction as found in the plasmid copy number control systems described above.

Further evidence that FinP interacts with *tra*J mRNA is provided by the fact that DNA fragments containing the *tra*J promoter and leader sequence but that lack the *fin*P promoter cause at least 100-fold derepression of R100 transfer when present in the cell at high copy num-ber [Dempsey, 1989a]. This would be expected if the increased concentration of *tra*J leader RNA titrates the antisense RNA present. Transcription of *tra*M that lies immediately up-stream of *tra*J may provide a natural system that can do precisely this. Both the full-length *tra*M transcript and a more abundant shorter transcript from a second promoter extend into the *tra*J open reading frame and thus can titrate FinP if the level is high enough [Dempsey, 1989b]. Transcription of *tra*M, which depends on TraJ, may thus amplify the derepression of *tra*J.

Antisense RNA–mediated control over expression of conjugative transfer functions differs significantly from control over plasmid replication. Even naturally repressed F-like transfer systems remain derepressed for a number of generations after transfer to plasmid-free bacteria. This is most likely explained by the fact that *fin*P RNA is transcribed from a very weak promoter and therefore a long time is required for a significant concentration of the inhibitor to accumulate. The eventual steady-state concentration will depend on how stable the FinP RNA is. Eventual repression by FinP depends on a functional *fin*O gene. It has been suggested that FinO is a short stretch of RNA transcript that interacts with FinP either to stabilize it directly or to modulate its structure [Dempsey, 1987]. The properties of *fin*P and *fin*O mutants described recently are consistent with the idea that FinO catalyzes the formation of loop I of FinP, stabilizes FinP, and stimulates its interaction with its target [Frost et al., 1989]. The idea of one RNA modulating the inhibitory activity of a second RNA molecule on a third RNA is unique among the systems described in this review.

III. ANTISENSE RNA FOR TRIGGERING HOST-LETHAL GENES

A novel role for antisense RNA, as a natural timer that switches on gene function only after the DNA template for the gene has been lost, is illustrated by the *par*B locus of plasmid R1, which consists of at least three genes, *hok*, *mok*, and *sok* [Gerdes et al., 1986b; reviewed by Gerdes et al., 1990a]. The product of the plasmid *hok* gene is lethal to *E. coli* [Gerdes et al., 1986a] but its effect is not normally observed because translation of *hok* mRNA is indirectly repressed by *sok* antisense RNA, which is produced at high levels but is unstable. The *hok* mRNA is quite stable, with a half-life of approximately 20 minutes, but is synthesized at a low rate [Gerdes et al., 1988]. When the plasmid template is lost, the *sok* RNA is degraded more rapidly than *hok* RNA, allowing translation of *hok* and thus killing of the bacteria (Fig. 10) [Gerdes et al., 1988]. However, as with many other examples, more detailed analysis has revealed greater complexity than originally suspected. First, the *hok–sok* dsRNA hybrid is cleaved by RNase III, a process that could be essential for either activation or inactivation of *hok*. In addition, the initial *hok* transcript encodes another gene, *mok*, which

starts before and overlaps with *hok* [Gerdes, et al., 1990a,b]. Introduction of mutations into the *mok* open reading frame that cause termination of translation prior to the overlap with *hok* interfere with *hok* expression, suggesting some sort of translational coupling between *hok* and *mok*. In addition, the structure of the initial transcript is highly folded, and the mRNA is not translated into Hok. Hok is only produced after further mRNA processing, including cleavage at the 3′ end (possibly by RNase E), which appears to occur in cells after loss of the plasmid [Gerdes et al., 1990b]. The requirement of this locus for multiple events before the lethal effects of Hok are revealed may ensure that there are no deleterious effects of *hok* until after the plasmid encoding it is lost. It thus appears to be very well adapted to ensuring the stable inheritance of the plasmid carrying it [Gerdes, 1988].

The *hok/sok* locus of R1 seems to be part of a family of related functions that are found on both plasmids and chromosomes [Gerdes et al., 1990a]. The most closely related of these functions is the *flm* locus of plasmid F, which appears to be a homolog of the R1 *par*B locus [Loh et al., 1988]. An interesting consequence of the posttranscriptional regulation of these loci is the formation of ghosts or lysed bacteria when transcription is inhibited by rifampicin, providing a useful phenotype for detecting the presence of such loci [Gerdes et al., 1990a].

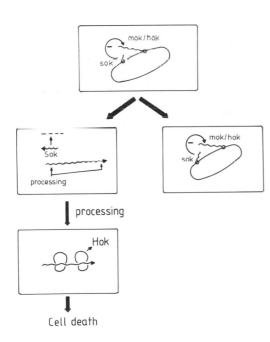

Fig. 10. *Killing of plasmid-free segregant bacteria by the* parB *region of R1.*

IV. ANTISENSE RNA IN CONTROL OF TRANSPOSITION

In IS*10*, the insertion sequence associated with the tetracycline resistance transposon Tn*10*, translation of the transposase from the pIN mRNA is repressed by the pOUT RNA. Formation of a duplex between these two species sequesters the ribosome-binding site into an inaccessible form [Simons and Kleckner,

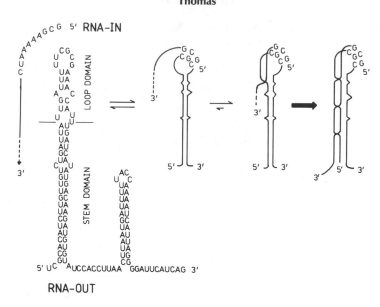

Fig. 11. *RNA–OUT of IS10 and its interaction with the 3' end of RNA–IN.*

1983; Ma and Simons, 1990]. While many of the structural features of RNA–OUT [Kittle et al., 1989] appear similar to stem–loop II of R1 CopA, there are significant differences in this system from the antisense RNAs involved in copy number control (Fig. 11).

First, RNA–OUT is very stable, having a half-life of 60 minutes. Its activity is dependent on this stability. Mutations that change the stem structure decrease its stability and abolish its inhibitory function while mutations restoring the native stem structure reverse this trend. Part of this stability is due to insensitivity to attack from ribonucleases such as RNase III, since some mutations affecting the stem allow attack [Case et al., 1989]. RNase III cuts about 15 bases away from the start of RNA duplexes. A mismatch in the RNA–OUT stem structure at position 15 (Fig. 11) may be crucial to make it a poor substrate for RNase III [Case et al., 1989]. In addition, the stability of the stem may protect against 3' exonucleolytic attack.

Second, RNA–OUT is transcribed from a weak promoter (a strong promoter could interfere with the activity of P_{IN}, the promoter responsible for transposase gene transcription [Case et al., 1988]. The combination of the high stability and the low rate of synthesis means that the concentration of RNA–OUT does not respond rapidly to changes in transposon copy number.

Third, the target for RNA–OUT, RNA–IN is complementary to only part of RNA–OUT. Initial "kissing" between inhibitor and target appears to involve the ssRNA loop of RNA–OUT, as with ColE1 and R1, and the 5' end of RNA–IN as indicated by the effect of point mutations in this region [Kittle et al., 1989].

Conversion to the full-length dsRNA duplex between RNA–OUT and RNA–IN appears to occur by propagation from the initial "kissing" complex rather than from interaction between ssRNA segments at the base of the stem structure (Fig. 11) [Kittle et al., 1989]. Although the interaction with RNA–OUT inhibits translation of transposase from RNA–IN, the inhibition is subsequently rendered irreversible by cleavage of the ds RNA by RNase III [Case et al., 1989].

V. ANTISENSE RNA IN CONTROL OF BACTERIOPHAGE DEVELOPMENT

Both *E. coli* phage λ and the related *Salmonella* phage P22 have been found to encode important antisense RNA molecules. In both phage, the early positive transcription factor (cII in λ, c1 in P22) not only stimulates transcription of genes required for lysogenic establishment, but also delays expression of late gene expression by stimulating transcription from promoter P_{aQ}. This promoter is within the Q gene but faces in the opposite direction and gives rise to an antisense RNA (thought to be 220 nucleotides in length) that inhibits expression of Q [Ho and Rosenberg, 1985; Hoopes and McClure, 1985]. The expression of cII itself appears to be modulated by an antisense RNA, OOP RNA (77 nucleotides), which complements a region of 55 nucleotides at the 3' end of the cII gene and 22 nucleotides in the intercistronic region between cII and O. OOP RNA inhibits cII expression apparently by associating with the cII mRNA, which is expressed as a di-cistronic message with O. This creates an RNase III– sensitive site that leads to cleavage and inactivation of the cII mRNA [Takayama et al., 1987; Krinke and Wulff, 1987]. The cleavage by RNase III occurs 13 nucleotides from the 3' end of the duplex region. By overproduction of OOP RNA the cleavage of the cII–O mRNA could essentially be driven to completion. In RNase III$^+$ bacteria it could be shown that the upstream sequences, encoding cII, are rapidly degraded after the initial cleavage. In RNase III$^-$ bacteria the dsRNA duplex was cleaved at different sites, with the result that it was the downstream section of mRNA, encoding gene O, that disappeared, the cII mRNA remaining reasonably stable [Krinke and Wulff, 1990]. This illustrates the specificity with which mRNA processing and degradation occurs and suggests that when we understand what creates the specificity for attack by all RNases it should then be possible to design antisense inhibitors that target just one part of an mRNA molecule for degradation.

One way in which phage P22 differs from λ is in encoding an antirepressor protein, Ant, that interferes with binding of phage repressors to DNA (it acts on both cI of λ and c2 of P22). Production of Ant is modulated by antisense RNA, Sar, which after an initial burst of Ant production (which may derepress any endogenous related prophages, allowing them to recombine with P22) serves to reduce Ant synthesis as infection progresses, allowing lysogeny to be established. Mutations in *sar* can lead to clear plaque mutants unable to establish lysogeny. The balance of *ant* expression versus *sar* inhibition is interesting. Initially *ant* is transcribed from P_{ANT}, and this transcription interferes with *sar* transcription, presumably the result of the waves of positive DNA supercoiling preceding RNA polymerase starting from P_{ANT}. Later transcription from P_{ANT} is repressed by protein Arc. This not only reduces *ant* transcription but also allows *sar* transcription, achieving inhibition of both transcription and translation of *ant* simultaneously [Wu et al., 1987]. This illustrates one of the most effective ways in which antisense RNA seems to be exploited both naturally and in designed expression vectors [O'Connor and Timmis, 1987]. The structure of *sar* consists of 71 nucleotides, which fold in a dumbbell structure, creating two single-stranded loops whose existence has been confirmed by RNase sensitivity. Initial interaction between *sar* and its *ant* mRNA target probably occurs through the ssRNA loop containing the anti Shine-Dalgarno sequences. Although *sar* and *ant* mRNA rapidly form a dsRNA duplex [Liao et al., 1987], the initial kissing between Sar and its target may be sufficient for inhibition, since this will compete directly with ribosomes for binding to the mRNA. The structure of Sar strengthens the idea that highly structured single-stranded loops are the most active form of antisense RNA.

A novel type of antisense RNA has recently been described in the closely related bacteriophages P1 and P7 [Citron and Schuster, 1990]. The c4 gene controls expression of the *ant* gene, which antagonizes the immunity system of these lysogenic bacteriophage. Both c4 and *ant* are encoded by the same mRNA, and both the absence of a protein product from c4 and the nature of mutations that affect c4 activity suggest that c4 consists of specific RNA sequences. These are proposed to interact with and block the ribosome-binding site of a small open reading frame (*orf*X), which precedes *ant* and whose translation is essential for *ant* expression (presumably translational coupling). This antisense RNA is therefore not a separate species but is part of the mRNA it regulates. How the action of c4 is modulated, since it does not cause constant repression of *ant*, is unclear. Therefore this type of situation seems difficult to exploit as a strategy for control of gene expression by artificial antisense RNA.

VI. ANTISENSE RNA IN CONTROL OF CHROMOSOMAL GENES

Although the majority of known examples of antisense RNA are found in extrachromosomal elements (plasmids, phage, transposable elements), two chromosomal systems have been studied in detail. The *mic*F RNA (93 nucleotides) was the first chromosomally encoded antisense RNA to be discovered, through its ability when cloned at high copy number to inhibit *omp*F expression. OmpF and OmpC are outer membrane porin proteins whose levels are inversely correlated; OmpC predominates at high osmolarity and OmpF at low osmolarity. The *mic*F locus is transcribed divergently from the *omp*C promoter region (47 minutes on the *E. coli* chromosome) but acts on the distant *omp*F locus (21 minutes on the *E. coli* chromosome), showing that inhibitor–target interactions do not require the antisense RNA to be transcribed from the complementary strand of the target gene itself. The sequence of *mic*F is complementary to the 5′ end of the *omp*F mRNA, including the Shine-Dalgarno sequence and the initiation codon, and this leads to the suggestion that it acts to inhibit *omp*F translation when *omp*C is activated (see Fig. 7, Murray and Crockett, this volume) [Mizuno et al., 1984; Andersen et al., 1987, 1989]. This hypothesis was confirmed by studies on new synthesis of OmpF after osmotic shift in strains with and without a functional *mic*F [Aiba et al., 1987]. Response to other forms of environmental stress such as temperature shift and ethanol stress also require *mic*F [Misra and

Reeves, 1987]. Analysis of RNA levels showed that *mic*F RNA levels increase in response to such changes whereas *omp*F mRNA levels decrease. However, while the increased *mic*F RNA level dramatically reduced OmpF production, it had a less marked effect on the level of *omp*F mRNA, suggesting that the primary effect of micF is to inhibit translation, degradation of *omp*F mRNA being a secondary effect that depends on a rate-limiting element such as a ribonuclease [Andersen et al., 1989].

A different mechanism is illustrated by the *crp* gene of *E. coli*, which encodes the cAMP receptor protein. A divergent transcript, *tic* RNA appears to associate with the beginning of the *crp* mRNA to create a structure resembling a transcriptional terminator and leads to attenuation of transcription [Okamoto and Freundlich, 1986]. Using a cloned *crp* gene that shows the same response to cAMP as the chromosomal gene, it has been shown in vivo that inactivation of the promoter for *tic* RNA abolishes regulation by cAMP [Okamoto et al., 1988].

More examples of antisense RNA in chromosomal genes are being discovered, partly because of the employment of more sophisticated techniques and partly because investigators are looking more closely for such phenomena. These new examples may well reveal novel features of antisense RNA. Detailed analysis of the *sul*A–*omp*A operon of *E. coli* has revealed a 353 nucleotide untranslated countertranscript, *isf*, complementary to *sul*A mRNA. Rather as suggested for *sok* of plasmid R1, this may ensure complete switch off of *sul*A when it is not induced, since any expression of *sul*A, whose product is an inhibitor of cell division, would be deleterious to bacterial growth. On SOS induction of *sul*A, transcription of *isf* is suppressed by the more active *sul*A transcription, which therefore effectively eliminates the translational inhibition, allowing full *sul*A expression [Cole and Honore, 1989].

Antisense RNA molecules can also arise from much longer transcriptional units. It has recently been shown that the *dic*B operon encodes a small repressor RNA, DicF (53 nucleotides), that arises from a long noncoding intercistronic region by a combination of transcriptional termination and RNA processing by RNase III and RNase E [Faubladier et al., 1990]. The target for this mRNA is not known but may well be a complementary sequence in the mRNA of *fts*Z (a gene essential for cell division), since overexpression of cloned *fts*Z can overcome the inhibition by DicF (P. Bouché, personal communication).

As with the c4 gene of bacteriophages P1 and P7, sequences internal to polycistronic mRNAs have also been implicated in control without processing. Thus sequences within the *gnd* mRNA, which encodes 6-phosphogluconate dehydrogenase, appear to pair with the translational initiation signals at the beginning of the operon to regulate gene expression [Carter-Muenchau and Wolf, 1989].

Not all countertranscripts necessarily act simply as antisense RNAs. In some cases they may be translated into one or more proteins that play a role in function or control of the genes encoded on the opposite strand. This may be the case in *pts*H–*pts*I locus of the *E. coli* chromosome [Levy et al., 1989].

VII. ARTIFICIAL CONTROL BY ANTISENSE RNA IN BACTERIA

It was quickly recognized that there are many practical applications for antisense RNA in genetic manipulation [Coleman et al., 1984; Pestka et al., 1984]. Some of the initial results were very encouraging and showed that plasmids could be constructed to make high levels of small antisense RNAs that could inhibit expression of chromosomal genes. Pestka et al. [1984] placed a 421 bp DNA fragment of *E. coli lac*Z in inverse orientation downstream of the λ P_L promoter, which is controlled by the λ temperature-sensitive repressor cI^{ts}. The fragment was not complementary to the ribosome-binding site or start codon but included the sequence from codons 6 to 140. However, it should be borne in mind that, since it is estimated that the ribosome can cover up to 40 nucleotides of the RNA [Steitz, 1969; Platt and Yanofsky, 1975], it is possible that a duplex starting at nucleotide 19 of the mRNA could affect ribosome binding. Heat induction of P_L resulted in 98% inhibition of *lac*Z expression as measured by β-galactosidase activity. It also gave 80% and 55% inhibition of the downstream genes galactoside permease and transacetylase, respectively. These effects on downstream genes could simply be the result of polarity, i.e., the fact that *lac*Z translation is inhibited, but in view of the sensitivity of dsRNA to specific nucleases it could include an mRNA destabilizing effect. These results are summarized in Figure 12.

Similar experiments were carried out by Ellison et al. [1985], who again tested inhibition of chromosomal β-galactosidase with antisense RNA produced from the thermoinducible λ P_L promoter. Using an antisense 2,990 bp fragment covering almost all of the coding region of *lac*Z except for the first 24 and last 45 bp, they obtained 80% inhibition, whereas with deletions of this fragment the inhibition was reduced (57% with 815 bp from the 5′ end of the same fragment and 25% with 367 bp from the 3′ end).

Coleman et al. [1984] constructed a general vector for production of antisense RNA, taking into account the likely need to optimise stability of the inhibitory RNA. This vector pJDC402 (Fig. 13) contains an *Xba*I restriction site flanked upstream by tandemly placed *E. coli* lipoprotein (*lpp*) and *lac* promoters and downstream by the *lpp rho*-independent transcriptional terminator. The RNA expressed by this vector therefore has defined upstream and downstream stem–loop structures, between which any antisense RNA can be built into this system. With the anti-*lpp* gene, which includes the ribosome-binding site and the start of the open reading frame, there was 94% inhibition. A similarly efficient inhibition (95%) was observed with anti-*omp*C (outer membrane protein C) RNA. For the inhibition of *lpp* expression, Northern analysis indicated that *lpp* mRNA was significantly less stable when the anti-*lpp* was induced, suggesting that the effect on *lpp* is not simply due to inhibition of translation but may also be the result of accelerated mRNA

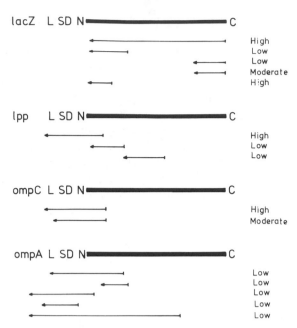

Fig. 12. *Schematic summary of effects of artificial antisense RNA in systematic studies. The genes shown are not to scale but serve to illustrate the different patterns of inhibition found for different genes. The target RNA is shown by the thick line, and the various antisense RNAs tested are shown below. L, leader RNA; SD, Shine-Dalgarno sequence; N, C, N and C termini of each open reading frame, respectively. The* lacZ *results are from Pestka et al. [1984] (example 1) and Ellison et al. [1985] (examples 2, 3, and 5). Example 4 is from Daugherty et al. [1989], who obtained 73% inhibition, in comparison to the 25% obtained in example 3, using the same antisense fragment transcribed from the same promoter. The only difference between the antisense constructs is the inclusion by Daugherty et al. [1989] of an artificial Shine-Dalgarno sequence at the 5' end of the antisense RNA, which they found considerably increased its stability and effectiveness. Note that even greater inhibition was obtained by these authors by targeting the 5' end of the gene (not shown here). These results of Daugherty et al. are discussed in more detail by Murray and Crockett (this volume).*

degradation due to cleavage by ribonucleases such as RNase III.

Not all attempts to use antisense RNA have been successful. For example, no interference with tetracycline resistance was observed for antisense transcription of the 346 bp *Bam*HI to *Hin*dIII fragment of pBR322 from the *lac* promoter on a high copy number vector [Zabarovskii, 1986]. Similarly, with the *E. coli gal*K gene and λ N gene, expression was found not to be inhibited by up to a 50-fold excess of antisense RNA produced either from the whole gene or its N-terminal portion [Hasan et al., 1988]. Coleman et al. [1984] found that in contrast to the successful inhibition of *lpp* and *omp*C (described above), *omp*A expression was hardly affected by antisense RNA, about 50% inhibition being achieved at maximum. The *omp*A

mRNA is highly structured and very stable, and this may be a major factor in the lack of inhibition.

Attempts to determine systematically what is required for successful inhibition have been made, and some of this information is summarized in Figure 12. The particularly detailed experiments of Daugherty et al. [1989], which studied the effect of various anti-*lac*Z RNAs of different lengths and representing different parts of the *lac*Z coding region, are discussed in some detail by Murray and Crockett (this volume). They found that targeting the Shine-Dalgarno sequence and AUG initiation codon was most effective, but even a section of only 367 bases hybridizing to the C-terminal segment of *lac*Z had some effect. This could be due to blocking of ribosome movement or to

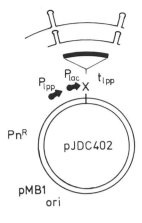

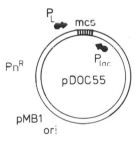

Fig. 13. *Structure of two vectors for control of gene expression by antisense RNA. Fragments from the target gene are inserted into pJDC402 in reverse orientation, allowing antisense RNA to be made against any part of the gene to be regulated [Coleman et al., 1984]. Plasmid pDOC55 employs antisense RNA to repress translation of residual mRNA from a cloned gene inserted at the multiple cloning site [O'Conner and Timmis, 1987].*

stimulation of mRNA cleavage by RNase III and thus irreversible inactivation of the mRNA. It was indeed noticed that the level of endogenous *lac*Z mRNA was inversely proportional to the amount of antisense RNA produced.

These findings are consistent with the earlier observation [Coleman et al., 1984] that the most effective inhibition of gene expression of endogenous *E. coli* genes was achieved when the antisense RNA included the ribosome-binding site of the target gene. However, there is no absolute requirement for inclusion of an anti Shine-Dalgarno sequence as long as the

duplex between mRNA and antisense RNA is formed rapidly and is of considerable length and stability [Pestka et al., 1984; Ellison et al., 1985].

The effectiveness of the anti Shine-Dalgarno sequence depends very much on its context. Hirashima et al. [1989] found that the most effective RNA repressor of expression of coliphage SP RNA was a 30 nucleotide RNA containing the anti Shine-Dalgarno sequence. This included 13 nucleotides upstream of the sequence that were essential for effective inhibition. Addition of downstream sequences had little effect, but addition of extra upstream sequences reduced inhibition. A pragmatic approach must therefore be adopted.

Another obvious use of antisense RNA is as a means of inhibiting translation of mRNA corresponding to a heterologous cloned gene whose expression may be deleterious to the bacterial host, thus preventing selection against bacteria carrying the desired gene. A general cloning vector employing antisense RNA, pDOC55, was described by O'Connor and Timmis [1987] (see Fig. 13). In this vector cloned DNA can be inserted between convergently oriented λ P_L and *E. coli* P_{lac} promoters, such that transcription from P_L gives sense RNA while transcription from P_{lac} when P_L is repressed both interferes with residual transcription from P_L (probably because of the wave of positive supercoiling that moves forward in front of RNA polymerase) and produces an antisense RNA that associates with any small amount of mRNA that is still produced and prevents its translation. The system worked well for both *tra*T, an F plasmid transfer gene, and the gene encoding *Eco*RI restriction enzyme, which would be extremely toxic to the host bacterium in the absence of *Eco*RI methylase if expressed at a significant level. Quantitative estimates suggested that there was greater than 95% inhibition of translation with no detectable accumulation of gene product.

In at least two cases where such convergent transcription occurs naturally, giving rise to sense and antisense RNA, the control has been artificially manipulated by replacing the natural sense promoter by the strong, regulated promoter P_L of bacteriophage λ [Larsen et al., 1984; Yarranton et al., 1984]. In both these cases, switching on the regulated promoter not only increases sense transcription but also blocks antisense transcription, leading to runaway gene expression and, in both cases, runaway plasmid replication.

An interesting vector described more recently incorporates an N-terminal fragment of the *htp*R gene oriented so that antisense RNA is transcribed from λ P_R on shift to 42°C. This anti-*htp*R inhibits expression of *htp*R, which encodes the heat shock response sigma factor σ^{32}, and therefore does not turn on the genes for proteases Lon, GroEL, and GroES, whose transcription depends on σ^{32}. The reduced production of proteases at 42°C results in greater stability of the polypeptide products of genes inserted in this plasmid, which are also induced by the shift to higher temperature [Kiselev and Tarasova, 1988].

Sometimes eukaryotic DNA cloned in bacteria can be deleterious because of the fortuitous expression of genes it encodes. Antisense RNA can be useful under such circumstances, as illustrated by its use to inhibit the production of a deleterious DNA binding protein from a section of cauliflower mosaic virus genome that was interfering with maintenance of the recombinant plasmid [Futterer et al., 1988].

VIII. CONCLUSIONS

From the above examples it can be seen that RNA functions naturally as a repressor in a number of ways. It can act as a modulator (inhibitor/stimulator) of RNA processing, as an inhibitor of transcription by causing premature termination of transcription, and as an inhibitor of translation by sequestering the ribosome-binding site for a gene. These natural examples thus illustrate the ways that artificial regulatory RNAs can be made to function. Two extreme scenarios exist for antisense RNA action.

1. The RNA may function to provide a rapid change in the level of gene expression. In this case it can be synthesized at a high rate and is degraded rapidly. Its concentration depends on whether its promoter is fully active and on the copy number of the gene encoding it. This is most applicable to tight regulation of cloned genes whose expression must be switched on rapidly.

2. The RNA may provide a mechanism for long-term silencing of gene expression and in this case it is generally found that the antisense RNA is highly stable but is synthesized at a low rate. This may be most applicable for providing immunity to phage in a way that does not impose a serious energetic drain on the bacteria.

Whichever strategy is employed, an effective level of antisense RNA must be reached. Since most bacterial mRNAs are translated as soon as transcription has created an mRNA segment long enough to be recognized by the ribosomes, the degree of inhibition achieved by an antisense RNA depends on its rate of interaction with its target, not the equilibrium that could be achieved eventually. The rate will depend on the concentration, and thus the relative rate of synthesis and degradation, of both RNA species and their reactivity toward each other. Most natural antisense RNAs are highly structured, and the evidence cited above suggests that maximum reactivity (i.e., highest rate of interaction with target) is achieved by strict adherence to relatively rigid single-stranded loop sizes that are involved in the initial "kissing" between repressor and target. Although not very easy to achieve, artificial regulators should aim to mimic this as far as possible. Alternatively, since most artificial antisense RNAs are relatively unreactive, it is normally necessary for them to be made in large quantities [e.g., Daugherty et al., 1989].

To control expression of a gene in its natural context, a starting point for choosing an effective inhibitor is to examine the nucleotide sequence around the start of translation, searching for potential secondary structure that could interfere with, or promote, RNA duplex formation. Ideally, any secondary structure that could be exploited should contain a single-stranded loop of five to seven bases whose stem should be flanked by single-stranded tails of at least 15 bases on either side. Insertion of such a segment into a vector such as pJDC402 should be used to add stabilizing secondary structure on either side. However, the sequence of the predicted product should be checked to ensure that it does not contain sequences whose juxtaposition favors the formation of unexpected structures.

Since the exact window of antisense RNA chosen can influence its effectiveness, initial attempts should include a number of trials. For example, a 70 base segment centered around the Shine-Dalgarno sequence could be chosen along with two other variants with 10 or 20 extra bases upstream or downstream. These could easily be created by using four different primers for a polymerase chain reaction (PCR), as illustrated in Figure 14. Any trends in terms of loss or gain of inhibitory activity with the addition of upstream or downstream sequences could provide a clue as to the most effective modifications to obtain a more effective inhibitor. If antisense RNA to the translation initiation signals is ineffective, then the rest of the gene should be searched for sequences resembling natural antisense inhibitors and their targets, and these could provide the starting point for further attempts. Alternatively, DNA of the gene could be fragmented randomly and overlapping fragments inserted and then tested for their effectiveness in achieving the desired inhibition. However, this strategy depends on there being an easy phenotypic screen for the inhibition desired.

The most effective strategy for using antisense RNA to control expression of cloned genes seems to be that incorporated into the vector developed by O'Connor and Timmis [1987]. Here a regulated promoter is arranged to transcribe the complementary strand of the gene to be cloned. Antisense transcription is maximal only when transcription of the gene to be

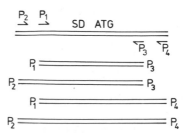

Fig. 14. *Use of four primers to generate four possible inhibitory segments by polymerase chain reaciton.*

controlled is repressed. It is switched off when the cloned gene is induced. Therefore translation is only inhibited when the level of mRNA is very low. Thus the antisense RNA serves to amplify the repression or derepression initiated at the level of transcription.

The effectiveness of antisense RNA can be modulated by other macromolecules in the cell. In bacteria we know of at least two examples of proteins (Rom from ColE1 and π from R6K) as well as one likely RNA species (FinO from F) that facilitate RNA–RNA interactions. The PcnB protein may sequester tRNA-like antisense molecules and reduce their effectiveness.

Finally, the susceptibility of the dsRNA molecules formed by interaction between antisense RNA and its target to RNase activity can critically influence the fate of the products of the interaction. A fuller understanding of the molecular basis of these effects may allow the principles they illustrate to be applied to other situations. If this is possible then not only will bacteria have provided the initial examples of antisense RNA in action but they may also have provided further tools to modulate the efficiency with which antisense RNAs can be exploited.

ACKNOWLEDGMENTS

I am grateful to the following people for provision of reprints, preprints, and unpublished information: K.N. Nordström, R.P. Novick, J.-I. Tomizawa, G. Wagner, and D.D. Womble. I thank Jim Murray for help in ensuring that the topic is covered thoroughly and Grazyna Jagura-

Burdzy, Cathy Rowlinson, and Deepan Shaw for critical reading of the manuscript.

IX. REFERENCES

Aiba H, Matsuyama S, Mizuno T, Mizushima S (1987): Function of *micF* as an antisense RNA in osmoregulatory expression of the *ompF* gene in *Escherichia coli*. J Bacteriol 169:3007–3012.

Alonso JC, Tailor RH (1987): Initiation of plasmid pC194 replication and its control in *Bacillus subtilis*. Mol Gen Genet 210:476–484.

Andersen J, Delihas N, Ikenaka K, Green PJ, Pines O, Ilercil O, Inouye M (1987): The isolation and characterization of RNA coded by the *micF* gene in *Escherichia coli*. Nucleic Acids Res 15:2089–2101.

Andersen J, Forst SA, Zhao K, Inouye M, Delihas N (1989): The function of *micF* RNA is a major factor in the thermal regulation of OmpF protein in *Escherichia coli*. J Biol Chem 264:17961–17970.

Banner DW, Kokkinidis M, Tsernoglou D (1987): Structure of the ColE1 Rop protein at 1.7Å resolution. J Mol Biol 196:657–675.

Blomberg P, Wagner EGH, Nordström K (1990): Control of replication of plasmid R1: The duplex between the antisense RNA, CopA and its target, CopT, is processed specifically in vivo and in vitro by RNase III. EMBO J 9:2331–2340.

Carleton S, Projan SJ, Highlander SK, Moghazeh S, Novick RP (1984): Control of pT181 replication. II. Mutational analysis. EMBO J 3:2407–2414.

Carter-Muenchau P, Wolf RE (1989): Growth-rate–dependent regulation of 6-phosphogluconate dehydrogenase level mediated by an anti-Shine-Dalgarno sequence located within the *Escherichia coli gnd* structural gene. Proc Natl Acad Sci USA 886:1138–1142.

Case CC, Roels SM, Gonzalez JE, Simons EL, Simons RW (1988): Analysis of the promoters and transcripts involved in IS*10* antisense RNA control. Gene 72: 219–236.

Case CC, Roels SM, Jensen PD, Lee J, Kleckner N, Simons RW (1989): The unusual stability of the IS*10* antisense RNA is critical for its function and is determined by the structure of its stem-domain. EMBO J 8:4297–4305.

Case CC, Simons EL, Simons RW (1990): The IS*10* transposase mRNA is destabilized during antisense RNA control. EMBO J 9:1259–1266.

Castagnoli L, Scarpa M, Kokkinidis M, Banner DW, Tsernoglou D, Cesareni G (1989): Genetic and structural analysis of the ColE1 Rop (Rom) protein. EMBO J 8:621–629.

Cesareni G, Muesing MA, Polisky B (1982): Control of ColE1 DNA replication: The *rop* gene product negatively affects transcription from the replication primer promoter. Proc Natl Acad Sci USA 79:6313–6317.

Cheah K-C, Skurray R (1986): The F plasmid carries an

IS*3* insertion within *finO*. J Gen Microbiol 132: 1141–1156.

Citron M, Schuster H (1990): The c4 repressors of bacteriophages P1 and P7 are antisense RNAs. Cell 62:591–598.

Cole ST, Honore N (1989): Transcription of the *sulA–ompA* region of *Escherichia coli* during the SOS response and the role of an antisense RNA molecule. Mol Microbiol 3:715–722.

Coleman J, Green PJ, Inouye M (1984): The use of RNAs complementary to specific mRNAs to regulate expression of individual bacterial genes. Cell 37: 429–436.

Crosa JH (1980): Three origins of replication are active in vivo in the R plasmid RSF 1040. J Biol Chem 255:11075–11077.

Dasgupta S, Masukata H, Tomizawa J-I (1987): Multiple mechanisms for initiation of ColIE1 DNA replication: DNA synthesis in the presence and absence of ribonuclease H. Cell 51:1113–1122.

Daugherty BL, Hotta K, Kumar C, Ahn YH, Zhu JD, Pestka S (1989): Antisense RNA: Effect of ribosome binding sites, target location, size, and concentration on the translation of specific RNA molecules. Gene Anal Tech 6:1–16.

Dempsey WB (1987): Transcript analysis of the plasmid R100 *traJ* and *finP* genes. Mol Gen Genet 209:533–544.

Dempsey WB (1989a): Derepression of conjugal transfer of the antibiotic resistance plasmid R100 by antisense RNA. J Bacteriol 171:2886–2888.

Dempsey WB (1989b): Sense and antisense transcripts of *traM*, a conjugal transfer gene of the antibiotic resistance plasmid R100. Mol Microbiol 3:561–570.

Devereux J, Haeberli P, Smithies O (1984): A comprehensive set of sequence analysis programs for the VAX. Nucleic Acids Res 12:387–395.

Dong X, Womble DD, Rownd R (1987): Transcriptional pausing in a region important for plasmid NR1 replication control. J Bacteriol 169:5353–5363.

Dooley TP, Polisky B (1987): Suppression of ColE1 RNA–RNA mismatch mutations in vivo by the ColE1 Rop protein. Plasmid 18:24–34.

Eguchi Y, Tomizawa J-I (1990): Complex formed by complementary RNA stem-loops and its stabilization by a protein: Function of ColIE1 Rom protein. Cell 60:199–209.

Ellison MJ, Kelleher RJ, Rich A (1985): Thermal regulation of β-galactosidase synthesis using antisense RNA directed against the coding portion of the mRNA. J Biol Chem 260:9085–9087.

Faubladier M, Cam K, Bouché J-P (1990): *Escherichia coli* cell division inhibitor DicF–RNA of the *dicB* operon. Evidence for its generation in vivo by transcription termination and RNase III and RNase E–dependent processing. J Mol Biol 212:461–471.

Finlay BB, Frost LS, Paranchych W, Willetts NS (1986): Nucleotide sequences of five IncF plasmid *finP* alleles. J Bacteriol 167:754–757.

Fitzwater T, Zhang X-Y, Elble R, Polisky B (1988): Conditional high copy number ColE1 mutants: Resistance to RNA I inhibition in vivo and in vitro. EMBO J 7:3289–3297.

Frost L, Lee S, Yanchar N, Paranchych W (1989): *finP* and *fisO* mutations in FinP antisense RNA suggest a model for FinOP action in the repression of bacterial conjugation by the Flac plasmid JCFLO. Mol Gen Genet 218:152–160.

Futterer J, Gordon K, Pfeiffer P, Hohn T (1988): The instability of a recombinant plasmid, caused by a prokaryotic-like promoter within the eukaryotic insert, can be alleviated by expression of antisense RNA. Gene 15:141–145.

Gerdes K (1988): The *parB* (*hok/sok*) locus of plasmid R1: A general purpose plasmid stabilization system. Biotechnology 6:1402–1405.

Gerdes K, Bech FW, Jorgensen ST, Lobner-Oleson A, Rasmussen PB, Atlung T, Karlstrom O, Molin S, Von Meyenburg K (1986a): Mechanism of postsegregational killing by the *hok* gene product of the *parB* system of plasmid R1 and its homology to the *relF* gene product of the *E. coli relB* operon. EMBO J 5:2023–2029.

Gerdes K, Helin K, Christensen OW, Lobner-Oelsen A (1988): Translational control and differential RNA decay are key elements regulating postsegregational expression of the killer protein encoded by the *parB* locus of plasmid R1. J Mol Biol 203:119–129.

Gerdes K, Poulsen LK, Thisted T, Nielsen AK, Martinussen J, Andreasen PH (1990a): The *hok* killer family in gram-negative bacteria. New Biologist 2:946–956.

Gerdes K, Rasmussen PB, Molin S (1986b): Unique type of plasmid maintenance function: Postsegregational killing of plasmid free cells. Proc Natl Acad Sci USA 83:3116–3120.

Gerdes K, Thisted T, Martinussen J (1990b): Mechanism of post-segregational killing by the *hok/sok* system of plasmid R1: *sok* anti-sense RNA regulates the formation of a *hok* mRNA species correlated with killing plasmid free cells. Mol Microbiol 4:1807–1818.

Green PJ, Pines O, Inouye M (1986): The role of antisense RNA in gene regulation. Annu Rev Biochem 55:569–597.

Hasan N, Somaskhar G, Szybalski W (1988): Antisense RNA does not significantly affect expression of the *galK* gene of *Escherichia coli* or the N gene of coliphage lambda. Gene 72:247–252.

Helmer-Citterich M, Anceschi MM, Banner DW, Cesareni G (1988): Control of ColE1 replication: Low affinity specific binding of Rop (Rom) to RNAI and RNAII. Cell 7:557.

Highlander SK, Novick RP (1990): Mutational and physiological analyses of plasmid pT181 functions expressing incompatibility. Plasmid 23:1–15.

Hirashima A, Sawaki S, Inokuchi Y, Inouye M (1986): Engineering of the mRNA-interfering complementary RNA immune system against viral infection. Proc Natl Acad Sci USA 38:7726–7730.

Hirashima A, Sawaki S, Mizuno T, Houba-Herin N, Inouye M (1989): Artificial immune system against viral infection involving antisense RNA targeted to the 5'-terminal non-coding region of coliphage SP RNA. J Biochem (Tokyo) 106:163–166.

Ho YS, Rosenberg M (1985): Characterization of a third, cII-dependent, coordinately activated promoter on phage lambda involved in lysogenic development. J Biol Chem 260:11838–11844.

Hoopes BC, McClure WR (1985): A cII-dependent promoter is located within the Q gene of bacteriophage-lambda. Proc Natl Acad Sci USA 82:3134–3138.

Inouye M (1988): Antisense RNA: Its functions and applications in gene regulation—A review. Gene 72:25–34.

Inouye M, Delihas N (1988): Small RNAs in prokaryotes: A growing list of diverse roles. Cell 53:5–7.

Ippen-Ihler K, Minkley EG (1986): The conjugation system of F, the fertility factor of *Escherichia coli*. Annu Rev Genet 20:593–624.

Itoh T, Tomizawa J-I (1980): Formation of an RNA primer for initiation of replication of ColE1 DNA by ribonuclease H. Proc Natl Acad Sci USA 77:2450–2454.

Jacobson AB, Good L, Simonetti J, Zuker M (1984): Some simple computational methods to improve the folding of large RNAs. Nucleic Acids Res 12:45–52.

Kiselev VI, Tarasova IM (1988): Regulation of proteolysis in *Escherichia coli* cells by antisense RNA of *htpR* gene. Biotechnol Appl Biochem 10:59–62.

Kittle JD, Simons RW, Lee J, Kleckner NC (1989): Insertion sequence IS*10* antisense pairing initiates by an interaction between the 5' end of the target RNA and a loop in the antisense RNA. J Mol Biol 210:561–572.

Konings DAM, Hogeweg P (1989): Pattern analysis of RNA secondary structure. Similarity and consensus of minimal-energy folding. J Mol Biol 207:597–614.

Krinke L, Wulff DL (1987): OOP RNA, produced from multicopy plasmids, inhibits lambda cII gene expression through an RNase III–dependent mechanism. Genes Dev 1:1005–1013.

Krinke L, Wulff DL (1990): RNase III–dependent hydrolysis of lambda cII-O gene mRNA mediated by lambda OOP antisense RNA. Genes Dev 4:2223–2233.

Kumar CC, Novick RP (1985): Plasmid pT181 replication is regulated by two countertranscripts. Proc Natl Acad Sci USA 82:638–642.

Lacatena RM, Cesareni G (1981): Base pairing of RNA I with its complementary sequence in the primer precursor inhibits ColE1 replication. Nature 294:623–626.

Lacatena RM, Cesareni G (1983): Interaction between RNA I and the primer precursor in the regulation of ColE1 replication. J Mol Biol 170:635–650.

Larsen JEL, Gerdes K, Light J, Molin S (1984): Low copy–number plasmid-cloning vectors amplifiable by derepression of an inserted foreign promoter. Gene 28:45–54.

Levy S, De Reuse H, Danchin A (1989) Antisense expression at the *pts*H–*pts*I locus of *Escherichia coli*. FEMS Microbiol Lett 48:35–38.

Liao SM, Wu TH, Chiang CH, Susskind MM, McClure WR (1987): Control of gene expression in bacteriophage P22 by a small antisense RNA. I. Characterization in vitro of the P_{sar} promoter and the sar RNA transcript. Gene Dev 1:197–203.

Light J, Molin S (1983): Post-transcriptional control of expression of the *rep*A gene of plasmid R1 mediated by a small RNA molecule. EMBO J 2:93–98.

Lin-Chao S, Bremer H (1987): Activities of the RNAI and RNAII promoters of plasmid pBR322. J Bacteriol 169:1217–1222.

Liu J, Parkinson JS (1989): Genetic and sequence analysis of the *pcn*B locus, an *Escherichia coli* gene involved in plasmid copy number control. J Bacteriol 171:1254–1261.

Loh SM, Cram DS, Skurray RA (1988): Nucleotide sequence and transcriptional analysis of a third function (Flm) involved in F-plasmid maintenance. Gene 66:259–268.

Lopilato J, Bortner S, Beckwith J (1986): Mutations in a new chromosomal gene of *Escherichia coli* K-12, *pcn*B, reduce plasmid copy number of pBR322 and its derivatives. Mol Gen Genet 205:285–290.

Ma C, Simons RW (1990): The IS10 antisense RNA blocks ribosome binding at the transposase translation initiation site. EMBO J 9:1267–1274.

Masai H, Arai K-I (1988): RepA protein- and *ori*R-dependent initiation of R1 plasmid replication: identification of a *rho*-dependent transcription terminator required for *cis*-action of *rep*A protein. Nucleic Acids Res 16:6493–6514.

Masters M, March JB, Oliver IR, Collins JF (1989): A possible role for the *pcn*B gene product of *Escherichia coli* in modulating RNA:RNA interactions. Mol Gen Genet 220:341–344.

Masukata H, Dasgupta S, Tomizawa J-I (1987): Transcriptional activation of ColE1 DNA synthesis by displacement of the nontranscribed strand. Cell 51: 1123–1130.

Masukata H, Tomizawa J-I (1986): Control of primer formation for ColE1 plasmid replication: Conformational change of the primer transcript. Cell 44:125–136.

Misra R, Reeves PR (1987): Role of *mic*F in the *tol*C-mediated regulation of OmpF, a major outer membrane protein of *Escherichia coli* K-12. J Bacteriol 169: 4722–4730.

Mizuno T, Chou M-Y, Inouye M (1984): A unique mechanism regulating gene expression: Translation inhibition by a complementary RNA transcript (micRNA). Proc Natl Acad Sci USA 81:1966–1970.

Nordström K, Wagner EGF, Persson C, Blomberg P, Öhman M (1988): Translational control by antisense RNA in control of plasmid replication. Gene 72: 237–240.

Novick RP (1989): Staphylococcal plasmids and their replication. Annu Rev Microbiol 43:537–565.

Novick RP, Iordanescu S, Projan SJ, Kornblum J, Edelman I (1989): pT181 plasmid replication is regulated by a countertranscript-driven transcriptional attenuator. Cell 59:395–404.

O'Connor CD, Timmis KN (1987): Highly repressible expression system for cloning genes that specify potentially toxic proteins. J Bacteriol 169:4457–4462.

Öhman M, Wagner EG (1989): Secondary structure analysis of the RepA mRNA leader transcript involved in control of replication of plasmid R1. Nucl. Acids Res. 17:2557–2579.

Okamoto K, Freundlich M (1986): Mechanism for the autogenous control of the *crp* operon: transcriptional inhibition by a divergent RNA transcript. Proc Natl Acad Sci USA 83:5000–5004.

Okamoto K, Hara S, Bhasin R, Freundlich M (1988): Evidence in vivo for autogenous control of the cyclic AMP receptor protein gene (*crp*) in *Escherichia coli* by divergent RNA. J Bacteriol 170:5076–5079.

Patel I, Bastia D (1986): A replication origin is turned off by an origin "silencer" sequence. Cell 47: 785–792.

Patel I, Bastia D (1987): A replication initiator protein enhances the rate of hybrid formation between a silencer RNA and an activator RNA. Cell 51, 455–462.

Persson C, Wagner EGH, Nordström KN (1988): Control of replication of plasmid R1: kinetics of in vitro interaction between the antisense RNA, CopA, and its target, CopT. EMBO J 7:3279–3288.

Persson C, Wagner EG, Nordstrom K (1990a): Control of replication of plasmid R1: formation of an initial transient complex is rate-limiting for antisense RNA-target RNA pairing. EMBO J 9:3777–3785.

Persson C, Wagner EG, Nordstrom K (1990b): Control of replication of plasmid R1: structures and sequences of the antisense RNA, CopA, required for its binding to the target RNA, CopT. EMBO J 9: 3767–3775.

Pestka S, Daugherty RK, Jung V, Hotta K, Pestka RK (1984): Anti-mRNA: Specific inhibition of translation of single mRNA molecules. Proc Natl Acad Sci USA 81:7525–7528.

Platt T, Yanofsky C (1975): An intercistronic region and ribosome-binding site in bacterial messenger RNA. Proc Natl Acad Sci USA 72:2399–2403.

Polisky B (1988): ColE1 replication control circuitry: Sense from antisense. Cell 55:929–932.

Polisky B, Zhang X-Y, Fitzwater T (1990): Mutations affecting primer RNA interaction with the replication repressor RNA I in plasmid ColE1: Potential folding pathway mutants. EMBO J 9:295–304.

Riise E, Molin S (1986): Purification and characterization of the CopB replication protein, and precise mapping of its target site in the R1 plasmid. Plasmid 15:163–171.

Robertson HD (1990): *Escherichia coli* RNase III. Methods Enzymol 181:189–202.

Simons RW (1988): Naturally occurring antisense RNA control—A brief review. Gene 72:35–44.

Simons RW, Kleckner NC (1983): Translational control of IS10 transposition. Cell 34:683–691.

Simons RW, Kleckner N (1988): Biological regulation by antisense RNA in prokaryotes. Annu Rev Genet 22:567–600.

Steitz JA (1969): Polypeptide chain initiation: Nucleotide sequence of the three ribosome-binding sites in bacteriophage R17 RNA. Nature 224:957–964.

Summers D, Sherratt D (1984): Multimerization of high copy number plasmids causes instability: ColE1 encodes a determinant essential for plasmid monomerization and stability. Cell 36:1097–1103.

Takayama KM, Houba-Herin N, Inouye M (1987): Overproduction of an antisense RNA containing the OOP RNA sequence of bacteriophage lambda induces clear plaque formation. Mol Gen Genet 210:184–186.

Takayama KM, Inouye M (1990): Antisense RNA. Crit Rev Biochem Mol Biol 25:155–184.

Tamm J, Polisky B (1983): Structural analysis of RNA molecules involved in plasmid copy number control. Nucleic Acids Res 11:6381–6397.

Tinoco I, Borer PN, Dengler B, Levine MD, Uhlenbeck OC, Crothers DM, Gralla J (1973): Improved estimation of secondary structure in ribonucleic acids. Nature [New Biol] 246:40–41.

Tomizawa J-I (1984): Control of plasmid ColE1 replication: The process of binding of RNA I to the primer transcript. Cell 38:861–870.

Tomizawa J I (1986): Control of ColE1 plasmid replication: Binding of RNA I to RNA II and inhibition of primer formation. Cell 47:89–97.

Tomizawa J-I (1990a): Control of ColE1 plasmid replication. Intermediates in the binding of RNA I and RNA II. J Mol Biol 212:683–694.

Tomizawa J-I (1990b): Control of ColE1 plasmid replication. Interaction of Rom protein with an unstable complex formed by RNA I and RNA II. J Mol Biol 212:695–708.

Tomizawa J-I, Itoh T, Selzer G, Som T (1981): Inhibition of ColE1 RNA primer formation by a plasmid-specified small RNA. Proc Natl Acad Sci USA 78:1421–1425.

Tomizawa J-I, Som T (1984): Control of plasmid ColE1 replication: Enhancement of binding of RNA I to the primer transcript by the Rom protein. Cell 38:871–878.

Wagner EGH, Nordström KN (1986): Structural analysis of an RNA molecule involved in replication control of plasmid R1. Nucleic Acids Res 14:6381–6397.

Wagner EGH, von Heijne J, Nordström K (1987): Control of replication of plasmid R1: Translation of the 7K reading frame in a RepA mRNA leader region counteracts the interaction between CopA RNA and CopT RNA. EMBO J 6:515–527.

Wagner EGH, Persson C, Öhman M, Nordström K (1988): Antisense RNA in replication control of the IncFII plasmid R1. Nucleosides Nucleotides 7:559–564.

Womble DD, Dong X, Wu R-P, Luckow VA, Martinez AF, Rownd R (1984): IncFII plasmid incompatibility product and its target are both RNA transcripts. J Bacteriol 160:28–35.

Womble DD, Rownd RH (1988): Genetic and physical map of plasmid NR1: Comparison with other IncFII antibiotic resistance plasmids. Microbiol Rev 52:433–451.

Wong EM, Polisky B (1985): Alternative conformations of the ColE1 replication primer modulate its interaction with RNA I. Cell 42:959–966.

Wu TH, Liao SM, McClure WR, Susskind MM (1987): Control of gene expression in bacteriophage P22 by a small antisense RNA. II. Characterization of mutants defective in repression. Genes Dev 1:204–212.

Yarranton GT, Wright E, Robinson MK, Humphreys GO (1984) Dual-origin plasmid vectors whose origin of replication is controlled by coliphage lambda promoter P_L. Gene 28:293–300.

Zabarovskii ER (1986): The function of the *tet* gene in the plasmid pBR322 is not inhibited by expression of an anti-*tet* gene. Mol Biol [Mosk] 20:639–645.

ABOUT THE AUTHOR

CHRISTOPHER M. THOMAS is Professor of Molecular Genetics in the School of Biological Sciences at the University of Birmingham (England), where he teaches bacterial molecular genetics. After receiving his B.A. in biochemistry from the Queen's College, Oxford, in 1974, he carried out research for his D.Phil. under Keither Dyke in the sub-department of microbiology in the Department of Biochemistry at Oxford University. This work concentrated on developing an in vitro DNA replication system for *Staphylococcus aureus* and analyzing the relationship between plasmid and chromosome replication. Dr. Thomas then pursued postdoctoral research in the laboratory of Professor Donald Helinski, at the University of California-San Diego, La Jolla, funded by a Medical Research Council Travelling Fellowship. His study of broad host range plasmids was started at this time and continued when he took up a permanent position in the Department of Genetics at the University of Birmingham. These studies now cover all aspects of plasmid functions, including replication, partitioning, and conjugative transfer and control of the genes required for these processes. His research papers have appeared in such journals as *Nature*, *EMBO Journal*, the *Journal of Molecular Biology*, and *Gene*. He has edited one book, *Promiscuous Plasmids of Gram Negative Bacteria*. He is currently an editor of the *Journal of General Microbiology*. His current research is supported by the Medical Research Council, the Science and Engineering Research Council, and the Wellcome Trust. He received a Ciby-Geigy ACE Award for the 1991–1993 sessions to allow international collaboration on analyzing the complete sequence (60 kb) of broad host range IncP plasmids.

Antisense RNA and DNA:
1992 Wiley-Liss, Inc.

Interference of Gene Expression by Antisense RNAs in *Drosophila*

Rekha Patel and Marcelo Jacobs-Lorena

I. INTRODUCTION

The use of antisense RNA or DNA fragments to block the expression of selected genes, and thereby assess their function, is a powerful new tool for the molecular biologist [Weintraub, 1990]. It is an approach that could be particularly useful in higher eukaryotes, which lack the broad spectrum of genetic tools available in bacteria and yeast. In recent years, antisense RNA has been shown to interfere effectively with the expression of specific genes in several biological systems [Walder, 1988; Izant, 1989; Van der Krol et al., 1988]. Despite many successes, this technique is not yet universally applicable, and various aspects of antisense inhibition remain to be clarified. For instance, in some systems inhibition requires the antisense RNA to be in vast excess (50-fold or higher) over the endogenous mRNA [Izant and Wein-traub, 1984, 1985; Melton, 1985; Xiao and Rank, 1988], while in others a one- to five-fold excess is sufficient [Crowley et al., 1985; Izant and Weintraub, 1985; McGarry and Lindquist, 1986; Pecorino et al., 1988; Rothstein et al., 1987]. Whereas in one case the hybrids of sense and antisense RNA were detected in the nucleus [Kim and Wold, 1985], in other cases such hybrids were present in the cytoplasm [Pecorino et al., 1988]. In some systems the mRNA appears to be destabilized, indicating that the cell has mechanisms for selectively degrading the hybrid RNAs or that lack of translation reduces the stability of that particular mRNA [Crowley et al., 1985; Knecht and Loomis, 1987; Strickland et al., 1988; Smith et al., 1988] (see also Munir et al., this volume). It appears that in teratocarcinoma cells the hybrid region may be stable while the rest of the mRNA is targeted for degradation [Pecorino et al., 1988]. *Xenopus* embryos, cultured mammalian cells, and certain other systems have an activity that unwinds the hybrids, thus potentially relieving inhibition of gene expression [Bass and Weintraub, 1987; Rebagliati and Melton, 1987; Wagner and Nishikura, 1988], although the discovery that unwinding is associated with base modification of the RNA complicates this interpretation. This double-stranded RNA unwinding/modifying activity is discussed by Bass (this volume). Many regions of mRNA have been targeted for hybrid formation with diverse results. Successes and failures have been noted for antisense experiments that utilize antisense RNA complementary to the 5' or 3' untranslated sequences, the coding regions, and intron–exon boundaries in the mRNA [Kim and Wold, 1985; Knecht and Loomis, 1987; Melton, 1985; Wormington, 1986].

This diversity of results indicates that there may be differences in the cellular responses to the presence of antisense RNAs. These differences may depend on the biological system and on the specific target mRNA. A clearer understanding of how antisense RNAs interfere with gene expression will require a step past the description of phenotypes and a deeper probing into the mechanisms involved.

The interference of gene expression by antisense RNAs in *Drosophila* is dealt with in this chapter. The experiments are discussed in three categories, according to the experimental approach utilized. Experiments that used antisense RNA injection into embryos to confirm the identity of a cloned gene are described in Section II.A. Experiments performed in *Drosophila* cultured cells in which RNAs complementary to various genes were expressed from inducible promoters are summarized in Section II.B, and experiments conducted in the whole organism by stable transformation of antisense genes into the germ line of the fly are reported in Section 11.C. Section III concludes this chapter with an evaluation of the results described and with an assessment of the status of antisense research in *Drosophila*.

II. INTERFERENCE OF GENE EXPRESSION BY ANTISENSE RNAs IN *DROSOPHILA*
A. Injection of Antisense RNA Into Embryos

The antisense approach has been successfully used to identify genes involved in early pattern formation in the *Drosophila* embryo. Pattern-forming genes are particularly suitable for this approach in part because, as indicated by genetic analysis, a complete loss of wild-type activity is usually not required. For most genes hypomorphic (partial-loss-of-function) alleles have been described, and their phenotype can be ordered in a graded series that represents different levels of wild-type activity. A reduction in wild-type activity by antisense RNA can therefore be evaluated by comparison of phenotypes of the different hypomorphic mutants and the phenocopies produced by antisense RNA. This approach has been successfully undertaken for the following genes: *Krüppel* [Rosenberg et al., 1985], *snail* [Boulay et al., 1987], *wingless* [Cabrera et al., 1987], *spalt* [Schuh and Jäckle, 1989], *knirps* [Nauber et al., 1988], and *pecanex* [LaBonne et al., 1989].

The *Krüppel* (*Kr*) gene is first transcribed at the early blastoderm stage [Preiss et al., 1985] and produces a rare transcript (0.01% of the total polyadenylated RNA). Rosenberg et al. [1985] injected wild-type embryos at the syncytial blastoderm stage with in vitro transcribed *Kr* antisense RNA (and sense RNA as controls). Embryos injected with *Kr* antisense RNA developed lethal phenocopies at a frequency of up to 30%, while sense RNA had no specific effects. Significantly, extreme *Kr* phenocopies resembling the amorphic *Kr* phenotype were never obtained, indicating that Kr^+ activity was not completely abolished by antisense RNA. The frequency and strength of the phenocopies were dependent on the concentration of injected antisense RNA. The strongest phenotypes were observed when 10^8 molecules of antisense RNA were injected per embryo, which represents a 1,000-fold excess over the endogenous *Kr* transcripts. The weakest detectable response required more than 50-fold excess of antisense RNA over the endogenous transcripts. Even the injection of antisense *Kr* RNA into heterozygous *Kr* embryos (which have 50% of the wild-type *Kr* activity) did not produce any amorphic *Kr* phenotypes.

Boulay et al. [1987] used antisense RNA to obtain biological proof that they had correctly identified the cDNA for *snail* (*sna*), a gene required for normal formation of the dorsal–ventral pattern. Injection of antisense RNA, but not of buffer or sense RNA, resulted in more than 60% of viable embryos developing a weak or intermediate *sna* phenocopy. As in the case of *Kr*, extreme phenotypes were not observed. Antisense RNA therefore proved valuable in providing a link between cloned DNA and the phenotype produced by mutation, confirming that the open reading frame identified did indeed correspond to the gene being sought.

The *wingless* (*wg*) gene is transcribed at the blastoderm stage in a narrow stripe of cells at the posterior border of each parasegment [Baker, 1987]. Sense and antisense *wg* RNA were injected into early embryos [Cabrera et al., 1987]. While sense RNA injections produced no specific phenotype, antisense RNA injections produced *wg* phenocopies in about 50% of the embryos. These phenocopies varied from the minimum detectable to a global *wg* phenotype, intermediate phenotypes being the most commonly observed. In most of the positive cases, the effect was seen in the region of the egg where the RNA was injected.

A similar approach was taken for the gene *spalt* (*sal*) [Schuh and Jäckle, 1989]. In this case only 20% of the injected embryos exhibited a weak *sal* phenotype at the first instar larval stage. The *sal* phenocopies were obtained only if antisense RNA was injected in a 1,000,000-fold molar excess over the endogenous *sal* transcript. One reason for the lower effectiveness of the *sal* antisense RNA could be that *sal* transcripts remain in the embryo for the first 12 hours of development [Frei et al., 1988], while *Kr* transcripts are not detectable after 5 hours of development. Similarly, *sna* transcripts disappear after 6 hours. Thus *sal* activity may be required for a longer period during embryogenesis than *Kr* and *sna*. Most of

the injected antisense RNA could be degraded rapidly soon after injection; thus a vast excess of *sal* antisense RNA is probably necessary to inhibit the late *sal*+ activity. Alternatively, the *sal* phenotype may be less dose sensitive than *Kr* and *sna*.

Knirps (*kni*) is a zygotic gene required for segmentation of the abdomen, and it is normally expressed in early blastoderm in an anterior and a posterior domain. In *kni* mutants, adjacent segments that arise from this posterior region are deleted. The gene was cloned by Nauber et al. [1988], who used injected antisense RNA to produce *kni* phenocopies. This confirmed that they had indeed cloned a cDNA for the *kni* gene, since control injections with sense RNA did not affect abdominal segmentation. However, only 5% of antisense embryos developed *kni* phenocopies, and these all resembled weak *kni* alleles. As in the case of *sal*, this may reflect the fact that expression of *kni* peaks at 4–6 hours into embryogenesis, but remains detectable until 8–24 hours, by which time most of the antisense RNA may have been degraded.

LaBonne et al. [1989] obtained *pecanex* (*pcx*) phenocopies by injection of antisense *pcx* RNA into 0–30-minute-old embryos. The production of *pcx* phenocopies was proportional to the amount of antisense *pcx* RNA injected. When 10^4 copies were injected per embryo, no phenocopies were produced. Injection of 10^5–10^8 copies of antisense *pcx* RNA produced phenocopies in up to 70% of the surviving embryos.

Recently, injection of antisense RNA has been used to confirm the identity of yet another cloned gene, *teashirt*, which was originally identified by use of the "enhancer trap" approach [O'Kane and Gehring, 1987]. Injection of a large excess of antisense *teashirt* RNA led to developmental effects at a higher frequency than that obtained with injection of sense RNA [Fasano et al., 1991]. However, in comparison to some of the examples mentioned above, the effect of the antisense RNA was less clear cut, giving only a quantitative difference. At least two factors may have contributed to this. First, the mutant *teashirt* phenotype is less specific and may not be as easy to distinguish from nonspecific defects caused by the injection process itself. Second, RNA blots from embryos and larvae at different developmental stages reveals that *teashirt* mRNA remains abundant from about 2 hours of embryogenesis onward, at least until the end of larval development. This observation raises the possibility that *teashirt* may be transcribed during embryogenesis, thus decreasing the effectiveness of the injected RNAs as discussed above.

The developmental program of the *Drosophila* embryo sharply limits the type of genes that can serve as targets for antisense RNA injection. The *Drosophila* embryo develops as a syncytium for the first 2.5 hours of development, meaning that during this time the entire embryo is accessible to the injected materials. Injection of antisense RNA after cellularization of the blastoderm is probably ineffective, since the injected RNA would not be able to reach its target. Moreover, other data suggest that antisense RNA is likely to be relatively unstable [Qian et al., 1988; Patel and Jacobs-Lorena, submitted for publication]. In summary, physical limitations imposed by the formation of cell membranes and the possible instability of antisense RNA limits the usefulness of the injection approach to genes that are expressed early in development. It has, however, proved valuable for confirming the identity of clones of such genes isolated by chromosomal walks, bypassing the need to rescue the mutant phenotype with a cloned copy of the wild-type gene. For genes that are expressed at other developmental times, the antisense RNA would have to be expressed from genes stably integrated into the genome and driven by strong and inducible promoters. This approach is considered in the next two sections.

B. Stable Transformation of Cultured Cells With Antisense Genes

McGarry and Lindquist [1986] have examined *Drosophila* tissue culture cells stably transformed with a gene encoding a heat-inducible RNA complementary to the mRNA for hsp26, one of the small heat shock proteins. Upon heat shock, these cells produced much less hsp26 than untransformed controls. The inhibition was highly specific: Expression of the closely related heat shock proteins hsp22, hsp23, and hsp28 was unaffected. By varying the copy number of the antisense gene, the degree of inhibition varied over a broad range. Reducing the rate of heat shock protein synthesis did not affect the synthesis of any other protein during heat shock or recovery.

Bunch and Goldstein [1989] have examined the potential of two antisense genes, whose transcription is driven by a *Drosophila* metallothionein promoter, to inhibit the expression of alcohol dehydrogenase (ADH) or a microtubule-associated protein (205 kD MAP) in cultured *Drosophila* cells. Expression of ADH was significantly reduced upon induction of the anti-ADH gene, which produced a four- to sevenfold excess of antisense ADH mRNA over endogenous ADH mRNA. The ADH mRNA formed hybrids with the antisense RNA but was not destabilized. Hybrids of both spliced and unspliced ADH RNA were detected. However, the anti-205-kD MAP gene was completely ineffective even though the antisense RNA was expressed in very large excess over the 205 kD MAP mRNA (which is itself far less abundant than ADH mRNA).

C. Antisense Genes Stably Integrated Into the Genome of the Fly

1. Disruption of oogenesis by expression of RNA complementary to ribosomal protein rpA1. The gene coding for the acidic ribosomal protein rpA1 is located on polytene chromosome band 53CD [Qian et al., 1987], in the same region where the *Minute* mutation *M(2)S7* has been genetically mapped. As reviewed in Section C.2, *Minute* is a group of mutations believed to affect ribosomal protein genes. Qian et al. [1988] constructed hybrid DNAs that placed the rpA1 gene in either sense or antisense orientation under control of the hsp70 promoter. These constructs were transformed into the germ line of *Drosophila* by P-element–mediated transformation. Even very frequent antisense gene induction throughout postembryonic development failed to produce some of the characteristic *Minute* phenotypes. However, antisense rpA1 expression severely disrupted oogenesis. Following a heat pulse, antisense females laid eggs that were small, abnormal in appearance, and infertile. The strength of this phenotype was dependent on the strength of antisense RNA induction. Because of leakiness of the heat shock promoter, females raised at 25°C were less fertile than flies raised at 18°C. When the antisense genes from two independent lines were genetically combined into a single fly, the females had only rudimentary ovaries and were almost completely sterile at 25°C. Microscopic examination of ovaries dissected from heat-shocked antisense females showed abnormalities that helped to explain the small-egg phenotype induced by antisense expression. Normally, the vast majority of the egg constituents are synthesized by the nurse cells and transferred to the oocyte at the end of oogenesis via cytoplasmic bridges. After this transfer has occurred the follicle cells surrounding the oocyte secrete the egg shell. In antisense rpA1 flies the nurse cells frequently fail to transfer their contents to the oocyte. The follicle cells seem to secrete the egg shell prematurely, resulting in the disruption of the nurse cell–oocyte connection. Because of this blockage of nurse cell-to-oocyte cytoplasmic transfer, a smaller oocyte is produced. It appears that the coordination between the developmental programs of the germ cells and somatic cells is disrupted by antisense rpA1 expression.

2. Phenocopying the *Minute* phenotype by expression of RNA complementary to the mRNA for ribosomal protein rp49.

The *Minute* loci of *Drosophila* represent a class of about 50 phenotypically similar, dispersed, haploinsufficient, cell autonomous mutations that are believed to affect protein synthesis [reviewed by Kay and Jacobs-Lorena, 1987]. The *Minute* phenotype includes several or all of the following traits: prolonged larval and pupal development, short and thin bristles, reduced viability and fertility, rough eyes, small body size, and etched tergites. In homozygous condition, *Minute* causes lethality in late embryo-

genesis or early first instar larvae. Even though it has been proposed for a long time that *Minute* codes for components of the protein synthesis machinery, this relationship has been conclusively demonstrated only in one case. Kongsuwan et al. [1985] have shown by germline transformation experiments that the ribosomal protein rp49 gene is able to suppress the phenotype of the *Minute (3)99D* locus. Not all transformants suppressed the *99D Minute* phenotype to an equal extent. Some of the transformant lines yielded flies with an intermediate phenotype, presumably because when the rp49 gene is integrated at certain chromosomal positions not enough product is made to rescue the mutation fully. Another indication that the organism is exceptionally sensitive to changes in levels of rp49 expression is that the 50% reduction of gene dosage in *Minute (3)99D* results in a very strong *Minute* phenotype. This exquisite sensitivity to levels of gene expression makes rp49 an ideal target for antisense mutagenesis. This is because a visible phenotype is expected to result even after modest interference of rp49 expression by antisense RNAs.

The rp49 gene was placed in sense and antisense orientation under the control of the strong heat-inducible promoter hsp70 [Patel and Jacobs-Lorena, submitted for publication]. These "sense" and "antisense" rp49 constructs were transformed into the germ line of *Drosophila* by P-element–mediated transformation. Exposing these flies to elevated temperatures (37°C) for 1 hour resulted in accumulation of the sense or antisense RNAs to levels three- to fourfold over the endogenous rp49 mRNA. To determine whether expression of antisense rp49 RNA has any effect on development, the progeny of the "antisense" flies was heat treated at different stages of their life cycle. Frequent induction of the antisense gene caused a slight developmental delay. Moreover, a significant proportion (10%–15%) of the "antisense" flies that hatched from heat-treated pupae (adult structures form during pupal stages) had thin and short bristles, characteristic of the *Minute* phenotype (Fig. 1). This phenotype was never

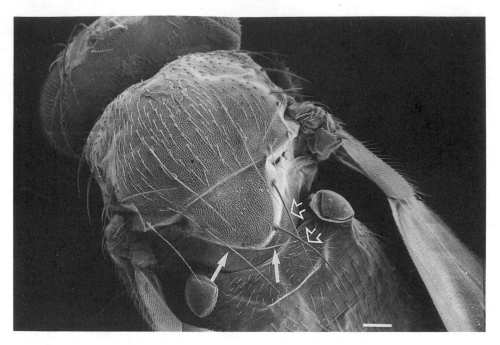

Fig. 1. *Induction of the* Minute *phenotype by expression of antisense ribosomal protein rp49 RNA from an integrated gene. Pupae carrying a heat-shock-promoter-driven antisense rp49 transgene were heat treated for 1 hour at 37°C at each of the first 3 days of pupal development. Rearing at all other times was at 25°C. Ten to 15% of the flies emerging from the treated pupae displayed thin and/or short bristles, character-istic of the* Minute *phenotype. The fly in this scanning electron micrograph emerged from a heat-treated "antisense" pupa. Closed arrows indicate abnormal* Minute-*type bristles; open arrows indicate the normal contralateral bristles. Control pupae carrying a "sense" hsp–rp49 gene and treated identically never yielded flies with abnormal bristles. Bar = 100 μm.*

observed in "sense" flies that had been subjected to identical treatment. Antisense rp49 gene expression also affected female fertility. There was a drastic drop in the number of eggs laid when females carrying the antisense gene were subjected to a heat pulse. Females carrying the sense construct were not affected by an identical treatment. This drop in fertility was temporary; the antisense flies recovered their normal fertility within 2 days of rearing at normal temperature. In short, conforming to expectations, antisense rp49 gene expression induced many of the characteristic *Minute* phenotypes.

III. EVALUATION AND PERSPECTIVES

Although interference of gene expression by antisense RNAs holds great promise as a research tool, presently this approach cannot be universally applied to *Drosophila*. In addition to the "success" cases reported in the preceeding sections, there is anecdotal information about attempts that have not yielded positive results. For instance, an hsp70-driven antisense *white* gene covering an intron–exon boundary plus a few hundred nucleotides of 3′ mRNA sequences, has been transformed into flies. Induction of this antisense gene had no effect on eye color, even when a hypomorphic mutation (*white^apricot*) was used (V. Pirrotta, personal communication). Note that *white* mRNA is quite rare. Thus, after heat induction the antisense RNA is expected to be present in large excess over the endogenous transcript. Similarly, an antisense *ovarian tumor* (*otu*) gene driven by an hsp83 promoter (in contrast to the hsp70, this promoter has high levels of constitutive activity in ovaries) was transformed into flies. However, no measurable phenotype caused by antisense *otu* expression could be observed (V. Schweickart and C. Laird, personal communication). The *otu* mRNA is rather abundant in ovaries, and in the best experiments a ratio of only 1:1 sense to antisense RNA was achieved. In our laboratory an attempt at obtaining a phenotype from flies expressing antisense RNA complementary to the *Drosophila* homolog of the *raf* oncogene was unsuccessful (Qian, Perrimon and Jacobs-Lorena, unpublished observations). For both *white* and *otu*, antisense gene expression in hypomorphic mutants of the same gene had no effect. However, an effect might be expected only if the hypomorphic mutation is the result of decreased expression of a wild-type gene. If the hypomorphic mutant expresses normal levels of an altered protein, it is conceivable that even a partial reduction of mRNA amount by the antisense RNA will have no measurable effect. It should be stressed, however, that these are all negative results and that there could be trivial reasons for the lack of effect. Possibly antisense induction was not given at the appropriate developmental time, or the antisense constructs did not cover the critical region of the gene. Thus, at this point, one cannot make a definite statement that antisense interference of *white*, *otu*, or *raf* does not work.

The mechanisms of antisense interference have hardly been evaluated in *Drosophila*. One important factor is the stability of the antisense RNA. Obviously, more stable antisense RNAs are expected to have longer lasting effects. In our experience, antisense RNAs are relatively unstable in whole flies, only very low levels being observed at 5 hours after induction. We have used the hsp70 polyadenylation signal to construct the antisense *rp49* genes. This sequence may have mediated the rapid degradation of the antisense RNA after return to normal temperatures. In the case of antisense rpA1 constructs, the *fushi tarazu* (*ftz*) 3'-untranslated region was used for the polyadenylation signal. The *ftz* mRNA is known to be unstable,

and this may have contributed to its rapid decay. However, the stability of the antisense transcript in different tissues of the fly may vary considerably, and extrapolations may not be valid. We also note that, unexpectedly, the stability of antisense rpA1 RNA changed with the age of the stock. Qian et al. [1988] observed that soon after obtaining the transformant flies the antisense rpA1 RNA was essentially stable (little or no decrease during the first 4 hours after induction). However, after maintaining the same strain of flies for a few months, the rpA1 antisense RNA became highly unstable. We cannot offer an explanation for these observations. It seems that with the passing generations the transgenic flies gained the ability to degrade the antisense RNA.

Interference of gene expression may depend on the formation of double-stranded RNA molecules. That such molecules accumulate in cultured cells has been shown by Bunch and Goldstein [1989]. An extensive search for double-stranded RNAs upon expression of the antisense rpA1 RNA in the ovary was unsuccessful (Qian, Hongo, and Jacobs-Lorena, unpublished observations). It is possible that in the latter case double-stranded RNAs never formed, that the duplex RNAs are very unstable and thus undetectable, or that the RNA "unwindase" similar to the one identified in *Xenopus* (see Bass, this volume) is responsible.

When considering the likelihood of success of the antisense approach, frequently invoked parameters are the abundance of the target mRNA and the magnitude of antisense RNA excess. As reviewed in this chapter, for each individual sense–antisense combination a clear relationship existed between the extent of antisense RNA expression and the magnitude of its effect. However, no generalizations can be made as to the ratio of sense to antisense RNA mass required to observe an effect. For instance, anti-ADH but not anti-MAP expression was effective, even though both antisense genes were expressed from the same promoter in the same type of cells. Ironically, ADH mRNA was much more abundant than MAP mRNA. Antisense RNA for two abundant mRNAs (rpA1 and rp49) expressed from an hsp70 promoter was effective, whereas antisense RNA for two relatively rare mRNAs (*white* and *raf*) expressed from the same promoter was ineffective (however, consider the caveats mentioned above).

At this point, all available evidence is compatible with the following working model. Antisense RNA, whether expressed endogenously or injected, is quite inefficient in reducing the expression of the target gene. Thus *the best candidates for experiments of antisense interference are genes for which there is a priori evidence that their expression is dosage sensitive*. For instance, it is unlikely that the antisense approach will succeed for a gene that produces no phenotype when its levels of expression are reduced by about one-third or one-half or for one whose activity is known to be extensively regulated posttranscriptionally. In the same vein and as discussed above, hypomorphic mutants that express normal levels of an altered protein may not enhance the effect of antisense gene expression. Other good candidates for antisense interference are genes (maternal or zygotic) that are required only transiently during early embryogenesis. These genes can be effectively inactivated during the syncytial stages of embryonic development by injection of a very large excess of antisense RNA. Injection of similar amounts of an unrelated RNA is an important control that needs to be done to test for nonspecific effects. Cellularization of the blastoderm and the possible instability of antisense RNAs preclude the use of this injection approach for genes expressed later than after 3 hours of embryonic development. In cases of genes for which a reduction of expression yields a phenotype without inducing lethality or sterility, the use of a strong constitutive promoter should be considered. In this connection, the hsp83 promoter may be quite appropriate because it has a very high basal level of expression approaching that of abundant genes (V. Schweickart and C. Laird, personal communication).

In summary, there are a number of genes in *Drosophila* for which interference of expression by antisense RNAs has been successfully used. At present, the antisense approach cannot be used for all genes. Genes that are known to be exceptionally dosage sensitive and genes that are required during early embryonic development are likely to be the best targets. A more systematic investigation of the mechanisms of antisense interference and technical improvements (e.g., use of different promoters) may widen the spectrum of genes amenable to this type of analysis.

ACKNOWLEDGMENTS

We thank V. Pirrotta, C. Laird, and V. Schweickart for allowing us to cite their unpublished observations. The work cited from our laboratory was supported by a grant from the National Institutes of Health.

IV. REFERENCES

Baker NE (1987): Molecular cloning of sequences from *wingless*, a segment polarity gene in *Drosophila*: The spatial distribution of a transcript in embryos. EMBO J 6:1765–1773.

Bass BL, Weintraub H (1987): A developmentally regulated activity that unwinds RNA duplexes. Cell 48:607–613.

Boulay JL, Dennefeld C, Alberga A (1987): The *Drosophila* developmental gene *snail* encodes a protein with nucleic acid binding fingers. Nature 330:395–398.

Bunch TA, Goldstein LSB (1989): The conditional inhibition of gene expression in cultured *Drosophila* cells by antisense RNA. Nucleic Acids Res 17:9761–9782.

Cabrera CV, Alonso MC, Johnston P, Phillips RG, Lawrence PA (1987): Phenocopies induced with antisense RNA identify the *wingless* gene. Cell 50:659–663.

Crowley TE, Nellen W, Gomer RH, Firtel RA (1985): Phenocopy of discoidin I-minus mutants by antisense transformation in *Dictyostelium*. Cell 43:633–641.

Fasano L, Röder L, Coré N, Alexandre E, Vola C, Jacq B, Kerridge S (1991): The gene *teashirt* is required for the development of *Drosophila* embryonic trunk segments and encodes a protein with widely spaced zinc finger motifs. Cell 64:63–79.

Frei E, Schuh R, Baumgartner S, Burri M, Noll M, Jurgens G, Seifert E, Nauber U, Jäckle H (1988): Molecular characterization of *spalt*, a homeotic gene required for head and tail development in *Drosophila* embryo. EMBO J 7:197–204.

Izant JG (1989): Antisense "pseudogenetics." Cell Motil Cytoskeleton 14:81–91.

Izant JG, Weintraub H (1984): Inhibition of thymidine kinase gene expression by antisense RNA: A molecular approach to genetic analysis. Cell 36:1007–1015.

Izant JG, Weintraub H (1985): Constitutive and conditional suppression of exogenous and endogenous genes by anti-sense RNA. Science 229:345–352.

Kay MA, Jacobs-Lorena M (1987): Developmental genetics of ribosome synthesis in *Drosophila*. Trends Genet 3:347–351.

Kim S, Wold BJ (1985): Stable reduction of thymidine kinase activity in cells expressing high levels of antisense RNA. Cell 42:129–138.

Knecht DA, Loomis WF (1987): Antisense RNA inactivation of myosin heavy chain gene expression in *Dictyostelium discoidium*. Science 236:1081–1091.

Kongsuwan K, Yu Q, Vincent A, Frisardi MC, Rosbash M, Lengyel JA, Merriam J (1985): A *Drosophila Minute* gene encodes a ribosomal protein. Nature 317:555–558.

LaBonne SG, Sunitha I, Mahowald AP (1989): Molecular genetics of *pecanex*, a maternal-effect neurogenic locus of *Drosophila melanogaster* that potentially encodes a large transmembrane protein. Dev Biol 136:1–16.

McGarry TJ, Lindquist S (1986): Inhibition of heat shock protein synthesis by heat-inducible antisense RNA. Proc Natl Acad Sci USA 83:399–403.

Melton DA (1985): Injected antisense RNAs specifically block messenger RNA translation in vivo. Proc Natl Acad Sci USA 82:144–148.

Nauber U, Pankratz MJ, Kienlin A, Seifert E, Klemm U, Jäckle H (1988): Abdominal segmentation of the *Drosophila* embryo requires a hormone receptor-like protein encoded by the gap gene *knirps*. Nature 336:489–492.

O'Kane C, Gehring WJ (1987): Detection in situ of genomic regulatory elements in *Drosophila*. Proc Natl Acad Sci USA 84:9123–9127.

Pecorino LT, Rickles RJ, Strickland S (1988): Anti-sense inhibition of tissue plasminogen activator production in differentiated F9 teratocarcinoma cells. Dev Biol 129:408–416.

Preiss A, Rosenberg UB, Kienlin A, Seifert E, Jäckle H (1985): Molecular genetics of *Krüppel*, a gene required for segmentation of the *Drosophila* embryo. Nature 313:27–32.

Qian S, Hongo S, Jacobs-Lorena M (1988): Antisense ribosomal protein gene expression specifically disrupts oogenesis in *Drosophila melanogaster*. Proc Natl Acad Sci USA 85:9601–9605.

Qian S, Zhang J-Y, Kay M, Jacobs-Lorena M (1987): Structural analysis of the *Drosophila* rpA1 gene, a member of the eucaryotic "A" type ribosomal protein family. Nucleic Acids Res 15:987–1003.

Rebagliati MR, Melton DA (1987): Antisense RNA injections in fertilized frog eggs reveal an RNA duplex unwinding activity. Cell 48:599–605.

Rosenberg UB, Preiss A, Seifert E, Jäckle H, Knipple DC (1985): Production of phenocopies by *Krüppel* antisense RNA injection into *Drosophila* embryos. Nature 313:703–706.

Rothstein SJ, DiNaio J, Strand M, Rice D (1987): Stable and heritable inhibition of the expression of nopaline synthase in tobacco expressing antisense RNA. Proc Natl Acad Sci USA 84:8439–8443.

Schuh R, Jäckle H (1989): Probing *Drosophila* gene function by antisense RNA. Genome 31:422–425.

Smith CJS, Watson CD, Ray J, Bird CR, Morris PC, Schuh W, Grierson D (1988): Antisense RNA inhibition of polygalacturonase gene expression in transgenic tomatoes. Nature 334:724–726.

Strickland S, Huarte J, Belin D, Vassali A, Rickles RJ, Vassali J-D (1988): Antisense RNA directed against the 3' noncoding region prevents dormant mRNA activation in mouse oocytes. Science 241:680–684.

Van der Krol AR, Mol JNM, Stuitje AR (1988): Antisense genes in plants: An overview. Gene 77:45–50.

Wagner RW, Nishikura K (1988): Cell cycle expression of RNA duplex unwindase activity in mammalian cells. Mol Cell Biol 8:770–777.

Walder J (1988): Antisense DNA and RNA: Progress and prospects. Genes Dev 2:502–504.

Weintraub HM (1990): Antisense RNA and DNA. Sci Am 262:40–46.

Wormington WM (1986): Stable repression of ribosomal protein L1 synthesis in *Xenopus* oocytes by microinjection of antisense RNA. Proc Natl Acad Sci USA 83:8639–8643.

Xiao W, Rank GH (1988): Generation of an *ilv* bradytrophic phenocopy in yeast by antisense RNA. Curr Genet 13:283–289.

ABOUT THE AUTHORS

REKHA PATEL is a research associate in the Molecular Biology Department of the Cleveland Clinic Foundation. She received her B.Sc. from the University of Bombay and her M.Sc. from M.S. University of Baroda, India. She received her Ph.D. under H. Sharatchandra at the Indian Institute of Science, where she worked on translational regulation of maternal mRNA in *Drosophila* oocytes. Dr. Patel then pursued postdoctoral research at Case Western Reserve University in the laboratory of Dr. Marcelo Jacobs-Lorena, working on regulation of ribosomal protein mRNA translation in early embryogenesis of *Drosophila*. At present she is working in the laboratory of Dr. Ganes Sen, studying the role of interferon-inducible genes in mediating immunity to viral infections.

MARCELO JACOBS-LORENA is Associate Professor of Genetics at Case Western Reserve University where he teaches molecular genetics and developmental biology. After receiving his B.A. in Chemistry from the University of São Paulo, Brazil, in 1964, he received his M.S. from Osaka University, Japan, working with A. Tsugita on the genetic code in bacteria. Dr. Jacobs-Lorena received the Ph.D. degree from the Massachusetts Institute of Technology, working with C. Baglioni on mechanisms of protein synthesis in eukaryotes and on the characterization of maternal mRNAs in sea urchins. He trained as a postdoctoral fellow with M. Crippa at the University of Geneva, Switzerland, working on the regulation of gene expression during oogenesis in *Drosophila*. Dr. Jacobs-Lorena's current research interests include regulation of mRNA translation and stability in early development of *Drosophila* and gene expression in the gut of hematophagous insects. His work has been published in the *Proceedings of the National Academy of Sciences USA*, *Science*, *Developmental Biology*, and other journals. Dr. Jacobs-Lorena served as a member of the National Institutes of Health Genetics Study Section and received the J. Diekhoff Award for Distinguished Graduate Teaching from Case Western Reserve University.

Antisense RNA and DNA: 87–96
© 1992 Wiley-Liss, Inc.

Use of Antisense RNA to Study Gene Expression in the Mammalian Preimplantation Embryo

Brian Levy, Robert P. Erickson, and Arturo Bevilacqua

I. INTRODUCTION

A. Antisense RNA

Antisense RNA can inhibit the normal expression of specific targeted genes. Thomas (this volume) discusses the natural use of antisense RNA by prokaryotes such as *E. coli* to regulate various processes including plasmid replication [Tomizawa and Itoh, 1981; Rosen et al., 1981] and membrane structure [Mizuno et al., 1983, 1984]. Stimulated by the studies of this natural role of antisense RNA, scientists have learned to use artificial antisense RNA to inhibit the expression of numerous exogenous and endogenous genes. There are, however, many questions that still need to be answered regarding the mechanism of antisense RNA inhibition of gene expression. Sequence-specific antisense RNA, transcribed from nuclear antisense genes, may hybridize with the heterogeneous nuclear RNA, preventing normal mRNA processing, including splicing and transport to the cytoplasm [Solnick and Lee, 1987]. Antisense RNA may also hybridize with cytosolic mRNA and disrupt function by a number of possible mechanisms: RNases can degrade the antisense RNA–mRNA duplex [Wormington, 1986[, the antisense RNA can interfere with ribosome binding by hybridizing to the binding site [Liebhaber et al., 1984], the antisense RNA can prevent posttranscriptional activation of dormant mRNAs [Strickland et al., 1988], and/or the antisense RNA may result in the modification of the mRNA by converting adenosines to inosines, as discussed by Bass (this volume).

The antisense RNA may be directly injected into cells, or it may be transcribed from a plasmid DNA expression vector with an inducible promoter. The latter technique permits more control over the quantity and timing of the antisense RNA expression. In addition, to optimize the inhibition, one must consider a number of factors. Antisense RNA directed at 5′ untranslated regions appear to give the greatest inhibition, although targeting 3′ translated and untranslated sequences also reduces expression of certain genes [Ch'ng et al., 1989]. Certain systems appear to be more susceptible to the RNA duplex unwinding activity originally discovered in *Xenopus* and discussed in more detail by Bass (this volume). This unwindase enzyme appears to unwind partially the antisense RNA–mRNA duplex while also converting adenosines to inosines. Considerable mRNA secondary structure may also disrupt antisense RNA hybridization. Finally, capping the antisense RNA with CH_3GpppG or linking the antisense RNA with stabilizing sequences may help to increase stability.

B. Antisense RNA and Embryos

Because of the dramatic developmental changes that occur, embryos are especially suited for antisense RNA studies. Often the inhibition of an important gene expressed during embryogenesis leads to significant morphological and sometimes lethal defects. Several examples from *Drosophila* are discussed by Patel and Jacob-Lorena (this volume), includ-

ing the use of in vitro synthesized antisense RNA to study the *Krüppel* gene, which affects pattern formation in the *Drosophila* embryo [Cohen et al., 1988]. Antisense RNA injections complementary to noncoding regions of the *Krüppel* gene altered the segmentation pattern of the embryo. Injections of antisense RNA to the 3′ half of the *wingless* gene message in *Drosophila* (an *int*-1 homolog) also resulted in an altered segmentation pattern that appeared similar to defects seen with a *wingless* mutation [Cabrera et al., 1987]. In the *Xenopus* system, antisense RNA inhibited expression of a membrane skeleton protein, causing abnormal retinal development [Giebelhaus et al., 1988].

We believe that the mouse preimplantation embryo also serves as a useful model to study the inhibition of selected genes. The results of experiments in which the endogenous genes for murine β-glucuronidase and connexin (involved in gap junctions) were inhibited with antisense RNA are discussed in this review. In addition, negative results with antisense RNA targeted to the murine protooncogene *int*-1 are contrasted, providing a basis for a more general discussion of the ''pros'' and ''cons'' of this approach.

II. INHIBITION OF THE EXPRESSION OF β-GLUCURONIDASE

A. Background

β-Glucuronidase, a lysosomal and microsomal hydrolase, is a good target gene for study with the antisense RNA technique, because a microfluorometric assay allowing single embryo assays has been developed, and its activity undergoes a dramatic 100-fold increase during mouse preimplantation development [Wudl and Chapman, 1976]. Extraction of total RNA from preimplantation embryos showed a corresponding 13-fold increase in β-glucuronidase mRNA between the four-cell and blastocyst stages [Bevilacqua et al., 1988]. Interestingly, β-glucuronidase mRNA did not significantly vary quantitatively between the one-cell and two-cell stages; however, a possible reduction in mRNA was seen at the four-cell stage. This reduction could be due to the degradation of maternal RNA. Also, crosses of C57BL/6J and C3H/HeJ strains, which carry variants of the β-glucuronidase gene, had shown that the paternal allele is expressed during the period of increased mRNA production at the eight-cell stage, thus excluding the possibility of activation of maternally stored mRNAs [Wudl and Chapman, 1976].

B. Results With Coding Region-Directed Antisense RNA

We synthesized antisense RNA to β-glucuronidase using a cDNA clone complementary to a large portion of the coding region of the gene, but lacking approximately 500 bases in the 5' and 600 bases in the 3' region of the gene. β-glucuronidase activity in the cells injected with the antisense RNA showed only a 45% reduction in activity compared with control embryos, even though the amount of antisense RNA injected was equivalent to the total amount of poly(A$^+$)RNA in the embryo [Bevilacqua et al., 1988]. To increase the efficiency of inhibition, a number of variables were tested to explain the initial low degree of inhibition. Because a double-stranded RNA melting activity had been observed in early *Xenopus* embryos [Bass and Weintraub, 1987], the preimplantation mouse embryos were assayed for the presence of a similar activity. After injecting a ^{32}P-labeled, synthetic sense–antisense hybrid obtained from a 0.5-kb fragment of the mouse β-glucuronidase genomic clone into the cytoplasm, no melting activity was observed after 5 hours. Further experiments with ^{35}S- or ^{32}P-labeled antisense RNA injections revealed that approximately 80% of the RNA was degraded after 36 hours. During the first 36 hours of development, most of the maternal RNAs undergo degradation as embryonic RNA begins to appear. To overcome the problem of degradation, we tried antisense RNA injections at the four-cell stage and increased inhibition to 68%. Also capping the injected RNA with CH$_3$GpppG greatly enhanced stability. Injections of capped antisense RNA

at the four-cell stage to this β-glucuronidase coding region sequence led to 75% inhibition [Bevilacqua et al., 1988].

C. Results With Antisense RNA Overlapping the Starting Codon

The greatest inhibition (89%) of β-glucuronidase activity occurred when an antisense RNA complementary to the 5' region of the gene was used [Bevilacqua and Erickson, 1989]. Initially, to isolate the 5' region of the β-glucuronidase gene, we prepared antisense RNA to subfragments of a mouse β-glucuronidase genomic close, cosBGUS1. The antisense RNA was injected into embryos at the four-cell stage, and the embryos were cultured to the blastocyst stage before being assayed for β-glucuronidase activity. We assumed that hybridization to the 5' region would result in the highest degree of inhibition. Because antisense RNA to several original, unmapped subfragments only inhibited activity by 47%, a supplementary approach was used to locate the 5' end of the gene. Using a 20 nucleotide 5' sequence conserved in the human and rat gene as a probe, we screened the cosBGUS1 subclones. A 0.35 kb long PstI fragment hybridized with the probe, so this fragment was used to prepare antisense RNA. Injections of this antisense RNA resulted in 89% inhibition of activity. Subsequent data [Funkenstein et al., 1988; D'Amore et al., 1988] showed that, although the 20 nucleotide sequence is only 90% conserved, the last 15 nucleotides are all identical. As predicted, the 0.35 kb long PstI fragment contains the initiation codon. Thus cytosolic injections of antisense RNA that cover the initiation codon and ribosomal binding site give greater inhibition than injections of antisense RNA to coding regions that do not include translation initiation sites [Bevilacqua and Erickson, 1989].

III. GAP JUNCTIONS
A. Background

The preimplantation embryo requires gap junctions to facilitate intercellular communication [Kidder, 1987]. These gap junctions,

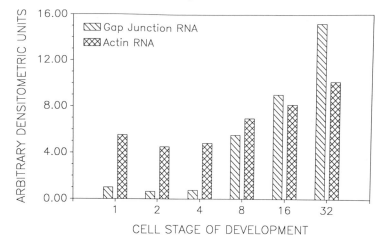

Fig. 1. *Twenty-one-day-old (C57BL/6J x SJL/J) F₁ female mice were primed with hormones as described elsewhere [Hogan et al., 1986] and mated with F₁ males of the same cross. The following midnight was taken as the time of coitus. At the times indicated embryos were collected from the oviducts of the plugged females, and total RNA from 300 embryos was extracted by the guanidinium isothiocyanate/cesium chloride method [Maniatas et al., 1982; Chirgwin et al., 1979]. The RNA was denatured, transferred to a DuPont NEN GeneScreen Plus nylon membrane, and hybridized with labeled riboprobe at 5 x 10⁵ cpm/ml at 68°C in 6 x SSC, 0.5% SDS, 5 x Denhardt's solution, 100 μg/ml Escherichia coli tRNA, and 0.01% EDTA (1 x SSC is 0.15 M NaCl/0.015 M sodium citrate, pH 7.0; 1 x Denhardt's is 0.02% polyvinylpyrrolidone, 0.02% Ficoll, 0.02% bovine serum albumin). Filters were washed at room temperature in 0.1 x SSC, 0.1% SDS for 30 minutes and three times at 70°C in the same solution. They were subjected to autoradiography for 1 to 3 days at −70°C with an intensifying screen. The 1.5 kb cDNA for rat liver 27/32 kD gap junction protein was obtained in pGEM-3 as an EcoRI insert from Dr. D. Paul. For riboprobe preparation, ³²P-UTP was added to the reaction mixture and the RNA synthesized as described by Melton et al. [1984] with the SP6 promoter, and HindIII was used to linearize the vector and terminate the transcript.*

Chicken β-actin riboprobe was used as a control for RNA recovery. pA1 was digested with PstI, and the 2.0 kb fragment was subcloned in the PstI site of pGEM-4 Blue. The orientation of the cDNA was determined by restriction mapping. The construct was linearized with EcoRI and ³²P-labeled riboprobe synthesized as described above using T7 RNA polymerase. The filters were stripped by washing four or five times for 2 minutes in 0.01% SDS, 0.01 × SSC at 95°C. They were subsequently rehybridized with the actin riboprobe under the same conditions as above. Densitometric analysis was performed using a Zeineh Soft Laser scanning densitometer and Apple data analyzer.

which specifically allow the exchange of small molecules and ions, first appear during the late eight-cell compaction stage [Magnuson et al., 1977; Kidder et al., 1987]. Injection of an antibody to the gap junction protein resulted in blastomere exclusion and delays in blastulation [Lee et al., 1987]. We found that inhibition of the function of gap junction mRNA with antisense RNA led to similar developmental abnormalities.

B. Appearance of Connexin 32 mRNA

In experiments with connexin, antisense RNA was prepared from a 1.5 kb cDNA for the rat liver 27 kD (connexin 32) gap junction protein [Bevilacqua et al., 1989]. Analysis of total embryonic RNA with the radiolabeled antisense RNA used as a hybridization probe to connexin 32 demonstrated that gap junction mRNA is present in one-cell embryos. Figure 1 shows that the gap junction mRNA is reduced in two- and four-cell embryos, then increases 12-fold at the eight-cell stage, 18-fold by the 16-cell stage, and 30-fold by the 32-cell stage. An actin probe used as a control shows good recovery of RNA at all stages. In addition, Northern analyses of blastocyst RNA showed two gap junction transcripts of 2.0 and 1.7 kb (Fig. 2).

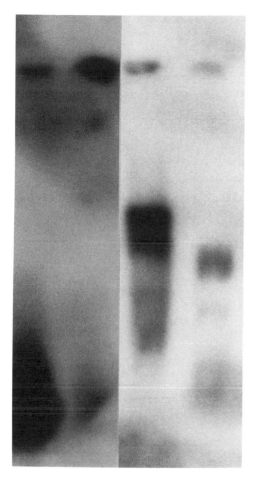

Fig. 2. *Northern analysis of total RNA from 1,000 blastocysts probed with antisense RNA riboprobe to GJ protein. Total RNA was isolated from 1,000 blastocysts by the guanidinium isothiocyanate/cesium chloride method [Chirgwin et al., 1979; Maniatis et al., 1982]. It was electrophoresed on a 0.9% agarose gel containing 2.2 M formaldehyde after denaturation in 2.2 M formaldehyde, 50% formanide at 55°C for 15 minutes [Maniatis et al., 1982]. A 1 kb ladder was used as a size marker. After transfer, the filter was baked 2 hours at 80°C and then prehybridized at 65°C in 6 x SSC, 5 x Denhardt's solution, 0.5% SDS, 10 μg/ml ssDNA (4 hours). Probe was added with 10 mM (final) EDTA and hybridized at 65°C overnight. The filter was then washed in 1 x SSC, 0.1% SDS at 68°C for 2 hours and exposed for 2 days at − 70°C with intensifying screen. Various amounts of sense RNA made from the pGEM-cDNA clone [Krieg and Melton, 1984] were also electrophoresed. Lane 1, 1,000 pg GJ sense RNA; lane 2, 100 pg GJ sense RNA; lane 3, total RNA from 1,000 blastocysts; lane 4, 1 kb ladder.*

C. Biological Effects of Injections of Antisense RNA to Connexin 32

Injections of the gap junction antisense RNA had dramatic effects on development [Bevilacqua et al., 1989]. First, the antisense RNA caused a greater percentage of degenerated embryos compared with the control β-glucuronidase RNA injections. Second, antisense RNA injected at the 2- or four-cell stage significantly inhibited compaction, while antisense RNA injected at the eight-cell stage inhibited the development of the blastocoel cavity in most of the blastomers. In other experiments, injections of a single blastomere from four-cell embryos resulted in embryos in which either one large cell or two smaller ones were excluded from the rest of the compacted embryo. Coinjections of gap junction antisense RNA and rhodamine-dextran into a single blastomere demonstrated that the excluded blastomeres were indeed the ones injected with the antisense RNA. When the rhodamine-dextran was present in two excluded cells, the injected blastomere is thought to have divided into two daughter cells. Finally, the fluorescent dye lucifer yellow (LY) provided more insight into the effects of antisense RNA on gap junction intercellular communication in compacted embryos. LY injected into one blastomere in control embryos and LY injected 10 minutes or less after antisense RNA injections spread to all of the blastomeres. However, LY injected at least 30 minutes after a gap junction antisense RNA injection did not transfer to all of the blastomeres. The dye was generally restricted to one or two cells or excluded from one or two cells [Bevilacqua, et al., 1989]. These dye-coupling results suggest that gap junctions in 8- and 16-cell embryos are replaced every 30–60 minutes, whereas gap junction proteins in tissues such as liver demonstrate a somewhat slower turnover rate [Yancey et al., 1981; Revel et al., 1984].

IV. *INT*-1

A. Background

The protooncogene *int*-1, when activated by insertion of the mouse mammary tumor virus, contributes to the development of neoplasms in mouse mammary glands [Nusse and Varmus, 1982]. int-1 was of interest to study because of its diverse pattern of expression in murine, Drosophila, and Xenopus systems. In the mouse, int-1 mRNA has been detected between days 8.5 and 14.5 of the postimplantation fetus in specific regions of the neural plate, spinal cord, diencephalon, midbrain, and hindbrain [Wilkinson et al., 1987]. In addition, it is expressed postmeiotically in testes. Shackleford and Varmus [1987] demonstrated the expression of int-1 in the round spermatids of adult mouse testis after 25 days of age (postnatal). In Drosophila, Rijsewijk et al. [1987] discovered that the Drosophila homolog of int-1, Dint-1, is identical to the segment polarity gene wingless. Heterozygote carriers of the wingless mutation develop an extra notum instead of a wing, whereas the homozygous condition causes lethality [Sharma, 1973; Sharma and Chapra, 1976]. Cabrera et al. [1987] replicated the wingless lethal phenotype by injecting antisense RNA to Dint-1 into young Drosophila embryos. Finally, the Xenopus homolog of int-1 is first detected at early neurula stages [Nordermeer et al., 1989]. Interestingly, injection of fertilized Xenopus eggs with murine int-1 results in bifurcation of the neural tube [McMahon and Moon, 1989].

Recent evidence demonstrates that int-1 encodes a secreted protein associated with the extracellular matrix [Bradley and Brown, 1990]. Thus int-1 may play a role in cell–cell communication, possibly assisting in directing cell fate. In fact, the experiments mentioned above demonstrate the profound morphogenic changes in cell fate induced with both coding and antisense RNA injections to int-1 in Drosophila and Xenopus. Therefore we decided to inject antisense RNA (and sense RNA as a control) into murine one-cell embryos to see if a phenotypic effect would result. Given the strong evolutionary conservation of int-1 and the very early embryonic lethality of null mutations at the wingless locus, we felt a more thorough search for int-1 expression in the early mouse embryo should be performed. In addition, we also used radiolabeled antisense RNA probes (riboprobes) as a more sensitive test than conventional DNA probes to look for int-1 expression in various tissues and detected some transcript expression in 7 and 14 day (postnatal) testis.

B. Biological Effects of Injection of Antisense RNA to int-1

We prepared antisense RNA to a 2.1 kb int-1 cDNA clone that was inserted into the vector

TABLE I. Injections of Antisense and Sense RNA to int-1 Into Mouse Preimplantation Embryos*

	Antisense (No. surviving/total)	Sense	Uninjected	χ^2 (antisense vs. sense)	χ^2 (antisense vs. control)
Early experience	5/22 (22.7)		24/82 (29.3)		$P > 0.05$
	11/22 (50.0)		17/50 (34.0)		$P > 0.05$
Total	16/44 (36.4)		41/132 (31.1)		$P > 0.05$
Later experience	11/36 (30.6)	0/7	35/47 (74.5)	$P > 0.05$	$P < 0.01$
	13/26 (50.0)		92/135 (68.1)		$P > 0.05$
	0/10	2/11 (18.2)	45/90 (50.0)	$P > 0.05$	$P < 0.01$
	0/7	5/12 (41.7)	17/36 (47.2)	$P > 0.05$	$P > 0.05$
Total	24/79 (30.4)	7/30 (23.3)	189/308 (61.4)	$P > 0.05$	$P < 0.01$

*The development of one-cell mouse embroys to the blastocyst stage following injection of antisense and sense int-1 capped RNA. Antisense and sense RNA were prepared as described in Figure 1. The synthesized RNA was visualized on a 0.9% agarose minigel prepared with diethylpyrocarbonate (DEPC)–treated water and resuspended in DEPC water. Approximately 20 pg of capped antisense or sense RNA was injected into the cytoplasm of one-cell mouse embryos (C57BL/6J × SJL/J) using a micromanipulator equipped with a Nikon microscope. Embryos were obtained from superovulated pregnant females and cultured in M16 media (without glucose) at 37°C, 5% CO_2 until the blastocyst stage. The embryos were handled in M2 media [Hogan et al., 1986]. Only embryos that survived injection were cultured until the control embryos reached the blastocyst stage. Values in parentheses are percentages.

pGEM 3-Z. Sense RNA to *int*-1 was prepared from the same cDNA clone. Table I shows the results of sense and antisense injections into one-cell embryos. Statistical analyses of the data show no significant effects of antisense RNA on embryo survival in early experiments. In two later experiments, antisense RNA injections did significantly reduce embryo survival; however, sense injections did so as well. Also, the first batch of embryos injected during a session did better than embryos injected later. Thus antisense RNA to *int*-1 seems to have little effect on embryonic development to the blastocyst stage.

C. Searching for Early Embryonic *int*-1 Transcripts

We also used a radiolabeled antisense *int*-1 construct to probe a dot blot of mRNA isolated from various tissues. As shown in Figure 3, *int*-1 is detected in 12 day embryos, corresponding to its expression in the developing nervous system [Wilkinson et al., 1987]. However, we were unable to detect *int*-1 expression in embryoid bodies. Thus we found no evidence for earlier expression of *int*-1, which may explain the absence of a specific phenotype induced by antisense *int*-1 RNA.

V. CONCLUSION

We used the mouse preimplantation embryo as a model to study the effects of antisense RNA on three important genes expressed during embryogenesis. Our studies with β-glucuronidase demonstrated the importance of the 5′ noncoding region and the timing of injections in inhibiting gene activity. These studies also failed to show a cytoplasmic RNA unwinding activity in the early mouse embryo. Learning from these experiments, we decided to use antisense RNA to inhibit the formation of gap junctions in preimplantation embryos. Injections of the gap junction antisense RNA significantly inhibited compaction, development of the blastocele cavity, and, with single blastomere injections, resulted in embryos in which one or two cells were excluded from the remaining compacted cells. Studies with the dye LY

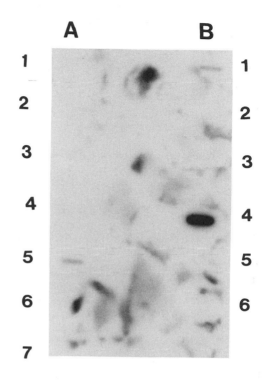

Fig. 3. *Dot blots containing poly(A⁺) mRNA from various tissue probes with an antisense* int-*1 construct.* **A:** *Slots contain approximately 0.25 μg of each of the following: 1, 7 day postnatal testis; 2, 14 day postnatal testis; 3, 21 day postnatal testis; 4, adult testis; 5, sense* int-*1 RNA; 6, antisense* int-*1 RNA; 7, yeast RNA.* **B:** *Slots contain approximately 0.83 μg of each of the following: 1, 12 day postnatal embryo; 2, embryoid bodies; 3, mouse liver; 4, sense* int-*1 RNA; 5, antisense* int-*1 RNA; 6, yeast RNA. The mRNA was extracted from the various tissues using the Fast Tract mRNA Isolation Kit^R supplied by Invitrogen Corporation. The tissue was frozen on dry ice immediately after extracting and then homogenized and/or passed through an 18 gauge syringe. We estimated the mRNA concentration by pentose analysis [Schneider, 1945]. This mRNA was then transferred to a DuPont NEN GeneScreen Plus nylon membrane and hybridized with an antisense* int-*1 RNA probe. This probe was prepared as in Figure 1. The filter was prehybridized for 3 hours and then hybridized overnight in a Collaborative Research hybridization solution lacking dextran sulfate but containing 200 μg/ml of yeast RNA. Finally, the filter was exposed overnight at −70°C before development.*

also clearly demonstrated the inhibition of gap junction formation and suggested that gap junctions in 8- and 16-cell embryos are replaced every 30–60 minutes. Finally, antisense RNA to *int*-1 did not appear to impair significantly the formation of blastocysts, correlating with our inability to detect early embryonic expression of *int*-1. We thus feel that antisense RNA serves as an excellent tool for inhibition of gene expression in the mouse preimplantation embryo.

ACKNOWLEDGMENTS

Work from our laboratory was supported by N.I.H. grant HD26454. We thank Mrs. Linda Yalkowsky for secretarial assistance.

VI. REFERENCES

Bass BL, Weintraub H (1987): A developmentally regulated activity that unwinds RNA duplexes. Cell 48:607–613.

Bass BL, Weintraub H (1988): An unwinding activity that covalently modifies its double-stranded RNA substrate. Cell 55:1089–1098.

Bevilacqua A, Erickson RP (1989): Use of antisense RNA to help identify a genomic clone for the 5′ region of mouse β-glucuronidase. Biochem Biophys Res Commun 160:937–941.

Bevilacqua A, Erickson RP, Hieber V (1988): Antisense RNA inhibits endogenous gene expression in mouse preimplantation embryos: Lack of double-stranded RNA ''melting'' activity. Proc Natl Sci USA 85: 831–835.

Bevilacqua A, Loch-Caruso R, Erickson RP (1989): Abnormal development and dye coupling produced by antisense RNA to gap junction protein in mouse preimplantation embryos. Proc Natl Acad Sci USA 86:5444–5448.

Bradley RR, Brown AMC (1990): The proto-oncogene *int*-1 encodes a secreted protein associated with the extracellular matrix. EMBO J 9:1569–1575.

Cabrera CV, Alonso MC, Johnston P, Phillips RG, Lawrence PA (1987): Phenocopies induced with antisense RNA identify the *wingless* gene. Cell 50:659–663.

Chirgwin JM, Przybyla AE, MacDonald RJ, Rutter WJ (1979): Isolation of biologically active ribonucleic acid from sources enriched in ribonuclease. Biochemistry 18:5294–5299.

Ch'ng JL, Mulligan RC, Schimmel P, Holmes EW (1989): Antisense RNA complementary to 3′ coding and noncoding sequences of creatine kinase is a potent inhibitor of translation in vivo. Proc Natl Acad Sci USA 86:10006–10010.

Cohen SM, Nauber U, Schuh R, Seifert E, Jackle H (1988): Phenocopies produced by antisense RNA identify genes required for pattern formation in the *Drosophila* embryo. In Melton DA (ed): Antisense RNA and DNA. Cold Spring Harbor, NY: Cold Spring Harbor Laboratory, pp 65–69.

D'Amore MA, Gallagher PM, Korfhagen TR, Ganschow RE (1988): Complete sequence and organization of the murine β-glucuronidase gene. Biochemistry 27: 7131–7140.

Funkenstein B, Leary SL, Stein JC, Catterall JF (1988): Genomic organization and sequence of the Gus-S α allele of the murine β-glucuronidase gene. Mol Cell Biol 8:1160–1168.

Giebelhaus DH, Eib DW, Moon RT (1988): Antisense RNA inhibits expression of membrane skeleton protein 4.1 during embryonic development of *Xenopus*. Cell 53:601.

Hogan B, Costantini F, Lacy E (1986): Manipulating the Mouse Embryo. Cold Spring Harbor, NY: Cold Spring Harbor Laboratory.

Kidder GM (1987): Intercellular communication during mouse embryogenesis. In Bavister BD (ed): The Mammalian Preimplantation Embryo: Regulation of Growth and Differentiation in Vitro. New York: Plenum, pp 43–64.

Kidder GM, Rains J, McKeon J (1987): Gap junction assembly in the preimplantation mouse conceptus is independent of microtubules, microfilaments cell flattening, and cytokinesis. Proc Natl Acad Sci USA 84:3718–3722.

Krieg PA, Melton DA (1984): Functional messenger RNAs are produced by SP6 in vitro transcription of cloned cDNAs. Nucleic Acids Res 12:7057–7070.

Lee S, Gilula NB, Warner AE (1987): Gap junctional communication and compaction during preimplantation stages of mouse development. Cell 51:851–860.

Liebhaber SA, Cash FE, Shakin SH (1984): Translationally associated helix-destabilizing activity in rabbit reticulocyte lysate. J Biol Chem 259:15597–15602.

Magnuson T, Densey A, Stackpole CW (1977): Characterization of intercellular junctions in the preimplantation mouse embryo by freeze-fracture and thin-section electron microscopy. Dev Biol 61:252–261.

Maniatis T, Fritsch EF, Sambrook J (1982): Molecular Cloning: A Laboratory Manual. Cold Spring Harbor, NY: Cold Spring Harbor Laboratory.

McMahon AP, Moon RT (1989): Ectopic expression of the proto-oncogene *int*-1 in *Xenopus* embryos leads to duplication of the embryonic axis. Cell 58: 1075–1084.

Melton DA, Krieg PA, Rebagliati MR, Maniatis T, Zinn K, Green WR (1984): Efficient in vitro synthesis of biologically active RNA and RNA hybridization probes from plasmids containing a bacteriophage SP6 promoter. Nucleic Acids Res 12:7035–7056.

Mizuno T, Chou M-Y, Inouye M (1983): A comparative

study on the genes for three porins of the *Escherichia coli* outer membrane: DNA sequence of the osmoregulated omp C gene. Proc Jpn Acad Sci 59: 335–338.

Mizuno T, Chou M-Y, Inouye M (1984): A unique mechanism regulating gene expression: Translational inhibition by a complementary RNA transcript (mic RNA). Proc Natl Acad Sci USA 81:1966–1970.

Nordermeer J, Meijlink F, Verrijzer P, Rijsewijk F, Destree O (1989): Isolation of the *Xenopus* homologue of *int*-1 and *wingless* and expression during neurula stages of early development. Nucleic Acids Res 17:11–18.

Nusse R, Varmus HE (1982): Many tumors induced by the mouse mammary tumor virus contain a provirus integrated in the same region of the host genome. Cell 31:99–109.

Revel J-P, Nicholson BJ, Yancey SB (1984): Molecular organization of gap junctions. Fed Proc Fed Am Soc Exp Biol 43:2672–2677.

Rijsewijk F, Schuermann M, Wagenaar E, Parren P, Weigel D, Nusse R (1987): The *Drosophila* homolog of the mouse mammary oncogene *int*-1 is identical to the segment polarity gene *wingless*. Cell 50:649–657.

Rosen J, Ryder T, Ohtsubo H, Ohtsubo E (1981): Role of RNA transcripts in replication incompatibility and copy number control in antibiotic plasmid derivatives. Nature 290:794–799.

Rosenberg UB, Preiss A, Seifert E, Jäckle H, Knipple DC (1985): Production of phenocopies by *Krüppel* antisense RNA injection into *Drosophila* embryos. Nature 313:703–706.

Schneider WC (1945): Phosphorus compounds in animal tissues in extraction and estimation of deoxypentose nucleic acid and of pentose nucleic acid. J Biol Chem 161:293–303.

Shackleford GM, Varmus HE (1987): Expression of the proto-oncogene *int*-1 is restricted to postmeiotic male germ cells and the neural tube of midgestational embryos. Cell 50:89–95.

Sharma RP (1973): *Wingless*, a new mutant in *D. melanogaster*. Dros Inf Service 50:134.

Sharma RP, Chapra VL (1976): Effect of *wingless* (*wg*) mutation on wing and haltere development in *Drosophila melanogaster*. Dev Biol 48:461–465.

Solnick D, Lee SI (1987): Amount of RNA secondary structure required to induce an alternative splice. Mol Cell Biol 7:3194–3198.

Strickland S, Huarte J, Belin D, Vassalli A, Rickles RJ, Vassalli JD (1988): Antisense RNA directed against the 3' noncoding region prevents dormant mRNA activation in mouse oocytes. Science 241: 680–684.

Tomizawa JI, Itoh T (1981): Inhibition of ColE1 RNA primer formation by a plasmid-specified small RNA. Proc Natl Acad Sci USA 78:6096–6100.

Wilkinson DG, Bailes JA, McMahon AP (1987): Expression of the protooncogene *int*-1 is restricted to specific neural cells in the developing mouse embryo. Cell 50:79–88.

Wormington WM (1986): Stable repression of ribosomal protein L1 synthesis in *Xenopus* oocytes by microinjection of antisense RNA. Proc Natl Acad Sci USA 83:8639–8643.

Wudl L, Chapman V (1976): The expression of β-glucuronidase during preimplantation development of mouse embryos. Dev Biol 48:104–109.

Yancey SB, Nicholson BJ, Revel J-P (1981): The dynamic state of liver gap junctions. J Supramol Struct Cell Biochem 16:221–232.

ABOUT THE AUTHORS

BRIAN LEVY received his B.S. from the University of Michigan. He performed the *int*-1 antisense experiments described in this paper for his honor's thesis. He is now a medical student at the University of Michigan.

ROBERT P. ERICKSON is the Holsclaw Family Professor of Human Genetics and Inherited Diseases in the Departments of Pediatrics and Molecular and Cellular Biology at the University of Arizona, where he teaches pediatric genetics and molecular biology. After receiving his B.A. from Reed College in 1960, he received his M.D. at Stanford University School of Medicine, where he also actively pursued research in the Department of Genetics. Dr. Erickson's internship at Cornell University School of Medicine and residency at Albert Einstein College of Medicine were followed by two years at NIH where he was a researcher in the laboratory of Dr. Christian B. Anfinsen. This was followed by another year of postdoctoral fellowship with N.A. Mitchison at the National Institute of Medical Research, Mill Hill, London, where he commenced his studies on the role of cell surface antigens in development. This research avenue was continued during a sabbatical with François Jacob at the Pasteur Institute in 1975–1976, while a shift in techniques to those of molecular genetics occurred with a sabbatical at the Imperial Cancer Research Fund Laboratories, London, in 1983–1984. Dr. Erickson's current research involves the use of molecular genetic techniques to study aberrations of sexual determination and differentiation in mice and man and the use of antisense techniques to study gene expression in the preimplantation embryo and during spermatogenesis. He is author or co-author of over 150 research papers and is on the editorial boards of *Molecular Reproduction and Development* and *Antisense Research and Development*. He has been the recipient of a research career development award from NIH, a Guggenheim Fellowship, and an Eleanor Roosevelt Cancer Research Fellowship.

ARTURO BEVILACQUA is a Research Associate at the University ''La Sapienza'' of Rome where he lectures in General Biology. After he received his ''Laurea'' degree in Biological Sciences from the University of L'Aquila in 1985, studying the heat shock response of mouse oogenetic cells, he pursued postdoctoral research at the University of Michigan in the laboratory of Robert P. Erickson. There he studied the inhibition of gene expression by antisense RNA in early mammalian embryos and produced transgenic mice. Dr. Bevilacqua then returned to Italy, where he received his Ph.D. from the University of L'Aquila under Franco Mangla. His thesis concentrated on the study of mammalian developmental genetics and the development of techniques for genetic manipulation of early mouse embryos. Dr. Bevilacqua's current research focuses on the analysis of dependence of gene expression on chromatin structure during mouse oogenesis and preimplantation embryo-genesis. His research papers have appeared in such journals as the *Proceedings of the National Academy of Sciences USA*, *Developmental Biology*, *Gamete Research*, and *Molecular Reproduction and Development*. In 1991 he won the Alberto Monroy Memorial Award for research.

Antisense RNA and DNA: 97–108
© 1992 Wiley-Liss, Inc.

Antisense RNA Production in Mammalian Fibroblasts and Transgenic Mice

M. Idrees Munir, Belinda J.F. Rossiter, and C. Thomas Caskey

I. INTRODUCTION

The possibility of preventing the expression of specific genes within an organism has important implications in the generation of animal models of human disease, in the determination of gene function, and in the treatment of certain diseases. Many human disorders result from a simple genetic defect but closer biochemical study and the development of potential treatments have been hindered by the lack of animal models. The techniques of molecular biology have enabled the rapid isolation of genes, some of whose functions are unknown, and the ability to ''turn off'' such an unidentified gene in an organism offers the possibility of determining its function. The ability to inhibit specifically the expression of genes such as viral genes or oncogenes as a form of therapy for diseases such as AIDS and cancer would obviously have enormous potential for improving the means of treatment currently available.

The methods employed to inhibit specific genes in mice may be divided into three categories. The first involves random inactivation of genes and screening for the desired result, as in the generation of hypoxanthine phosphoribosyltransferase (HPRT E.C. 2.4. 2.8) deficient mice [Hooper et al., 1987; Kuehn et al., 1987]. Second, targeting of foreign DNA into particular genes by using homologous recombination has resulted in the inactivation of HPRT in embryonic stem cells [Mansour et al., 1988; Thomas and Capecchi, 1987], from which genetically altered mice can be derived. Third, antisense RNA inhibition of gene expression, which is described here, has been used in our and in other laboratories in attempts to turn off gene expression. This technique has the advantage of targeting specific genes and also the possibility of partial inhibition in cases where the complete inactivation of a gene might be lethal to the organism.

Our reason for initially choosing the HPRT enzyme, which is involved in purine metabolism, for attempts at antisense inhibition of gene expression is its involvement in two human disorders. A partial deficiency in HPRT activity results in gouty arthritis, and complete deficiency results in the severe Lesch-Nyhan syndrome (for which there is currently no treatment), with symptoms of spasticity, choreoathetosis, mental retardation, and compulsive self-mutilation. There are obviously neurological disturbances involved in Lesch-Nyhan syndrome, but it is not at all clear how these result from a disruption in purine metabolism; an animal model of this disease might therefore provide insights into resolving the problem. Although there have been recent successes in generating HPRT-deficient mice (which, incidentally, display no symptoms of gout or Lesch-Nyhan syndrome) by means other than antisense inhibition [Hooper et al., 1987; Kuehn et al., 1987], we thought that the HPRT gene could still be used as a model in the investigation of the phenomenon of antisense inhibition of gene expression. In addition, selection of cultured cells with or without HPRT activity is straightforward, and normal levels of HPRT protein are low (about 0.01% of total protein), making it a reasonable choice of target for antisense inhibition.

Stout and Caskey [1987] demonstrated transient and stable inhibition of HPRT activity in COS cells using antisense constructs containing the adenovirus major late promoter. These experiments suggested that antisense constructs containing splice sites from the 5' portion of the HPRT gene were as effective at inhibiting gene expression as those containing cDNA sequences alone. Munroe [1988] has also shown that antisense RNA derived from intron or exon sequences can inhibit in vitro splicing of β-globin mRNA. Although other examples of antisense inhibition using genomic fragments are known [Edwards et al., 1988; Nishikura and Murray, 1987], the majority reported in the literature are fragments derived from cDNA, suggesting that interference with RNA splicing may not be the only mechanism of antisense inhibition.

An obvious extension to the expression of antisense genes in cultured cells is the generation of transgenic mice carrying antisense genes. Such animals have the potential of revealing the function of a particular gene by observing the phenotype of an animal inhibited in its expression. The feasibility of this approach has been shown by Katsuki and coworkers [1988] (also Katsuki et al., this volume), who have used antisense technology to inhibit myelin basic protein expression in transgenic mice and thus induce a *shiverer* phenotype.

We describe here the generation of antisense HPRT RNA in cultured cells and transgenic mice. Whereas antisense expression in cells could lead to the reduction of HPRT activity to undetectable levels, this effect was not observed in the transgenic animals. The possible reasons why the antisense approach to inhibition of HPRT appears to give different results in vitro and in vivo are discussed.

II. RESULTS
A. Antisense RNA Production In Vitro

Figure 1 shows the structures of the antisense constructs used to reduce HPRT activity in cells in culture [Stout and Caskey, 1987, 1990]. Antisense molecules containing only coding sequences were effective in reducing or eliminating HPRT activity in mouse or human fibroblasts, as were sequences from the junction of the first exon and first intron, and even sequences from within the first intron close to, but not including, any coding sequence. Intron sequences further away from the first exon were not effective in inhibiting HPRT activity. The fact that intron sequences close to exon 1, which are important in splicing, were as effective in mediating antisense inhibition of HPRT gene expression as the coding region led us to propose that in this system the mechanism of inhibition involved interference with RNA processing and with splicing in particular.

Further experiments showed that there was a threshold level of antisense RNA required for inhibition of HPRT activity to be observed, using constructs containing the first exon–intron junction; this is summarized in Figure 2. An inhibition of HPRT activity was seen only if there was at least a 20-fold excess of antisense

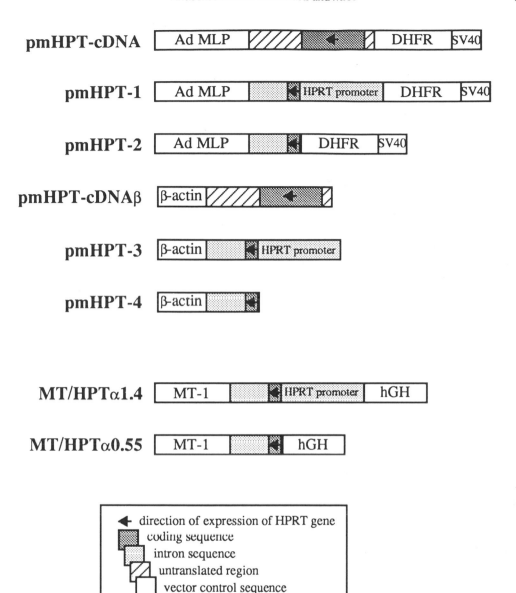

Fig. 1. *Schematic of antisense constructs. The upper group of six antisense constructs were used in the in vitro experiments and utilize the adenovirus major late promoter (Ad MLP) or the β-actin promoter. The presence of the DHFR (dihydrofolate reductase) gene is not relevant to these studies; the adenovirus major late promoter construct employs an SV40 polyadenylation signal. In each of these sets of vectors three different portions of the HPRT gene were expressed in the antisense orientation, one being the cDNA including some untranslated sequence (pmHPT-cDNA and pm-HPT-cDNAβ) and the others being two genomic re-gions around the first exon pmHPT-1, pmHPT-3, and MT/HPTα1.4 contain the first exon and about 400 bp of downstream intron sequence, and pmHPT-2, pmHPT-4, and MT/HPTα0.55 also contain about 960 bp of 5' flanking sequence, including the GC-rich promoter sequence from the HPRT gene. The lower pair of antisense constructs were used in the in vivo experiments and utilize the metallothionein-1 promoter (MT-1). The two constructs contain different-sized genomic regions of the HPRT gene around the first exon; fragments identical to these were used in the in vitro experiments. hGH, human growth hormone.*

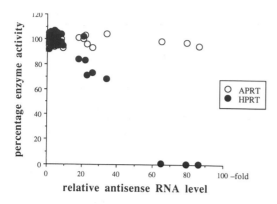

Fig. 2. *Inhibition of fibroblast HPRT activity with antisense RNA production. Cloned cell lines containing the antisense constructs described in Figure 1 were assayed for antisense RNA levels (relative to normal HPRT RNA levels) and HPRT and APRT enzyme activities (relative to control cell lines). The three cell lines with virtually no HPRT activity were able to survive 6-thioguanine selection and had no detectable HPRT mRNA.*

over normal HPRT mRNA levels, and complete inactivation was only observed in the presence of about a 60-fold excess of antisense RNA. Adenine phosphoribosyltransferase (APRT E.C. 2.4.2.8), a similar enzyme also involved in purine metabolism activity, is not affected by the presence of antisense HPRT RNA, indicating that the inhibition appears to be specific. In the instances in which HPRT activity was reduced to almost zero, the amount of HPRT mRNA was also reduced to undetectable levels, suggesting a degradation or inhibition of formation of mature message. It is interesting to note that the vast majority of transformed cell lines had only a 1- to 10-fold excess of antisense RNA over normal HPRT RNA and no detectable inhibition of HPRT activity. The latter detail may have relevance regarding the effectiveness of the in vivo experiments.

B. Antisense RNA Production In Vivo

Two antisense constructs were used in transgenic mice [Munir et al., 1990], differing in the length of genomic HPRT sequence expressed; these are indicated in Figure 1. The

region of the HPRT gene included the first exon and about 400 bp of downstream sequence; the longer construct also contained about 960 bp of 5′ flanking sequence. Both transgenes were under the control of the mouse metallothionein-1 (MT-1) promoter, which, although supposedly inducible with heavy metals, was found to have constitutive expression in the second series of animals. Both constructs were expressed in transgenic mice, most often in central nervous system tissue, but also in heart and liver (second series of mice).

1. The pMT/HPTα1.4 construct. Six transgenic mice were generated, each carrying 2-45 copies of the pMT/HPTα1.4 construct (Fig. 3).

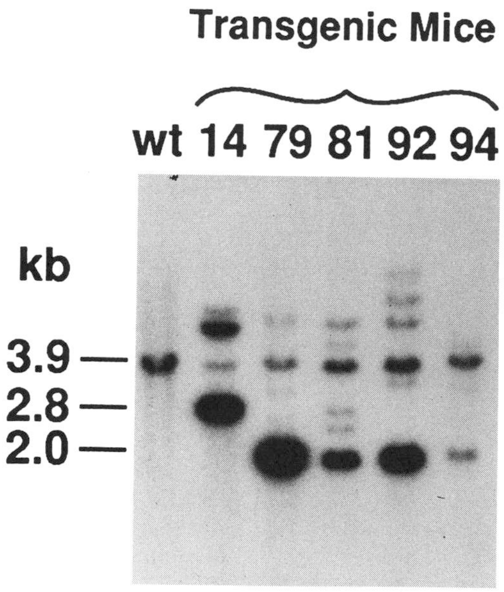

Transgenic Mice

wt 14 79 81 92 94

kb

3.9 —

2.8 —

2.0 —

Fig. 3. *Southern analysis of transgenic mice. Genomic DNA from each of the five antisense HPRT transgenic mouse lines was digested with EcoRI and probed with a mouse MT-1 fragment. Mouse No. 14 carries the MT/HPTα1.4 transgene (the F-A site of integration, 35 copies of the transgene); mice Nos. 79, 81, 92, and 94 carry the MT/HPTα0.55 transgene (25, 3, 12, and 1 copies, respectively). On digestion with EcoRI, the endogenous MT-1 fragment is 3.9 kb, the MT/HPTα1.4 construct yields a 2.8 kb fragment, and the MT/HPTα0.55 transgene results in a 2.0 kb fragment. Additional fragments correspond to multiple copies of the transgene, presumably resulting from inactivation of the flanking EcoRI site(s). [From Munir et al., 1990, with permission of the publisher.]*

Five of these transgenic mice did not produce detectable quantities of antisense RNA by Northern analysis. The sixth mouse (No. 14, F-A/B), carrying approximately 45 copies of the transgene, had two unlinked sites of integration that were separated by breeding. From this founder animal two separate mouse lines, each with one of the sites, were established and shown to be carrying approximately 35 (F-A) and 10 (F-B) copies of the transgene.

A mouse carrying both transgene sites (F-A/B) was treated with heavy metals as described, and RNA from various organs was assayed by Northern analysis for expression of antisense RNA. Antisense HPRT expression was only seen in central nervous system tissue, and this expression was seen only after induction by $ZnSO_4$ and/or $CdSO_4$ (Fig. 4). RNA was also isolated from the brains of induced mice carrying the two sites of insertion independently. Mouse F-A produced antisense RNA, whereas mouse F-B displayed no antisense expression after induction with both heavy metals (Fig. 4).

No significant or repeatable inhibition of HPRT activity in the brains of F-A mice was observed; therefore additional brains were further dissected into five parts (basal ganglia, brain stem, cerebellum, cortex, and hippocampus). RNA was extracted from these different brain regions in order to locate more precisely the antisense expression. Different levels of antisense RNA expression were observed in each of the five regions of the brain by Northern blot analysis (Fig. 5A). The basal ganglia, cortex, and hippocampus regions showed the highest levels of expression, whereas brain stem

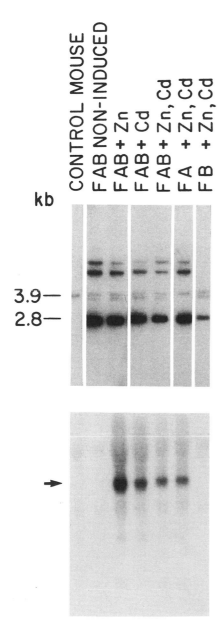

Fig. 4. *Induction of MT/HPTα1.4 transgene expression with heavy metals. An F-A/B transgenic founder animal was generated carrying multiple copies of the MT/HPTα1.4 transgene at two independent sites of integration that were separated by breeding to F-A and F-B. Control (nontransgenic), F-A/B, F-A, and F-B mice were treated with zinc (Zn) or cadmium (Cd) as described before the mice were sacrificed and DNA and RNA isolated. The upper panel shows Southern analysis of the DNA probed with mouse MT-1. In addition to the endogenous and transgene fragments (3.9 and 2.8 kb, respectively), three additional bands are visible in the F-A/B mouse; the lower of these is associated with the F-B site of integration and the other two lie within the F-A site (compare the right-hand two lanes with the others). The lower panel shows Northern analysis of RNA from the brains of the same mice, probed with a single-stranded "sense" HPRT RNA probe that hybridizes to the antisense transcript but not the endogenous HPRT message. The size of the antisense RNA is 2.1 kb. [From Munir et al., 1990, with permission of the publisher.]*

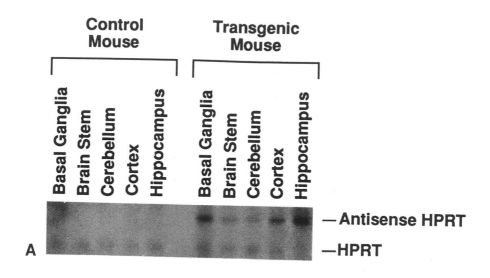

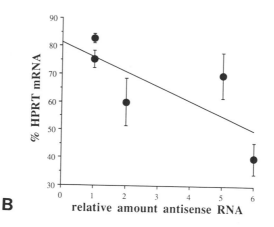

A

B

Fig. 5. *Effect of MT/HPTα1.4 antisense RNA on endogenous HPRT mRNA levels.* **A:** *Northern analysis of brain region RNA from a control mouse and from an MT/HPTα1.4 transgenic mouse, probed with an HPRT exon 1 fragment that hybridizes with both sense and antisense RNA species. The size of the antisense RNA is 2.1 kb, and the size of the HPRT RNA is 1.3 kb.* **B:** *Densitometry plots of the relative intensity of the bands from three such Northerns were used to generate the values plotted in the lower panel. The "relative amount antisense RNA" is defined as the level of the antisense RNA compared with the normal level of HPRT mRNA in the same brain region. The "% HPRT mRNA" is defined as the level of HPRT mRNA in the transgenic mouse compared with the same tissue in a normal mouse. The values determined by this method are in agreement with others obtained by independent comparison of antisense and HPRT mRNA bands (detected by different single-stranded probes) with β-actin bands. The error bars represent 1 SD (sample number = 3); the line was fitted by the Cricket Graph computer program (Cricket Software Inc., Philadelphia, PA). [From Munir et al., 1990, with permission of the publisher.]*

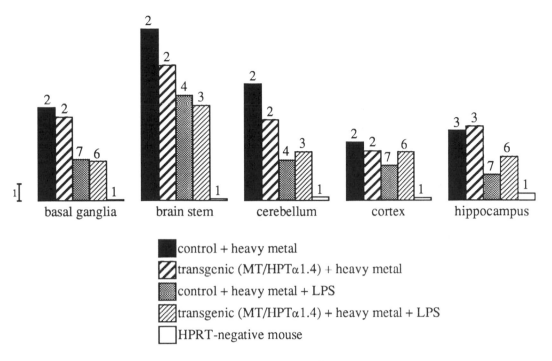

Fig. 6. *HPRT activities in control and MT/HPTα1.4 transgenic mouse brain regions. The height of each bar is proportional to the HPRT enzyme activity relative to APRT activity in the appropriate tissue; the scale bar on the left indicates a value of one. The number above each bar indicates the number of mice used to determine the overall value, which is the mean of the measurements observed. The shading key indicates the nature of the mice in each group and their treatment. Heavy metal, zinc and cadmium treatment; LPS, lipopolysaccharide treatment; HPRT-negative mouse, the Hprt-b^{m2} strain of Kuehn et al. [1987]. [From Munir et al., 1990, with permission of the publisher.]*

and cerebellum contained little antisense RNA; endogenous HPRT mRNA was detectable in all tissues expressing antisense RNA, but at reduced levels. The reduced level of HPRT mRNA was plotted against the amount of antisense RNA observed (Fig. 5B) and showed an approximate relationship between the amount of antisense RNA present (one- to sixfold excess over HPRT mRNA) and the degree of reduction of HPRT mRNA (20%–50%).

Since levels of endogenous HPRT mRNA were partly reduced as the result of expression of the MT/HPTα1.4 transgene, we wished to determine if there was a corresponding reduction in the HPRT enzyme activity. Measurement of HPRT enzyme activities in different regions of brain from the mouse F-A carrying

35 copies of the transgene did not show any significant difference from wild-type levels (Fig. 6). Control and transgenic mice were treated with heavy metals to induce the expression of the transgene, and some mice were given lipopolysaccharide (LPS) to further increase the induced expression [Searle et al., 1984]. Since LPS appears to have the effect of reducing HPRT activity in control animals lacking any transgene, it is only possible to compare control and transgenic animals that have received the same treatment with respect to LPS. Although some variation is seen between the HPRT levels in tissues from control and transgenic animals, no effect was perceived that was significant, consistent, or showed any relationship with the expression of antisense RNA.

2. The pMT/HPTα0.55 construct. The MT/HPTα0.55 transgene construct was constructed to counter some of the possible problems arising from the MT/HPTα1.4 construct, which contains part of the mouse HPRT promoter region. It is conceivable that the longer construct might produce short pieces of "sense" HPRT RNA from the transgene HPRT promoter, which could then potentially inhibit any antisense effect on the cell by competing with the native HPRT mRNA for hybridization with the genuine antisense transcript. In addition, the presence of the GC-rich promoter region expressed as a portion of the antisense transcript might have some nonspecific effect on other GC-rich parts of the genome or transcribed RNA.

Four out of six transgenic mice generated with the MT/HPTα0.55 construct expressed antisense RNA detectable by Northern analysis. Mouse No. 79, which carries 25 copies of the injected sequence, expressed the antisense transcript in heart and liver. Transgenic mice Nos. 81, 92, and 94 have 3, 12, and 1 copies of the injected DNA, respectively, and expressed antisense HPRT RNA in brain, heart, and liver. Native HPRT mRNA was also present in the tissues expressing antisense RNA. In contrast to the first transgenic mouse described above, the expression from the transgene in this series of animals was constitutive and not increased by treatment with heavy metals. When normalized to β-actin RNA expression, the levels of endogenous HPRT mRNA were not found to be significantly altered in tissues expressing antisense HPRT RNA.

Assays of HPRT activity in tissues from transgenic mice expressing antisense HPRT RNA in brain, heart, and liver again indicated that there was no inhibition of HPRT activity. Even heart, which had the lowest levels of endogenous HPRT activity and the highest levels of antisense HPRT RNA production, was unaffected in terms of HPRT activity.

III. DISCUSSION

We have successfully inhibited HPRT activity in mammalian cells by transfection of con-structs driving the expression of parts of the HPRT gene in the reverse orientation. Antisense constructs containing regions of the gene important for the correct splicing of RNA were as effective in mediating inhibition as constructs containing only coding sequences. This would suggest that the inhibition of HPRT gene expression may be occurring through the mechanism of interference with RNA splicing.

The use of antisense technology in transgenic mice is potentially a very powerful way of inducing dominant negative mutations. We demonstrate here the expression of HPRT antisense sequences in various tissues of transgenic mice with, in some cases, corresponding reductions in the levels of native HPRT mRNA. Nevertheless, it has not been possible to demonstrate any reproducible decrease in HPRT enzyme activity.

The choice of sequences to be used for antisense inhibition of gene expression has focused on regions of the RNA molecule that have functional importance and probably interact with proteins. It is possible to envisage the disruption of processing or transport of a single-stranded molecule that might occur when the appropriate domain is made double stranded. For this reason, investigators have most often chosen the cap site, ribosome binding site, initiation codon, splice acceptor and donor sites, coding region, or polyadenylation signals to be represented in the antisense molecule. Although there are several examples of successful antisense inhibition of gene expression in cells, there still have not been enough experiments performed to define "guidelines" as to the best design of an antisense construct in the whole organism. Experiments with oligonucleotides may have the potential to assist in the definition of these rules, if there are to be any, since it should be possible to tell more quickly what the important target regions are, although it should be noted that the mechanism of inhibition by antisense oligonucleotides may differ from that of antisense RNA. Daugherty et al. [1989] engineered a series of β-galactosidase antisense RNA expression constructs for use in *Escherichia coli* and determined that the most effective molecule in this system included a functional ribosome binding site and 5' sequences from the *lacZ* gene in the reverse orientation; shorter antisense species were produced more abundantly and were therefore more effective.

Experiments in our laboratory [Stout and Caskey, 1990] using cultured fibroblasts suggested a nuclear site of antisense inhibition and that perturbation of pre-mRNA processing was sufficient to inhibit HPRT gene expression. It has also been reported by Munroe [1988] that antisense intron or exon sequences can interfere with an in vitro splicing reaction. The fact that even exon sequences greater than 80 bp from a β-globin splice acceptor site can inhibit splicing raises the intriguing possibility that coding sequence antisense constructs could still have some effect on mRNA splicing. We chose to use antisense constructs in transgenic mice that include a splice donor site, a 5' coding sequence, and the translation start site in order to take advantage of multiple ways in which antisense inhibition might be effective in this system.

Successful inhibition of gene expression in cultured cells resulting from expression of an antisense gene has been reported on numerous occasions [van der Krol et al., 1988] (see also Murray and Crockett, this volume), often yielding valuable information about the function of the inhibited gene product. The use of antisense inhibition in whole organisms has been less fruitful in general, although there are some notable cases of its success, particularly in *Drosophila* (see Patel and Jacobs-Lorena, this volume), where, for example, injection of antisense *Krüppel* (*Kr*) RNA into wild-type *Drosophila* embryos resulted in the production of phenocopies of the *Kr* mutant [Rosenberg et al., 1985], and in plants (Rodermel and Bogorad, this volume; Bejarano et al., this volume). To date there is only one successful example of antisense inhibition of gene expression in adult animals, reported by Katsuki et al. [1988] (see Katsuki et al., this volume). This group was able to reduce the level of myelin basic protein mRNA in mice to 20% of normal, which was sufficient to induce the phenotype of the *shiverer* mutation. It may be significant that there is a scarcity of reports relating to the successful use of antisense in vivo, perhaps indicating differences between in vitro and in vivo systems in this respect, as yet uncharacterized.

In 1987 there were two reports of "unwinding" or "helicase" activities in *Xenopus* oocytes [Bass and Weintraub, 1987; Rebagliati and Melton, 1987] that were originally thought to counteract any antisense effect by unravelling the double stranded RNA molecule; this activity was not observed in mouse embryos [Bevilacqua et al., 1988] (see also Levy et al., this volume). As originally shown by Bass and Weintraub [1988] and discussed by Bass (this volume), it is now known that this process may assist rather than hinder the disruption of mRNA by chemically modifying the substrate molecule. It is possible that there may be other as yet uncharacterized enzymes present or absent in different systems that could moderate or enhance the antisense effect.

One obvious explanation for our lack of success in reducing HPRT activity in transgenic mice, compared with the in vitro experiments, is the amount of antisense RNA generated. In cultured cells it was shown that a reduction of HPRT activity was observed when antisense RNA was present at levels approximately 20-fold greater than the normal levels of endogenous HPRT mRNA; at a 60-fold excess, HPRT activity was undetectable, as was sense HPRT mRNA. Antisense studies in cells would suggest that the inhibitory effect is proportional to the amount of antisense RNA present [Kim and Wold, 1985], and this is supported by results in other systems, including *Drosophila* (see Patel and Jacobs-Lorena, this volume).

Whether or not antisense inhibition works may not be as simple as a function of the quantity of antisense RNA, as there have been other reports of failure, such as that of Kerr et al. [1988], for whom even a 1,000-fold excess of chloramphenicol acetyltransferase (CAT) antisense RNA would not inhibit expression of a transfected, interferon-inducible CAT gene. It is quite likely that there are differences between cultured cell systems and whole organisms in terms of biochemical activity within the cell. These differences might perhaps affect stability of the antisense–sense RNA–RNA hybrid or either of its components, or secondary structures of the RNA molecules. It has often been the case that in vitro experiments provide a good model and basis for in vivo studies, but in the antisense HPRT system this appears not to have been so.

If there is too little antisense RNA being generated in the cell, it may be possible to increase that amount by designing expression constructs using more powerful promoters, such as viral promoters, although it might be necessary to sacrifice inducibility to obtain maximum expression. It should also be possible to make the antisense RNA more effective by introducing into it a ribozyme sequence [Haseloff and Gerlach, 1988] (see also Joyce, this volume; Koizumi and Ohtsuka, this volume) such that the target mRNA is not only inhibited by antisense–sense interaction but also cleaved by the action of the ribozyme (and thus presumably degraded by cellular enzymes). Such constructs have been shown to reduce HIV-1 *gag* RNA levels in cultured cells [Sarver et al., 1990].

The process of generating transgenic mice by microinjection of DNA into fertilized embryos and implantation of them into pseudopregnant females can be slow and inefficient. The method of manipulating embryonic stem (ES) cells before their injection into developing blastocysts [Bradley et al., 1984; Robertson et al., 1986] and the generation of transgenic mice via chimeric animals is also not trivial, but has the advantage that only cells that have been modified genetically in the required way need to be used to generate the chimeric mice. The method thus has the potential of being more efficient. Selection protocols allowing only the growth of HPRT-deficient cells are available [Caskey and Kruh, 1979], and such selection of ES cells carrying effective antisense HPRT constructs would be an obvious improvement to the approach described in this report.

Although we have been unable thus far to demonstrate a decrease in HPRT activity as a result of antisense inhibition, it is clearly possible to generate transgenic mice that express antisense HPRT constructs. The most likely explanation for the lack of antisense effect at the protein level in our opinion is generation

of inadequate levels of antisense RNA. The use of strong constitutive promoters and selection of HPRT-deficient ES cells prior to the generation of transgenic mice hold the promise of a more effective approach. The incorporation of ribozyme sequences into antisense constructs is an exciting possibility for the enhancement of inhibition using antisense technology.

Despite the limited success with antisense inhibition of HPRT in vivo, the phenomenon of antisense inhibition of gene expression has enormous potential, as can be seen from other discussions in this volume. Instances in which antisense technology has been less successful may still be useful in defining important parammeters of the system.

ACKNOWLEDGMENTS

M.I.M. was supported as a Howard Hughes Medical Institute Associate, B.J.F.R. was supported as an Arthritis Foundation Fellow, and C.T.C. is a Howard Hughes Medical Institute Investigator. Additional support was provided by PHS grant DK31428.

IV. REFERENCES

Bass BL, Weintraub H (1987): A developmentally regulated activity that unwinds RNA duplexes. Cell 48:607–613.

Bass BL, Weintraub H (1988): An unwinding activity that covalently modifies its double-stranded RNA substrate. Cell 55:1089–1098.

Bevilacqua A, Erickson RP, Hieber V (1988): Antisense RNA inhibits endogenous gene expression in mouse preimplantation embryos: Lack of double-stranded RNA "melting" activity. Proc Natl Acad Sci USA 85:831–835.

Bradley A, Evans M, Kaufman MH, Robertson E (1984): Formation of germ-line chimaeras from embryo-derived teratocarcinoma cells. Nature 309:255–256.

Caskey CT, Kruh GD (1979): The HPRT locus. Cell 16:1–9.

Daugherty BL, Hotta K, Kumar C, Ahn YH, Zhu J, Pestka S (1989): Antisense RNA: Effect of ribosome binding sites, target location, size, and concentration on the translation of specific mRNA molecules. Gene Anal Tech 6:1–16.

Edwards SA, Rundell AYK, Adamson ED (1988): Expression of c-*fos* antisense RNA inhibits the differentiation of F9 cells to parietal endoderm. Dev Biol 129:91–102.

Haseloff J, Gerlach WL (1988): Simple RNA enzymes with new and highly specific endoribonuclease activities. Nature 334:585–591.

Hooper M, Hardy K, Handyside A, Hunter S, Monk M (1987): HPRT-deficient (Lesch-Nyhan) mouse embryos derived from germline colonization by cultured cells. Nature 326:292–295.

Katsuki M, Sato M, Kimura M, Yokoyama M, Kobayashi K, Nomura T (1988): Conversion of normal behavior to *shiverer* by myelin basic protein antisense cDNA in transgenic mice. Science 241:593–595.

Kerr SM, Stark GR, Kerr IM (1988): Excess antisense RNA from infectious recombinant SV40 fails to inhibit expression of a transfected, interferon-inducible gene. Eur J Biochem 175:65–73.

Kim SK, Wold BJ (1985): Stable reduction of thymidine kinase activity in cells expressing high levels of antisense RNA. Cell 42:129–138.

Kuehn MR, Bradley A, Robertson EJ, Evans MJ (1987): A potential animal model for Lesch-Nyhan syndrome through introduction of HPRT mutations into mice. Nature 326:295–298.

Mansour SL, Thomas KR, Capecchi MR (1988): Disruption of the proto-oncogene *int*-2 in mouse embryo-derived stem cells: A general strategy for targeting mutations to non-selectable genes. Nature 336:348–352.

Munir MI, Rossiter BJF, Caskey CT (1990): Antisense RNA production in transgenic mice. Somat Cell Mol Genet 16:383–394.

Munroe SH (1988): Antisense RNA inhibits splicing of pre-mRNA in vitro. EMBO J 7:2523–2532.

Nishikura K, Murray JM (1987): Antisense RNA of proto-oncogene c-*fos* blocks renewed growth of quiescent 3T3 cells. Mol Cell Biol 7:639–649.

Rebagliati MR, Melton DA (1987): Antisense RNA injections in fertilized frog eggs reveal an RNA duplex unwinding activity. Cell 48:599–605.

Robertson E, Bradley A, Kuehn M, Evans M (1986): Germ-line transmission of genes introduced into cultured pluripotential cells by retroviral vectors. Nature 323:445–448.

Rosenberg UB, Preiss A, Seifert E, Jäckle H, Knipple DC (1985): Production of phenocopies by *Krüppel* antisense RNA injection into *Drosophila* embryos. Nature 313:703–706.

Sarver N, Cantin EM, Chang PS, Zaia JA, Ladne PA, Stephens DA, Rossi JJ (1990): Ribozymes as potential anti-HIV-1 therapeutic agents. Science 247: 1222–1225.

Searle PF, Davison BL, Stuart GW, Wilkie TM, Norstedt G, Palmiter RD (1984): Regulation, linkage, and sequence of mouse metallothionein I and II genes. Mol Cell Biol 4:1221–1230.

Stout JT, Caskey CT (1987): Antisense RNA inhibition of endogenous genes. Methods Enzymol 151: 519–530.

Stout JT, Caskey CT (1990): Antisense RNA inhibition of HPRT synthesis. Somat Cell Mol Genet 16:369–382.

Thomas KR, Capecchi MR (1987): Site-directed mutagenesis by gene targeting in mouse embryo-derived stem cells. Cell 51:503–512.

van der Krol AR, Mol JNM, Stuitje AR (1988): Modulation of eukaryotic gene expression by complementary RNA or DNA sequences. BioTechniques 6:958–976.

ABOUT THE AUTHORS

M. IDREES MUNIR is a research scientist in the Cell Biology Department at Baylor College of Medicine. After receiving his master's degree from Quaid-i-Azam University, Pakistan, he joined Glasgow University, Scotland, where he completed his Ph.D. with Robin Leake. His thesis work was on steroid regulation of growth in the mammalian reproductive tract. Dr. Munir then pursued postdoctoral research at Baylor College of Medicine in Houston with Dr. Stanley Glasser in the area of reproductive biology, studying the implantation process in mammals. Then he moved to the Institute for Molecular Genetics within Baylor and worked with Dr. C. Thomas Caskey as a Howard Hughes Medical Institute Associate, developing an interest in gene transfer studies. He contributed to a number of projects involved with transgenic mice. He corrected a genetic defect at the OTC locus in the *sparse fur* mouse line. He showed partial inhibition of endogenous message by antisense RNA in transgenic mice. Currently Dr. Munir is studying the process of muscle development. His interest is toward developing animal models of human genetic diseases and the eventual goal of developing strategies for gene replacement therapy. Using the antisense RNA approach and homologous recombination techniques, he is elaborating the role of muscle-specific genes during early mouse development.

BELINDA J.F. ROSSITER is an Instructor at the Baylor College of Medicine, Houston, Texas, where she works in the Institute of Molecular Genetics and the Human Genome Program Center. After receiving her B.A. from the University of Cambridge, England, in 1983, she received her Ph.D. in 1987 under Margaret Fox at the Paterson Institute for Cancer Research, University of Manchester, England, where she studied the molecular structure of wild-type and mutant Chinese hamster hypoxanthine guanine phosphoribosyltransferase

(HPRT) genes. Dr. Rossiter then pursued postdoctoral research with C. Thomas Caskey at the Baylor College of Medicine, studying further the Chinese hamster HPRT gene, performing gene transfer of human HPRT into cells from Lesch-Nyhan syndrome patients, and investigating antisense inhibition of mouse HPRT in vivo. Dr. Rossiter currently serves a number of writing and editing functions in the fields of medical genetics and genome research, including the generation of review articles, meeting reports, and educational materials.

C. THOMAS CASKEY received his M.D. from Duke University, Durham, North Carolina, in 1963. Board certified in internal medicine, clinical genetics, and biochemical and molecular genetics, he is Professor of Medicine, Biochemistry, and Cell Biology, Henry and Emma Meyer Professor in Molecular Genetics, Director of the Institute for Molecular Genetics, and Investigator of the Howard Hughes Medical Institute, all at Baylor College of Medicine in Houston, Texas. His most recent honors include the Distinguished Service Professor Award at Baylor College of Medicine and the Distinguished Alumnus Award at Duke University. He is past president of the American Society of Human Genetics ('90), a member of the Institute of Medicine of the National Academy of Sciences USA, a member of the Department of Energy Advisory Committee on Mapping the Human Genome, and a member of the Advisory Panel on Mapping the Human Genome of the U.S. Congress Office of Technology Assessment. He is also a liaison member of the Program Advisory Committee on the Human Genome and a member of the Human Genome Steering Committee of the National Institutes of Health. He serves on several national research review panels and editorial boards and is a member of several societies. His research interests include inherited disease and mammalian genetics.

Antisense RNA and DNA: 109–120
© 1992 Wiley-Liss, Inc.

Manipulation of Myelin Formation in Transgenic Mice

Minoru Kimura, Masahiro Sato, and Motoya Katsuki

I. INTRODUCTION

Transgenic mice have become useful tools for studies of complicated biological phenomena such as developmental processes of central nervous system (CNS) formation. We present our results from recent studies of myelin formation in the CNS of transgenic mice.

Myelin is a lamellar structure that surrounds and insulates axons. These specialized myelin membranes aid in the conduction of nerve impulses. Myelin basic protein (MBP) is a major component of CNS myelin [Raine 1989; Braun 1989]. At least four types of MBPs with molecular weights of 21.5, 18.5, 17, and 14 kilodaltons (kD) have been identified in the myelin of adult mice [Barbarese et al., 1978]. The respective ratios of their amounts are 1:5:2:10 in young mice and 1:10:3.5:35 in adults. The presence of a fifth type of MBP in young mice has been also reported [Newman et al., 1987]. All of these MBP are encoded by a single gene, and each of them is translated from separate mRNAs derived by simple alternative splicing of the primary transcript from the single MBP gene [Takahashi et al., 1985; deFerra et al., 1985; Kimura et al., 1986]. These studies have been carried out using mutant mice, which are invaluable in the elucidation of the developmental processes of myelin formation.

Mutations affecting MBP synthesis have been found in the hypomyelinating mutant mice *shiverer* (*shi*) [Biddle et al., 1973; Chernoff, 1981] and *myelin-deficient* (*mld*) [Doolittle and Schweikart, 1977]. They are allelic [Bourre et al., 1980; Neuman et al., 1989] and show a very similar phenotype with vigorous tremors and convulsions of progressively increasing severity, leading to a short life span of 3–6 months. However, at the root of these two apparently similar phenotypes lie completely different types of mutations in the MBP gene. In *shi* mice, the MBP gene is largely deleted from exons 3–7, resulting in the complete absence of all four types of MBPs from its myelin and severe hypomyelination in homozygous mutant mice [Kimura et al., 1985; Roach et al., 1983; Molineaux et al., 1986]. From these results, it was suggested that MBP is a necessary component for myelin formation in the CNS of the mouse.

On the other hand, *mld* mice have a normal MBP gene indistinguishable from the wild-type gene by Southern blot analysis. However, they also carry an additional rearranged MBP gene upstream of the normal MBP gene as a possible consequence of tandem gene duplication. The rearrangement of this extra MBP gene involves an inversion of the regions of the gene from exons 3–7. Since the rearranged gene is transcribed, antisense transcripts corresponding to exons 3–7 are detected in the nuclei of *mld* oligodendrocytes [Akowitz et al., 1987; Okano et al., 1987, 1988a,b; Popko et al., 1987, 1988]. Therefore it appears that, although MBP transcripts are normally expressed in *mld* mice, the steady-state levels of MBP mRNA are reduced to 2%–5% of those of wild-type mRNA because of the inhibitory effect of the antisense transcripts expressed from the upstream inverted gene.

If it is true that the antisense transcripts in *mld* mice exhibit an inhibitory effect on the sense MBP mRNA synthesis, or make the endogenous MBP mRNA unstable through generating a heteroduplex between these RNA molecules, the *mld* mutation might be expected to be dominant over the wild-type MBP locus. However, although heterozygous (*mld*/ +) mice do not seem to be affected, the amount of MBP mRNA is one-half that of the wild-type mice [Roch et al., 1986, 1987]. These results, and recent in situ hybridization analysis of MBP gene expression in *mld* oligodendrocytes, strongly suggest that the expression of the downstream wild-type MBP gene must be repressed by the readthrough transcripts of the rearranged upstream MBP gene [Fremeau and Popko, 1990; Tosic et al., 1990]. It was thought that the *mld* mutation would provide a naturally occurring antisense regulation system for elucidation. However, it might be the unique structure of tandemly duplicated MBP genes with an inverted gene in *mld* mice that is primarily responsible for the reduction of MBP synthesis, even if the antisense MBP transcripts are expressed, causing heteroduplex formation between antisense transcripts and endogenous sense transcripts [Fremeau and Popko, 1990; Tosic et al., 1990].

To elucidate the role of MBP gene expression in CNS myelin formation, we performed experiments to manipulate myelin formation by introducing MBP minigenes into mouse zygotes, thereby generating transgenic mice in two ways (Fig. 1). First, the sense MBP cDNA constructed under the control of the MBP promoter region was introduced into *shi* mouse zygotes to investigate whether myelin formation can be restored in hypomyelinating *shi* mouse brains by a single type of MBP expressed by transgenes [Kimura et al., 1989]. Second, the antisense MBP cDNA was introduced into wild-type mouse zygotes to determine if they act as a dominant gene to reduce endogenous MBP synthesis [Katsuki et al., 1988].

II. RESTORATION OF MYELIN FORMATION IN TRANSGENIC *SHIVERER* MICE

The structure of the DNA fragment introduced into homozygous *shi* mouse zygotes consists of mouse MBP cDNA under the control of the 1.3 kb promoter region of the mouse MBP gene, which contains the cap site of MBP mRNA and 5' noncoding sequences. Poly(A) addition signals are provided by polyadenylation sites taken from the rabbit β-globin and simian virus 40 (SV40) early genes. The second intron of the rabbit β-globin gene was also placed between the cap site and MBP cDNA for possible use in splicing (Fig. 2A). Two minigenes containing mouse cDNA, each of which codes for the smallest or the second smallest type of MBP, were constructed.

As mentioned earlier, the smallest type of MBP is the most abundant [Barbarese et al., 1978]. The amount of the second smallest type is one-tenth of that of the smallest type. It has already been reported that transgenic mice carrying the full length of the mouse genomic MBP gene, myelination of which was completely restored, looked normal and no longer displayed a mutant phenotype [Readhead et al., 1987]. If all types of MBPs were necessary for myelin formation, transgenic *shi* mice with the single type of MBP cDNA would not be rescued from the mutant phenotypes even when MBP is expressed from the transgenes.

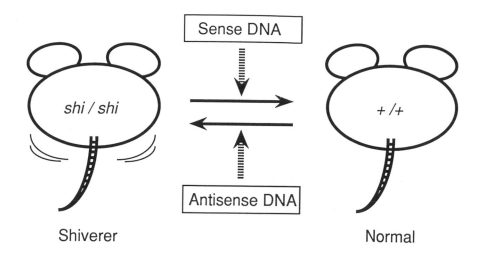

Fig. 1. *Manipulation of myelin formation in transgenic mice in two ways.*

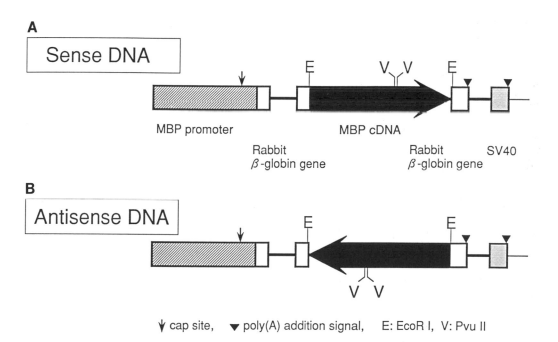

↓ cap site, ▼ poly(A) addition signal, E: EcoR I, V: Pvu II

Fig. 2. *Structures of DNA microinjected into mouse zygotes. See text for further details. **A:** Sense MBP cDNA is inserted in the expression vector under control of mouse MBP promoter. **B:** Antisense MBP cDNA constructed in the same vector as in A, except for the orientation of the cDNA. ↓, Cap site; ▼, poly(A) addition signal; E, EcoRI, V, PvuII.*

TABLE I. Summary of Data From Seven (Five Rescued and Two Not Rescued) Transgenic *shi* Mice

Animal	Copy No.*	Life span (days)[†]	MBP mRNA in brain (%)[‡]	MBP in cerebellum	Myelin formation[§]	Shivering behavior
561-7	50	>223	ND	+	+	−
617-5	60	>115	40	+	+	−
625-5	30	>134	ND	+	+	−
643-7	100	>100	3	+	+	−
654-2	100	>78	15	+	+	−
551-9[¶]	10	>55	10	±[‖]	−	+
563-1	3	>78	1	±[‖]	−	+
shi/shi	0	60–90	0	−	−	+
+/+	0	1.5–2.5 yr	100	+	+	−

*Copy numbers of the transgene were estimated from intensity of the DNA blot analysis with MBP cDNA used as a probe.
[†]For transgenic mice the number shows the day of sacrifice after birth.
[‡]MBP mRNA expression determined by RNA blot hybridization compared with total MBP mRNA from a +/+ mouse of same age. ND, not determined.
[§]Recovery of myelin formation was evaluated by staining with toluidine blue and electron microscopic analyses.
[¶]This mouse had pMP405 DNA, and other transgenic *shi* mice had pMP302 DNA.
[‖]Regional expression was recognized.
(Reproduced from Kimura et al., 1989, with permission of the publisher.)

Twenty-one transgenic mice were obtained after introduction of the construct that contains the smallest MBP cDNA. Five of these mice showed no tremors or only weak shivering, which was apparently different from that of mutant mice. Moreover, they showed no signs of tonic convulsions and no change in their normal behavior for more than 3 months. All these rescued mice may have been the direct consequence of the MBP minigene being integrated in their chromosomes.

We next examined the transgene expression in these rescued mice. In wild-type mice, it is already known that the MBP mRNA are expressed exclusively in the brain. Therefore, one-half of each transgenic mouse brain was subjected to RNA blot analysis and the other half to histological and histochemical analyses. As shown in the Table I, not only in rescued mice but also in nonrescued transgenic mice, transgenes were expressed at various levels. Except for one mouse (No. 551-9), the transgenes were expressed more in the rescued mice than in the nonrescued mice (including the other 14 nonrescued transgenic mice for which data are not shown). No hybridizing mRNA were detected in the nontransgenic *shi* mice. Expression of the MBP transgenes was detected only in the brain. This indicates that the 1.3 kb MBP promoter region may contain the sequences responsible for the tissue-specific expression.

To identify the protein products of MBP transgenes and myelin formation, we performed an immunochemical analysis of paraffin-embedded specimens using the antimouse MBP antibodies as well as histological observations. As shown in Table I, restoration of MBP synthesis was correlated with myelin formation and the disappearance of shivering behavior in the rescued transgenic mice.

We have also obtained seven transgenic mice carrying MBP cDNA that encodes the second smallest MBP. Two of them were rescued from the mutant phenotypes: absence of MBP, hypomyelination, and shivering behavior. From these results, we conclude that only one type of the MBP synthesized in oligodendrocytes may be sufficient to support myelin formation if it is correctly expressed in the brain [Kimura et al., 1989].

III. CONVERSION FROM NORMAL TO SHIVERING BEHAVIOR BY MBP ANTISENSE cDNA IN TRANSGENIC MICE

This section of the book includes many examples in which antisense RNA has been shown to repress the expression of specific genes, including in mammalian systems (see Murray and Crockett, this volume; Munir et al., this volume). In *mld* mice, partial antisense transcripts were detected that were expressed from the duplicated gene, which has an inverted region extending from exons 3–7. However, it is not clear whether these antisense transcripts repress the expression of the endogenous normal MBP gene directly, since, as discussed above, no effect is observed in heterozygous (*mld*/+) mice. To clarify the effect of MBP antisense RNA in vivo, we generated transgenic mice carrying genes that express MBP antisense cDNA.

We have demonstrated that the 1.3 kb mouse MBP promoter region has the ability to control the tissue-specific expression of MBP minigenes in transgenic mice. Therefore, we tried to generate transgenic mice with antisense MBP cDNA, using the same construct of the sense MBP cDNA, except for the orientation. It therefore included the native mouse MBP promoter, rabbit β-globin exons and introns, and poly(A) signal sequences (Fig. 2B). The antisense MBP minigene was injected into eggs heterozygous for the *shi* mutation (*shi*/+). It has been shown that the amount of MBP mRNA in heterozygous (*shi*/+) mice is one-half that in the wild type (+/+) (see Fig. 4B, below) [Barbarese et al., 1983], and therefore inhibition of the endogenous MBP sense mRNA expression by antisense RNA would be expected to be more pronounced in heterozygous (*shi*/+) than in the wild type.

Five transgenic mice were obtained. After gathering their offspring, all of the founder mice were sacrificed, and their brains were subjected to RNA blot analysis and hybridized with three different probes (Fig. 3). First, total RNAs were probed with MBP cDNA. In this experiment, not only antisense MBP mRNA but also endogenous sense mRNAs are detected. However, the expected size of endogenous MBP mRNA, 2.3 kb, was larger than that of transgenic mRNA, 1.9 kb. Therefore, if both mRNAs were expressed at almost the same level, we could determine the antisense MBP mRNA by this experiment. As shown in Figure 3A, the endogenous MBP mRNA was expressed 10 times more than the antisense MBP mRNA. If we take a very close look at Figure 3A, it is possible to observe that the antisense mRNA as well as the sense mRNA is expressed in the lane of the 664-9 (AS100) mouse. After stripping the MBP cDNA probe from the filter, the same filter was probed with rabbit β-globin exon 3 DNA fragment, which exists only in the transgenes. Therefore, it was determined that the transgenic MBP mRNA are expressed in the AS100 mouse brain (Fig. 3B). To confirm the antisense mRNA expression in the AS100 mouse brain, freshly prepared total RNA of AS100 mouse brain were hybridized with sense MBP RNA as a single-stranded riboprobe that therefore hybridized only with antisense MBP mRNA. As shown in Figure 3C, antisense mRNA were expressed in the AS100 transgenic mouse brain. In several other tissues examined, neither sense nor antisense MBP mRNA were detected by RNA blot analysis. It was again proved that the 1.3 MBP promoter region retains the sequences responsible for the control of tissue-specific MBP expression. We have never noticed any mutant phenotypes in AS100 founder mouse, but it is possible that it is too weak to be recognized.

The AS100 male mouse (*shi*/+) was mated with a wild-type female mouse (+/+), and 50 offspring were obtained (AS100-1 to AS100-50). The transgene of the AS100 mouse was transmitted to 21 of them with different genetic backgrounds, *shi*/+ and +/+.

Very surprisingly, 10 of 21 transgenic offspring of the founder transgenic mouse AS100 began to show shivering about 2 weeks after birth. The question that then arose was, why were not all of the transgenics shivering if the MBP antisense RNA was expressed in them? To determine the reason for these differences in the manifestation of shivering behavior among

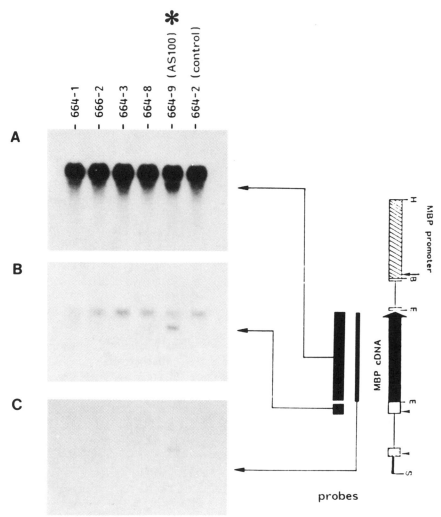

Fig. 3. *RNA blot analyses of MBP antisense transgenic founder mouse brains. Total RNA were isolated, and RNA blot analyses were performed.* **A:** *Filter hybridized with MBP cDNA as a probe. Both endogenous (2.3 kb) and antisense (1.9 kb) RNA could be detected.* **B:** *After stripping the MBP cDNA probe, the same filter was rehybridized using rabbit β-globin exon 3 DNA fragment, a probe that is specific for the introduced antisense construct. Transgene RNAs could only be detected in the brain.* **C:** *Total RNAs were hybridized with sense MBP mRNA synthesized in the SP6 riboprobe (Promega) system as a probe. Only the antisense MBP RNA could be detected.*

21 transgenic mice, we examined the genetic backgrounds of all of these mice to find out if the shivering phenotype was attributable to heterozygosity (*shi*/ +), which could be detected by the presence of the truncated restriction fragment from the *shi* mutant allele. Three of the nine transgenic mice with the wild-type genetic background (+ / +) appeared to be shivering phenotypes, whereas 5 of the 12 with a heterozygous (*shi*/ +) genetic background appeared to be normal. This not only indicates that the genetic backgrounds in the case of the *shi* mutation could not explain the differences in the manifestation of tremors but also that the MBP

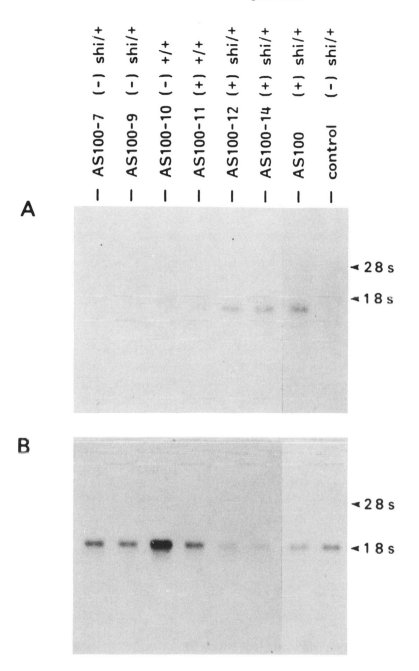

Fig. 4. *Antisense and endogenous sense MBP mRNA expression in AS100 transgenic mice. +, Transgenic animals containing the antisense construct; −, non-transgenic littermates. The* shi *genotype is shown as* shi/+ *(heterozygous) or* +/+ *(wild type). Only AS100-12 and -14 exhibit a* shi *phenotype.* **A:** *RNA blot analysis of the AS100 founder mouse and that of the offspring using sense MBP RNA as the probe. Only antisense RNA is detected.* **B:** *RNA analysis of the same mice with double-stranded MBP cDNA as the probe. The difference in MBP expression between* +/+ *and* +/shi *genotypes is also apparent (compare AS100-10, with -7, -9, and the control). [From Katsuki et al., 1988, with permission of the publisher.]*

TABLE II. Correlation Between MBP Expression and the Tremor Phenotype in Transgenic Mice

Mouse	Genetic back-ground	Trans-gene*	Antisense MBP mRNA[†]	Endogenous MBP mRNA (%)[‡]	MBP in cere-bellum[§]	Tremor pheno-type[‖]
AS100-10	+/+	−	−	100	+	−
AS100-9	shi/+	−	−	50	+	−
AS100-11	+/+	B type	+	50	+	−
AS100-12	shi/+	A type	+ +	30	± (m)	±
AS100-14	shi/+	B type	+ +	20	± (m)	+

*A and B type represent the types of transgene transmitted. Two types of transgenes were observed, B type, which gave only the expected 1.2 kb band, and A type, which had B type and further additional bands. These band types were stably inherited.

−, A nontransgenic mouse.

[†]Antisense MBP mRNA in the brain was detected by RNA blot analysis with sense MBP RNA as the probe.

[‡]Relative amount of MBP mRNA in the brain, the wild type (+/+) being taken as 100%.

[§]+, The MBP was detected uniformly in the cerebellar tissue by anti-MBP; ±, relatively sparse distribution of MBP; ±, intermediate distribution of MBP between − and ±; and m, mosaic expression of MBP.

[‖]−, Normal behavior; +, tremor phenotype; and ±, the faint tremor phenotype.

(Reproduced from Katsuki et al., 1988, with permission of the publisher.)

antisense RNA expressed in the transgenic mice can act in a dominant manner to repress the endogenous MBP sense mRNA.

We next measured the amounts of endogenous MBP mRNA and antisense RNA expressed from the transgenes. Total RNA was isolated from the brains of transgenic mice and subjected to RNA blot analysis. As shown in Figure 4A, antisense RNAs, which were hybridized with radiolabeled MBP sense RNA as riboprobes, were detected in the transgenic mice AS100-11, -12, -14 and the founder mouse AS100. RNA blot analyses of these transgenic mouse brains probed with MBP cDNA showed that they had reduced amounts of endogenous MBP mRNA (Fig. 4B) when compared with their nontransgenic littermates AS100-7, -9, and -10.

It is still not clear why there are differences in the manifestation of tremors among antisense transgenic mice, but the reduced amount of MBP mRNA seems to be correlated with the appearance and severity of the shivering phenotype and also with the amount of antisense RNA expressed (Table II). Therefore it appeared that these differences might be due to differences in myelination among antisense

transgenic mice. To examine these possibilities, we performed immunohistochemical and histological analyses of the cerebella of transgenic and nontransgenic mice. As mentioned earlier, MBP and myelin formation are not observed in the shi mouse cerebellum. In the antisense MBP transgenic mouse cerebellum, myelination was also not observed (Fig. 5B), although it has the endogenous MBP gene.

The antisense MBP mRNA were expressed tissue specifically, and they repressed the expression of MBP sense mRNAs. In these mice, the MBP and myelination of the CNS were also reduced, resulting in the same tremor behavior and tonic convulsions as in the shi mouse that has a deletion of the MBP gene. Recently, we discovered that not only the smallest type of MBP, but also all other types of MBP in the antisense transgenic mice, were reduced or almost abolished by the expression of only the smallest type (14 kD) of MBP antisense mRNAs. These other MBP are normally the result of alternative splicing from the transcript of the single MBP gene. Moreover, antisense mRNA expressed at less than 10% of the wild-type MBP mRNA level are capable of repressing the more than 60% of endog-

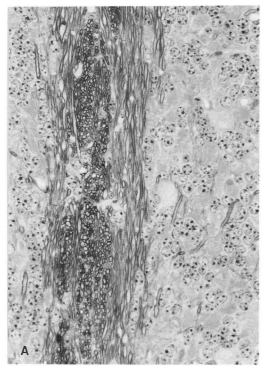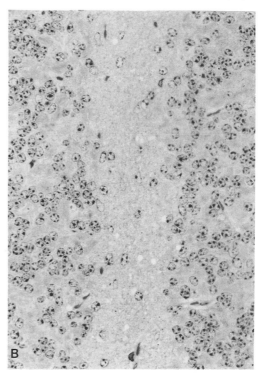

Fig. 5. *Hypomyelination in the antisense transgenic mouse cerebellum.* **A:** *Nontransgenic wild-type mouse cerebellum. Myelin is clearly recognized.* **B:** *Antisense* *transgenic mouse cerebellum. Myelin disappears in the white matter of the cerebellum.*

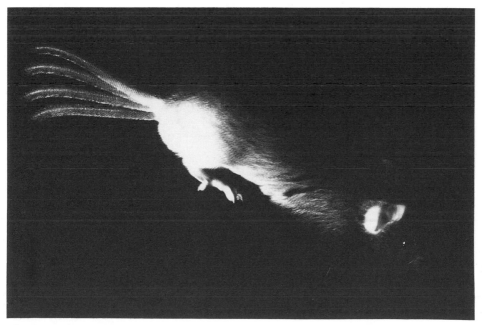

Fig. 6. *Shivering antisense transgenic mouse. Shivering behavior of the antisense transgenic mouse is almost the same as that of* shi *mutant mouse. Both* *mice lack the CNS myelin, a phenotype produced by complete different genetic disorders.*

enous MBP mRNA synthesis, and the antisense RNA has complete dominance over the endogenous MBP gene, leading to the shivering phenotype (Fig. 6).

This is in contrast to the *mld* mutation in which a phenotype is only observed in *mld/mld* homozygotes. This difference may possibly be explained by the apparent requirement for endogenous MBP RNA to be reduced to around 30% or less of the wild-type level for a *shi* phenotype to be observed, as seen by examining the phenotypes of AS100-12 and -14 in Table II. Thus in *shi/+* or *mld/+* heterozygotes [Roch et al., 1986, 1987], as in the mouse AS100-11 (Table II), no phenotype is observed as MBP mRNA is still present at 50% of the level found in wild-type mice.

We have shown here that, by manipulating the expression levels of MBP mRNA, we can control the levels of myelination and shivering behavior by introducing sense MBP minigenes into *shi* mice or antisense MBP minigenes into wild-type mice. We have therefore demonstrated that it is feasible to use the antisense method for elucidation of biological phenomena that are controlled by stoichiometric proteins like MBP and that it may be feasible to manipulate such phenomena by antisense gene expression.

IV. REFERENCES

Akowitz AA, Barbarese E, Scheld K, Carson JH (1987): Structure and expression of myelin basic protein gene sequences in the *mld* mutant mouse: Reiteration and rearrangement of the MBP gene. Genetics 116: 447–464.

Barbarese E, Carson JH, Braun PE (1978): Accumulation of the four myelin basic proteins in mouse brain during development. J Neurochem 31:779–782.

Barbarese E, Nielson ML, Carson JH (1983): Effect of the *shiverer* mutation on myelin basic protein expression in homozygous and heterozygous mouse brain. J Neurochem 40:1680–1686.

Biddle F, March E, Miller JR (1973): Research news. Mouse News Lett 48:24.

Bourre JM, Jacque C, Delassalle A, Nguyen-Legros J, Dumont O, Lachapelle F, Raoul M, Alvarez C, Baumann N (1980): Density profile and basic protein measurements in the myelin range of particulate material from normal developing mouse brain and from neurological mutants (*jumpy: quaking: trembler:*

shiverer and its *mld* allele) obtained by zonal centrifugation. J Neurochem 35:458–464.

Braun PE (1989): Molecular organization of myelin. In Morell P (ed): Myelin. New York: Plenum Press, pp 97–116.

Chernoff GF (1981): *Shiverer:* An autosomal recessive mutant mouse with myelin deficiency. J Hered 72:128.

deFerra F, Engh H, Hudson L, Kamholz J, Puckett C, Molineaux S, Lazzarini RA (1985): Alternative splicing accounts for the four forms of myelin basic protein. Cell 43:721–727.

Doolittle DP, Schweikart KM (1977): *Myelin deficient*: A new neurological mutant in the mouse. J Hered 68:331–332.

Fremeau RT Jr, Popko B (1990): In situ analysis of myelin basic protein gene expression in myelin-deficient oligodendrocytes: Antisense hnRNA and readthrough transcription. EMBO J 9:3533–3538.

Katsuki M, Sato M, Kimura M, Yokoyama M, Kobayashi K, Nomura T (1988): Conversion of normal behavior to shiverer by myelin basic protein antisense cDNA in transgenic mice. Science 241:593–595.

Kimura M, Inoko H, Katsuki M, Ando A, Sato T, Hirose T, Takashima H, Inayama S, Okano H, Takamatsu K, Mikoshiba K, Tsukada Y, Watanabe I (1985): Molecular genetic analysis of myelin-deficient mice: *Shiverer* mutant mice show deletion in gene(s) coding for myelin basic protein. J Neurochem 44:692–696.

Kimura M, Katsuki M, Inoko H, Ando A, Sato T, Hirose T, Inayama S, Takashima T, Takamatsu K, Mikoshiba K, Tsukada Y, Watanabe I (1986): Structure and expression of the myelin basic protein gene in mouse. In Tsukada Y (ed) The Ninth Taniguchi Symposium on Brain Science: Molecular Genetics in Developmental Neurobiology. Utrecht, the Netherlands: VNU Science, pp 125–133.

Kimura M, Sato M, Akatsuka A, Nozawa-Kimura S, Takahashi R, Yokoyama M, Nomura T, Katsuki M (1989): Restoration of myelin formation by a single type of myelin basic protein in transgenic *shiverer* mice. Proc Natl Acad Sci USA 86:5661–5665.

Molineaux SM, Engh H, deFerra F, Hudson L, Lazzarini RA (1986): Recombination within the myelin basic protein gene created the dysmyelinating *shiverer* mouse mutation. Proc Natl Acad Sci USA 83:7542–7546.

Neuman P, Applegate C, Ganser A, Heller A, Kirschner D, Rauch S, Rosen K, Villa-Komaroff L, White F (1989): Research news. Mouse News Lett 83:157.

Newman S, Kitamura K, Campagnoni AT (1987): Identification of a cDNA coding for a fifth form of myelin basic protein in mouse. Proc Natl Acad Sci USA 84:886–890.

Okano H, Ikenaka K, Mikoshiba K (1988a): Recombination within the upstream gene of duplicated myelin basic protein genes of myelin deficient shi^{mld} mouse results in the production of antisense RNA. EMBO J 7:3407–3412.

Okano H, Miura M, Moriguchi A, Ikenaka K, Tsukada Y, Mikoshiba K (1987): Inefficient transcription of the myelin basic protein gene possibly causes hypomyelination in myelin-deficient mutant mice. J Neurochem 48:470–476.

Okano H, Tamura T, Miura M, Aoyama A, Ikenaka K, Oshimura M, Mikoshiba K (1988b): Gene organization and transcription of duplicated MBP genes of myelin deficient (shi^{mld}) mutant mouse. EMBO J, 7:77–83.

Popko B, Puckett C, Hood L (1988): A novel mutation in myelin-deficient mice results in unstable myelin basic protein gene transcripts. Neuron 1:221–225.

Popko B, Puckett C, Lai E, Shine HD, Readhead C, Takahashi N, Hunt SW, Sidman RL, Hood L (1987): Myelin deficient mice: Expression of myelin basic protein and generation of mice with varying levels of myelin. Cell 48:713–721.

Raine CS (1989): Molecular organization of myelin. In Morell P (ed): Myelin. New York: Plenum Press, pp 1–50.

Readhead C, Popko B, Takahashi N, Shine HD, Saavedra RA, Sidman RL, Hood L (1987): Expression of myelin basic protein gene in transgenic *shiverer* mice: Correction of the dysmyelinating phenotype. Cell 48: 703–712.

Roach A, Boylan S, Horvath S, Prusiner L, Hood L (1983): Characterization of cloned cDNA representing rat myelin basic protein: Absence of expression in brain of *shiverer* mice. Cell 34:799–806.

Roch J-M, Brown-Luedi M, Cooper BJ, Matthieu J-M (1986): Mice heterozygous for the *mld* mutation have intermediate levels of myelin basic protein mRNA and its translation products. Mol Brain Res 1:137–144.

Roch J-M, Cooper BJ, Ramirez M, Matthieu J-M (1987): Expression of only one myelin basic protein allele in mouse is compatible with normal myelination. Mol Brain Res 3:61–68.

Takahashi N, Roach A, Teplow DB, Prusiner SB, Hood L (1985): Cloning and characterization of the myelin basic protein gene from mouse: One gene can encode both 14Kd and 185Kd MBPs by alternate use of exons. Cell 42:139–148.

Tosic M, Roach A, de Rivaz J-C, Dolivo M, Matthieu J-M (1990): Post-transcriptional events are responsible for low expression of myelin basic protein in myelin deficient mice: Role of natural antisense RNA. EMBO J 9:401–406.

ABOUT THE AUTHORS

MINORU KIMURA is Associate Professor at the Department of DNA Biology, Tokai University School of Medicine, where he researches and teaches genetics and molecular biology. After receiving his B.Sc. from Kyoto University in 1976, he received his Ph. D. in 1981 under Prof. Takashi Yura at Kyoto University, where he concentrated in the molecular genetics of bacterial DNA replication. Dr. Kimura then moved to the laboratory of Prof. Itaru Watanabe in Keio University as an instructor and started transgenic mouse experiments in 1982 in collaboration with Dr. Katsuki. He first cloned the gene of mouse myelin basic protein (MBP) and its cDNA, then rescued the tremor phenotype of the shiverer mice whose MBP gene is deleted. Transgenic mouse experiments with MBP antisense cDNA showed the reciprocal results. Drs. Kimura and Katsuki have also constructed and analyzed the transgenic mice with genes playing important roles in the nervous system, the immune system, and tumorigenesis. His research papers have appeared in such journals as the *Journal of Bacteriology,* the *Journal of Molecular Biology,* the *Journal of Neurochemistry,* the *Proceedings of the National Academy of Sciences USA, Science,* and *Oncogene.* Dr. Kimura's current research involves oncogenesis and neurogenesis using genetically manipulated animals.

MASAHIRO SATO is a researcher of molecular biology at Hoechst Japan Co. Ltd. After receiving his B.A. from Tohoku University in 1978, he received his Ph.D. in 1985 under Professor Takashi Muramatsu at Kagoshima University School of Medicine, where he concentrated on the analysis of various kinds of cell surface antigens expressed in murine embryonic cells using monoclonal antibodies and lectins as probes. His research papers have appeared in such journals as the *Journal of Experimental Embryology and Experimental Morphology* (now *Development*), *Differentiation, Developmental Biology,* and the *Journal of Reproductive Immunology.* Soon after joining Hoechst Japan in 1985, he studied at Tokai University for two years, where he learned techniques of making transgenic mice and their analysis under Professor Motoya Katsuki. He contributed to the work on transgenic mice with antisense MBP cDNA that was published in *Science.* Dr. Sato's current research involves constructing mouse models for bone diseases using antisense gene technology.

MOTOYA KATSUKI was until recently Professor of DNA Biology at Tokai University School of Medicine, where he taught genetics and molecular biology. He was recently appointed Professor of Molecular and Cellular Biology at the Medical Institute of Bioregulation at Kyushu University. After receiving his B.Sc. from University of Tokyo in 1967, he studied molecular genetics of spore formation in *B. subtilis* under Professor Ikeda at the University of Tokyo and under Professor Sekiguchi at Kyushu University. Dr. Katsuki then moved to the laboratory of Professor Itaru Watanabe in Keio University as an instructor; here his research focused on molecular genetics of behavior in *B. mori*. He received his Ph.D. from Kyushu University in 1980. Since 1981 he has concentrated on the research of mammalian embryology, using chimeric mice, uniparental embryos, and transgenic mice. In the course of this research, he studied under Dr. Peter Hoppe at the Jackson Laboratory for about two years as a visiting investigator. He is a pioneer of the transgenic mice experiments in Japan and has made tremendous contributions in this field. His research papers have appeared in such journals as *Nature*, the *Proceedings of the National Academy of Sciences USA*, *Science*, *Oncogene*, *Developmental Biology*, and the *Journal of Neurochemistry*. Dr. Katsuki's current research involves oncogenesis and neurogenesis using genetically manipulated embryonic stem cells of mice as well as transgenic mice.

Antisense RNA and DNA: 121–135
© 1992 Wiley-Liss, Inc.

Antisense mRNA Inhibition of Ribulose Bisphosphate Carboxylase—The Most Abundant Protein in Photosynthetic Cells

Steven R. Rodermel and Lawrence Bogorad

I. INTRODUCTION

Most multimeric protein complexes in chloroplasts are composed of subunits encoded by genes in the nucleus and chloroplast [reviewed by Bogorad, 1982]. These complexes include components of the photosynthetic apparatus, the protein synthesis machinery of the plastid, and enzymes involved in carbon fixation. Free subunit pools of these complexes do not accumulate, despite differences in copy number between the nuclear and the organelle genomes, which can range up to a factor of 10^4 [reviewed by Bendich, 1987]. These observations indicate that mechanisms exist to coordinate subunit biosynthesis in the two compartments.

The mechanisms that integrate nuclear and plastid gene expression are poorly understood; one reason is that there is a dearth of mutants available for study that display specific lesions affecting nuclear–plastid interactions, because classic selection procedures are generally ineffective for generating mutations in genes that are members of multigene families (most nuclear genes) or in genes that are present in multiple copies per cell (all plastid genes). To avoid the difficulties of classic selection procedures, we have used antisense RNA technology as an alternative approach to gain entrance into the complex regulatory circuitry governing nuclear–chloroplast interactions [Rodermel et al., 1988].

For our model system we have chosen to study the biosynthesis of ribulose bisphosphate carboxylase (RUBISCO)—the key regulatory enzyme of photosynthetic carbon metabolism. This enzyme is well suited for studies of nuclear–chloroplast interactions because it is a relatively simple example of a plastid multimeric protein complex; i.e., it is composed of a single type of nuclear-encoded subunit (small subunit proteins) and a single type of plastid-encoded subunit (large subunit proteins). In addition, molecular biological analyses are facilitated because RUBISCO is the most abundantly transcribed and expressed protein in photosynthetic cells, constituting up to 50% of the total soluble leaf protein in some species [Ellis, 1979]. A final advantage of using RUBISCO as a model system is that this enzyme is perhaps the most intensively studied plant protein, and hence there is a wealth of information available concerning its biosynthesis.

In this review we describe the generation of transgenic tobacco plants expressing antisense mRNA sequences for the nuclear-encoded small subunit of RUBISCO. We discuss those elements that we initially considered important to our experimental plan, as well as those elements we presently consider important for the successful generation of antisense mutants in higher plants. Although our principal aim was to use the RUBISCO antisense mutants to study nuclear–chloroplast interactions, these mutants are proving to be an ideal system in which to address other fundamental problems in plant biochemistry, physiology, and ecology. These topics are also discussed.

The successful inhibition of RUBISCO expression by antisense mRNA offers a striking example of the effectiveness of antisense mRNA technology in generating hypomorphic mutants in higher plants. The fact that the expression of this abundant plant protein can be inhibited by antisense mRNA further suggests that it should be possible to use antisense genes to inhibit the expression of any plant gene, regardless of expression level or copy number in the genome.

II. RUBISCO BIOSYNTHESIS: A MODEL SYSTEM IN WHICH TO STUDY NUCLEAR–CHLOROPLAST INTERACTIONS

RUBISCO is localized in the chloroplasts of photosynthetic eukaryotic cells, where it catalyzes both the carboxylation of ribulose-1-5-bisphosphate in the Calvin Cycle and the oxygenation of the same substrate in the photorespiratory pathway [reviewed by Miziorko and Lorimer, 1983]. The holoenzyme is composed of eight small subunits (SS), encoded by a small nuclear multigene (rbcS) family, and eight catalytic large subunits (LS), encoded by a single gene (rbcL) on the multicopy chloroplast chromosome [reviewed by Miziorko and Lorimer, 1983]. A typical leaf cell of a higher plant consequently contains from 10 to 20 rbcS sequences and from 2,000 to 50,000 copies of the rbcL gene [Bendich, 1987]. Whereas rbcL mRNAs are translated on membrane-bound and free 70S plastid ribosomes [Hattori and Margulies, 1986], rbcS mRNAs are translated as precursors on cytoplasmic 80S ribosomes and then transported into the chloroplast posttranslationally [reviewed by Schmidt and Mishkind, 1986]. During this latter process, an amino-terminal transit sequence is clipped off the precursor to yield the mature SS polypeptide. Assembly of the RUBISCO holoenzyme from its subunits is mediated by the RUBISCO subunit–binding protein, a molecular chaperone containing two types of nuclear-encoded subunits [reviewed by Roy, 1989].

Studies in a variety of developmental systems in higher plants and eukaryotic algae have demonstrated that SS and LS protein accumulation is coordinated primarily at the level of rbcS and rbcL transcript pool sizes [reviewed by Tobin and Silverthorne, 1985]. For example, during the process of light-induced chloroplast differentiation (or the "greening" of pale-yellow, photosynthetically incompetent etioplasts in dark-grown plants into photosynthetically competent chloroplasts), coordinate increases in the levels of the *rbc*S and *rbc*L mRNAs are accompanied by corresponding increases in the amount of the RUBISCO holoenzyme. In addition, only stoichiometric amounts of the LS and SS proteins accumulate during this process.

The above observations raise the question: What are the mechanisms that coordinate the accumulation of *rbc*S and *rbc*L transcripts and SS and LS proteins during such a process as greening? Is such integration achieved by feedback from one compartment to the other? For example, do the transcription and/or translation products of one subunit exert feedback on the biosynthesis of the other subunit, e.g., to regulate transcription or posttranscriptional events? Numerous studies over the past 30 years have attempted to answer this latter question by perturbing the mRNA and/or protein levels of one subunit and then examining the effects of this perturbation on the transcript and protein levels of the other subunit. Most of these experiments have involved the use of inhibitors; as described below, these experiments are difficult to interpret. Mutants have also been used to address this question. However, their use has been limited, for reasons also discussed below.

A. Inhibitor Studies

Inhibitors of transcription and translation have been used in a variety of systems to determine whether alterations in the mRNA or protein levels of one RUBISCO subunit affect the production of mRNA or protein of the other subunit. These experiments have demonstrated that *rbc*S mRNA accumulation is dependent on plastid transcription and translation: In the presence of inhibitors of plastid transcription or translation, decreased *rbc*L mRNA or LS protein amounts are accompanied by decreased *rbc*S mRNA amounts [e.g., Radetzsky and Zetsche, 1987]. On the other hand, *rbc*L mRNA accumulation appears to be largely independent of events in the nuclear–cytoplasmic

compartment: In the presence of inhibitors of transcription or translation in the nuclear–cytoplasmic compartment, decreased *rbc*S or SS protein amounts do not affect *rbc*L mRNA production, at least in the short term [e.g., Radetzsky and Zetsche, 1987; Sasaki, 1986]. Considered together, these data suggest that *rbc*S and *rbc*L transcript pool sizes are coordinated by a signal from the plastid compartment that is generated in response to the levels of *rbc*L mRNA and/or LS protein [Radetzsky and Zetsche, 1987; Sasaki, 1986].

In addition to "coarse" controls that coordinate *rbc*S and *rbc*L transcription, inhibitor experiments have shown that "fine-tune" control of SS and LS protein accumulation may be mediated by signals generated in response to *rbc*S and *rbc*L transcription or translation products. For example, pulse-chase experiments performed in the presence of inhibitors of plastid transcription and translation have demonstrated that excess amounts of the imported SS are degraded within the plastid, presumably in response to a lack of LS proteins [Radetzsky and Zetsche, 1987; Schmidt and Mishkind, 1983]. On the other hand, similar experiments performed in the presence of inhibitors of nuclear–cytoplasmic transcription and translation have shown that the LS protein concentration is adjusted to that of the SS by translational control of *rbc*L mRNA expression [Radetzsky and Zetsche, 1987]. It is assumed that this latter effect is mediated by reductions in *rbc*S mRNA or SS protein.

The data from the inhibitor experiments are difficult to interpret because of the pleiotropic effects of these agents; i.e., even though they may inhibit a single activity (e.g., 70S translation), there can be secondary effects. For example, it would be anticipated that inhibition of nuclear transcription would block not only *rbc*S mRNA and SS protein production but also the production of any regulatory signals generated in response to the levels of these molecules that are dependent on nuclear transcription or translation. A similar qualification attends the interpretation of data from such experimental systems as heat-bleached rye seedlings, which lack plastid ribosomes [e.g., Biekmann and Feierabend, 1985], and plastid ribosome-deficient mutants of *Chlamydomonas* [Mishkind and Schmidt, 1983]. In both of these systems a failure of SS proteins to accumulate within the plastid has been assumed to be due to LS protein depletion. However, the data are equally consistent with the possibility that one or more factors required for SS stability cannot be translated within the plastid.

B. Mutant Analyses

Because of the drawbacks of experimental systems with broad pleiotropic effects, mutants displaying specific alterations in the amounts of transcription or translation products from either *rbc*S or *rbc*L would appear to provide the best opportunities for studying the effects of perturbation on the regulation of RUBISCO biosynthesis. However, very few mutants have been isolated that display these characteristics. This is because classic selection procedures are hampered by the presence of many members of the *rbc*S gene family and by many copies of the chloroplast chromosome, each with an *rbc*L gene. Thus a mutation in one gene would be likely to be complemented by the wild-type copies of the gene. In fact, no *rbc*S mutants have been reported, and only a few *rbc*L mutants have been described.

Mutants specifically perturbed in LS metabolism have been characterized in *Chlamydomonas* [Spreitzer and Ogren, 1983; Spreitzer et al., 1985] and *Oenothera* [Hildebrandt et al., 1984]. All of these mutants are deficient in RUBISCO content (and activity) because of nonsense mutations in the *rbc*L gene. These mutations result in the production of truncated forms of the LS protein that are subsequently degraded in *Chlamydomonas*, but not in *Oenothera*. However, the accumulation of *rbc*S mRNAs and (truncated) *rbc*L mRNAs is unaffected in the mutants. Whereas SS protein synthesis is also unaffected, imported SS proteins are degraded within the chloroplast. These findings demonstrate convincingly that a lack of LS proteins (or, in the case of *Oenothera*, a lack of "functional" LS proteins) serves as a signal to fine-tune the levels of the SS protein.

III. RUBISCO ANTISENSE MUTANTS

We wanted to address the question whether *rbc*S transcription or translation products in the nucleus–cytoplasm could directly (or indirectly) serve as a source of signals to regulate LS metabolism in the plastid. Our overall strategy was to use antisense RNA technology as an alternative approach to classic selection procedures for the generation of mutants specifically perturbed in the metabolism of the SS [Rodermel et al., 1988]. It was our hope that these mutants would display a much reduced accumulation of either *rbc*S mRNAs or SS proteins, such that the effect of this single alteration on *rbc*L mRNA and LS protein accumulation could be determined.

A. Antisense Mutant Generation

Our approach to generate SS antisense mutants involved the introduction of an *rbc*S antisense gene sequence into the nuclear genome of tobacco by *Agrobacterium*-mediated DNA transfer methods. At the outset, several elements were considered vital to the successful inhibition of RUBISCO expression by *rbc*S antisense mRNA and were incorporated into the experimental plan. Whether these elements, in retrospect, played a decisive role in the success of the experiments is discussed in subsequent sections.

1. Ratio of target mRNA to antisense mRNA. Successful applications of antisense RNA technology to plants had not yet been reported when the present experiments were initiated. However, the conventional wisdom from animal systems was that successful inhibition of gene expression by antisense RNA depended on the presence of excess amounts of the antisense vs. the target mRNA [reviewed by van der Krol et al., 1988b]. High levels of antisense RNA were also deemed crucial in our case, because we wanted to downregulate the expression of the most abundantly transcribed and expressed protein in photosynthetic cells. To achieve as high a level of antisense mRNA as possible, the cauliflower mosaic virus (CaMV) 35S promoter [Guilley et al., 1982; Odell et al., 1985]

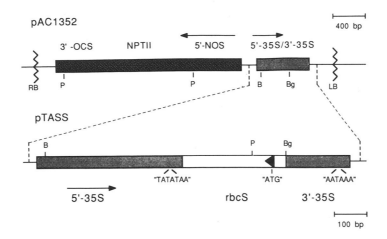

Fig. 1. *Construction of the* rbcS *antisense expression vector. The* rbcS *sequences used in the present experiments were derived from pSEM1, a* Nicotiana sylvestris rbcS *cDNA clone [Pinck et al., 1984], and they encompass a 322 bp fragment of DNA from the 5' region of the gene (from 22 bp upstream to 300 bp downstream of the initiation codon). This region of the N.* sylvestris *gene shares nearly 100% homology with the same region of two N.* tabacum rbcS *genes that account for the vast bulk of* rbcS *mRNA production in the leaves of this species [Mazur and Chui, 1985; O'Neal et al., 1987]. The N.* sylvestris rbcS *fragment was cloned in reverse orientation into the BglII* site of pAC1352 (A. Cheung and L. Bogorad, unpublished data), a cauliflower mosaic virus (CaMV) 35S promoter/terminator cassette in which a BglII site separates the 35S promoter and terminator elements. This vector also contains a copy of the NPTII gene (conferring kanamycin resistance) fused to nopaline synthetase (NOS) promoter and octopine synthetase (OCS) terminator elements to serve as a selectable marker in the transformation experiments. The resulting rbcS antisense vector was designated pTASS. [Reproduced from Rodermel et al., 1988, with permission of the publisher.]*

was chosen to drive the expression of *rbcS* antisense sequences (see Fig. 1); this promoter gives high level of expression of foreign genes in plant cells [e.g., Sanders et al., 1987]. It is also worth pointing out that, in contrast to *rbcS*, which is transcribed at high levels only in photosynthetic tissues [reviewed by Tobin and Silverthorne, 1985], the 35S promoter is constitutively expressed. As described later, constitutive expression of the *rbcS* antisense mRNAs may have played a role in the success of these experiments.

2. Region of the *rbcS* gene to target with antisense mRNA. From prokaryotic and animal systems it was known that antisense mRNA directed toward the 5' portion of the target mRNA successfully inhibited target mRNA expression in most, if not all, cases [reviewed by van der Krol et al., 1988b] (see also Murray and Crockett, this volume; Thomas, this

volume). In many of these cases, the antisense mRNA was found to interfere with target mRNA translation. Therefore, sequences overlapping the 5'-nontranslated region of the *rbcS* gene were thought to be especially important to include in the antisense gene (see Fig. 1).

3. Growth medium. We decided to grow potential *rbcS* antisense mutants on tissue culture medium supplemented with sucrose. Since RUBISCO is the key regulatory enzyme in carbon fixation, it was thought that any reductions in RUBISCO content (or activity) might be lethal for plant growth under photoautotrophic conditions (such as growth in soil in the greenhouse).

Taking the above factors into consideration, an *rbcS* antisense RNA expression vector was constructed (pTASS; see Fig. 1). pTASS was then introduced by the triparental mating procedure (see Fig. 1, Bejarano et al., this volume) into a strain of *Agrobacterium tume-faciens* harboring the disarmed Ti plasmid pGV2260 [Deblaere et al., 1985]; GJ23 served as the *Escherichia coli* mobilizing strain in these conjugations [Van Haute et al., 1983]. Exconjugants expressing the appropriate antibiotic resistances were used to infect leaf discs of *Nicotiana tabacum* (cv. SR1), and the infected discs were placed on tissue culture medium containing kanamycin [Horsch et al., 1985]. Following several weeks on selective medium, putative transformants were removed from the discs and propagated in tissue culture on non-selective B5 medium supplemented with 1% sucrose. As controls, nontransformed SR1 plants were germinated from seed and maintained in tissue culture under the same conditions as the transformed plants.

B. Characterization of the *rbc*S Antisense Mutants

Five plants transformed with *rbc*S antisense gene sequences (designated transformants 3, 5, 7, 23, and 25) and three nontransformed control plants were randomly selected for analysis. Southern hybridization experiments showed that four of these transformants contained a single copy of the antisense gene, but that transformant 5 contained at least four copies of the gene [Rodermel et al., 1988].

*rbc*S and *rbc*L transcript levels in the transformed and control plants were assessed by Northern hybridization experiments [Rodermel et al., 1988]. Total cell RNA was isolated from the top three (expanding) leaves of plants growing on sugar-containing tissue culture medium, and equal amounts (10 µg) of each RNA sample were electrophoresed through a 1.2% MOPS-formaldehyde gel and transferred to nylon filters. The filters were hybridized with nick-translated probes specific for the tobacco *rbc*S or *rbc*L gene (Fig. 2). As shown in Table I, densitometric analyses of autoradiographs obtained from replicate experiments revealed that the average *rbc*L mRNA levels in the transformants do not vary significantly from the control plant mean. On the other hand, the *rbc*S transcript levels are depressed an average of about fourfold in transformants 3, 7, 23 and

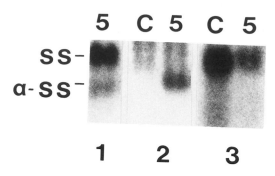

Fig. 2. *Identification of the* rbsS *antisense transcript in transformed plants. Northern blots of RNA extracted from transformant 5 and a control plant (C), showing the sense 950 nucleotide* rbcS *mRNA (SS) and the 750 nucleotide antisense* rbcS *transcript (α-SS).* **Lane 1**, *10 µg of RNA from transformant 5, hybridized to nick-translated* rbcS *DNA probe;* **lanes 2**, *hybridization of an RNA probe synthesized in vitro with SP6 RNA polymerase, specific for antisense transcripts;* **lanes 3**, *hybridization of an RNA probe synthesized in vitro with T7 RNA polymerase, specific for sense transcripts. Note the much greater steady-state level of* rbcS *mRNA than the antisense mRNA sequence (lane 1) and the reduction in the level of sense* rbcS *mRNA caused by the antisense RNA in transformant 5, when compared with the control plant in lanes 2 and 3. [Reproduced from Rodermel et al., 1988, with permission of the publisher.]*

25 and an average of 10-fold in transformant 5. (The variations in *rbc*L and *rbc*S mRNA levels among the control plants were about 25% for each.) These results indicate that the accumulation of *rbc*S and *rbc*L transcripts is not tightly coupled in this system.

To assess the LS and SS protein levels in the transformed and control plants, soluble protein fractions were isolated from the top three leaves of each plant, and equal amounts of protein from each sample were applied to a 15% discontinuous SDS-polyacrylamide gel (SDS-PAGE), which was stained with Coomassie blue [Rodermel et al., 1988]. The LS and SS bands on these gels were identified by Western blot analysis using antibodies generated against the tobacco SS and LS proteins. Densitometric analyses of replicate stained gels (Table I) revealed that the LS and SS protein levels in transformants 3, 7, 23, and 25 are each decreased about 40% compared with controls,

TABLE I. *rbc*S and *rbc*L mRNA and SS and LS Protein Amounts in the Transformed Versus Control Plants

Transformant	mRNA (% controls)		Protein (% controls)	
	*rbc*S	*rbc*L	SS	LS
3	33	79	53	55
5	12	101	37	38
7	25	100	61	65
23	26	86	63	71
25	26	96	49	60

Densitometric scans of autoradiographs or Coomassie blue–stained gels were made, and the areas under the peaks were measured. To ensure linearity, all gels contained a dilution series of one sample, and autoradiographs were exposed for varying amounts of time. Each value represents the mean from replicate experiments for each transformant expressed as a percentage of the control plant mean. In each experiment, the variation in the individual transformed plants was within the range of variation observed among the control plants (from 20% to 27%). (Reproduced from Rodermel et al., 1988, with permission of the publisher.)

whereas the levels of these proteins are each depressed about 60% in transformant 5. (The variations in SS and LS protein amounts among the control plants were about 25% for each.) These results were verified by comparisons of stained SDS-polyacrylamide gels in which the amount of soluble protein applied to the gel was standardized to equal amounts of chlorophyll (as a rough gauge of thylakoid membrane protein) rather than to equal amounts of soluble protein [Rodermel et al., 1988]. The fact that the LS and SS proteins are depressed to a similar extent in the transformed vs. control plants indicates that the stoichiometry of the two proteins is the same in the two sets of plants; i.e., excess LS or SS pools do not exist. This has been confirmed by electrophoresis of soluble proteins from the control and transformed plants on nondenaturing polyacrylamide gels: The amount of reduction in RUBISCO holoenzyme in each transformant is about the same as that observed for the LS and SS proteins.

Considered together, the data from the RNA and protein analyses indicate that the accumulation of RUBISCO is roughly correlated with *rbc*S mRNA pool sizes in the transformants. On the other hand, RUBISCO accumulation does not appear to be correlated with *rbc*L transcript pool sizes. This indicates that LS protein accumulation is regulated by posttranscriptional factors in the transformants.

C. Posttranscriptional Regulation of LS Protein Accumulation

To test whether LS protein accumulation is regulated at the translational level in the *rbc*S antisense gene mutants, transformed and control plants growing on sucrose-containing tissue culture medium were labeled with ^{35}S-methionine for 1 hour [Rodermel et al., 1988]. Equal counts per minute of soluble protein were then electrophoresed through a 15% SDS-polyacrylamide gel, and the gel was fluorographed. Examination of replicate fluorographs revealed that, with the exception of transformant 5, the amounts of label incorporated into the LS during the 1 hour pulse were about the same in the transformed and control plants. This indicates that *rbc*L mRNAs are translated with about the same efficiency in these plants, suggesting that excess LS is degraded in the transformants—perhaps in response to a lack of SS protein. On the other hand, transformant 5 incorporated considerably less ^{35}S into the LS during the 1 hour labeling period than did any of the other plants. Since this transformant also had the most severely depressed *rbc*S mRNA and SS protein levels of any of the transformants (see Table I), the data suggest that, under such conditions, restrictions on LS translation may come into play to regulate LS accumulation. Alternatively, the data are consistent with the hypothesis that there is enhanced LS degradation in transformant 5.

D. Conclusions

Two central findings emerge from our analyses of the *rbc*S antisense DNA mutants. The first of these concerns the "coarse control" mechanisms that coordinate *rbc*S and *rbc*L transcript pool sizes: rbcL mRNA accumulation is unaffected by rbcS mRNA or SS protein levels in the transformed plants. This observation is consistent with the data from short-term inhibitor studies suggesting that rbcL mRNA accumulation is generally independent of nuclear transcription and translation [Sasaki, 1986; Radetzky and Zetsche, 1987]. However, because of the specific means of perturbation achieved in the present study, it is clear that if a signal is transduced from the nuclear–cytoplasmic compartment to regulate rbcL mRNA amounts in the chloroplast, this signal is neither rbcS mRNA nor SS protein, nor is it a secondary messenger generated in response to the amounts of either of these molecules.

In the second place, our analyses demonstrate that rbcS transcript pool sizes may be the major factor defining RUBISCO levels in this system, but that the amounts of the LS and SS proteins are fine-tuned by other mechanisms. In particular, LS accumulation appears to be controlled primarily by posttranslational factors—perhaps by LS degradation in response to depleted amounts of the SS protein. Little is understood about plastid-localized proteolytic processes [e.g., Malek et al., 1984]. However, the present data and the observations of others that both LS and SS protein accumulation are controlled by posttranslational factors [e.g., Roy et al., 1982; Schmidt and Mishkind, 1983; Spreitzer and Ogren, 1983; Hildebrandt et al., 1984; Spreitzer et al., 1985; Cannon et al., 1986] clearly indicate that there must be a plastid-localized proteolytic system that is sensitive to "excess" levels of one or the other subunits of RUBISCO. The elucidation of how this system operates would be particularly interesting in light of the fact that RUBISCO assembly is mediated by the RUBISCO subunit–binding protein (RSBP)—a molecular chaperone that facilitates assembly of the holoenzyme by binding newly synthesized LS proteins, and perhaps newly imported SS proteins [reviewed by Roy, 1989]. Experiments are in progress to determine whether RSBP synthesis is coordinated with that of the SS in the RUBISCO antisense mutants.

Although the antisense mutant analyses showed that LS protein accumulation appears to be regulated primarily at the posttranslational level, the analyses also revealed that under conditions of severe rbcS mRNA or SS protein depletion, LS accumulation may also be regulated at the translational level. Experiments are in progress to determine whether this inhibition is at the level of translational initiation or at the level of translational elongation. Regardless of the precise mechanism, these findings are consistent with the data of others demonstrating that LS accumulation can be influenced by translational factors [e.g., Inamine et al., 1985; Berry et al., 1985, 1986; Steinbiss and Zetsche, 1986; Abbott and Bogorad, 1987; Nikolau and Klessig, 1987; Radetzky and Zetsche, 1987; Sasaki et al., 1987; Sheen and Bogorad, 1986].

IV. PHOTOAUTOTROPHIC GROWTH OF THE RUBISCO ANTISENSE MUTANTS

When grown on sucrose-containing tissue culture medium, some of the RUBISCO antisense mutants grow more slowly than control plants. Some of these plants are also more pale and have thinner stems and less broad leaves than the controls. To assess the growth characteristics of the mutant plants under photoautotrophic conditions, seeds derived from the selfing of each of the transformants were planted in vermiculite in a growth chamber (28°C, 14 hours/day, 10 hours/night, 500 μEinsteins M^2 sec^{-1}) along with seeds from nontransformed tobacco plants. In contrast to the control plants, the progeny plants displayed markedly varying rates of growth. For example, Figure 3 shows a control plant and four representative progeny plants from the selfing of transformant 5 after 6 weeks growth under photoautotrophic conditions: Whereas most of the control plants were flowering (about 120 cm high), the progeny plants ranged in height from 5 to 120 cm (also flowering).

Southern hybridization experiments revealed that antisense gene dosage is negatively correlated with the growth of these plants, i.e., the more antisense genes, the slower the plants grow

Fig. 3. *Growth of progeny plants of* rbcS *antisense mutant 5. Progeny plants were grown in a growth chamber under conditions described in the text. This photograph was taken about 6 weeks after planting and shows representative control (C) and progeny plants. The latter include a big plant (B), a medium-sized plant (M), a small plant (S), and a tiny plant (T). [Reproduced from Rodermel et al., 1988, with permission of the publisher.]*

[Rodermel et al., 1988]. RUBISCO content is also roughly correlated with the growth of these plants—i.e., the more RUBISCO, the faster the plants grow—although some mutants with much reduced RUBISCO levels grow as well as wild-type plants. Quick et al. [1991] have examined these latter plants and found that they perform photosynthesis as well as wild-type plants. In fact, examination of a spectrum of progeny antisense plants has revealed that RUBISCO content can be decreased about 40% before net carbon dioxide fixation is affected [Quick et al., 1991].

The mechanisms by which the leaf adjusts to maintain the same photosynthetic flux in plants with widely varying amounts of RUB-ISCO is currently under investigation. It is the long-term goal of these experiments to define the precise relationships between RUBISCO content, RUBISCO activity, photosynthesis, and plant growth. These relationships have proven refractory to understanding in the past, principally because of a dearth of informative mutants, e.g., mutants displaying a graded series of RUBISCO concentrations. It is hoped that knowledge gleaned from these analyses can be used to design appropriate molecular genetic strategies—centered around the modulation of RUBISCO content and/or activity—to improve the productivity of important crop species.

V. MECHANISM OF ANTISENSE RNA INHIBITION

Antisense mRNA has been found to inhibit gene expression in animal systems at the levels of target mRNA transcription, processing, transport from the nucleus, and translation [reviewed by van der Krol et al., 1988b] (see also Murray and Crockett, this volume). In some cases sense–antisense duplex RNAs have been observed, but in most cases their existence has only been inferred; such structures may be short lived because of rapid turnover.

The only generality to emerge from about a dozen successful reports of antisense mRNA inhibition of plant gene expression is that repression of protein amount and/or activity is accom-panied by decreased amounts of the target mRNA species. In some cases these decreases are stoichiometric [e.g., Robert et al., 1989; Smith et al., 1990], but in other cases they are not [e.g., Rothstein et al., 1987; van der Krol et al., 1988a; Cornelissen, 1989]. Amounts of the antisense RNA required for effective inhibition of gene expression also vary widely, from amounts that are in vast excess over the target mRNA [e.g., Ecker and Davis, 1986] to amounts that are significantly less than the target mRNA [e.g., Smith et al., 1988; van der Krol et al., 1990a]. Although it is presumed, as in other systems, that antisense RNA exerts its effect by forming duplexes with the target RNA, such duplexes have only been observed in cases in which the antisense RNA has been directed against cytoplasmically localized viral RNAs rather than against nuclear genes [Cuozzo et al., 1988; Hemenway et al., 1988].

Regions of certain genes, when in an anti-sense orientation, give rise to mRNAs that are more efficient than others in inhibiting target mRNA expression [e.g., Delauney et al., 1988; Sandler et al., 1988; van der Krol et al., 1990b]. However, there are apparently no "rules" regarding the best antisense sequence to use, since the most effective regions differ from gene to gene (e.g., 5′ vs. 3′ sequences). In this regard, it should be pointed out that the effectiveness of a given antisense sequence is likely to depend on its inherent stability, as well as on the rate and extent of hybridization of the antisense RNA with the target RNA. The three-dimensional conformations of the antisense and sense RNAs may thus play a crucial role in determining whether a given antisense sequence is effective in inhibiting the expression of a given gene.

There often appears to be variable expression of the target gene in plants transformed with an antisense gene construct, ranging from 0% to 99% [reviewed by van der Krol et al., 1988b]. This variability has been attributed to position effects arising from the random integration of antisense genes into host chromosomes. In addition, there appears to be no quantitative relationship between the steady-state level of a particular antisense mRNA and the degree to which the target mRNA or target protein is reduced in amount [e.g., Sheehy et al., 1988; van der Krol et al., 1990b]. For example, plants transformed with the same antisense sequence can have similar amounts of antisense mRNA but vastly differing amounts of the target mRNA.

A. Two Case Studies in Plants

Given the array of observations noted above, it might be anticipated that the mechanism of antisense mRNA inhibition would vary according to the plant system, target gene, and particular antisense gene construct used. Two cases in which the mechanism of antisense mRNA inhibition has been most fully examined in plants support this conclusion.

The first of these cases involves the inhibition by antisense mRNA of polygalacturonase (PG), a cell wall–degrading enzyme [Smith et al., 1988, 1990; Sheehy et al., 1988]. PG mRNA and protein are normally induced at the onset of tomato fruit ripening and are present only in orange fruit tissue; they are not expressed in green fruit tissue or other plant tissues. PG mRNA and protein levels are drastically reduced in the orange fruit tissue of transgenic plants carrying a CaMV 35S promoter/antisense PG DNA construct. However, excess amounts of the antisense mRNA do not appear to be necessary for these reductions. This conclusion is based on a comparison of the levels of the constitutively expressed antisense PG mRNAs, which are present in all tissues of the transformed plants, with the levels of the induced PG mRNAs, which are present at maximal levels only in orange fruit tissue of nontransformed plants. It is found that antisense PG mRNAs are about 10-fold less abundant than PG mRNAs, but are still able to bring about the large reduction in PG mRNA and protein levels.

Perhaps the simplest hypothesis compatible with this finding is a mechanism whereby antisense mRNA inhibits PG gene transcription. To test this hypothesis, nuclear run-off transcription experiments were performed, and it was observed that PG mRNAs are transcribed at the same rate in transformed and control

orange fruit tissues [Sheehy et al., 1988]. This indicates that antisense PG mRNA does not exert its effect at the transcriptional level, but rather at the posttranscriptional level—most likely by forming duplex structures that result in rapid degradation of the mRNA. Since such a proposed mechanism would require at least equal amounts of the antisense and target mRNAs, how does one reconcile this with the observed low steady-state levels of the antisense PG mRNA? The apparent answer to this question comes from other nuclear run-off transcription experiments, in which it was found that the antisense PG mRNAs are transcribed at a higher rate than PG mRNAs. It is therefore proposed that a pool of rapidly turning-over antisense mRNAs is present at the onset of ripening when PG expression normally starts and that duplex structures between sense and antisense PG mRNAs are formed and degraded. As ripening proceeds, so does this process, since the antisense mRNA is being produced at a higher rate than the target mRNA. PG mRNA is therefore unable to accumulate and little PG protein is produced. However, it should be pointed out that the situation may not be quite this simple, since it would be expected that the antisense and target mRNAs accumulate with complicated kinetics that not only are a function of the rates of production and degradation of these two mRNA species but also are a function of the rates of formation and degradation of antisense:sense RNA duplex structures.

A second case in which the mechanism of antisense mRNA inhibition has been partially characterized is in doubly transformed tobacco plants carrying both the bialophos resistance (*bar*) gene—encoding the enzyme phosphoinotricin acetyltransferase (PAT)—and a 35S promoter/antisense *bar* construct introduced by a second transformation [Cornelissen, 1989; Cornelissen and Vandewiele, 1989]. In plants carrying the antisense gene, there is an approximately 13-fold decrease in PAT enzyme activity, but only a fourfold decrease in *bar* mRNA. This indicates that there is an approximately threefold reduced synthesis of PAT per *bar* mRNA molecule because of the antisense gene in the transgenic plants. It is found that the proportion of *bar* mRNA in the cytoplasm (vs. the nucleus) was independent of the presence of the antisense *bar* gene. Furthermore, inhibitor experiments showed that the rate of turnover of *bar* mRNA in the cytoplasm is the same in plants carrying only the *bar* gene and in doubly transformed plants that carry in addition the antisense *bar* construct.

The total reduction in the PAT enzyme level that is observed is therefore apparently due to the fact that antisense *bar* mRNA exerts its inhibitory effect at two distinct cellular sites: It inhibits translation of *bar* mRNA in the cytoplasm by forming unstable duplex structures that are not degraded, and it also inhibits either processing or transport of *bar* mRNA in the nucleus by forming rapidly turned-over duplex structures.

B. Mechanism of *rbc*S Antisense mRNA Inhibition

To understand the mechanism by which antisense *rbc*S mRNA inhibits RUBISCO expression, the first point one should note is that the decreases in *rbc*S steady-state levels in the various transformants roughly correspond with the observed decreases in RUBISCO content (Table I). However, this correlation is not strictly proportional, since *rbc*S mRNA amounts are more severely depressed than are SS protein amounts. Consequently, it is possible that *rbc*S mRNA may be more efficiently translated or that the SS protein may be more slowly turned-over in the transformed plants.

Because the *rbc*S antisense cDNA fragment in pTASS includes 5′-flanking sequences as well as coding sequences that overlap the first intron–exon border of the gene, it is possible that antisense RNAs bind to *rbc*S transcripts and prevent their processing, their transport from the nucleus, or their translation in the cytoplasm. Blockage of any of these events could possibly result in degradation of the RNA. Alternatively, the data do not rule out the possibility that antisense RNAs bind to DNA and block *rbc*S transcription. As a first step toward understanding how the reductions in *rbc*S sense

mRNA occur, Northern hybridization experiments were carried out using strand-specific probes for *rbc*S sense or antisense mRNAs in the transformed plants [Rodermel et al., 1988]. A low abundance, 750 bp transcript was identified as the product of the *rbc*S antisense gene. An mRNA of this size falls within the expected size range of a transcript encoded by the antisense gene, assuming the CaMV 35S promoter and polyadenylation sequences of the vector are used (see Fig. 1). In other Northern experiments, antisense *rbc*S mRNAs have been observed in root tissues of the transformed plants at levels that vary among the transformants but that, for a given transformant, are similar to those levels found in leaf tissues (S. Rodermel, unpublished data).

The fact that the steady-state levels of the *rbc*S antisense mRNA are about the same in roots (which do not synthesize *rbc*S mRNA) and leaves (which synthesize *rbc*S mRNA) suggests that inhibition of RUBISCO expression does not require a steady-state excess of antisense over sense mRNA. In addition, these findings suggest that the antisense transcript is not degraded in leaves at rates higher than in roots. If so, this may indicate that antisense *rbc*S mRNA exerts its inhibitory effect at the level of *rbc*S transcription. Alternatively, the data are equally consistent with a mechanism such as that proposed in tomato for the inhibition of PG expression by antisense PG mRNA (above) that involves duplex formation and degradation between RNA species with differing stabilities, rates of transcription, and inducibility. Nuclear run-off transcription assays are in progress as a first step toward discriminating among these various possibilities.

VI. SUMMARY

The biosynthesis of RUBISCO provides an ideal model system for studying the mechanisms that coordinate gene expression in the nucleus and the chloroplast. For the express purpose of determining whether LS metabolism is affected by the abundance of SS mRNAs or protein, we generated SS antisense mutants of tobacco using *Agrobacterium*-mediated DNA transfer methods. We found that expression of the antisense sequences in the transgenic plants results in drastic reductions in the accumulation of both SS mRNA and protein and that these reductions are accompanied by corresponding reductions in LS protein, but not LS mRNA. This indicates that LS mRNA levels are not coordinated with SS transcript amounts by signals generated either directly or indirectly in response to the amounts of the SS mRNA or protein. We also observed that the accumulation of LS protein in these mutants is regulated primarily by posttranslational factors but that restrictions on LS translation come into play under conditions of severe SS mRNA or protein repression. This suggests that a hierarchy of posttranscriptional controls regulates LS accumulation.

The RUBISCO antisense DNA mutants have not only provided valuable information regarding the mechanisms that coordinate nuclear and plastid gene expression, but also are being used to answer fundamental questions in plant biochemistry, physiology, and ecology. For example, under photoautotrophic conditions these mutants display a wide spectrum of variation in both RUBISCO content and plant growth that makes them an ideal model system for analyzing the poorly understood relationships between RUBISCO content, RUBISCO activity, photosynthesis, and plant growth.

The mechanism by which antisense SS mRNA inhibits RUBISCO gene expression is not fully understood. The data are most compatible with an effect during transcription, processing, and/or transport of the target mRNA from the nucleus. Regardless of the precise mechanism, our results demonstrate that antisense RNA is capable of sharply reducing the accumulation of one of the most plentiful leaf mRNA species and the most abundant soluble leaf protein. Therefore our results show that in principle antisense RNA technology may be of general utility in inhibiting the expression of any plant gene, regardless of expression level or copy number in the genome. However, since few generalities have emerged from reports of successful antisense RNA inhibition in plants,

it is concluded that a good deal of trial-and-error may be required to inhibit successfully the expression of a gene of choice.

ACKNOWLEDGMENTS

The work cited from the authors' laboratory was funded by research grants from the National Institute of General Medical Sciences and the Maria Moors Cabot Foundation of Harvard University.

VII. REFERENCES

Abbott MS, Bogorad L (1987): Light regulation of genes for the large and small subunits of ribulose bisphosphate carboxylase in tobacco. In Biggins J (ed): Progress in Photosynthesis Research.'' Dordrecht, The Netherlands: Martinus Nijhoff Publishers, vol IV, pp 527–530.

Bendich AJ (1987): Why do chloroplasts and mitochondria contain so many copies of their genome? BioEssays 6:279–282.

Berry JO, Nickolau BJ, Carr JP, Klessig DF (1985): Transcriptional and post-transcriptional regulation of ribulose 1,5-bisphosphate carboxylase gene expression in light- and dark-grown amaranth cotyledons. Mol Cell Biol 5:2238–2246.

Berry JO, Nickolau BJ, Carr JP, Klessig DF (1986): Translational regulation of light-induced ribulose 1,5-bisphosphate carboxylase gene expression in amaranth. Mol Cell Biol 6:2347–2353.

Biekmann S, Feierabend J (1985): Synthesis and degradation of unassembled polypeptides of the coupling factor of photophosphorylation in 70S ribosome-deficient rye leaves. Eur J Biochem 152:529–535.

Bogorad L (1982): Regulation of intracellular gene flow in the evolution of eukaryotic genomes. In Schiff JA (ed): On the Origins of Chloroplasts. Amsterdam: Elsevier/North Holland, pp 277–295.

Cannon S, Wang P, Roy H (1986): Inhibition of ribulose bisphosphate carboxylase assembly by antibody to a binding protein. J Cell Biol 103:1327–1335.

Cornelissen M (1989): Nuclear and cytoplasmic sites for antisense control. Nucleic Acids Res 17:7203–7209.

Cornelissen M, Vandewiele M (1989): Both RNA level and translation efficiency are reduced by antisense RNA in transgenic tobacco. Nucleic Acids Res 17: 833–843.

Cuozzo M, O'Connel KM, Kaniewski W, Fang R-X, Chua N-H, Tumer NE (1988): Viral protection in transgenic tobacco plants expressing the cucumber mosaic virus coat protein or its antisense RNA. Bio/Technology 6:549–557.

Deblaere R, Bytebier B, DeGrave H, Bedoeck F, Schell J, Van Montagu M, Leemans J (1985): Efficient octopine Ti plasmid-derived vectors for *Agrobacterium*-mediated gene transfer to plants. Nucleic Acids Res 13:4777–4788.

Delauney AJ, Tabaeizadeh Z, Verma DPS (1988): A stable bifunctional antisense transcript inhibiting gene expression in transgenic plants. Proc Natl Acad Sci USA 85:4300–4304.

Ecker JR, Davis RW (1986): Inhibition of gene expression in plant cells by expression of antisense RNA. Proc Natl Acad Sci USA 83:5372–5376.

Ellis RJ (1979): The most abundant protein in the world. Trends Biochem Sci 4:241–244.

Guilley H, Dudley RK, Jonard G, Balazs E, Richards KE (1982): Transcription of cauliflower mosaic virus DNA: Detection of promoter sequences, and characterization of transcripts. Cell 30:763–773.

Hattori T, Margulies MM (1986): Synthesis of large subunit of ribulosebisphosphate carboxylase by thylakoid-bound polyribosomes from spinach chloroplasts. Arch Biochem Biophys 244:630–640.

Hemenway C, Fang R-X, Kaniewski WK, Chua N-H, Tumer NE (1988): Analysis of the mechanism of protection in transgenic plants expressing the potato virus coat protein or its antisense RNA. EMBO J 7: 1273–1280.

Hildebrandt J, Bottomley W, Moser J, Herrmann RG (1984): A plastome mutant of *Oenothera hookeri* has a lesion in the gene for the large subunit of ribulose-1,5-bisphosphate carboxylase/oxygenase. Biochim Biophys Acta 783:67–73.

Horsch RB, Fry JE, Hoffmann NL, Eichholtz D, Rogers SG, Fraley RT (1985): A simple and general method for transferring genes into plants. Science 227: 1229–1231.

Inamine G, Nash B, Weissbach H, Brot N (1985): Light regulation of the synthesis of the large subunit of ribulose-1,5-bisphosphate carboxylase in peas: Evidence for translational control. Proc Natl Acad Sci USA 82:5690–5694.

Malek L, Bogorad L, Ayers AR, Goldberg AL (1984): Newly synthesized proteins are degraded by an ATP-stimulated proteolytic process in isolated pea chloroplasts. FEBS Lett 166:253–257.

Mazur BJ, Chui C-F (1985): Sequence of a genomic DNA clone for the small subunit of ribulose bis-phosphate carboxylase-oxygenase from tobacco. Nucleic Acids Res 13:2373–2386.

Mishkind MC, Schmidt GW (1983): Posttranscriptional regulation of ribulose 1,5-bisphosphate carboxylase small subunit accumulation in *Chlamydomonas reinhardtii*. Plant Physiol 72:847–854.

Miziorko HM, Lorimer G (1983): Ribulose-1,5-bisphosphate carboxylase-oxygenase. Annu Rev Biochem 52:507–535.

Nikolau BJ, Klessig DF (1987): Coordinate, organ-specific and developmental regulation of ribulose 1,5-bisphos-phate carboxylase gene expression in *Amaranthus hypochondriacus*. Plant Physiol 85:167–173.

Odell JT, Nagy F, Chua N-H (1985): Identification of DNA sequences required for activity of the cauliflower mosaic virus 35S promoter. Nature 313:810–812.

O'Neal JK, Pokalsky AR, Kiehne KL, Shewmaker CK (1987): Isolation of tobacco SSU genes: Characterization of a transcriptionally active pseudogene. Nucleic Acids Res 15:8661–8677.

Pinck M, Guilley E, Durr A, Hoff M, Pinck L, Fleck J (1984): Complete sequence of one of the mRNAs coding for the small subunit of ribulose bisphosphate carboxylase of *Nicotiana sylvestris*. Biochimie 66:539–545.

Quick WP, Schurr U, Scheibe R, Schulze E-D, Rodermel SR, Bogorad L, Stitt M (1991): Decreased RUBISCO in tobacco transformed with ''antisense'' *rbc*S. I. Impact on photosynthesis in ambient growth conditions. Planta 183:542–554.

Radetzky R, Zetsche K (1987): Effects of specific inhibitors on the coordination of the concentrations of ribulose-bisphosphate-carboxylase subunits and their corresponding mRNAs in the alga *Chlorogonium*. Planta 172:38–46.

Robert LS, Donaldson PA, Ladaique C, Altosaar I, Arnison PG, Fabijanski SF (1989): Antisense RNA inhibition of β-glucuronidase gene expression in transgenic tobacco plants. Plant Mol Biol 13:399–409.

Rodermel SR, Abbott MS, Bogorad L (1988): Nuclear-organelle interactions: Nuclear antisense gene inhibits ribulose bisphosphate carboxylase enzyme levels in transformed tobacco plants. Cell 55:673–681.

Rothstein SJ, DiMaio J, Strand M, Rice D (1987): Stable and heritable inhibition of the expression of nopaline synthase in tobacco expressing antisense RNA. Proc Natl Acad Sci USA 84:8439–8443.

Roy H (1989) Rubisco assembly: A model system for studying the mechanism of chaperonin action. Plant Cell 1:1035–1042.

Roy H, Bloom M, Milos P, Monroe M (1982): Studies on the assembly of large subunits of ribulose bisphosphate carboxylase in isolated pea chloroplasts. J Cell Biol 94:20–27.

Sanders PR, Winter JA, Barnason AR, Rogers SG, Fraley RT (1987): Comparison of cauliflower mosaic virus 35S and nopaline synthase promoters in transgenic plants. Nucleic Acids Res 15:1543–1558.

Sandler SJ, Stayton M, Townsend JA, Ralston ML, Bedbrook JR, Dunsmiur P (1988): Inhibition of gene expression in transformed plants by antisense RNA. Plant Mol Biol 11:301–310.

Sasaki Y (1986): Effects of α-amanitin on coordination of two mRNAs of ribulose-bisphosphate carboxylase in greening pea leaves. FEBS Lett 204:279–282.

Sasaki Y, Nakamura Y, Matsuno R (1987): Regulation of gene expression of ribulose bisphosphate carboxylase in greening pea leaves. Plant Mol Biol 8:375–382.

Schmidt GW, Mishkind ML (1983): Rapid degradation

of unassembled ribulose 1,5-bisphosphate carboxylase small subunits in chloroplasts. Proc Natl Acad Sci USA 80:2632–2636.

Schmidt GW, Mishkind ML (1986): The transport of proteins into chloroplasts. Annu Rev Biochem 55:879–912.

Sheen J, Bogorad L (1986): Expression of the ribulose-1,5-bisphosphate carboxylase large subunit gene and three small subunit genes in two cell types of maize leaves. EMBO J 5:3417–3422.

Sheehy RE, Kramer M, Hiatt WR (1988): Reduction of polygalacturonase activity in tomato fruit by antisense RNA. Proc Natl Acad Sci USA 85:8805–8809.

Smith CJS, Watson CF, Ray J, Bird CR, Morris PC, Schuch W, Grierson D (1988): Antisense RNA inhibition of polygalacturonase gene expression in transgenic tomatoes. Nature 334:724–726.

Smith CJS, Watson CF, Morris PC, Bird CR, Seymour GB, Gray JE, Arnold C, Tucker GA, Schuch W, Harding S, Grierson D (1990): Inheritance and effect on ripening of antisense polygalacturonase genes in transgenic tomatoes. Plant Mol Biol 14:369–379.

Spreitzer RJ, Goldschmidt-Clermont M, Rahire M, Rochaix J-D (1985): Nonsense mutations in the *Chlamydomonas* chloroplast gene that codes for the large subunit of ribulosebisphosphate carboxylase-oxygenase. Proc Natl Acad Sci USA 82:5460–5464.

Spreitzer RJ, Ogren WL (1983): Rapid recovery of chloroplast mutations affecting ribulose bisphosphate carboxylase/oxygenase in *Chlamydomonas reinhardtii*. Proc Natl Acad Sci USA 80:6293–6297.

Steinbiss HJ, Zetsche K (1986): Light and metabolic regulation of the synthesis of ribulose-1,5-bisphosphate carboxylase/oxygenase and the corresponding mRNAs in the unicellular alga *Chlorogonium*. Planta 167: 575–581.

Tobin EM, Silverthorne J (1985): Light regulation of gene expression in higher plants. Annu Rev Plant Physiol 36:569–593.

van der Krol AR, Lenting PE, Veenstra J, van der Meer IM, Koes RE, Gerats AGM, Mol JNM, Stuitje AR (1988a): An anti-sense chalcone synthase gene in transgenic plants inhibits flower pigmentation. Nature 333:866–869.

van der Krol AR, Mol JNM, Stuitje AR (1988b): Modulation of eukaryotic gene expression by complementary RNA or DNA sequences. BioTechniques 6: 958–976.

van der Krol AR, Mur LA, de Lange P, Gerats AGM, Mol JNM, Stuitje AR (1990a): Antisense chalcone synthase genes in petunia: Visualization of variable transgene expression. Mol Gen Genet 220:204–212.

van der Krol AR, Mur LA, de Lange P, Mol JNM, Stuitje AR (1990b): Inhibition of flower pigmentation by antisense CHS genes: Promoter and minimal sequence requirements for the antisense effect. Plant Mol Biol 14:457–466.

van Haute E, Joos H, Maes M, Warren G, Van Montagu M, Schell J (1983): Intergeneric transfer and exchange recombination of restriction fragments cloned in pBR322: A novel strategy for the reversed genetics of the Ti plasmids of *Agrobacterium tumefaciens*. EMBO J 2:411–417.

ABOUT THE AUTHORS

STEVEN R. RODERMEL is an Assistant Professor of Botany at Iowa State University, where his current research involves the use of antisense gene mutants to dissect the regulatory processes involved in photosynthesis. After graduating from Yale with a B.A. in Philosophy in 1972, Dr. Rodermel received his M.S. in Zoology in 1976 at the University of Wyoming, where he worked under the guidance of Joan Sonneborn, studying the mechanisms of cellular aging in the ciliated protozoan *Paramecium*. He then joined the lab of Lawrence Bogorad at Harvard, where he completed his Ph.D. in 1986 and his postdoctoral work in 1990. Dr. Rodermel's doctoral work involved the identification and characterization of photoregulated chloroplast genes in maize, and his postdoctoral work focused upon the analysis of RUBISCO antisense DNA mutants of tobacco. His papers have appeared in such journals as *Cell*, *Genetics*, and the *Journal of Cell Biology*. Dr. Rodermel teaches a course in introductory biology for non-majors and also a graduate level course in plant cell biology.

LAWRENCE BOGORAD is Maria Moors Cabot Professor of Biology in the Department of Cellular and Developmental Biology at Harvard University. He teaches plant physiology and plant molecular biology. He received his B.S. and Ph.D. at the University of Chicago, where he concentrated on the physiology of greening in plants. Dr. Bogorad then took up a postdoctoral fellowship, working at the Rockefeller University under the direction of Sam Granick. There he studied the path of chlorophyll biosynthesis, using defective mutants of the single-celled alga *Chlorella*, and the enzymology of tetrapyrrole formation. Dr. Bogorad then joined the faculty of the University of Chicago, where he continued working on tetrapyrrole biosynthesis and later on the molecular biology of chloroplasts. At Harvard University his research has been concentrated largely on various facets of the latter subject. Dr. Bogorad is also Chairman of the Editorial Board of the *Proceedings of the National Academy of Sciences USA*.

Antisense RNA and DNA: 137–158
© 1992 Wiley-Liss, Inc.

Use of Antisense RNA Technology to Engineer Virus Resistance in Plants

Eduardo R. Bejarano, Anthony G. Day, and Conrad P. Lichtenstein

I. INTRODUCTION

Antisense RNA technology has been used very successfully to suppress gene expression in transgenic plants, and the example is discussed by Rodermel and Bogorad (this volume) of the use of such technology to inhibit expression of ribulose bisphosphate carboxylase. As these authors also review the work from other laboratories on the application of this technol-ogy to plants and discuss various models to account for the mechanism by which antisense RNA can suppress gene expression, we do not cover it here.

In this chapter we review, briefly, various strategies to engineer virus resistance in plants and focus on the use of antisense RNA. We also review the development of plant gene trans-fer technology using the ''natural plant genetic engineer'' soil bacterium *Agrobacterium tume-faciens*, as this technology has revolutionized the methods to improve plant resistance to pathogens.

Initially antisense RNA was used against plant RNA viruses, but perhaps because RNA viruses probably replicate in the cytoplasm these experiments have had limited success. The approach we took was to use *A. tumefaciens*-mediated gene transfer to construct transgenic tobacco plants carrying a genetic cassette including an antisense gene sequence of a virally encoded gene of the plant viral pathogen tomato golden mosaic virus (TGMV) [Day et al., 1991]. The virus is a member of the family of single-stranded (ss) DNA viruses called *geminiviruses* that replicate in the nucleus. The gene we chose encodes a protein absolutely required for TGMV DNA replication. These genetic cassettes also contained, on the same transcription unit, a gene encoding hygromycin resistance, allowing selection for concomitant expression of the antisense gene. Transgenic plants were challenged by infection with TGMV; the frequency of symptom development was very significantly reduced in a number of antisense lines and correlated broadly with the abundance of antisense RNA transcript and with a reduction in viral DNA harvested from infected leaf tissue. The expression of the antisense RNA inhibits the replication of the virus in leaves; thus the reduction in symptom development is presumably due mainly to inhibition of DNA replication.

II. PLANT GENE TRANSFER TECHNOLOGY USING *AGROBACTERIUM TUMEFACIENS*

Plant genetic engineering was begun with the discovery that *A. tumefaciens* could transfer DNA to plants [see Lichtenstein and Fuller, 1987; Memelink et al., 1987; Herrera-Estrella and Simpson, 1988; Zambryski, 1988; Hooykaas, 1989]. This bacterium causes crown gall tumors in wounded gymnosperms and dicotyledonous angiosperms (dicots). Oncogenic strains contain a single copy of a large (150–250 kb) tumor-inducing (Ti) plasmid. A region of

this plasmid DNA (the ''transfer'' or T-DNA) is transferred to the wounded plant cell and stably integrated into the nuclear genome. T-DNA–encoded genes, despite their bacterial origin, are expressed in infected plant cells (they contain eukaryotic signal sequences required for transcription). Expression of these genes results in the synthesis of auxins and cytokinins. It is these phytohormones that, when overexpressed, result in the proliferation of the infected plant tissue to produce the characteristic tumorous gall of crown gall disease.

At this stage, the bacteria are no longer required to maintain the tumorous state, but this state is restricted only to those plant cells originally infected and the T-DNA cannot be transferred from plant cell to plant cell. T-DNA also encodes the synthesis of ''opines,'' sugar and/or amino acid derivatives that are presumed to diffuse from the tumor into the surrounding soil where they may serve as a sole carbon and nitrogen source for agrobacteria harboring a Ti plasmid; Ti plasmids also encode genes for opine catabolism that are induced by the opines. Opines also induce the Ti plasmid *tra* operon, which allows conjugal transfer of the Ti plasmid to other agrobacteria.

Thus the Ti plasmid of *A. tumefaciens* has evolved an elegant genetic parasitism in which infection results in the hijacking of plant metabolic resources to produce food (opines) that are metabolizable only through bacterial products encoded by the Ti plasmid. This plasmid can spread only in the presence of infected tissue. Furthermore, infected tissue proliferates, thereby increasing the amount of opine available.

Various features of Ti plasmid biology have allowed the exploitation of Ti plasmids as vectors for plant genetic transformation. First, T-DNA does not encode gene products required for T-DNA transfer. Another region of the Ti plasmid, the so-called virulence (*vir*) regulon, does this. Second, the only T-DNA–encoded transfer functions are two *cis*-essential sequences that signal initiation and termination of T-DNA transfer. These are two 25 bp direct repeats that flank the T-DNA so-called left and right borders (LB and RB, respectively). Third, recombinant DNA inserted into the T-DNA region is expressed in plant tissue transformed with that T-DNA if the recombinant DNA is attached to the correct plant regulatory signals. Originally T-DNA–encoded phytohormone synthesis served as a dominant marker to select for transformed cells that can grow to form a callus in hormone-free medium in tissue culture. As phytohormone synthesis can interfere with plant regeneration or produce abnormal plants, alternative dominant selectable markers were subsequently developed, e.g., using bacterial genes encoding antibiotic resistance flanked by T-DNA transcription initiation and termination signals. Fourth, Ti plasmids themselves are too large for direct genetic manipulation, and it is fortunate that the *vir* region of Ti plasmids will act *in trans* on T-DNA carried on another plasmid.

These features, among others, have led to the development of the so-called binary vector system. Here, an *A. tumefaciens* strain with a "helper" Ti plasmid deleted in the T-DNA region supplies all *trans*-acting functions for DNA transfer. A small shuttle or binary vector, more amenable to the manipulation of recombinant DNA, can then be used. Many such vectors have been designed; the vector used in our laboratory, pGA482 [An, 1986], illustrates the properties (Fig. 1). This plasmid is 13.2 kb and contains the following salient features: a broad host range origin of DNA replication (*oriV*) allowing DNA replication in both *Escherichia coli* and *A. tumefaciens*, and an origin of conjugal transfer (*oriT*) allowing transfer from *E. coli* to *A. tumefaciens*. The mobilization (*mob*) and transfer (*tra*) functions are supplied *in trans* by a helper plasmid, usually in a second *E. coli* strain in a triparental mating. A second *oriV*, the high copy number origin of replication from plasmid ColE1, facilitates recovery of high plasmid DNA yields from *E. coli*. A bacterial tetracycline resistance gene allows selection in *E. coli* and *A. tumefaciens*. The left and right T-DNA borders flank a polylinker sequence for cloning foreign DNA and a chimeric kanamycin resistance gene that expresses in plants, allowing for selection of transformed plant tissue; this gene has the coding region of neomycin phosphotransferase II, from transposon Tn5, flanked by the nopaline (a T-DNA–encoded opine gene) promoter and polyadenylation signals. Thus plasmid constructs are propagated in *E. coli* and then conjugated to *A. tumefaciens*, ready to transfer to plant tissue.

The easiest method to generate transgenic plants is by leaf disc transformation. Plant leaves, cut into small pieces, are cocultivated with an *Agrobacterium* culture. Addition of the appropriate antibiotic selects for transformation, and addition of phytohormones allows either shoot or callus regeneration. Most plants regenerated from this tissue will be morphologically normal and will produce viable seed. Not all plant species are amenable to regeneration in this way; tobacco is particularly accommodating and for this reason is a popular model system. Once T-DNA is integrated into the plant genome it segregates stably in a mendelian fashion. However, although T-DNA appears to be faithfully inherited, levels of gene expression can vary.

III. ENGINEERING RESISTANCE TO PLANT RNA VIRUSES

The effects of viral infection on plants are more insidious than those of other pests such as insects, fungi, and nematodes. Although viruses often cause only mild disease symptoms and thus may go unnoticed, it is clear that they may also seriously affect crop yields. The vast majority (98%) of plant viruses are RNA viruses, and of these most are ss-positive RNA viruses [Francki et al., 1985; Francki, 1988; van Regenmortet and Fraenkel-Conrat, 1988; Koening, 1988; Milnie, 1988]. Typically RNA viruses encode a coat protein, a cell–cell movement factor, and a possible RNA-dependent RNA polymerase (or a subunit interacting with a host enzyme). Several host-encoded mechanisms of resistance exist, which target different stages in the viral infection cycle, e.g., replication or cell–cell spread. Additionally, a

Bejarano et al.

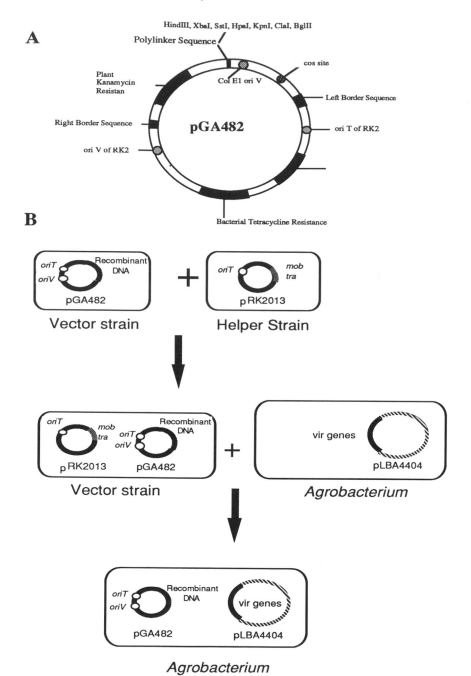

Fig. 1. A: *The binary vector system showing pGA482, the cloning vector.* **B:** *Triparental mating to transfer pGA482 recombinants to* Agrobacterium *strain harboring Ti plasmid pLBA4404 ready for plant transformation. The* E. coli *vector strain, the* E. coli *helper strain, and the* Agrobacterium *recipient are mixed together. The plasmid pRK2013 carried by the helper* strain enters the vector strain and provides conjugal transfer functions that transfer the pGA482 recombinant to Agrobacterium, *where by virtue of its broad host range origin,* oriV, *it can replicate. Though pRK2013 may also be transferred, it cannot replicate in* Agrobacterium *and is lost.*

hypersensitive response by the plant leads to death of the infected cells, thus containing the virus and limiting its spread. Such host-encoded resistance genes may be selected by classic plant breeding, but very few of these genes have been identified, and their mechanism(s) of action have not been elucidated. Thus the host response to viral infection is an area for future exploitation in engineering resistance.

A variety of approaches have been taken to engineer resistance to RNA viruses [for reviews, see Baulcombe, 1989; van den Elzen et al., 1989] that exploit natural mechanisms of resistance brought about by viral–viral interactions. Related plant viruses may interact with one another to limit infection, a phenomenon known as cross-protection. In some cases small extragenomic RNA species, satellite RNA, may interfere with viral replication.

A. Engineering Cross-Protection

In cross-protection a mild virus is used as a biological control to bring about resistance to a more virulent strain. However, the problem is that there is still a viral infection, and thus this approach led to the separation of the cross-protection function from the viral infection. Such approaches were first taken with tobacco mosaic virus (TMV). Upon infection, the coat protein of this virus disassembles and the RNA thus exposed is simultaneously translated. This process can be performed in vitro but is inhibited by addition of free coat protein, leading to the idea that this might be a component of the cross-protection observed in vivo. To test this hypothesis, transgenic plants expressing high levels of TMV coat protein were generated and were found to be resistant to viral infection. Protection is thought to occur by at least two steps: plants expressing the coat protein are protected from infection by the excess of coat protein in the cell, preventing the virus from being uncoated; the degree of resistance is dependent on the concentration of the viral inoculum and is only effective against intact virus and not against naked TMV RNA [Powell et al., 1986; Nelson et al., 1987]. If the viral RNA is uncoated, the infection can be estab-

lished, but there is a significant reduction in the rate of local spread within the inoculated leaves and the rate of systemic spread of virus from inoculated leaves to upper parts of the plant; that reduction is not overcome by inoculation with naked TMV RNA [Wisniewski et al., 1990].

Similar mechanisms could be responsible for the protection obtained by expressing the coat protein of other viruses, e.g., alfalfa mosaic virus (AIMV) [Tumer et al., 1987], cucumber mosaic virus (CMV) [Cuozzo et al., 1988], tobacco rattle virus (TRV) [van Dun and Bol, 1988], and potato virus X (PVX) [Hemenway et al., 1988] and even for viruses expressing a polyprotein requiring cleavage of the coat protein, e.g., potato virus Y (PVY) [Lawson et al., 1990]. Coat-mediated protection involves the coat protein itself and not its messenger RNA, since transgenic plants expressing an AIMV coat protein gene with a frameshift produce a normal amount of RNA that does not confer resistance to the virus [van Dun et al., 1988]. However, protection by coat protein is not the only mechanism involved in natural cross-protection; Gerber and Sarkar [1989] have reported that a coat protein-free mutant of TMV does confer cross-protection, and recently Golemboski et al. [1990] have shown that transgenic plants expressing a nonstructural gene of TMV are resistant to viral infection.

B. Satellite RNA

Some strains of certain plant RNA viruses contain satellite RNA; these are small RNA molecules, which require helper virus for replication [Semancik, 1987]. When they are present, viral symptoms may be attenuated, e.g., CMV and tobacco ringspot virus (TobRSV). Transgenic tobacco transcribing CMV [Baulcombe et al., 1986; Harrison et al., 1987] or TobRSV satellite RNAs [Gerlach et al., 1987] were shown to have reduced symptoms following infection by satellite-free CMV or TobRSV, respectively. The viruses replicate the satellite RNA, and this in turn inhibits viral replication, which might suggest competition for a limiting replicase. But the proposed mechanism is not as clear-cut, because infection of plants expressing CMV satellite RNA by a related virus, tomato aspermy virus (TAV), also leads to satellite RNA replication. In this case, however, there is no evidence of inhibition of TAV replication, despite a reduction in symptoms. Cross-protection with satellite RNA differs from coat protein protection in that it is independent of the concentration of the satellite RNA transcripts prior to infection (since they are replicated by the virus) and is also independent of the concentration the viral inoculum. There are some caveats to using expression of satellite RNA to confer resistance to viral infection, as some satellite–virus combinations are known to produce more severe symptoms than the virus alone. Moreover, there is always a risk of a mutation of the satellite sequence that could convert it from a benign satellite into a virulent one.

C. Antisense RNA

As antisense RNA acts at the transcript level, it is effective against multiple gene copies and, in addition, acts *in trans*. It would thus seem to be ideal for the suppression of viral infection; viruses invade the cell from outside and multiply to a high copy number. The reports of protection against viral infectivity in plants by antisense RNA to date [Baulcombe et al., 1987; Hemenway et al., 1988; Cuozzo et al., 1988; Rezaian et al., 1988; Powell et al., 1989] have focused on RNA viruses that probably replicate in the cytoplasm. This limits the points at which the virus can be blocked, since, if efficient antisense mechanisms rely on nuclear interactions, then suppression of viral infection by antisense RNA will be difficult. Perhaps because of this the results have in many cases not been encouraging. In three cases, the RNA viruses CMV [Cuozzo et al., 1988], PVX [Hemenway et al., 1988], and TMV [Powell et al., 1989], the gene targeted was the coat protein gene; this was essentially a control for a cross-protection experiment, as described above. Coat protein genes will not in general be good targets for suppression of viral infectivity by antisense technology, as they tend to

be expressed at higher levels than other viral genes and may sometimes be removed without abolishing replication and symptom development. In the above cases, protection was only observed at low levels of viral inoculum. Protection was assayed by symptom development or coat protein level and not viral RNA replication at the cellular level.

In other cases, antisense RNA was directed against different regions of CMV genome [Rezaian et al., 1988]. Little or no protection was observed. However, there is little symptom development with the host they used, and infection was assayed by RNA analysis, generally a more stringent test for infection than symptom development. Also in experiments with TRV transgenic plants expressing antisense RNAs to various regions of the viral genome have shown no resistance to viral infection [Baulcombe et al., 1987].

Recently Kawchuk et al. [1991] obtained transgenic potato plants that express the potato leafroll luteovirus (PLVR) coat protein gene in both sense and antisense orientations. When the plants were challenged with the virus, two of the transformants expressing the antisense RNA gave similar levels of resistance to those expressing the sense RNA. The amount of virus accumulated in these transgenic plants was less than 10% of the controls. When RNA and protein expression were analyzed, it was found that in the resistant transformants that contain the sense RNA the level of coat protein present was very low, although the level of RNA transcript was very high. If the coat protein is imparting resistance by itself (see above), then either it must be able to do so at extremely low levels or, alternatively, it is expressed at higher levels in the specific cells where the virus replicates, i.e., in phloem cells. The viral particle of PLVR contains a positive ssRNA genome. The fact that the antisense RNA provides a comparable level of protection as the sense RNA suggests that a similar mechanism of action may be sequestering the corresponding complementary positive or negative viral strand transcript. If this were the case, both sense and antisense RNA would work as an "antisense molecule.'' The construct used in this work contains both the coat protein and a region encoding a 17 kD protein that is thought to be a Vpg protein (a protein linked to the 5' end of the RNA genome used for the initiation of replication). The possible involvement of the Vpg protein or its RNA coding sequence in the protection against PLVR cannot be ruled out. Alternatively, this may be an example of the phenomenon of cosuppression, as discussed by Murray and Crockett (this volume).

IV. ENGINEERING RESISTANCE TO PLANT DNA VIRUSES

Plant DNA viruses divide into two classes, the caulimoviruses and the geminiviruses.

A. Caulimoviruses

The caulimoviruses are transmitted by aphids, and the host ranges of individual viruses of the group are restricted to only a few plant species. Their genome is a double-stranded (ds) circular DNA molecule that replicates through an RNA intermediate [Shepherd and Lawson, 1981; Bonneville et al., 1988]. The best-studied of these viruses is cauliflower mosaic virus (CaMV) [reviewed by Lichtenstein and Fuller, 1987; Gronenborn, 1987]. CaMV has a genome of approximately 8 kb and a host range limited mainly to cruciferous plants such as cauliflower and turnip. This virus encodes a powerful constitutive promotor, the 35S promotor, so-called because, following infection, its DNA genome is transcribed in the nucleus from this promotor to produce a 35S RNA transcript, which is greater than the length of the genome. This transcript serves as the template for synthesis of virion DNA by a virally encoded reverse transcriptase. In this sense, CaMV and the other caulimoviruses resemble the hepatitis B virus and retroviruses of vertebrates. The 35S promotor has been extensively used to drive expression of foreign DNA in transgenic plants.

B. Geminiviruses

Geminiviruses, viewed under the electron microscope, consist of twinned (hence the name) quasi-icosahedral particles of about 20 × 35 nm. Each particle contains a small, circular, ssDNA molecule of between 2.5 and 3.0 kb in size, encapsulated in a viral protein coat [reviewed by Stanley, 1985; Harrison, 1985; Lazarowitz, 1987; Davies, 1987; Davies et al., 1987; Davies and Stanley, 1989]. Geminiviruses are transmitted to plants by whitefly, infecting dicots, and leafhoppers, infecting predominantly monocotyledonous plants (monocots). The whitefly-transmitted geminiviruses have bipartite genomes in which each genome (typically designated DNA A and DNA B) is separately encapsulated. Thus successful infection requires that both genomes infect a plant. In contrast, leafhopper-transmitted geminiviruses are monopartite and, having only one genome, are thus approximately half the size of the bipartite ones.

Geminiviruses accumulate in the nuclei (where it is thought that they replicate), and most dicot-infecting geminiviruses are associated almost exclusively with the phloem cells, especially phloem parenchyma. Like all plant viruses described thus far, geminiviruses neither integrate into the plant genome nor are transmitted through the germ line.

Some members of this group of viruses cause significant diseases in crop plants, most notably in maize (e.g., maize streak virus [MSV]), wheat (wheat dwarf virus [WDV]), and other cereals (cassava latent virus [CLV], also known as African cassava mosaic virus [ACMV]) and in beet (beet curly top virus [BCTV]). Yet, when we began our work, there were no reports of engineering plants resistant to geminivirus infection.

The coat protein approach to engineering resistance, described above for some of the RNA viruses, has not been used to obtain resistance to geminivirus infection. This may be because the replication cycle of geminiviruses and the characteristics and functions of their coat proteins make it unlikely that overexpression of this protein in the plant would produce resistance. Additionally, there is no known example of natural cross-protection in geminiviruses.

Recently, Stanley et al. [1990] performed an experiment analogous to the satellite RNA approach. Plants infected by CLV can accumulate a family of related subgenomic (approximately half-size) components of DNA B. They constructed transgenic plants containing a tandem repeat of one such subgenomic DNA B. The subgenomic DNA was shown to excise from the plant genome and replicate following CLV infection. This led to reduction in replication of the full-length DNA A and DNA B and was associated with a decreased symptom severity. The mechanism of resistance is presumed to be by competition for replication.

V. ENGINEERING RESISTANCE TO GEMINIVIRUS

We decided to examine the potential of using antisense RNA to engineer resistance to a geminivirus. As it is believed that replication occurs in the nucleus, where the virus and cytopathological structures are found, we thought that there would be a greater likelihood of success than with an RNA virus replicating in the cytoplasm. Most models accounting for how antisense RNA suppresses gene expression in eukaryotes support this prejudice: These models include antisense RNA–DNA interaction interfering with transcription, interference of splicing of the duplex RNA, rapid degradation of the sense RNA–antisense RNA duplex, a block in the transport of duplex RNA into the cytoplasm, and direct inhibition of translation following duplex formation. Of these only the last must occur in the cytoplasm, though arguably duplex formation may be kinetically favored in the smaller volume of the nucleus than in the cytoplasm. The other models either require or strongly suggest a nuclear localized event.

A. Tomato Golden Mosaic Virus Genome Organization and Replication

We chose TGMV as our target. TGMV causes the disease "tomato golden mosaic" (Mosaico Dourado) of tomatoes, *Lycopersicum esculentum*, and is widespread in Brazil. As

suggested by its name (with some poetic license), the disease symptoms, necrotic leaf lesions, generate a mosaic of green and yellow areas (see Fig. 2). When we began our work this was one of the best-characterized of the geminiviruses: The complete nucleotide sequence of both genomes had been determined, there was some information on viral transcription patterns, and open reading frames had been assigned (and for some of these the genetic functions were known). Additionally, although tomatoes are the natural host, various species of *Nicotiana*, including tobacco, are also susceptible to infection.

Geminivirus research has been aided by several approaches: 1) DNA sequence comparisons of different viruses, drawing attention to common and perhaps necessary genetic components; 2) production of transgenic plants expressing isolated geminivirus genes, usually from constitutive promoters; and 3) analysis of geminivirus mutants obtained by site-directed mutagenesis. These genetic approaches require the successful introduction of such mutants into the plant. Although TGMV can be transmitted mechanically with naked DNA, usually plasmid DNA, carrying an at least partially duplicated insertion of geminivirus DNA, the efficiency is very low.

A better technique, used for the first time with CaMV [Grimsley et al., 1986], called *agroinoculation* or *agroinfection*, has enabled molecular genetic approaches to be taken to elucidate gene function. In this technique an *A. tumefaciens* strain is used to introduce, with high efficiency, a recombinant T-DNA containing an insertion of geminivirus DNA, typically by inoculation of the plant stem. In the case of the bipartite geminiviruses, e.g., TGMV, this can be done by mixed infection with *A. tumefaciens* strains carrying T-DNAs with genomes A and B [Hayes et al., 1988; Elmer et al., 1988b]. To produce infection by agroinoculation the only requirement is that both genomes must be present in at least a partial duplication. Following infection, the viral genomes excise from the T-DNA, circularize, and then enter the viral life cycle, leading to

Fig. 2. *Photograph showing the diagnostic symptoms of infection of* Nicotiana tabacum *by TGMV.*

systemic infection of the plant distant from the site of inoculation. The exact mechanism for the release of the viral monomer from the T-DNA is unknown. Two mechanisms have been proposed [Elmer et al., 1988b]: one via homologous recombination between the viral duplication and the other via replication that generates a ssDNA monomer.

These approaches have allowed infection, by agroinoculation, with mutated viral DNAs and complementation analyses to be performed. Agroinoculation has also allowed the development of geminivirus vectors [Davies and Stanley, 1989].

The genetic organization of TGMV is similar to that of all of the bipartite genome geminiviruses; the gene organization of the monopartite geminiviruses is related to the A genome of the bipartite viruses. Figure 3 shows a genetic map of TGMV DNA A and DNA B. DNA A alone is sufficient for DNA replication and encapsulation with coat protein; DNA

B alone cannot replicate but is required for cell–cell spread and symptoms. This was shown by using *A. tumefaciens* to clone dimers of DNA A into the nuclear genome of petunia plants (A2 plants): Freely replicating DNA A components were found [Rogers, et al., 1986] and shown to be encapsulated in virions of apparently normal morphology, but the plant had no symptoms. Trimers of TGMV B were similarly cloned into the chromosome of petunia (B3 plants), but no freely replicating DNA B molecules were found. However, crossing these A2 and B3 plants gave systemic infection in the F_1 heterozygotes. Also agroinoculation of A2 plants with *A. tumefaciens* strains containing B dimers gave complemented A2 plants, and agroinoculation of B3 plants with *A. tumefaciens* strains containing A dimers gave complementation and systemic infection [Sunter et al., 1987; Elmer et al., 1988b].

The two genomes have a noncoding region of about 200 base pairs that contains a poten-

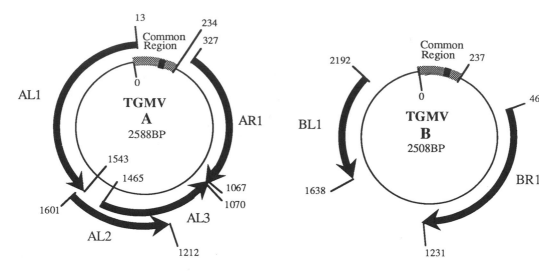

Fig. 3. *Maps of the DNA A and DNA B genomes of TGMV (TGMV A, TGMV B, respectively). Nucleotide positions of the six ORFs are marked. Only the coat protein genes AR1, and BR1 (transcribed in the rightward direction) are encoded on the viral plus strand, i.e., by the infecting viral ssDNA. The 200 bp region common to both TGMV A and TGMV B is shaded grey; within this, a palindromic sequence, yielding a potential stem–loop structure, is marked in black.*

tial stem–loop structure. This region, named the *common region*, is thought to be *cis*-essential for DNA replication, as its removal abolishes replication [Revington et al., 1989]. The common region also includes the promoters for transcription of two divergent transcripts. Though highly conserved between the two DNA components, the common region is not conserved between different geminiviruses, with the exception of the stem–loop structure.

The stem–loop structure consists of a highly G:C-rich stem flanking an A:T-rich loop; this structure is fairly well conserved in all geminiviruses sequenced thus far in which a nanonucleotide AT loop is invariant. It bears some similarity to a loop that is the recognition and cleavage site for the bacteriophage ØX174 gene A protein [Arai and Kornberg, 1981; Rogers et al., 1986]. Gene A protein is a site-specific nicking endonuclease essential for replication of ds RF I and the formation of ss from RF I and RF II by rolling circle replication. When total DNA is prepared from infected plant tissue, monomers of ss circular, ds open circular, and ds closed circular viral DNA forms are

found. Sometimes multimers, usually dimers, are also found. Thus, although the pathway for geminivirus replication has not been elucidated, these features suggest that it may resemble that of bacteriophage ØX174 and involve a rolling cycle mechanism. The stem–loop sequence was shown to be necessary for DNA replication, as an insertion of 8 bp into the loop in TGMV abolishes infectivity and drastically reduces DNA replication [Revington et al., 1989].

From Figure 3 it can be seen that DNA A encodes four open reading frames (ORFs). One rightward transcript, designated AR1, is encoded by the virion or plus strand DNA, while the leftward transcripts, AL1, AL2, and AL3, are encoded by the minus strand. DNA B encodes only two ORFs, one rightward and one leftward, designated BR1 and BL1, respectively. Both DNAs exhibit the complex bidirectional transcription pattern shown in Figure 4 [Sunter et al., 1989a,b; Sunter and Bisaro, 1989; Hanley-Bowdoin et al., 1989].

The AR1 ORF encodes the viral coat protein [Kallender et al., 1988]; deletion or muta-

TGMV A

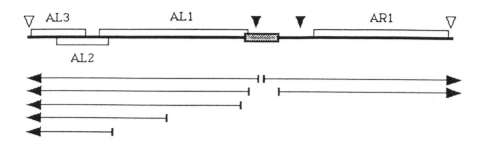

TGMV B

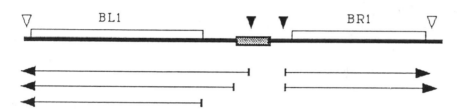

Fig. 4. *Transcription map of TGMV A and TGMV B showing the DNA map in a linear form and the viral transcripts (arrows) identified in the cells of infected plants. The common regions (dashed boxes), ORFs (open boxes), TATA boxes (closed triangles), and the polyadenylation signals (open triangles) are shown. The map is based in the data of Hanley-Bowdoin et al. [1988, 1989], Sunter and Bisaro [1989], and Sunter et al. [1989].*

tion of the coat protein gene causes no loss in infectivity, although symptom severity is diminished proportional to the length of the deletion. This has suggested that the efficiency of cell–cell movement is restricted by the size of the viral DNA [Gardiner et al., 1988; Hayes et al., 1988; Brough et al., 1988]. DNA molecules smaller than wild type would replicate in a plant cell, but transport from cell to cell would be delayed. Whether or not the coat protein of geminiviruses is involved in vascular spread in plants, there is increasing evidence that they play an important role in plant–plant spread by insects. The coat protein gene is well conserved at the amino acid level among the whitefly-transmitted viruses, but not between the white-

fly- and leafhopper-transmitted viruses [Stanley, 1985; Davies, 1987]. Indeed, the "missing link," BCTV, which is monopartite and leafhopper transmitted but infects dicots, has strong homology to component A of the whitefly-transmitted bipartite geminiviruses, except in the coat protein gene. This suggests that the coat protein may be predominantly involved in the determination of insect vector rather than host range. Direct evidence for this is shown by the substitution of the coat protein gene of ACMV (a whitefly-transmitted virus) by that of BCTV. This modified ACMV is now transmitted by the leafhopper [Briddon et al., 1990].

Viral DNAs, which are mutant in either BR1 or BL1, do not show systemic infectivity when inoculated [Hayes and Buck, 1989; Brough et al., 1988], suggesting that DNA B encodes the functions necessary for cell–cell spread and symptom development. When leaf disc inoculations of AL1/AL2 transgenic plants by mutant TGMV B molecules were carried out, it was observed that mutations in BL1 did not affect the amount of virus found, whereas mutations in BR1 led to a decreased yield of viral DNAs. This has been interpreted as suggesting that BR1 is necessary for cell–cell transmission (perhaps in addition to symptom development) and BL1 for symptom development alone [Hayes and Buck, 1989].

There is some homology between the coat protein gene, AR1, and BR1 in TGMV, CLV, and bean golden mosaic virus (BGMV), and it is possible that BR1 encodes functions necessary for systemic infection that are supplied by the coat protein in the monopartite geminiviruses. In support of this, in the monopartite viruses BCTV [Briddon et al., 1989] and MSV [Lazarowitz et al., 1989], the coat protein is essential for systemic infection, although it is dispensable in the dicot-infecting bipartite geminiviruses.

Viral replication is determined by the protein products of the AL1, AL2, and AL3 genes. These genes are all on the viral minus strand; therefore second strand synthesis must be initially carried out by host-encoded proteins. More detailed information on the functions of the AL1, AL2, and AL3 genes has come from the analysis of symptoms in plants and the amounts of viral DNA forms present in protoplasts, leaf discs, or whole plants infected by viral DNAs mutant in these genes [Elmer et al., 1988a; Gardiner et al., 1988; Brough et al., 1988; Hayes and Buck, 1989; Hanley-Bowdoin, et al., 1989; Sunter et al., 1990; Sunter and Bisaro, 1991].

The sequence of the AL3 gene can be disrupted without eliminating virus spread. However, there is a delay in symptom appearance that correlates with a decrease in ds and ss viral DNA accumulation after transfection of *Nicotiana tabacum* protoplasts [Sunter et al., 1990]. The function of AL3 protein remains unclear,

although a role in the activation of viral replication through interaction with the host replication proteins or AL1 protein has been proposed.

The products of the AL1 and AL2 genes are essential for systemic infection. However, AL2 mutants can replicate in protoplasts, producing the DNA forms typical of TGMV infection, although there is a 10-fold reduction in the viral ssDNA accumulation [Sunter et al., 1990]. The same results are obtained by infection of leaf discs, although the reduction of ssDNA is less pronounced [Elmer et al., 1988a,b]. The product of the AL2 gene could be an ssDNA binding protein that protects the viral ssDNA for its movement between cells. AL2 protein transactivates expression of the coat protein gene (AR1) and possibly one or both of the DNA B genes needed for viral spreading, giving another explanation for the phenotype of AL2 mutants [Sunter and Bisaro, 1991]. All AL1 mutants studied are unable to replicate in any of infection systems used.

These results and the coat protein deletion analysis suggest that the product of the AL1 gene and the common region are the only essential requirements for TGMV replication. It has been suggested that the AL1 product could have a similar function to gene A of ØX174, i.e., making a nick in the dsDNA (presumed to be the replicative form DNA) during viral replication by rolling-circle mechanism.

The homolog of the TGMV AL1 gene appears to be split into two exons in the monocot-infecting geminiviruses, as there is amino acid homology between two proteins apparently encoded by adjacent ORFs in these geminiviruses and the single protein encoded by the AL1 gene in the dicot-infecting geminiviruses (including the monopartite BCTV). Two ORFs in WDV [Schalk et al., 1989] are spliced by deletion of an intron and a frame shift to give a transcript encoding a single protein with homology to the product of the AL1 gene. It is possible that two polypeptides are expressed, one corresponding to the fused protein and another corresponding to the product of one of the unspliced mRNAs.

B. Targeting Tomato Golden Mosaic Virus With Antisense RNA

1. Choice of viral gene to target with antisense RNA. We chose the region encompassing the entire AL1 ORF but also including the 5' regions of two other ORFs, AL2 and AL3. At the time we made the choice, the evidence for the role of AL1 in replication was not as strong as it is now. It is fortunate that subsequent work has validated our choice of this target, because it is conserved in all geminiviruses, encodes a protein absolutely required for viral DNA replication, and is not abundantly transcribed [Elmer et al., 1988a; Hanley-Bowdoin et al., 1988, 1990].

2. Choice of signal sequence to initiate and terminate transcription. To suppress expression of a multicopy virus we felt that it would be necessary to obtain high levels of expression of the antisense RNA. Therefore, we used the strongest constitutive plant promoter available to maximize transcription, i.e., the 35S promoter of CaMV. This promoter contains an enhancer, and it has been shown that if this sequence is duplicated then expression levels increase by 3–10-fold [Kay et al., 1987]. Thus we used this enhancer duplication of the CaMV 35S promoter to increase the antisense transcript levels further. To terminate transcription, we used the polyadenylation and termination signal sequences of the T-DNA encoded octopine (an opine) synthase gene.

3. Choice of bifunctional construct. A common problem in transgenic plants is variations in the levels of expression of the input DNA. This is in part attributable to "position effects." However, levels of expression of unselected genes can vary considerably, even when the adjacent marker with which it is cotransferred, i.e., the antibiotic resistance gene used to select for transformation, is well expressed. To overcome this potential problem and to ensure further high levels of expression of the antisense RNA, we made bifunctional constructs in which the antisense sequence was fused to a drug resistance marker gene in a single transcription unit.

Such an approach was first taken by Kim and Wold [1985]. Mouse L cells were transformed with the herpes simplex virus (HSV) thymidine kinase (TK) gene and then retransformed with a bifunctional construct in which an antisense gene (complementary to the 3' end of the TK gene) was fused to the 5' end of a dihydrofolate reductase (DHFR) gene. The antibiotic methotrexate selects for expression of DHFR and hence for concomitant expression of anti-TK RNA on the same transcript. Cells resistant to increasing levels of methotrexate were selected by gradually increasing the concentration of methotrexate in the culture media. The increased resistance occurred by gene duplication, and high levels of methotrexate resistance resulted in high levels of antisense-containing transcripts. In their experiments, a high ratio of antisense–sense RNA was needed to obtain measurable levels of TK inhibition.

Concurrent with our work, Delauney et al. [1988] also made bifunctional constructs to overcome possible position effects in plants. Transgenic tobacco plants expressing chloramphenicol acetyltransferase (CAT) were retransformed with a bifunctional construct in which an anti-*cat* gene was fused 3' to the coding region of a hygromycin resistance gene (*hyg*); transcription of this bifunctional construct was driven by the CaMV 35S promoter. Between 30% and 50% of retransformed plants had decreased levels of CAT, in some cases to background level. A construct containing the whole of the CAT coding region appeared to be more efficient than one containing only the 5' terminal 172 bp. There was a strong correlation between mRNA levels and CAT levels. In the plants that showed no decrease in CAT activity, no antisense RNA was observed.

An additional rationale in our choice of a bifunctional construct was that, if the antisense region was cloned upstream of the drug marker used for selection of transformants, this would reduce the efficiency of translation of the latter; thus we would select for integration of the input DNA at "high expression sites." Experiments on mammalian cells show that a chimeric polycistronic gene comprising the DHFR coding region fused 3′ to a xanthine–guanine phosphoribosyl transferase (XGPRT) coding region gives 5–20-fold reduced translation relative to the monocistronic construct, depending on the construction [Peabody and Berg, 1986; Peabody et al., 1986; Kaufman et al., 1987]. As long as the downstream start codon is within 50 nucleotides (up or downstream) of the previous termination codon, the downstream gene is translated [Peabody et al., 1986]. However, if the distance is much greater than this, translation is decreased a further fivefold. When different upstream genes are used, the result is quantitatively different [Kaufman et al., 1987]. In this case downstream genes in di- or tricistronic constructs were expressed at between 100- and 300-fold less than when the genes were in monocistronic constructs or when they were the first coding region of a polycistronic construct. These results are consistent with the binding of ribosome to the cap site, with translation starting at the first AUG and continuing to the first stop codon, at which point the ribosome pauses and binds with reduced efficiency to a nearby AUG codon (if there is one) from which it translates a second polypeptide. The antisense sequence we chose contains 17 AUG start codons for very short reading frames. Thus this suggested that, if placed upstream of our selectable marker, the latter would be inefficiently translated.

C. Construction of Transgenic Plants Expressing an Antisense AL1 Gene

A 1,258 bp fragment from TGMV DNA A, comprising the complete coding region of AL1 and the 5′ regions of two other ORFs, AL2 and AL3, was cloned in either the sense or antisense orientation, both upstream and downstream, of the hygromycin resistance gene (hyg) in a plasmid plant transformation binary vector to yield the plasmids pP1AEN, pP5AEN, pP6AEN, and pP2AEN, shown in Figure 5. Tobacco leaf discs were transformed by A. tumefaciens. Surprisingly we were unable to obtain transformants with pP1AEN, despite repeated attempts; perhaps in this construct, where hyg is downstream of the sense AL1

region, it is not possible to get sufficient translation of the hyg gene to get resistance to hygromycin. It was also difficult to obtain transgenic plants from pP5AEN, as most transformants, while able to grow, failed to root on hygromycin selection. However, plant tissue transformed by pP5AEN did yield two antisense lines, P5AEN1 and P5AEN6, which were maintained for further analysis. Similarly, from pP2AEN five antisense plant lines were maintained, P2AEN1–5; and from pP6AEN two sense lines, P6AEN5 and P6AEN6, were maintained. Integration, structure, and copy number of the constructs in the transformed plants were examined by Southern blotting of genomic DNA; most lines appeared to have only one copy of the input DNA, which was inserted intact. However, P2AEN2 had 2–4 copies; P2AEN3, 6–10 copies (as possible head to tail dimers); P5AEN1, 6–10 copies; and P5AEN6, 4 copies. P6AEN5 and P6AEN6 appeared identical, so only P6AEN6 was analyzed further.

We next examined, by Northern blotting, the transcription of these chimeric bifunctional gene fusions. All transformed plants examined expressed a single full-length transcript of the predicted size (3.5 kb). This suggests that the transcript comprises the entire hyg-coding region and the complete AL1 (sense or antisense) sequence and that therefore the transcriptional signal sequences were used and there were no fortuitous signal sequences arising from the antisense orientation. Not surprisingly, different lines expressed different amounts of this RNA transcript. We quantitated this by RNA slot blot analysis using strand-specific probes to distinguish between sense and antisense transcripts and a probe against an endogenous tobacco transcript as an internal standard to calibrate expression. We observed a fivefold range in the expression of antisense RNA among the different transgenic plant lines.

1. Resistance to symptom development by agroinoculation of plant viruses. We challenged the transgenic lines with TGMV to see whether they had reduced susceptibility to infection. In our initial experiments, this was done

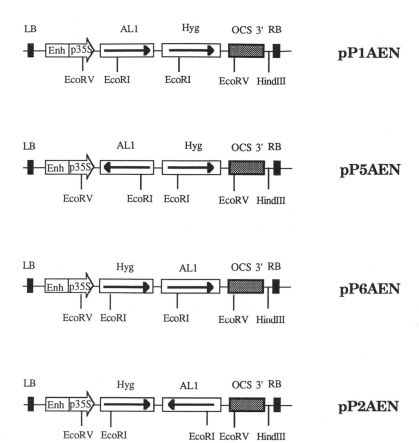

Fig. 5. *Physical map of the chimeric AL1 sense and AL1 antisense bifunctional* hyg *gene fusions in plasmids pP1AEN, pP2AEN, pP5AEN, and pP6AEN. RB and LB denote the flanking signal sequences required for transfer of T-DNA from* A. tumefaciens *to plant cells. In each plasmid, transcription of both the* hygromycin resistance gene, hyg, *and the AL1 antisense (or sense) gene is driven by the cauliflower mosaic virus (CaMV)–"enhanced" 35S promoter. The genes have the 3' transcriptional regulatory signals of the octopine synthase gene,* ocs. *Arrows indicate the sense and antisense orientations of the genetic units.*

by mechanical infection of plants with a viral pellet. This turned out to be a not very efficient method, as many of the controls failed to become infected. So we switched to using agroinoculation (see above) in a mixed infection with *Agrobacterium* strains carrying tandem repeats of both TGMV DNA A and DNA B genomes within T-DNA borders. Following infection of the stem, monomers of virion DNA, released from the T-DNA, replicate and spread systemically through the plant, giving rise to symptom development in leaves distant from the site of infection. This turned out to be a

considerably more efficient method for infection, as typically 70%–80% of control plants came down with viral symptoms. Necrotic leaf lesions were the symptoms we scored. These were fully developed 14 days after infection, as a second scoring carried out 7 days after this gave no further increase in symptom development or severity.

All the lines with the antisense AL1 downstream of *hyg* (P2AEN series) and one containing it upstream (P5AEN6) had significantly fewer symptom-bearing plants than the untransformed plants; moreover, the sense lines were as susceptible as the untransformed controls (Fig. 6). Some plants from antisense "resistant" lines did show symptoms, but these were typically of reduced severity compared with controls—fewer necrotic lesions per leaf and fewer leaves per plant with lesions.

2. DNA levels in infected plants. Symptoms develop when the viral titer reaches a threshold level. Below this threshold there may still be some systemic infection and hence viral DNA replication in infected tissue. As our antisense construct was designed to block such replication, we decided to evaluate the degree of suppression of replication. Viral DNA was quantitated from leaves of both plants that had developed symptoms and plants that had not. All the symptom-bearing plants contained high levels of all three viral DNAs typical of infection, essentially equivalent to the levels found both in untransformed controls and in the sense control. Analysis of viral DNAs from symptomless untransformed plants showed that 70% of these plants had viral DNAs present. The levels of these viral DNAs were, however, reduced to about 60% of those found in plants showing symptoms. Viral DNA was present in a similar percentage of symptomless plants from the P2AEN lines (antisense downstream of *hyg*); however, the amount of viral DNA was significantly less than that of the untransformed plants, ranging from about 20% to 38%. The P5AEN lines (antisense upstream of *hyg*) were found to be less resistant to developing symptoms. Interestingly, despite this, the suppression of viral DNA replication was found to be almost complete in the symptomless plants.

We also analyzed the data in another way: This was to calculate the average amount of viral DNAs summed together for all the individual plants of each line, i.e., plants both with and without symptoms. The average amount of viral DNAs of the untransformed control was set for reference at 100% to compare with the average amount of viral DNAs in the sense line and antisense lines. The data summarized in Figure 6 show a strong correlation between this average and the percentage of plants showing symptoms; in the more "resistant lines" (fewer

plants showing symptoms), there was correspondingly less viral DNA.

3. In vitro analysis of TGMV DNA replication using a leaf disc assay. Symptom development involves both cell–cell spread and viral DNA replication. To examine the effect on DNA replication in the absence of cell–cell spread, we performed in vitro leaf disc agroinoculation assays [Elmer et al., 1988a,b] with TGMV DNA A alone. This DNA can only replicate in the agroinoculated cell; it cannot spread without the DNA B–encoded functions. Analysis of untransformed leaf disc tissue gave three DNA species of free TGMV DNA A. In transgenic tissues of lines P2AEN1-4, no TGMV DNA A was detected; in P2AEN5, no TGMV DNA A was observed in one assay and very low levels in a duplicate assay. In P5AEN1 tissue, a slight reduction was observed.

D. Conclusions and Future Perspectives

Transgenic plant lines expressing an antisense AL1 transcript showed a statistically significant reduction in symptom development following TGMV agroinoculation. In the most resistant line, P2AEN4, only 5 of 57 plants (~9%) challenged displayed any symptoms at all, and in these 5 plants the symptoms were reduced; this compares with 76% of untransformed plants developing symptoms.

Our rationale in placing the antisense region upstream of the drug marker was that this should reduce the translational efficiency of the marker gene, and hence drug resistance should require correspondingly higher levels of RNA expression. Consistent with this, we obtained fewer transformants with this configuration than when the antisense AL1 gene was downstream of *hyg*, suggesting perhaps that many integration events resulted in too low a level of *hyg* expression to allow selection of hygromycin resistant transformants. It was thus an unexpected observation that the two lines expressing the most antisense (P2AEN3 and P2AEN4) RNA actually had the anti-AL1 downstream of *hyg*. However, our sample size of transgenic lines studied was too small to determine which configuration of anti-AL1 relative to *hyg* gives the high-

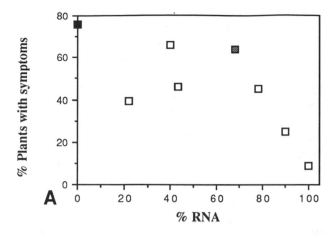

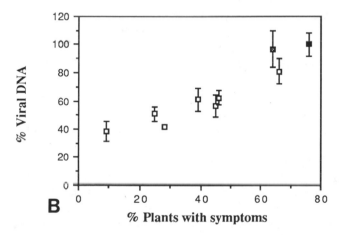

Fig. 6. *Correlation of antisense RNA levels with symptom development and accumulation of viral DNA. Percentage of plants that develop symptoms in untransformed (black square), sense (grey square), and antisense (empty squares) lines plotted versus the level of sense or antisense RNA found in the leaves (**A**) and versus the level of viral DNA present in the leaves (**B**). The levels of RNA are expressed as percentages relative to that of the highest expressing line. The levels of viral DNA are expressed as percentages relative to that of the untransformed plant.*

est level of antisense RNA expression, and so it is difficult to assess the significance of this observation.

We observed a broad correlation between antisense RNA levels and resistance (symptom development); those two lines expressing the most antisense RNA show greatest resistance (Fig. 6). Variation in gene expression between lines transformed by the same construct is well-documented in plant transformation experiments [Weising et al., 1988; Hobbs et al., 1990]; though there is no clear explanation, it is perhaps related to the transcriptional activity of the region surrounding the integration site of the input DNA, the so-called position effects described above. We observed no correlation between copy number of the integrated DNA and either degree of resistance or level of antisense RNA expression; though this may seem at first surprising, it suggests that position effects are more important in determining gene expression than gene dosage. Perhaps in those lines with extra copies not all were contributing to expression.

Presumably, a threshold level of virus must accumulate in infected tissue to give symptoms. Below that threshold, no symptoms may be evident despite the presence of virus. Thus it is not surprising that many of the plants that did not display symptoms did have some viral DNA replication. However there was a reduction in the amount of viral DNA accumulating in the leaf tissue of antisense plant lines, showing that reduced symptom development and the degree of reduction in viral DNA did correlate with the degree by which symptom development was reduced (Fig. 6). What was surprising was that the lines with the antisense AL1 region upstream of *hyg* were less resistant, with respect to symptom development, than those antisense lines with the AL1 region downstream of *hyg*, yet when we looked at the plants in each group without symptoms, the degree of suppression of viral DNA replication in the upstream lines'' was more pronounced. We have no explanation for this observation.

Though AL1 is required for DNA replication, the antisense constructs might confer resistance by prevention of cell–cell spread. The leaf disc agroinoculation experiments, using only the TGMV A genome, preclude cell–cell spread and allow examination of viral DNA replication only in the infected cells; these experiments show clearly that the antisense AL1 RNA blocks replication. Moreover, the degree of such blocking correlates with the degree of symptom development measured in whole plants. The block in viral DNA replication in the leaf disc assay was very much stronger (in most cases absolute) than in whole plants that were agroinoculated. We do not believe that this is an important discrepancy, as there are differences between the two situations; e.g., 1) there may be physiological differences between isolated leaf discs in tissue culture and whole plants that modulate their response to infection, and 2) as we agroinoculated whole plants with both TGMV genomes, viral DNA escaping from T-DNA,

following agroinoculation, is free to spread from cell to cell in contrast to agroinoculation of leaf disc with the DNA A alone.

Although we did not obtain complete resistance, the mode of infection we used, agroinoculation, is probably more stringent than natural infection by the whitefly. In the former mode, the integrated T-DNA of the transformed cells presumably provides a constant source of TGMV DNA, thus allowing multiple attempts by the virus to overcome the antisense RNA and enter the viral replication cycle. In the latter case, the TGMV DNA, introduced by the whitefly, does not integrate into the plant genome and thus is presumably present only transiently. Thus, if it initially fails to replicate, it probably disappears. We have not been able to address this difference more directly by evaluating the degree of resistance of the antisense lines to infection by whitefly; these whitefly are not endemic to the United Kingdom, and in any case the natural host for TGMV is tomato, not tobacco. However, we predict that antisense plants would be yet more resistant to infection by whitefly; thus we believe our results are very encouraging and suggest that the use of antisense RNA technology may find application in the control of economically important geminivirus diseases.

The work we have described leaves open a number of interesting questions that we are currently addressing:

1. We have no detailed information on the sequence of events by which antisense RNA confers virus resistance to the transgenic plants. This question can be addressed by analyzing the distribution and stability of the mRNA and antisense RNA of AL1 in TGMV-infected transgenic plant tissue at different times following infection and by comparing this to levels of viral DNA accumulation and symptom development. We also do not know what the major block to the viral life cycle is, i.e., whether it is the production of ssDNA or dsDNA and, if ssDNA, whether it is encapsulated virion DNA or systemically transferred

ssDNA that may not be encapsulated as virus for traffic within the plant.

2. What is the most effective region to target antisense RNA? We have not formally proved that it is the AL1 antisense region that confers resistance. We chose a region comprising almost the entire AL1 transcript, but this region also overlaps with two other TGMV genes, AL2 and AL3. Perhaps, given the role of AL2 and AL3 in determining levels of viral DNAs and the time of onset of symptoms, the AL2 and AL3 antisense regions expressed in our constructs are also contributing to a diminution of the yield of viral DNA. This can be addressed by generating plants containing isolated regions of the original construct.

3. What other factors contribute to the degree of virus resistance? This follows from the previous point: We believe that our bifunctional constructs allowed a selection for transgenic lines that express high levels of antisense RNA, either intrinsic to the *hyg* selection or by RNA stabilization. We have constructed monofunctional antiAL1 transgenic plant lines and plan to challenge these with TGMV, concurrently performing analyses of steady-state RNA levels and de novo transcription and comparing them to the original bifunctional constructs.

4. Can the TGMV AL1 antisense gene confer resistance to other geminiviruses? The AL1 gene is conserved in this family of plant viruses, so it may be that there is sufficient sequence homology to allow antisense RNA to suppress other viral AL1 gene expression and hence viral replication. This would not be without precedent, since the antisense chalcone synthase gene from petunia can suppress the endogenous tobacco gene in transgenic tobacco [van der Krol et al., 1988]. A variety of geminivirus DNAs are available and DNA sequences known, so it will be possible to compare the degree of DNA sequence homology shown to the TGMV AL1 gene and to correlate this with the degree of "virus resistance" monitored by the leaf disc assay and symptom development in whole plants.

5. Can our antisense constructs confer resistance to TGMV infection of tomatoes? Although

tomatoes are the natural host for TGMV, we chose tobacco as a convenient model system. It will be interesting, and indeed useful, to extend our observations to tomato plants and make them resistant to their natural pathogen.

ACKNOWLEDGMENTS

We acknowledge the support of the Science and Engineering Research Council (AGD) and the European Community and European Molecular Biology Organisation for senior fellowships (E.R.B.).

VI. REFERENCES

An G (1986): Development of plant promoter expression vectors and their use for analysis of differential activity of nopaline synthase promoter in transformed tobacco cells. Plant Physiol 81:86–91.

Arai K, Kornberg A (1981): Unique primed start of phage ØX174 DNA replication and mobility of the primasome in a direction opposite chain synthesis. Proc Natl Acad Sci USA 78:69–73.

Baulcombe D (1989): Strategies for virus resistance in plants. Trends Genet 5:56–60.

Baulcombe DC, Hamilton WDO, Mayo MA, Harrison BD (1987): Plant Resistance to Viruses. CIBA Foundation Symposium 133. New York: John Wiley, pp 170–184.

Baulcombe DC, Saunders GR, Bevan MW, Mayo MA, Harrison BD (1986): Expression of biologically active viral satellite RNA from the nuclear genome of transformed plants. Nature 321:446–449.

Bonneville JM, Hohn T, Pfeiffer P (1988): Reverse transcription in the plant virus, cauliflower mosaic virus. In Domingo E, Holland JJ (eds): RNA Genetics. Boca Raton, FL: CRC Press, pp 24–42.

Briddon RW, Pinner MS, Stanley J, Markham PG (1990): Geminivirus coat protein gene replacement alters insect specificity. Virology 177:85–94.

Briddon RW, Watts J, Markham PG, Stanley J (1989): The coat protein of beet curly top virus is essential for infectivity. Virology 172:628–633.

Brough CL, Hayes RJ, Morgan AJ, Coutts RHA, Buck KW (1988): Effects of mutagenesis in vitro on the ability of cloned tomato golden mosaic virus DNA to infect *Nicotiana benthamiana* plants. J Gen Virol 69:503–514.

Cuozzo M, O'Connell KM, Kaniewski W, Fang RX, Chua NH, Turmer NE (1988): Viral protection in transgenic tobacco plants expressing the cucumber mosaic virus coat protein or its antisense RNA. BioTechnology 6:549–557.

Davies JW (1987): Geminivirus genomes. MicroSciences 4:18–23.

156   Bejarano et al.

Davies JW, Stanley J (1989): Geminivirus genes and vectors. Trends Genet 5:77–81.

Davies JW, Stanley J, Donson J, Mullineaux PM, Boulton MI (1987): Structure and replication of geminiviruses genomes. J Cell Sci Suppl 7:95–107.

Day AG, Bejarano ER, Burrell M, Buck KW, Lichtenstein CP (1991): Expression of antisense viral gene in transgenic tobacco plants confers resistance to the DNA virus, tomato golden mosaic virus. Proc Natl Acad Sci USA 88:6721–6725.

Delauney A, Tabaeizadeh Z, Verma D (1988): A stable bifunctional antisense transcript inhibiting gene expression in transgenic plants. Proc Natl Acad Sci USA 85:4300–4304.

Elmer JS, Brand L, Sunter G, Gardiner WE, Bisaro D, Rogers SG (1988a): Genetic analysis of the tomato golden mosaic virus II. The product of the AL1 coding sequence is required for replication. Nucleic Acids Res 16:7043–7060.

Elmer JS, Sunter G, Gardiner WE, Brand L, Browing CK, Bisaro D, Rogers SG (1988b): *Agrobacterium*-mediated inoculation of plants with tomato golden mosaic virus DNAs. Plant Mol Biol 10:225–234.

Francki RIB (ed) (1985): The Plant Viruses. In Fraenkel-Conrat H, Wagner RR (eds): Viruses Series. New York: Plenum Press, vol I, pp. 1–15.

Francki RIB, Milne RG, Hatta T (1985): Atlas of Plant Viruses. Boca Raton, FL: CRC Press.

Gardiner WE, Sunter G, Brand L, Elmer JS, Rogers SG, Bisaro DM (1988): Genetic analysis of tomato golden mosaic virus: The coat protein is not required for systemic spread or symptom development. EMBO J 4:899–904.

Gerber M, Sarkar S (1989): The coat protein of tobacco mosaic virus does not play a significant role for cross protection. J Phytopathol 124:323–331.

Gerlach WL, Llewellyn D, Haseloff J (1987): Construction of plant disease resistance gene from the satellite RNA of tobacco ringspot virus. Nature 328:802–805.

Golemboski DB, Lomonossoff GP, Zaitlin M (1990): Plants transformed with a tobacco mosaic virus nonstructural gene sequence are resistant to the virus. Proc Natl Acad Sci USA 87:6311–6315.

Grimsley N, Hohn B, Hohn T, Walden R (1986): Agroinfection: An alternative route for viral infection of plants using Ti plasmid. Proc Natl Acad Sci USA 83:3282–3286.

Gronenborn B (1987): The molecular biology of cauliflower mosaic virus and its application as plant vector. In Hohn T, Schell J (eds): Plant DNA Infectious Agents. New York: Springer-Verlag, pp 1–30.

Hanley-Bowdoin L, Elmer JS, Rogers SG (1988): Transient expression of heterologous RNAs using tomato golden mosaic virus. Nucleic Acids Res 16:10511–10528.

Hanley-Bowdoin L, Elmer JS, Rogers SG (1989): Functional expression of the leftward open reading frames of the A component of tomato golden mosaic virus in transgenic tobacco plants. Plant Cell 1:1057–1067.

Hanley-Bowdoin L, Elmer JS, Rogers SG (1990): Expression of functional replication protein from tomato golden mosaic virus in transgenic tobacco plants. Proc Natl Acad Sci USA 87:1446–1450.

Harrison BD (1985): Advances in geminivirus research. Annu Rev Phytopathol 23:55–82.

Harrison BD, Mayo MA, Baulcombe DC (1987): Virus resistance in transgenic plants that express cucumber mosaic virus satellite RNA. Nature 328:799–802.

Hayes RJ, Buck KW (1989): Replication of tomato golden mosaic virus DNA-B in transgenic plants expressing open reading frames (ORFs) of DNA A: Requirement of ORF-A12 for production of single-stranded DNA. Nucleic Acids Res 17:10213–10222.

Hayes RJ, Coutts RHA, Buck KW (1988): Agroinfection of *Nicotiana* spp. with cloned DNA of tomato golden mosaic virus. J Gen Virol 69:1487–1496.

Hemenway C, Fang RX, Kaniewski WK, Chua NH, Turner NE (1988): Analysis of the mechanism of protection in transgenic plants expressing the potato virus X coat protein or its antisense RNA. EMBO J 7:1273–1280.

Herrera-Estrella L, Simpson J (1988): Foreign gene expression. In Shaw CHS (ed): Plant Molecular Biology. Oxford: IRL Press, pp 131–158.

Hobbs SLA, Kpodar P, DeLong MO (1990): The effect of T-DNA copy number, position and methylation on reporter gene expression in tobacco transformants. Plant Mol Biol 15:851–864.

Hooykaas PJJ (1989): Tumorigenicity of *Agrobacterium* in plants. In Hopwood DA, Chater KE (eds): Genetics of Bacterial Diversity. London: Academic Press, pp 373–391.

Kallender H, Petty ITD, Stein VE, Panico M, Blench IP, Etienne AT, Morris HR, Coutts RHA, Buck KW (1988): Identification of the coat protein gene of tomato golden mosaic virus. J Gen Virol 69:1351–1357.

Kaufman J, Murtha P, Davies M (1987): Translation efficiency of polycistronic mRNAs and their utilization to express heterologous genes in mammalian cells. EMBO J 6:187–193.

Kawchuk LM, Martin RR, McPherson J (1991): Sense and antisense RNA-mediated resistance to potato leafroll virus in russet burbank potato plants. Mol Plant Microbe Interact 4:247–253.

Kay R, Chan A, Daly M, McPherson J (1987): Duplication of CaMV 35S promoter sequences creates a strong enhancer for plant genes. Science 236:1299–1302.

Kim S, Wold B (1985): Stable reduction of thymidine kinase activity in cells expressing high levels of antisense RNA. Cell 42:129–138.

Koening R (1988): Plant Viruses. New York: Plenum Press, vol III.

Lawson C, Kaniewski W, Haley L, Rozman R, Newell C, Sanders P, Tumer NE (1990): Engineering resistance to multiple virus infection in a commercial potato cultivar: Resistance to potato virus X and potato virus Y in transgenic russet burbank potato. BioTechnology 8:127–134.

Lazarowitz SG (1987): The molecular characterization of geminiviruses. Plant Mol Biol Rep 4:177–192.

Lazarowitz SG, Pinder AJ, Damsteegt VD, Rogers SG (1989): Maize streak virus genes essential for systemic spread and symptom development. EMBO J 8: 1023–1032.

Lichtenstein C, Fuller SL (1987): Vectors for the genetic engineering of plants. In Rigby PWJ (ed): Genetic Engineering 6. London: Academic Press, pp 104–171.

Memelink JS, Harry J, Hoge C, Schilproort RA (1987): T-DNA hormone biosynthetic genes: Phytohormones and gene expression in plants. Dev Genet 8:321–337.

Milne RG (1988): Plant Viruses. New York, Plenum Press, vol IV.

Nelson RS, Powell AP, Beachy RN (1987): Lesions and viral accumulation in inoculated transgenic tobacco plants expressing the coat protein of tobacco mosaic virus. Virology 158:126–132.

Peabody D, Berg P (1986): Termination-reinitiation occurs in the translation of mammalian cell mRNAs. Mol Cell Biol 6:2695–2703.

Peabody D, Subramani S, Berg P (1986): Effect of upstream reading frames on translational efficiency in SV40 recombinants. Mol Cell Biol 6:2074–20711.

Powell PA, Nelson RS, Hoffman N, Rogers SG, Fraley RT, Beachy RN (1986): Delay of disease development in transgenic plants that express the tobacco mosaic virus coat protein gene. Science 232:738–743.

Powell PA, Stark DM, Sanders PR, Beachy RN (1989): Protection against tobacco mosaic virus in transgenic plants that express tobacco mosaic virus antisense RNA. Proc Natl Acad Sci USA 86:6949–6952.

Revington GN, Sunter G, Bisaro D (1989): DNA sequences essential for replication of the B genome component of tomato golden mosaic virus. Plant Cell 1:985–992.

Rezaian MA, Skene KGM, Ellis JG (1988): Anti-sense RNAs of cucumber mosaic virus in transgenic plants assessed for controls of the virus. Plant Mol Biol 11:463–471.

Rogers SG, Bisaro DM, Horsch RB, Fraley RT, Hoffmann NL, Brand L, Elmer JS, Lloyd AM (1986): Tomato golden mosaic virus A component DNA replicates autonomously in transgenic plants. Cell 45:593–600.

Schalk HJ, Matzeit V, Schiller B, Schell J, Gronenborn B (1989): Wheat dwarf virus, a geminivirus of graminaceous plants needs splicing for replication. EMBO J 8:359–364.

Semancik JS (1987): Viroids and Viroid-Like Pathogens. Boca Raton, FL: CRC Press.

Shepherd RJ, Lawson RH (1981): Caulimoviruses. In Kurstak E (ed): Handbook of Plant Virus Infections

and Comparative Diagnosis. Amsterdam: Elsevier, pp. 847–877.

Stanley J (1985): The molecular biology of geminiviruses. Adv Virus Res 30:139–177.

Stanley J, Frischmuth T, Ellwood S (1990): Defective viral DNA ameliorates symptoms of geminivirus infection in transgenic plants. Proc Natl Acad Sci USA 87: 6291–6295.

Sunter G, Bisaro DM (1989): Transcription map of the B genome component of tomato golden mosaic virus and comparison with A component transcripts. Virology 173:1–9.

Sunter G, Bisaro DM (1991): Transactivation in a geminivirus: AL2 gene product is needed for coat protein expression. Virology 180:416–419.

Sunter G, Gardiner WE, Bisaro DM (1989a): Identification of tomato golden mosaic virus–specific RNAs in infected plants. Virology 170:243–250.

Sunter G, Gardiner E, Bisaro DM (1989b): Identification of tomato golden mosaic virus-specific RNAs in infected plants. Virology 170:243–250.

Sunter G, Gardiner WE, Rushing AE, Rogers SG, Bisaro DM (1987): Independent encapsidation of tomato golden mosaic virus A component DNA in transgenic plants. Plant Mol Biol 8:477–484.

Sunter G, Hartitz MD, Hormuzdi SG, Brough CL, Bisaro DM (1990): Genetic analysis of tomato golden mosaic virus—ORF-AL2 is required for coat protein accumulation while ORF-AL3 is necessary for DNA replication. Virology 179:69–77.

Tumer NE, O'Connell KM, Nelson RS, Sanders PR, Beachy RN, Fraley RT, Shah DM (1987): Expression of alfalfa mosaic virus coat protein gene confers cross-protection in transgenic tobacco and tomato plants. EMBO J 6:1181–1188.

van den Elzen PJM, Huisman MJ, Willink DPL, Jongedijk E, Hoekema A, Cornelissen BJC (1989): Engineering virus resistance in agricultural crops. Plant Mol Biol 13:337–346.

van der Krol AR, Lenting PE, Veenstra J, van der Meer I, Koes RE, Gerats AGM, Mol JNM, Stuitje KR (1988): An antisense chalcone synthase gene in transgenic plants inhibits flower pigmentation. Nature 333:866–869.

van Dun CMP, Bol JF (1988): Transgenic tobacco plants accumulate tobacco rattle virus coat protein, resist infection with tobacco rattle virus and pea early browing virus. Virology 161:649–652.

van Dun CMP, Overduin B, van Vloten-Doting L, Bol JF (1988): Transgenic tobacco expressing tobacco streak virus or mutated alfalfa mosaic virus coat protein does not cross-protect against alfalfa mosaic virus infection. Virology 164:383–389.

van Regenmortet MHV, Fraenkel-Conrat H (1988): Plant Viruses. New York: Plenum Press, vol II.

Weising K, Schell J, Kahl G (1988): Foreign genes in plants: Transfer, structure, expression and applications. Annu Rev Genet 22:421–177.

Wisniewski LA, Powell PA, Nelson RS, Beachy RN
(1990): Local and systemic spread of tobacco mosaic
virus in transgenic tobacco. Plant Cell 2:559–567.

Zambryski P (1988): Basic processes underlying *Agrobacterium*-mediated DNA transfer to plant cells. Annu
Rev Genet 22:1–30.

ABOUT THE AUTHORS

EDUARDO R. BEJARANO is Associate Lecturer in the Department of Genetics at Malaga University
(Spain) where he teaches general and molecular genetics. He received his B.A. in Biology from the University
of Córdoba in 1982. He then moved to Seville University to work under Enrique Cerdá-Olmedo. He worked in
fungal genetics and carotene biosynthesis and obtained his Ph.D. in 1988. Dr. Bejarano then pursued postdoctoral research in the laboratory of Conrad Lichtenstein at the Imperial College of Science, Technology and
Medicine in London. There he came in contact with plant molecular biology and the use of the antisense RNA
technology to engineer virus resistance in plants. Dr. Bejarano's current research involves the improvement of
plant protection against virus and the analysis of plant genomic DNA sequences homologous to viral genes.

ANTHONY G. DAY is a postdoctoral fellow in the laboratory of Professor Alan R. Fersht at Cambridge
University. He is studying the enzymes ribulose bisphosphate carboxylase/oxygenase and barnase (a small
ribonuclease), using site-directed mutagenesis to probe structure and function. After receiving his B.Sc. from
Portsmouth Polytechnic in 1981, he received an M.Sc. under the supervision of Steven Withers at the University of British Columbia, working on the kinetic characterization of a β-glucosidase. Following this, Dr. Day
embarked on a Ph.D. under the supervision of Conrad Lichtenstein at the Imperial College of Science, Technology and Medicine, where he engineered tobacco plants to express antisense RNA directed against tomato
golden mosaic virus and successfully conferred resistance to this virus on the transgenic plants. Dr. Day is the
proud owner of three lizards, a royal python, and approximately 20,000 honeybees.

CONRAD P. LICHTENSTEIN is Lecturer in Genetic Engineering at the Imperial College of Science,
Technology and Medicine, University of London, where he teaches prokaryotic molecular genetics and plant
molecular biology. After receiving his B.Sc. (Hons.) from the University of Sussex in 1975, he received his
Ph.D. under Sydney Brenner at the MRC laboratory of Molecular Biology, Cambridge, England, where he
studied bacterial transposons. Dr. Lichtenstein then pursued postdoctoral research in the laboratory of Eugene
Nester at the University of Washington, Seattle. Here he studied the tumor-inducing genes of *Agrobacterium
tumefaciens*, the causative agent of crown gall disease in plants. Dr. Lichtenstein's current research involves
the development of antisense RNA technology and homologous recombination as tools for reverse genetics,
both to study plant development and to engineer crop improvements. He is also studying the mechanism of
DNA transposition in bacteria. His research papers have appeared in such journals as *Nature*, *Cell*, the *Journal Molecular Biology*, *Nucleic Acids Research*, and the *Proceedings of the National Academy of Sciences
USA*. Dr. Lichtenstein is News and Views editor of *Plant Molecular Biology*.

Antisense RNA and DNA: 159–174
© 1992 Wiley-Liss, Inc.

The Double-Stranded RNA Unwinding/ Modifying Activity

Brenda L. Bass

I. INTRODUCTION

Several years ago, a novel biological activity was discovered in eggs and embryos of the South African clawed toad, *Xenopus laevis* [Rebagliati and Melton, 1987; Bass and Weintraub, 1987]. The activity requires a double-stranded RNA (dsRNA) substrate and acts to convert adenosine (A) residues within its substrate to inosine (I) residues (Fig. 1) [Bass and Weintraub, 1988; Wagner et al., 1989]. Since IU base pairs are less stable than AU base pairs [Kawase et al., 1986; Lomant and Fresco, 1975], the dsRNA substrate becomes increasingly single-stranded during the reaction [Bass

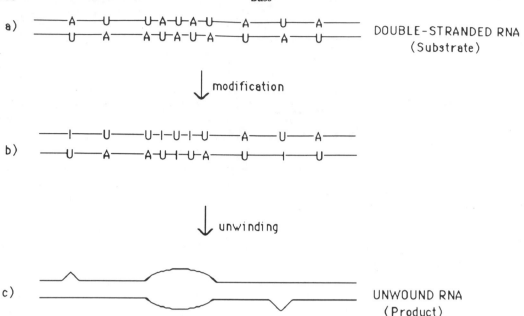

Fig. 1. *Diagram outlining the modification of a dsRNA molecule by the unwinding/modifying activity. Only the adenosine and uridine residues of a random sequence duplex (a) are shown. After the modification has occurred (b), the unstable IU base pairs are thought to result in an increase in the single-stranded charac-ter of the RNA. It has been proposed (see Section V) that the final product (c) consists of internal single-stranded loops (highly modified regions) interspersed with sequences that remain largely double stranded (sparsely modified regions).*

and Weintraub, 1988; Wagner et al., 1989]. The ability of the activity to "unwind" its dsRNA substrate locally was realized prior to the detection of the modified base; for this reason, the activity was originally referred to as a *dsRNA unwinding activity* [Rebagliati and Melton, 1987; Bass and Weintraub, 1987]. At present, the biological role of the activity is unknown. Although the role of the modification may be to unwind the dsRNA, the discovery that the activity covalently modifies its RNA substrate raises the possibility that the primary function of the activity is to modify rather than unwind; in light of these recent data the activity has been called an *unwinding/modifying activity* [Bass and Weintraub, 1988].

Since antisense techniques rely on the formation of a stable duplex, the existence of the activity in *Xenopus* early embryos was thought to explain why antisense RNA techniques are

unsuccessful in these cells [Bass and Weintraub, 1987; Rebagliati and Melton, 1987]. The discovery that adenosines are modified to inosines during the reaction makes this interpretation questionable. It is known that inosines are read as guanosines during translation [Basilio et al., 1962]; the modification would be expected to change the coding capacity of the RNA drastically and therefore produce a nonfunctional protein. Thus it seems more likely that the activity would contribute to the success of antisense RNA techniques. In fact, as I will discuss later, certain data obtained in the course of antisense RNA experiments are consistent with this hypothesis. At present, the reason why antisense experiments are unsuccessful at certain stages during early *Xenopus* embryogenesis remains a mystery.

My goal is to clarify what is currently known about the unwinding/modifying activity. In particular, I have tried to cover those aspects that may aid persons attempting to demonstrate the involvement of the unwinding/modifying activity in their particular system. The relationship of the activity to the mechanism of antisense RNA inhibition is discussed at the end.

II. LOCALIZATION

A. Interspecies

The unwinding/modifying activity was originally detected by microinjecting radiolabeled dsRNA into the cytoplasm of *Xenopus* eggs or pre-mid-blastula transition (pre-MBT) embryos [Bass and Weintraub, 1987; Rebagliati and Melton, 1987]. At that time it was also shown that the activity was present in whole cell extracts prepared from these cells [Bass and Weintraub, 1987]. Subsequently, the activity was observed in whole cell extracts prepared from *Caenorhabditis elegans* embryos (M. Krause, unpublished results), *Drosophila melanogaster* embryos (B. Bass, unpublished results), *Drosophila melanogaster* tissue culture cells (J. Yin, personal communication), and numerous mammalian tissues as well as cultured cells [Wagner and Nishikura, 1988; Wagner et al. 1990]. Attempts to detect the activity in *Saccharomyces cerevisae* have failed (B. Bass, unpublished results), and, to my knowledge, extracts derived from a prokaryote or a plant have not yet been assayed. Thus, at present, the unwinding/modifying activity appears to be ubiquitous among the various phyla of the animal kingdom, but has not been detected elsewhere.

B. Intracellular

Although the activity was originally observed in the cytoplasm of *Xenopus* embryos, it appears that at many stages of *Xenopus* development, if not the majority, the activity is localized to the nucleus [Bass and Weintraub, 1988; B. Bass, unpublished results]. The intracellular localization of the activity has been determined at various stages during early *Xenopus* devel-

opment, and the existing data are summarized schematically in Figure 2. Microinjection experiments indicate that the activity resides in the nucleus (germinal vesicle) of the stage 6 oocyte, is released into the cytoplasm during meiotic maturation, and remains in the cytoplasm during fertilization and early embryogenesis. It is not yet known exactly when the activity returns to the nucleus, but subsequent to the MBT the activity cannot be detected in the cytoplasm [Rebagliati and Melton, 1987]. Embryos between stages 17 and 30 have been fractionated and the nuclear and cytoplasmic components assayed; the activity is detectable only in the nuclear fraction (B. Bass, unpublished results).

Since breakdown of the oocyte nucleus is a requisite event of meiotic maturation, the detection of the activity in the cytoplasm of the early embryo may merely reflect a lag in its relocalization to the nucleus. It is possible that the unwinding/modifying activity is typical of a number of other proteins found in the oocyte germinal vesicle that, after release into the cytoplasm during meiotic maturation, return to the nucleus at characteristic developmental stages [Dreyer and Hausen, 1983]. It remains to be seen whether the activity will exert its effect primarily in the cytoplasm, the nucleus, or throughout the cell.

The nuclear localization of the activity has led to some confusion. For example, based on results from microinjecting dsRNA into the cytoplasm of *Xenopus* cells, it was concluded that the activity was present in eggs and early stages of embryogenesis, but not in oocytes or post-MBT embryos [Bass and Weintraub, 1987; Rebagliati and Melton, 1987]. This led to the suggestion that the expression of the activity was developmentally regulated. It is now known that the activity is present even in oocytes and post-MBT embryos (see Fig. 2), but is localized to the nucleus and thus not detected by microinjection into the cytoplasm.

Initially, mammalian cells were assayed using whole cell extracts [Wagner and Nishikura, 1988; Wagner et al., 1989]. However, more recently, extracts prepared from nuclei have been

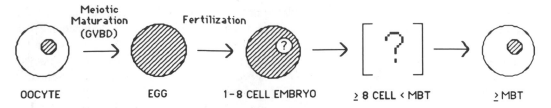

Fig. 2. *Schematic representation of the intracellular localization of the unwinding/modifying activity during early* Xenopus *development. Diagonal lines indicate localization of the activity at various developmental stages. A question mark is used to indicate stages or organelles of a particular stage that have not yet been assayed. See text for details.*

assayed [Wagner et al., 1990]. The activity has been detected in the nuclear fraction of extracts prepared from 15-day-old rat brain, 14-day-old rat astrocyte cultures derived from 1-day-old rat cerebra, and three cell lines: the rat Schwann cell line D6P2T and the glial cell lines hamster HJC and rat C6. It has been speculated that this trend will continue, and in mammals the activity will be localized to the nucleus in all somatic cells [Wagner et al., 1990]. As in the work with *Xenopus*, the nuclear localization initially led to some confusion [Wagner et al., 1990]. For example, assays of whole cell extracts led to the conclusion that certain mammalian cell lines had low or undetectable levels of the activity. It was subsequently determined that the activity is easily detected in nuclear extracts prepared from the same cells. Apparently, in certain cell lines the activity cannot be extracted from the nucleus with methods for preparing whole cell extracts.

Experiments in which dsRNA was microinjected into the cytoplasm of mouse embryos at the one cell stage led to the conclusion that these cells do not contain the unwinding/modifying activity [Bevilacqua et al., 1988] (see also Levy et al., this volume). This result is inconsistent with more recent studies in which the activity is clearly detectable in whole cell extracts prepared from mouse tissues or cultured cells [Wagner et al., 1990]. It seems possible that, as with early experiments in *Xenopus*, the activity was overlooked because of its nuclear localization. Although experiments with *Xenopus* cells indicate that the unwinding/modifying activity is still in the cytoplasm at the eight cell stage [Bass and Weintraub, 1987], it is possible that relocalization to the nucleus is temporally different in the mouse. In fact this seems probable, since early development occurs on a very different time scale in the mouse.

III. BIOCHEMICAL REQUIREMENTS

The unwinding/modifying activity has not yet been purified, and very little is known about its biochemical requirements. In crude whole cell extracts, the activity is sensitive to proteinase K and does not exhibit a requirement for ATP or for any other nucleoside triphosphate energy source [Bass and Weintraub, 1987; Wagner and Nishikura, 1988]. The activity is stimulated by the addition of magnesium, but, surprisingly, it is also stimulated by EDTA [Bass and Weintraub, 1987; Wagner and Nishikura, 1988]. In light of the EDTA results it has been concluded that there is no divalent cation requirement. However, it seems prudent to wait until the protein is purified before drawing conclusions about this or any other characterization performed in the crude extract. The optimal temperature of the amphibian activity is about 25°C [Bass and Weintraub, 1987], while that of the mammalian activity is 37°C [Wagner and Nishikura, 1988]; in all other respects the biochemical requirements of the amphibian and mammalian activities appear to be identical.

IV. SUBSTRATE SPECIFICITY

A. Double-Stranded RNA

Competition experiments indicate that the unwinding/modifying activity greatly prefers to bind dsRNA. For example, in *Xenopus* extracts, the reaction of a radioactively labeled, intermolecular RNA duplex is inhibited by adding a 30-fold molar excess of a second unlabeled RNA duplex, but it is not inhibited by the addition of a similar amount of ssRNA, ssDNA, dsDNA or a DNA–RNA hybrid [Bass and Weintraub, 1987, and unpublished results]. Similar results have been obtained using mammalian extracts [Wagner and Nishikura, 1988]. The ssRNAs tested as inhibitors in these experiments were not denatured and presumably contained double-stranded regions; thus it has been speculated that the unwinding/modifying activity may also distinguish between intermolecular and intramolecular duplex interactions [Bass and Weintraub, 1988; Wagner et al., 1989]. However, more recent studies suggest that certain intramolecular duplexes can function as substrates [Luo et al., 1990; Sharmeen et al., 1991; Nishikura et al., 1991]. Compared with intermolecular duplexes, intramolecular duplexes seem to function less well as substrates, exhibiting fewer inosines in the final product or requiring higher concentrations of extract protein to reach the final product. One interpretation of the latter result is that the unwinding/modifying activity has a lower affinity for intramolecular duplexes. If this interpretation is correct, intramolecular duplexes should function as competitive inhibitors if assayed at higher relative concentrations than previously tested.

B. Termini

One difference between intermolecular and intramolecular duplexes is the number of termini. Although a sensitivity to the number of ends cannot be ruled out until a circular substrate is tested, it is known that the unwinding/modifying activity does not exhibit a strict requirement for a particular structure at the termini of its substrate (B. Bass, unpublished results). dsRNA substrates that have blunt ends, 5′ overhangs, 3′ overhangs, and noncomplementary 5′ and 3′ overhangs (''frayed ends'') work well in the reaction. One substrate tested had a long 5′ overhang of approximately 250 bases, while the overhangs of other substrates were much shorter, i.e., approximately 20 bases. dsRNA substrates capped at their 5′ end work as well as uncapped molecules [Bass and Weintraub, 1987] and substrates in which the 3′ termini have been oxidized to aldehydes work as well as those ending in a phosphate or hydroxyl (B. Bass, unpublished results).

C. Length Requirement

Although the preference of the activity for intermolecular substrates may have a structural basis, it is also possible that it simply reflects a requirement for a certain number of contiguous base pairs. An examination of RNAs for which secondary structures have been characterized reveals that intramolecular helical regions contain at most 8–10 contiguous base pairs. In most cases the substrates used to assay the unwinding/modifying activity have been quite long, ranging in length from 300 to 800 base pairs. Consistent with the hypothesis that optimal binding requires a minimal length of contiguous base pairs is the observation that an intermolecular duplex of shorter length, i.e., 36 base pairs, like intramolecular duplexes, exhibits a reduced affinity for the unwinding/modifying activity (B. Bass, M. Paul and A. Polson, unpublished results). It should be noted that, although very few dsRNA binding proteins have been characterized, those that have been studied seem to require a substrate with a relatively long, perfect, or nearly perfect double-stranded region. The bacterial enzyme RNase III requires 20 base pairs with at least 90% homology [Robertson, 1982, 1990] while the dsRNA-binding proteins involved in the interferon response, 2′,5′-oligo(A) synthetase and the associated protein kinase, are fully active only with duplexes longer than 65–80 contiguous base pairs. A duplex containing a mismatch approximately every 35 base pairs is 20 times more active in promoting synthesis of 2',5'-oligo(A) than one that contains a mismatch every 7 base pairs [Minks et al., 1979]. Thus, among proteins exhibiting a rigid specificity for dsRNA, a precedent exists for a requirement of a substrate with a proscribed number of contiguous base pairs.

D. Sequence Specificity

Although the unwinding/modifying activity is quite specific for dsRNA, it does not exhibit a strict sequence specificity. dsRNA molecules prepared from synthetic sense and antisense transcripts of β-globin or c-*myc* [Wagner and Nishikura, 1988], chloramphenicol acetyltransferase (CAT) [Bass and Weintraub, 1987], or the *Xenopus* genes AN-1 and *Xhox* 1B [Rebagliati and Melton, 1987] are all modified by the activity in vitro. A synthetically prepared hepatitis delta virus transcript [Luo et al., 1990], as well as a 4.6 kb dsRNA isolated from yeast L-A viral particles (B. Bass, unpublished results) were also substrates in vitro. Furthermore, the reaction of a CAT duplex can be competitively inhibited by poly(A)–poly(U) duplexes, as well as by poly(G)–poly(C) and poly(I)–poly(C) duplexes (B. Bass, unpublished results). The fact that even duplexes that do not contain adenosines can inhibit the reaction suggests that binding may occur separately from modification.

Experiments have also been performed to determine whether adenosines are chosen for modification with regard to their sequence context [Bass, unpublished results]. For example, sites of inosine in a modified CAT duplex were mapped by primer extension in the presence of dideoxynucleotides. With this method adenosine to inosine changes can be detected by the loss of ddT bands and the concomitant appearance of ddC bands. These studies suggested that adenosine residues in which the 5' base is adenine or uracil are preferentially modified. Many, but not all, of the adenosines 3' to an A or a U were modified, while modification of adenosines 3' to a G or a C was not observed. In these studies no evidence for additional

requirements in the bases surrounding the modification site was observed. However, work by others suggests a 3' nearest neighbor preference such that G = C ≥ U > A [Wagner et al., 1989].

It is possible that the unwinding/modifying activity specifically recognizes the base 5' to the modification site. An alternative explanation is that the modification reaction occurs preferentially in regions of local instability. The latter model is attractive in light of the fact that the functional group on the adenosine that must be modified is involved in base pairing. The amino group of an adenosine that occurs 3' to the thermodynamically more stable GC base pair may be less accessible to the unwinding/modifying protein(s) and thus less often modified.

V. REACTION PRODUCTS
A. Inosine Content

During the modification reaction adenosines within the dsRNA substrate are covalently modified to inosines. Thus, when compared with the substrate at zero time, the reaction products show a decrease in the number of adenosine residues and a corresponding increase in inosine content. It has been determined that adenosines in both strands of an intermolecular duplex are modified, but at the end of the reaction not all of the adenosines have been modified [Bass and Weintraub, 1988; Wagner et al., 1989]. In fact, the reaction stops when 30%–50% of the adenosines have been modified; the exact number seems to depend on the particular substrate and extract used [Bass and Weintraub, 1988; Wagner et al., 1989; Kimelman and Kirschner, 1989]. The observation that a significant number of adenosines remain unmodified is consistent with the idea that only adenosines within a certain sequence context can be modified. However, as discussed, the activity greatly prefers to bind dsRNA, and thus an alternate explanation is that after 30%–50% of the adenosines have been modified the RNA is too single stranded to serve as an efficient substrate. The analysis of modified basic fibro-blast growth factor RNA (see Section VII) emphasizes the specificity for dsRNA; single-stranded overhangs, even though contiguous with double-stranded regions, do not show base changes [Kimelman and Kirschner, 1989].

B. Single-Stranded Character

Since the modification results in the substi-tution of a stable AU base pair by the less sta-ble IU base pair, the reaction products have properties of unwound molecules. For example, the products exhibit an increase in suscepti-bility to single-strand–specific ribonucleases during the reaction [Rebagliati and Melton, 1987; Bass and Weintraub, 1988; Wagner et al., 1989]. It is important to note that this is a characteristic of modified RNA exposed to ribo-nucleases in vitro and that it remains to be seen if IU base pairs lead to degradation within the cell (see Section VII). Although the RNA sub-strates increase in single-stranded character dur-ing the reaction, the complementary strands do not completely separate [Bass and Weintraub, 1988; Wagner et al., 1989].

Given that approximately 50% of the aden-osines in both strands are modified, it can be estimated that in the final product one out of every four base pairs is an IU mismatch. Although the actual structure of the reaction products remains unknown, it has been pro-posed that the final product consists of a mol-ecule with internal loops interspersed with regions of dsRNA [Bass and Weintraub, 1988] (Fig. 1). The loops would occur in regions that were highly modified so that near-by normal base pairs would be disrupted, because the ther-modynamic stability they provided could not compete with the favorable free energy change of a helix–coil transition; these regions would explain the RNase sensitivity of the modified RNA. The double-stranded regions would occur in sparsely modified regions (such as GC-rich regions); these regions would explain the fact that the molecules never completely separate.

VI. ASSAYS

A. Categories

Several assays have been used to detect the unwinding/modifying activity. The assays fall into two categories: those that directly moni-tor the presence of inosine within the RNA sub-strate and those that detect inosine indirectly by its ability to change the base-pairing prop-erties of the substrate. At present, very little is known concerning RNAs that may be natu-ral substrates for the activity (see Section VII). Although I briefly comment on the feasibility of applying various assays to the identification of putative endogenous substrates, it should be understood that the assays described below have been developed for the detection of the unwind-ing/ modifying activity with synthetically pre-pared substrates.

B. Chromatography

Inosine can be monitored directly by digest-ing the modified RNA to completion with ribonuclease, followed by a chromatographic analysis of the resulting mononucleotides or nucleosides. Thin-layer chromatography [Bass and Weintraub, 1988; Wagner et al., 1989] and high performance liquid chromatography (HPLC) [Wagner et al., 1989; Polson et al., 1991] have been used in previous studies. Care should be taken when choosing a chromatographic system, since guanosine and inosine, as well as the corresponding monophosphates, often comigrate or migrate very similarly [Bass and Weintraub, 1988; Nishimura and Kuchino, 1983; Gehrke and Kuo, 1990]. It is neces-sary to use an isotopically labeled substrate to distinguish the digestion products from other nucleic acids that may be present in the crude extracts, as well as to increase the sensitivity of detection. The latter is neces-sary since, at present, quantities of protein, and therefore modified substrate, are limit-ing. The sense and antisense transcripts are radiolabeled during in vitro transcription by including in the transcription mixture α^{32}P-ATP or an ATP molecule labeled with ^{3}H or ^{14}C in its purine ring. Then, if the RNA is digested with the appropriate ribonuclease, the appearance of radiolabeled inosine will be accompanied by the disappearance of adenosine. For obvious reasons, when α^{32}P-ATP is utilized, the RNA must be digested with a nuclease that leaves a 5' phosphate (e.g., P1 nuclease). Detection of inosine among the nucleotide constituents of the substrate is by far the most convincing assay, and thus this method should always be used for the initial verification of the presence of the unwinding/modifying activity. However, since this is not the fastest assay, after initial verification it is often prudent to use one of the indirect methods discussed below for routine assays. Of course, for an endogenous substrate, different labeling protocols must be employed, and even then the direct detection of inosine may be difficult or impossible if the transcript is rare or hard to separate from other cellular RNAs.

C. Electrophoresis

Several assays are based on the fact that the modification changes the base-pairing properties of the substrate. For example, the reaction can be monitored electrophoretically, on a native polyacrylamide gel. A time course of a reaction using a CAT duplex is shown in Figure 3; as the reaction proceeds a decrease in gel mobility is observed. Although the change in electrophoretic mobility is not well understood, it is known to correlate with the increase in inosine [Wagner et al., 1989] and presumably reflects the change in conformation as the RNA increases in single-stranded character. Although for me this assay has proved to be the fastest, a few precautions should be noted. First, this assay is by no means quantitative. Although a decrease in gel mobility signals an increase in inosine content, the modification of a precise number of adenosines does not equate with a particular quantitative change in gel mobility. This is best understood by considering that the change in electrophoretic mobility correlates with a change in the con-

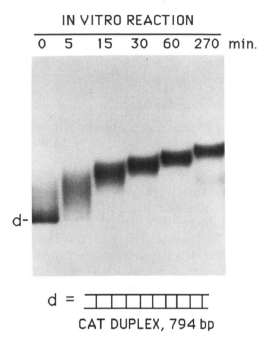

Fig. 3. *In vitro modification of a ^{32}P-labeled CAT duplex in a whole cell* Xenopus *embryo extract. After incubation, the RNA was deproteinized and electrophoresed on a 4% polyacrylamide native gel [for reaction conditions, see Bass and Weintraub, 1988].*

formation of the duplex. The change in conformation cannot be predicted and will depend on the sequence context of the modified base; a mismatch in a GC-rich region may have a relatively small effect on the conformation, whereas a mismatch in an AU-rich region may destabilize the helix to produce an internal loop.

A further precaution to be attached to the electrophoretic mobility assay relates to the ability of other proteins in the extract to change the electrophoretic mobility of the dsRNA. In the original protocol, after the RNA was modified in the extract, it was prepared for the electrophoretic analysis by a proteinase K digestion followed by a phenol/chloroform extraction. In more recent assays we use a quick protocol that eliminates the phenol/chloroform extraction step (R. Hough and B. Bass, unpublished results). We have found that under certain conditions this results in the change of duplex mobility because of binding of a relatively proteinase K-resistant protein. This RNA-binding protein does not seem to cofractionate with the modifying activity in more pure preparations and in fact seems to inhibit the modification reaction (R. Hough and B. Bass, unpublished results). Phenol extraction does of course remove this protein from the RNA and should be performed if the reason for the mobility change is questionable.

Finally, it should be noted that for the electrophoretic assay, which does not require the digestion of the RNA to mononucleotides, the RNA substrate can also be endlabeled posttranscriptionally, with T4 RNA ligase [Uhlenbeck and Gumport, 1982] or T4 polynucleotide kinase [Sambrook et al., 1989]. However, since phosphatase activities have been observed in crude extracts prepared from *Xenopus* cells (B. Bass, unpublished results), labeling with kinase should be avoided until extracts have been monitored for such activities.

D. Ribonuclease

Another assay based on the ability of the unwinding/modifying activity to change the base-pairing properties of its substrate involves the use of single-strand–specific ribonucleases. As the reaction proceeds, the substrate increases in single-stranded character and thus increases in its sensitivity to single-strand–specific ribonucleases. This was one of the assays used in the original detection of the activity [Rebagliati and Melton, 1987]. At that time the sensitivity of the RNA to single-strand–specific ribonuclease was thought to be the result of a typical unwinding activity, and the modification had not yet been detected. To my knowledge a ribonuclease assay has not been developed for use as a routine assay of the unwinding/ modifying activity.

E. Primer Extension

As discussed, primer extension in the presence of dideoxynucleotides has been used to map sites of inosine within modified substrates (see Section V). Again, because of the time involved, this is probably not the method of choice for routine assays. However, this method may be very helpful in future studies for monitoring the modification of endogenous substrates. Of course, some sequence must be known so that an appropriate primer can be synthesized. A problem with primers complementary to transcripts that may be modified is that, even in the most GC-rich regions, it is usually necessary to include a few thymidines within the primer; thus, if the adenosines in the complementary strand have been modified, the primer may not hybridize well. A valid worry would be that a population of RNAs that has not been modified in the region complementary to the primer would be overrepresented while modified sequences would be selected against. Of course, if the region of complementarity between the sense and antisense transcripts does not extend the entire length of the transcripts, then the primer can be chosen in an adjacent single-stranded region.

VII. BIOLOGICAL ROLE

The biological role of the unwinding/modifying activity is unknown. Given the ubiquity of double-stranded regions within cellular RNAs, many potential substrates and roles can be imagined. It has been speculated that the activity plays a regulatory role in the cell by modulating or reversing RNA–RNA interactions, and, since the activity has the potential to change the meaning of codons within a mRNA, it has also been suggested that the activity may function in the posttranscriptional regulation of gene expression [Bass and Weintraub, 1988; Wagner et al., 1989]. Finally, it has been proposed that the role of the activity may be to degrade dsRNA formed in the cell [Bass and Weintraub, 1988; Wagner et al., 1989]. The possibilities seem endless. However, several RNAs have recently been implicated as substrates for the activity in vivo, thus providing the first clues as to ways the activity may function in the cell. These studies will be described below.

A. Fibroblast Growth Factor

Several years ago Kimelman and Kirschner [1989] discovered that *Xenopus* oocytes contain an antisense transcript complementary to the maternal mRNA encoding basic fibroblast growth factor (bFGF). They speculated that, if an RNA duplex formed between the antisense transcript and the bFGF mRNA, it might serve as a substrate for the unwinding/modifying activity. Since the unwinding/modifying activity is localized to the nucleus in the oocyte and does not appear in the cytoplasm until the oocyte nuclear envelope breaks down during meiotic maturation, they reasoned that modification would occur subsequent to germinal vesicle breakdown (GVBD). To test their hypothesis, the authors isolated RNA from stage 6 oocytes, as well as from oocytes at various stages of meiotic maturation. Reverse transcriptase was used to prepare cDNA from the various RNA populations, and the cDNA was amplified by the PCR, using primers that bound just outside the overlap region between the sense and the antisense RNA. The amplified DNA was cloned and sequenced. Although the cDNA prepared from oocyte RNA had the sequence predicted by the protein coding mRNA, that prepared from oocytes 4 hours after GVBD showed extensive A to G changes. Such transitions within the cDNA are consistent with the involvement of the unwinding/modifying activity since reverse transcriptase would treat inosines as guanosines. Five clones were sequenced; on average, 47% of the adenosines in the region predicted to hybridize with the antisense transcript were modified. Although the modified bases were not at identical sites in each clone, as with the reaction in vitro (see Section IV) adenosines 3' to adenine and uracil bases were preferentially modified. Since the mRNA was clearly detectable 4 hours after GVBD, but absent by 6 hours after GVBD, the authors have speculated that the modification targets the bFGF message for degradation. It is intriguing to imagine that the unwinding/modifying activity is part of a more extensive degradation pathway and that other cellular RNAs are targeted for degradation by modification. However, one caveat to the hypothesis that the modification of adenosines to inosines triggers RNA degradation is the apparent stability of RNA substrates after microinjection into *Xenopus* embryos or maturing oocytes; although modified and unwound, the RNAs do not seem less stable than microinjected, single-stranded control RNAs [Bass and Weintraub, 1987].

B. RNA Viruses

Some evidence indicates that certain RNA viruses may be substrates for the unwinding/modifying activity in vivo. For example, several years ago, a measles virus matrix gene transcript, isolated from the brain of a patient who had died of the disease measles inclusion body encephalitis (MIBE), was sequenced, and it was discovered that 132 of 266 uridine (U) residues had been changed to cytidine (C) residues [Cattaneo et al., 1988]. It was proposed that these "hypermutations" resulted from A to I changes by the action of the unwinding/modifying activity on the viral RNA template [Bass et al., 1989]. Indeed, depending on whether the genomic or antigenomic strand of a viral RNA was modified, the unwinding/modifying reaction could explain U to C mutations as well as A to G mutations. Other examples of hypermutation events have now been identified in measles virus genomes [Billeter and Cattaneo, 1991]. Hypermutations of the U to C variety have been observed in matrix gene transcripts isolated from different measles virus strains [Wong et al., 1989; Cattaneo et al., 1989], and A to G type changes have been observed in another measles virus gene, the H gene [Cattaneo et al., 1989]. Finally, in contrast to the examples where the unwinding/modifying activity is thought to produce hypermutations, it has been speculated that the unwinding/modifying activity is responsible for a specific base change observed in hepatitis delta virus [Luo et al., 1990]. The change is dependent on replication of the RNA and results in the substitution of the amber termination codon (UAG) by the codon UGG. The base change correlates with the appearance of an extended protein; it has not been determined if both proteins are required for viral replication.

The hypermutations described above would be expected to result in a nonfunctional protein; thus these data are consistent with a role of the unwinding/modifying activity in a cellular response to infection by RNA viruses. However, the fact that the activity is localized in the nucleus at most developmental stages, while most RNA viruses are cytoplasmic, combined with the fact that the unwinding/modifying activity does not seem to be inducible, argues against such a role. Furthermore, it is not clear how the various RNA viruses provide the double-stranded structure required by the unwinding/modifying activity. It has been pointed out that there is no evidence that dsRNA is an intermediate in transcription or replication of ssRNA viruses [Weissman, 1989]. Rather, the template and nascent strand are thought to be wrapped in nucleocapsid proteins at all times, thus preventing the base pairing of the complementary strands. dsRNA has been observed to form during Q-β replication, but only when there is a failure of the mechanism that keeps the strands apart.

Compelling evidence that the unwinding/modifying activity is *not* part of an antiviral response comes from recent work by Morrissey and Kirkegaard [1991]. These authors found that the unwinding/modifying activity is actually inhibited in HeLa cells induced to an antiviral state with poly I:C. One of the antiviral effects elicited when mammalian cells are treated with poly I:C is the activation of a dsRNA-dependent protein kinase [see review by Pestka et al., 1987]. An exception to the latter is observed in adenovirus-infected cells; i.e., adenovirus codes for two small RNAs, VA1 and VA2, that block the activation of the dsRNA-dependent protein kinase [O'Malley et al., 1986; Kitajewski et al., 1986]. Morrissey and Kirkegaard [1991] observed that poly I:C is ineffective in inhibiting the unwinding/modifying activity in cells infected with adenovirus and thus speculate that the inhibition of the unwinding/modifying activity may be mediated by the dsRNA-dependent kinase. Although pre-

liminary, this work hints that the unwinding/modifying activity may be regulated in a complicated, yet interesting, manner.

In summary, the existing data argue against the idea that the unwinding/modifying activity is part of a cellular viral defense mechanism. It seems most likely that the unwinding/modifying activity serves an unrelated host cell function and that, coincidentally, certain RNA viruses may be substrates for the activity.

VIII. RELATED PROTEINS

Until the biological function of the unwinding/modifying activity has been determined it will remain unclear as to what particular category of biological activity it belongs. Although the activity clearly has some properties of unwinding, its mechanism of unwinding differs significantly from previously described ones that unwind their substrates either by utilizing energy produced by ATP hydrolysis or by virtue of their ability to bind single strands preferentially [see review by Matson and Kaiser-Rogers, 1990]. It has been proposed that the unwinding/modifying activity may exemplify a third type of unwinding activity, one that produces a change in base-pairing properties by covalently modifying its substrate [Bass and Weintraub, 1988]. This putative class of proteins would be characterized by its ability to unwind its substrates permanently and irreversibly.

Several previously characterized enzymes are known to catalyze the conversion of adenosine to inosine in the cell, notably, the enzymes involved in purine catabolism, adenosine deaminase, and adenylic deaminase [reviewed by Zielke and Suelter, 1971] and the enzyme that converts adenosine to inosine in the anticodon of many tRNAs [Elliott and Trewyn, 1984; Grosjean et al., 1987]. The substrate specificities of these previously characterized enzymes are quite different from that of the unwinding/modifying activity [Bass and Weintraub, 1988; Polson et al., 1991]. Neither adenosine deaminase nor adenylic deaminase can modify dsRNA, and likewise the unwinding/modifying activity cannot modify adenosine or adenosine monophosphates, the normal substrates of the previously characterized deaminases. In addition, the unwinding/modifying reaction is not inhibited by deoxycoformycin, a powerful inhibitor of both adenosine and adenylic deaminase.

The modification of adenosines to inosines in the anticodon loop of tRNAs is thought to occur by a base replacement mechanism rather than a deamination [Elliott and Trewyn, 1984; Grosjean et al., 1987]. According to the existing model, the phosphodiester backbone is not cleaved; rather the adenine is removed by cleaving the glycosyl bond and inserting hypoxanthine (the inosine base). Experiments with the unwinding/modifying activity have demonstrated that modification of a dsRNA substrate labeled with ^{14}C or ^{13}C in the purine ring of adenosine results in a product containing ^{14}C- or ^{13}C-labeled inosine [Wagner et al., 1989; Polson et al., 1991]. This result is not consistent with a base replacement mechanism, suggesting again that the unwinding/modifying activity is a previously uncharacterized activity.

Recent reports provide a number of examples in which mRNAs have been found to contain nontemplated nucleotides that presumably arise by posttranscriptional RNA editing mechanisms [reviewed in Weissmann et al., 1990; Scott, 1989]. In most cases these processes result in the addition or deletion of ribonucleotides from the RNA rather than the substitution type modification that occurs in the unwinding/modifying reaction. However, in one type of editing, exemplified by apolipoprotein B mRNA [Powell et al., 1987; Driscoll et al., 1989; Chen et al., 1987, 1990] and certain plant mitochondrial mRNAs [Gualberto et al., 1989; Covello and Gray, 1989; Schuster et al., 1990], single cytidine residues are replaced, or perhaps modified, to uridine. Like the unwinding/modifying activity, these conversions would be predicted to involve a deamination reaction. The modification of apolipoprotein B is tissue specific and produces a stop codon that results in a truncated form of the protein. The editing of the plant mitochondrial mRNAs also has a defined function, producing altered codons that result in proteins containing different amino acids.

IX. RELATIONSHIP TO THE MECHANISM OF ANTISENSE RNA TECHNIQUES

The mechanism by which antisense RNA inhibits gene expression is not well understood. It is presumed that inhibition involves the hybridization of the antisense RNA with its target mRNA, and in some cases the appropriate duplex has been detected [Melton, 1985; Wormington, 1986, Kim and Wold, 1985; Yokoyama and Imamoto, 1987]. The mystery lies in the reason that the duplex is inhibitory. Some experiments indicate that the duplex is inhibitory at the level of translation [Harland and Weintraub, 1985; Melton, 1985; Ch'ng et al., 1989], while other data also suggest an effect on transcription (Yokoyama and Imamoto, 1987] (see also Yokoyama, this volume), transport from the nucleus [Kim and Wold, 1985] (see also discussion of Cornelissen [1989] by Rodermel and Bogorad, Section V.A, this volume), or RNA stability [McGarry and Lindquist, 1986; Crowley et al., 1985]. It is possible that antisense RNA can exert its effect in a number of ways, the exact mechanism depending on the organism and perhaps additional factors such as the particular developmental stage of the cell or whether hybridization occurs in the nucleus or cytoplasm.

Given that the substrate for the unwinding/modifying activity is dsRNA and furthermore that the activity has been detected in many cell types that have proved amenable to antisense inhibition (see Section II), the question is raised as to whether the unwinding/modifying activity is involved in the observed inhibition. Clearly the activity has the potential to inhibit gene expression. As discussed, translation of an mRNA in which a number of adenosine residues had been converted to inosines would produce an altered or nonfunctional protein (see Section I). The idea that the unwinding/modifying activity is involved in antisense inhibition is attractive in light of the fact that the modification would result in the permanent destruction of a functional RNA.

A recurring observation in many successful antisense experiments is the disappearance of the target mRNA [see review by Takayama and Inouye, 1990]. A common hypothesis offered to explain the absence of the message is that the antisense–sense hybrid is degraded by a dsRNA specific nuclease. In most cases, the antisense transcript is detectable; this is as expected, since the antisense RNA is usually present in excess relative to the mRNA and only an amount stoichiometric to the message would be subject to hybridization and thus degradation. Indeed, in a recent study in which expression of an antisense RNA in a tomato plant resulted in the disappearance of the target mRNA, it was observed that the antisense transcript disappeared as well when the plant was exposed to conditions known to induce higher levels of expression of the target gene [Hamilton et al., 1990]. It has been speculated that the degradation of the *Xenopus* bFGF mRNA during oocyte maturation is triggered by modification by the unwinding/modifying activity (see Section VII). If this hypothesis is correct, it seems possible that degradation of the hybrid formed during antisense RNA experiments may also be mediated by modification.

Proof that the unwinding/modifying activity is involved in antisense RNA experiments awaits careful analyses directed at the detection of inosine within the sense and antisense transcripts. Of course, if degradation follows modification as in the bFGF case, such experiments may be hindered by the necessity of isolating the RNA after modification yet prior to degradation. It should be pointed out that in many previous antisense experiments modified RNAs may have eluded detection. In fact, the methods used to detect the target mRNA may not have distinguished between the absence of the transcript and its modification. For example, in an S1 analysis, probes designed to protect the target mRNA would not be expected to form a perfect duplex with a modified message; mismatches created by unpaired inosine residues would lead to the degradation of the RNA by S1. It would thus be concluded that the RNA was absent when in fact it was present in a modified form. Similarly, the lack of hybridization in a Northern analysis could indicate a modified mRNA rather than degradation of the transcript. Depending on the number of adenosines modified, probes designed to hybridize to the nascent transcript may not detect a modified transcript. One way in which these discrepancies can be avoided is to design a probe for the mRNA that lies outside the region targeted by the antisense RNA. Such a probe has been utilized in at least one antisense experiment, and in that case the message is clearly degraded [Giebelhaus et al., 1988].

If the unwinding/modifying activity does not modify hybrids formed by base pairing of sense and antisense RNA, how are these RNAs protected from modification? One obvious explanation is that the dsRNA is not accessible to the unwinding/modifying activity because it is localized to a different region or compartment in the cell. Clearly the latter is relevant to antisense RNA experiments in the *Xenopus* oocyte. Microinjection of antisense RNA into the cytoplasm of *Xenopus* oocytes results in the formation of hybrids between the antisense RNA and its target, and expression of the message is inhibited [Harland and Weintraub, 1985; Melton, 1985; Wormington, 1986]. There is no evidence for modification of dsRNA in the oocyte cytoplasm, consistent with the localization of the activity to the nucleus of these cells [Bass and Weintraub, 1987, 1988]. Similarly, antisense and sense transcripts of bFGF are not modified in the cytoplasm of stage 6 *Xenopus* oocytes [Kimelman and Kirschner, 1989]. A perplexing issue for future studies is how the endogenous transcripts, which presumably exist together in the nucleus during their synthesis, make it to the cytoplasm without being modified.

X. FUTURE STUDIES

What is the specific biological function of the activity? Which endogenous RNAs are substrates for the activity in vivo? Do the natural substrates exemplify a particular type of cellular RNA? These questions, pertaining to the biology of the unwinding/modifying activity, will certainly be the focus of studies in the immediate future. However, since examples of proteins with a rigid specificity for dsRNA are rare, it seems likely that in the long term, studies of the unwinding/modifying activity may also provide important information about RNA–protein interactions. Furthermore, it is possible that the unwinding/modifying activity may be useful in studies of the role structure plays in the function of cellular RNAs. Since the activity is specific for dsRNA, the demonstration that a cellular RNA can be modified by the activity in vivo would be convincing evidence for a specific double-stranded structure within the cell. Alternatively, the unwinding/modifying activity could be utilized for in vitro structure– function studies. AU base pairs within an RNA could be modified to IU base pairs and the requirement for stable base paired regions determined by monitoring the activity of the RNA before and after modification.

Although the discovery of a previously unrecognized biological activity is an exciting event, it is the subsequent studies, designed to elucidate the relationship of the activity to normal cell metabolism, that often provide the more interesting story. It seems likely that this will hold true for the unwinding/modifying activity.

ACKNOWLEDGMENTS

I thank K. Nishikura, L. Morrissey, and K. Kirkegaard for communicating unpublished results and the members of my laboratory for their comments and suggestions regarding this manuscript. Work conducted in my laboratory was supported by a grant from NIH (GM44073) and by the Pew Scholars Fund.

XI. REFERENCES

Basilio C, Wahba AJ, Lengyel P, Speyer JF, Ochoa S (1962): Synthetic polynucleotides and the amino acid code. Proc Natl Acad Sci USA 48:613–616.

Bass BL, Weintraub H (1987): A developmentally regulated activity that unwinds RNA duplexes. Cell 48:607–613.

Bass BL, Weintraub H (1988): An unwinding activity that covalently modifies its double-stranded RNA substrate. Cell 55:1089–1098.

Bass BL, Weintraub H, Cattaneo R, Billeter MA (1989): Biased hypermutation of viral RNA genomes could be due to unwinding/modification of double-stranded RNA [letter]. Cell 56:331.

Bevilacqua A, Erickson RP, Hieber V (1988): Antisense RNA inhibits endogenous gene expression in mouse preimplantation embryos: Lack of double-stranded RNA "melting" activity. Proc Natl Acad Sci USA 85:831–835.

Billeter MA, Cattaneo R (1991): Molecular biology of defective measles viruses persisting in the human central nervous system. In Kingsbury DW (ed): The Paramyxoviruses. New York: Plenum Press, pp 323–345.

Cattaneo R, Schmid A, Eschle D, Baczko K, ter Meulen V, Billeter MA (1988): Biased hypermutation and other genetic changes in defective measles viruses in human brain infections. Cell 55:255–265.

Cattaneo R, Schmid A, Spielhofer P, Kaelin K, Baczko K, ter Meulen V, Pardowitz J, Flanagan S, Rima BK, Udem SA, Billeter MA (1989): Mutated and hypermutated genes of persistent measles viruses which caused lethal human brain diseases. Virology 173:415–425.

Chen S-H, Habib G, Yang C-Y, Gu Z-W, Lee BR, Weng S-A, Silberman SR, Cai S-J, Deslypere JP, Rosseneu M, Gotto Jr. AM, Li W-H, Chan L (1987): Apolipoprotein B-48 is the product of a messenger RNA with an organ-specific in-frame stop codon. 238:363–366.

Chen S-H, Li XX, Liao WS, Wu JH, Chan L (1990): RNA editing of apolipoprotein B mRNA. Sequence specificity determined by in vitro coupled transcription editing. J Biol Chem 265:6811–6816.

Cornelissen M (1989): Nuclear and cytoplasmic sites for anti-sense control. Nucleic Acids Res 17:7203–7209.

Covello PS, Gray MW (1989): RNA editing in plant mitochondria. Nature 341:662–666.

Crowley TE, Nellen W, Gomer RH, Firtel RA (1985): Phenocopy of discoidin I–minus mutants by antisense transformation in Dictyostelium. Cell 43:633–641.

Davies MS, Wallis SC, Driscoll DM, Wynne JK, Williams GW, Powell LM, Scott J (1989): Sequence requirements for apolipoprotein B RNA editing in transfected rat hepatoma cells. J Biol Chem 264:13395–13398.

Dreyer C, Hausen P (1983): Two-dimensional gel analysis of the fate of oocyte nuclear proteins in the development of Xenopus laevis. Dev Biol 100:412–425.

Driscoll DM, Wynne JK, Wallis SC, Scott J (1989): An in vitro system for the editing of apolipoprotein B mRNA. Cell 58:519–525.

Elliott MS, Trewyn RW (1984): Inosine biosynthesis in

transfer RNA by an enzymatic insertion of hypoxanthine. J Biol Chem 259:2407–2410.

Gehrke CW, Kuo KC (1990): Ribonucleoside analysis by reversed-phase high performance liquid chromatography. In Gehrke CW, Kuo KC (eds): Chromatography and Modification of Nucleosides, Vol. 45A. Amsterdam: Elsevier, pp A3-A71.

Giebelhaus DH, Eib DW, Moon RT (1988): Antisense RNA inhibits expression of membrane skeleton protein 4.1 during embryonic development of *Xenopus*. Cell 53:601–615.

Grosjean H, Haumont E, Droogmans L, Carbon P, Fournier M, de Henau S, Doi T, Keith G, Gangloff J, Kretz K, Trewyn R (1987): A novel approach to the biosynthesis of modified nucleosides in the anticodon loops of eukaryotic transfer RNAs. In Bruzik KS, Stec WJ (eds): Biophosphates and Their Analogues-Synthesis, Structure, Metabolism and Activity. Amsterdam: Elsevier Science Publishers BV, pp 355–378.

Gualberto JM, Lamattina L, Bonnard G, Weil JH, Grienenberger JM (1989): RNA editing in wheat mitochondria results in the conservation of protein sequences. Nature 341:660–662.

Hamilton AJ, Lycett GW, Grierson D (1990): Antisense gene that inhibits synthesis of the hormone ethylene in transgenic plants. Nature 346:284–287.

Harland RM, Weintraub H (1985): Translation of mRNA injected into *Xenopus* oocytes is specifically inhibited by antisense RNA. J Cell Biol 101:1094–1099.

Kawase Y, Iwai S, Inoue H, Miura K, Ohtsuka E (1986): Studies on nucleic acid interactions I. Stabilities of mini-duplexes ($dG_2A_4XA_4G_2 \cdot dC_2T_4YT_4C_2$) and self-complementary d(GGGAAXYTTCCC) containing deoxyinosine and other mismatched bases. Nucleic Acids Res 14:7727–7736.

Kim SK, Wold BJ (1985): Stable reduction of thymidine kinase activity in cells expressing high levels of antisense RNA. Cell 42:129–138.

Kimelman D, Kirschner MW (1989): An antisense mRNA directs the covalent modification of the transcript encoding fibroblast growth factor in *Xenopus* oocytes. Cell 59:687–696.

Kitajewski J, Schneider RJ, Safer B, Munemitsu SM, Samuel CE, Thimmappaya B, Shenk T (1986): Adenovirus VAI RNA antagonizes the antiviral action of interferon by preventing activation of the interferon-induced eIF-2 alpha kinase. Cell 45: 195–200.

Lomant AJ, Fresco JR (1975): Structural and energetic consequences of noncomplementary base oppositions in nucleic acid helices. In Cohn WE (ed): Progress in Nucleic Acid Research and Molecular Biology. New York: Academic Press, pp 185–218.

Luo G, Chao M, Hsieh S-Y, Sureau C, Nishikura K, Taylor J (1990): A specific base transition occurs on replicating hepatitis delta virus RNA. J Virol 64: 1021–1027.

Matson SW, Kaiser-Rogers KA (1990): DNA helicases. Ann Rev Biochem 59:289–329.

McGarry TJ, Lindquist S (1986): Inhibition of heat shock protein synthesis by heat-inducible antisense RNA. Proc Natl Acad Sci USA 83:399–403.

Melton DA (1985): Injected anti-sense RNAs specifically block messenger RNA translation in vivo. Proc Natl Acad Sci USA 82:144–148.

Minks MA, West DK, Benvin S, Baglioni C (1979): Structural requirements of double-stranded RNA for the activation of 2',5'-oligo(A) polymerase and protein kinase of interferon-treated HeLa cells. J Biol Chem 254: 10180–10183.

Morrissey LM, Kirkegaard K (1991): Regulation of a double-stranded modification activity in human cells. Mol Cell Biol 11: 3719–3725.

Nishikura K, Yoo C, Kim U, Murray JM, Estes PA, Cash FE, Leibhaber SA (1991): Substrate specificity of the dsRNA unwinding/modifying activity. EMBO J 10:3523–3532.

Nishimura S, Kuchino Y (1983): Characterization of modified nucleosides in tRNA. In Weissman SM (ed): Methods of DNA and RNA Sequencing. New York: Praeger Publishers, pp 235–260.

O'Malley RP, Mariano TM, Siekierka J, Mathews MB (1986): A mechanism for the control of protein synthesis by adenovirus VA RNAI. Cell 44:391–400.

Pestka S, Langer JA, Zoon KC, Samuel CE (1987): Interferons and their actions. Annu Rev Biochem 56:727–777.

Polson AG, Crain PF, Pomerantz SC, McCloskey JA, Bass BL (1991): The mechanism of adenosine to inosine conversion by the double-stranded RNA unwinding/modifying activity: A high-performance liquid chromatography–mass spectrometry analysis. Biochemistry 30:11507–11514.

Powell LM, Wallis SC, Pease RJ, Edwards YH, Knott TJ, Scott J (1987): A novel form of tissue-specific RNA processing produces apolipoprotein-B48 in intestine. Cell 50:831–840.

Rebagliati MR, Melton DA (1987): Antisense RNA injections in fertilized frog eggs reveal an RNA duplex unwinding activity. Cell 48:599–605.

Robertson HD (1982): *Escherichia coli* ribonuclease III cleavage sites. Cell 30:669–672.

Robertson HD (1990): *Escherichia coli* ribonuclease III. Methods Enzymol 181:189–202.

Sambrook J, Fritsch EF, Maniatis T (1989): Ligases, kinases, and phosphatases. In Nolan C (ed): Molecular Cloning: A Laboratory Manual. Cold Spring Harbor, NY: Cold Spring Harbor Laboratory Press, pp 5.61–5.72.

Schuster W, Wissinger B, Unseld M, Brennicke A (1990): Transcripts of the NADH-dehydrogenase subunit 3 gene are differentially edited in *Oenothera* mitochondria. EMBO J 9:263–269.

Scott J (1989): Messenger RNA editing and modification. Curr Opinion Cell Biol 1:1141–1147.

Sharmeen L, Bass B, Sonenberg N, Weintraub H, Groudine M (1991): Tat-dependent adenosine-to-inosine modification of wild-type transactivation response RNA. Proc Natl Acad Sci USA 88:8096–8100.

Takayama KM, Inouye M (1990): Antisense RNA. Crit Rev Biochem Mol Biol 25:155–184.

Uhlenbeck OC, Gumport RI (1982): T4 RNA ligase. In Boyer PD (ed): The Enzymes, Vol 15. New York: Academic Press, pp 31–58.

Wagner RW, Nishikura K (1988): Cell cycle expression of RNA duplex unwindase activity in mammalian cells. Mol Cell Biol 8:770–777.

Wagner RW, Smith JE, Cooperman BS, Nishikura K (1989): A double-stranded RNA unwinding activity introduces structural alterations by means of adenosine to inosine conversions in mammalian cells and Xenopus eggs. Proc Natl Acad Sci USA 86: 2647–2651.

Wagner RW, Yoo C, Wrabetz L, Kamholz J, Buchhalter J, Hassan NF, Khalili K, Kim SU, Perussia B, McMorris FA, Nishikura K (1990): Double-stranded RNA unwinding and modifying activity is detected ubiquitously in primary tissues and cell lines. Mol Cell Biol 10:5586–5590.

Weissmann C (1989): Scientific correspondence: Single-strand RNA. Nature 337:415–416.

Weissmann C, Cattaneo R, Billeter MA (1990): RNA editing. Sometimes an editor makes sense [news]. Nature 343:697–699.

Wong TC, Ayata M, Hirano A, Yoshikawa Y, Tsuruoka H, Yamanouchi K (1989): Generalized and localized biased hypermutation affecting the matrix gene of a measles virus strain that causes subacute sclerosing panencephalitis. J Virol 63:5464–5468.

Wormington WM (1986): Stable repression of ribosomal protein L1 synthesis in Xenopus oocytes by microinjection of antisense RNA. Proc Natl Acad Sci USA 83:8639–8643.

Yokoyama K, Imamoto F (1987): Transcriptional control of the endogenous MYC protooncogene by antisense RNA. Proc Natl Acad Sci USA 84:7363–7367.

Zielke CL, Suelter CH (1971): Purine, purine nucleoside, and purine nucleotide aminohydrolases. In Boyer PD (ed): The Enzymes, Vol 4. New York: Academic Press, pp 47–78.

ABOUT THE AUTHOR

BRENDA L. BASS is an Assistant Professor of Biochemistry at the University of Utah, where she teaches nucleic acid biochemistry. Dr. Bass received her B.A. in chemistry from Colorado College and subsequently obtained a Ph.D. at the University of Colorado in the laboratory of Dr. Thomas Cech. Her graduate work involved characterization of the kinetics and guanosine binding site of the *Tetrahymena* ribozyme. As a postdoctoral fellow in Dr. Harold Weintraub's laboratory, Dr. Bass discovered the dsRNA unwinding/modifying activity that she continues to study in her laboratory at the University of Utah. Dr. Bass's research publications have appeared in *Nature*, *Biochemistry*, and *Cell*. She is the recipient of awards from the Pew Scholars Program and the David and Lucile Packard Foundation.

Antisense RNA and DNA: 175–194
© 1992 Wiley-Liss, Inc.

Artificial Regulation of Gene Expression by Complementary Oligonucleotides— An Overview

Jean-Jacques Toulmé

I. INTRODUCTION

The regulation of gene expression by nucleic acids depends on the formation of hybrids between the sense sequence—the one that bears the genetic information—and the complementary one, the so-called antisense sequence. The formation of the sense–antisense duplex interferes with one of the steps leading from the information contained in double-stranded DNA to the encoded protein. The inhibition of gene expression, which results from hydrogen bonding between nucleic acid–base from the sense and antisense strands, is highly specific. This mode of regulation can be considered for any gene whose sequence is known: The antisense sequence is deduced in a straightforward manner from the sense information. Its simplicity, specificity, and elegance make this strategy extremely attractive.

The first example of what we would now call an antisense oligonucleotide dates to 1967: An alkylating oligoribonucleotide was designed to modify selectively a target nucleic acid [Belikova et al., 1967]. In the 1970s several examples of specific interactions between a nucleic acid and complementary sequences that resulted in the blockage of a biologic process were reported. Short oligoribonucleotides and modified oligodeoxyribonucleotides have been targeted to transfer [Barrett et al., 1974; Miller et al., 1977] and ribosomal [Eckhardt and Lührmann, 1979; Taniguchi and Weissmann, 1978; Trudel et al., 1981] RNAs. Oligonucleotides complementary to messenger [Hastie and Held, 1978; Paterson et al., 1977] or viral [Stephenson and Zamecnik, 1978; Zamecnik and Stephenson, 1978] RNAs have been used as well. The discovery of natural gene regulation via antisense RNA in prokaryotes and in eukaryotes has led to a blossoming of studies in the field [Inouye, 1988; Simons, 1988; Simons and Kleckner, 1988; Takayama and Inouye, 1990; van der Krol et al., 1988b]. Antisense sequences can be engineered either genetically (antisense RNA) or chemically (antisense olignucleotides). Even if conceptually identical, these two approaches have different requirements and uses. This chapter will review only the synthetic oligonucleotides.

Major developments in molecular biology and advances in nucleic acid chemistry have made oligonucleotides readily available to almost any laboratory. Whereas numerous applications in molecular biology require standard unmodified oligomers (primers, linkers), the improvement of antisense properties in biological systems has led to the design of chemically modified molecules. The backbone of the molecules has been made resistant to nucleases. Various ligands have been linked at the end of the oligonucleotide to improve its binding affinity or its cellular uptake. We now face a panoply of molecules with their advantages and their inherent drawbacks. Numerous genes have been successfully targeted either in cell-free extracts or in intact cells.

In the last few years a number of reviews have been devoted to chemical [Goodchild, 1990; Uhlmann and Peyman, 1990], therapeutic [Zon, 1988], and technical [Marcus-Sekura, 1988] aspects, as well as to biological properties of antisense oligonucleotides [Hélène and Toulmé, 1990; Stein and Cohen, 1988, 1989; Toulmé and Hélène, 1988; van der Krol et al., 1988a]. In this chapter, I will focus on two key questions: Which target for complementary oligomers? Which antisense oligonucleotide?

II. WHICH TARGET?

Any nucleic acid sequence can be chosen as a target for antisense oligonucleotides; consequently gene expression might be affected at different steps (Table I). At the early stage only single-stranded binding sites (any kind of RNA) have been selected. Recently the formation of triple-stranded structures, at least in particular regions, has also been considered, making it possible to target the precursor DNA molecule itself. The sequence-dependent control of gene expression by oligonucleotides has been further extended to the design of traps for nucleic acid–binding factors, leading to what I will call the *sense approach*.

Not all these targets exhibit the same potential interest with respect to the regulation of gene expression. As mentioned, the first attempts were made against tRNAs or rRNAs. In the first case oligonucleotides complementary to the CCA stem or to the anticodon region were used [Miller et al., 1981]. In the second case the ribosome-binding site, the so-called Shine-Dalgarno sequence of the 16S *Escherichia coli* rRNA, was chosen as a target [Eckhardt and Lührmann, 1979; Jayaraman et al., 1981; Taniguchi and Weissmann, 1978; Trudel et al., 1981]. In all cases the investigators observed an inhibition of translation. Later,

TABLE I. Targets for Antisense Oligonucleotides and Steps of Gene Expression That Might Be Subsequently Affected

DNA	Replication
	Transcription
U snRNA	Splicing
Pre-mRNA	Splicing
	Polyadenylation
	Decay
tRNA	Translation
Retroviral RNA	Reverse transcription
mRNA	Transport
	Decay
	Translation
Proteins	Replication
	Transcription
	Maturation
	Translation

a similar approach was used with oligonucleotides complementary to U RNAs, the nucleic acid components of snRNPs, which are involved in the splicing of pre-mRNA. Both unmodified and 2'-O-methyl analogs of oligonucleotides have been tested in cell-free extracts and in injected *Xenopus* oocytes. The accessibility and the function of different regions of these RNAs were assessed in several species (human, yeast, *Xenopus*): The splicing reaction was inhibited by oligomers complementary to the 5' end of U1, U2, U4, and U6 snRNAs [Berget and Robberson, 1986; Black et al., 1985; Krainer and Maniatis, 1985; Krämer et al., 1984]. Such studies demonstrated the potential of these nucleic acids. Despite their interest as tools in molecular biology, no specificity can be expected from the use of antisense oligonucleotides complementary to such widespread targets as tRNA, rRNA or U snRNA: The expression of all genes within the cell (or in a cell-free mixture) will be affected by a single oligonucleotide sequence.

In contrast, effects restricted to a single gene can be obtained with oligomers targeted to mRNA, pre-mRNA, or DNA sequences. At present this represents a powerful means of downregulating the expression of a preselected gene. Despite the fact that antisense science is still in its infancy, a few guidelines for using antisense oligonucleotides successfully in preventing a gene from being translated, spliced, or transcribed have begun to emerge. I will now describe the effects of antisense oligonucleotides with respect to the location of their target.

A. Cytoplasmic Targets

1. Translation. Antisense oligonucleotides have been shown to prevent translation of mRNA of both prokaryotic and eukaryotic origin. Two possible modes of action for passive oligomers (which do not modify the target mRNA) are indicated in Figure 1. The double-stranded oligonucleotide–mRNA hybrid can act through steric hindrance (Fig. 1a) and prevent ribosomes from binding to or scanning the message. Alternatively, inhibition of translation will result from the degradation of the target mRNA

by RNase H (Fig. 1b), an enzyme that hydrolyzes the RNA moiety of RNA–DNA complexes.

Clear-cut information obtained from experiments performed in either cell-free extracts or microinjected cells has shown that the relative contribution of these two mechanisms depends on both the biological system under investigation and the location of the target along the mRNA (see also Minshull and Hunt, this volume). Oligonucleotides complementary to the coding region are not inhibitory unless the mRNA is cleaved by RNase H [Cazenave et al., 1987b; Minshull and Hunt, 1986; Shuttleworth and Colman, 1988; Walder and Walder, 1988]. Consequently, derivatives such as α-oligonucleotides that do not elicit RNase H activity do not inhibit protein synthesis if targeted downstream of the AUG [Cazenave et al., 1989; Gagnor et al., 1987; see Morvan et al., 1990, for review of α-oligomers]. Moreover, oligonucleotides that induce cleavage of the target mRNA are more efficient at inhibiting polypeptide chain elongation in wheat germ extract than in rabbit reticulocyte lysate [Blake et al., 1985a; Boiziau et al., 1991b; Minshull and Hunt, 1986]. In the former system, RNase H is fairly active under in vitro translation conditions. Indeed, RNA cleavage products have been identified following incubation of mRNA in the presence of a complementary oligomer in wheat germ extract [Cazenave et al., 1987b]. In contrast, in reticulocyte lysate the RNase H activity has been reported to vary from a low level to zero, at least in commercial extracts. In a few cases, no inhibition was observed unless exogenous RNase H was added [Maher and Dolnick, 1988; Minshull and Hunt, 1986]. Moreover, in the reticulocyte lysate system a strand-separating activity [not to be confused with the double-stranded RNA unwinding/ modifying activity discussed by Bass (this volume)] associated with the elongating ribosome has been characterized that is responsible for duplex dissociation [Liebhaber et al., 1984; Shakin and Liebhaber, 1986]. As discussed later, the location of the target sequence on the mRNA is less important for reactive oligonu-

a)

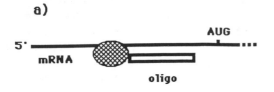

b)

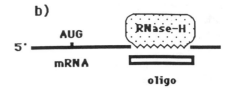

c)

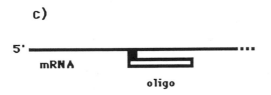

d)

Fig. 1. *Different mechanisms for inhibition of translation by antisense oligomers.* **a:** *Antisense oligonucleotides targeted upstream of the initiation codon AUG; antisense can block the initiation complex through physical arrest.* **b:** *Oligonucleotides complementary to the coding region prevent the polypeptide chain elongation only if the mRNA is cleaved by RNase H unless irreversible damage, crosslink* **(c)**, *or cleavage* **(d)** *are introduced by antisense oligomers linked to active groups such as photosensitizers or metal complexes. (Adapted from Toulmé et al., 1990, with permission of the publisher.)*

cleotides than for passive ones: With the former the damaged mRNA (either cleaved or crosslinked) can no longer support protein synthesis (Fig. 1c,d).

Inhibition of protein synthesis by oligonucleotides complementary to the 5′ leader of mRNA (from the cap to a few nucleotides downstream of the AUG initiation codon) occurs independently of RNase H activity [Maher and Dolnick, 1988]. α-Oligomers complementary to either the cap site or the AUG initiation region of the rabbit β-globin mRNA selectively blocked β-globin synthesis [Bertrand et al., 1989; Boiziau et al., 1991b]. Nevertheless, the situation is not yet clear: Walder and Walder [1988] have concluded that the RNase H pathway was predominant even for oligomers targeted to the 5′ end of mRNAs, whereas others came to the opposite conclusion, since α-oligomers complementary to the cap region of the rabbit β-globin message were more efficient inhibitors than homologous β-oligonucleotides in rabbit reticulocyte lysate [Bertrand et al., 1989; Boiziau et al., 1991b].

RNase H has been shown to mediate the antisense effect both in *Xenopus* [Cazenave et al., 1989; Dash et al., 1987; Jessus et al., 1988; Shuttleworth and Colman, 1988] and mouse [O'Keefe et al., 1989; Paules et al., 1989] oocytes. There is no direct proof of its involvement in other eukaryotic cells, but inhibition of protein synthesis by oligonucleotides complementary to the coding region of mRNA [Bories et al., 1989] suggests that this might be a general mechanism of inhibition if one assumes that arrested elongation cannot occur with an intact message. Moreover, in a few cases a decrease in the steady-state level of the target RNA has been reported following treatment of cultured cells with oligonucleotides [Gewirtz et al., 1989; Harel-Bellan et al., 1988].

In a number of other cases successful inhibition of gene expression in mammalian cell lines has been reported with oligonucleotides targeted to the AUG initiation region of the mRNA, but it is not known whether the effects are mediated by RNase H (e.g., Matsukura, this volume; Wickstrom, this volume).

2. Reverse transcription. One of the first events following the entry of a retrovirus into the host cell is the synthesis of a cDNA copy of the RNA genome by a viral RNA-dependent DNA polymerase (reverse transcriptase). This is accomplished in several steps involving an association between complementary nucleic acid sequences (see To, this volume, for details). Therefore there is room for competition with antisense molecules. The inhibition of reverse transcription by synthetic oligonucleotides has been achieved through different processes: competition with the primer, arrest of cDNA elongation, binding to the polymerase.

The use of a primer analog that is not recognized by the reverse transcriptase will prevent DNA synthesis. Such an approach has been described by Gagnor et al. [1989]: An α-oligomer, bound in a parallel orientation to its complementary sequence, was not elongated by the reverse transcriptase of the avian myeloblast virus. Competition between such a derivative and a conventional primer led to decreased cDNA synthesis. Another possibility to block reverse transcription is to use an oligonucleotide bound to the template RNA downstream of the primer. In a similar manner to what occurs for translation, the hybrid could block the scanning of the RNA. We have investigated such a possibility using the AMV reverse transcriptase and various hybrids. An unmodified 17-mer, complementary to the 5' end of the rabbit β-globin mRNA, led to the synthesis of a truncated cDNA fragment the size of which was consistent with an abortive polymerization at the level of the oligonucleotide–RNA complex [Loreau et al., 1990]. The arrested synthesis did not arise from a competition between the stopper oligonucleotide and the enzyme but resulted from the cleavage of the template RNA by the RNase H activity associated with the polymerase. Indeed, the 17-mer methylphosphonate and α analogs, which do not induce RNA cleavage, did not prevent cDNA synthesis [Boiziau et al., 1992a]. However, hybrid-arrested re-

verse transcription was observed with an alpha antisense oligomer adjacent to the primer binding site. Oligonucleotides that interact directly with the enzyme might also interfere with reverse transcription. This has been described with several modified oligomers: phosphorothioates [Majumdar et al., 1989], polylysine conjugates [Chu and Orgel, 1989], α-analogs [Bloch et al., 1988], and acridine-linked oligonucleotides [Loreau et al., 1990]. If these interactions display some specificity compared with other polymerases this could lead to interesting properties: The formation of oligonucleotide–reverse transcriptase complexes might be responsible for the inhibition of HIV development in cultured cells by phosphorothioate oligomers such as $S-(dC)_{28}$ [Matsukura et al., 1987] (see also Matsukura, this volume).

B. Nuclear Targets

Penetration of the nuclear membrane by oligonucleotides is a prerequisite to targeting nuclear sequences. A few examples indicate that this is actually the case. ^{32}P-labeled oligomer, associated with its target mRNA, has been extracted from the nucleus of rat PC12 cells following incubation with antisense oligonucleotide [Teichman-Weinberg et al., 1988]. It has also been demonstrated that fluorescence accumulates in the nucleus of Swiss 3T3 cells that have been microinjected with fluorescein- or rhodamin-linked oligomers [Leonetti et al., 1991]. This is true for both phosphodiester and phosphorothioate conjugates, the latter being detected for a longer time (up to 24 hours) than the former ones. We came to a similar conclusion in the case of procyclic trypanosomes incubated in the presence of an acridine-linked 6-mer with an unmodified backbone [Verspieren et al., 1987]. Lastly, the biological results described below definitely demonstrate that nuclear nucleic acid sequences can be targeted by antisense oligonucleotides.

Therefore, in intact cells, antisense oligonucleotides can bind to their complementary sequence not only on mRNA but also on pre-mRNA and even DNA. By so doing, they can potentially interfere with nuclear events such as transcription, splicing, or the export of mature transcripts to the cytoplasm. Up to now this latter possibility has been documented only with antisense RNA [Kim and Wold, 1985]. Other possibilities cannot be ruled out, as examples of such effects have been described either in cell-free systems or in intact cells.

1. Splicing. Besides binding to U snRNAs, antisense oligonucleotides can prevent maturation of mRNA if targeted to pre-mRNA sequences. Both unmodified and modified oligomers complementary to splice junctions have been used, particularly in the case of viruses. The acceptor splice site of an immediate early gene of herpes simplex virus (HSV 1) has been selected as a target for methylphosphonate oligonucleotides by Miller, Ts'o, and coworkers (see Miller, this volume): Inhibition of viral protein synthesis and decrease of virus titer in cultured Vero cells have been observed when oligonucleotides were added at the time of infection or postinfection [Kulka et al., 1989; Smith et al., 1986]. Antiviral properties have been improved by using a methylphosphonate analog coupled to a psoralen derivative. The formation of covalent adducts between the antisense and the pre-mRNA sequences, following UV irradiation, led to inhibition at concentrations 100-fold lower than nonconjugated oligomer [Kulka et al., 1989]. Several donor and acceptor splice sites of the HIV genome have been targeted with unmodified [Zamecnik et al., 1986] or modified oligonucleotides, such as methylphosphonates [Sarin et al., 1988; Zaia et al., 1988], phosphorothioates, phosphoramidates [Agrawal et al., 1989], and oligomers linked to various ligands [Letsinger et al., 1989; Stevenson and Iversen, 1989]. These types of modifications are illustrated by Murray and Crockett (Fig. 2, this volume). All these compounds prevented viral development, reverse transcriptase activity, and syncytia formation in newly infected cells, to varying extents.

2. Transcription. At first sight there is no room for antisense oligonucleotide-mediated inhibition at the level of transcription, as this would require an interaction with either DNA or with proteins. However, open regions are formed upon binding of the RNA polymerase

to the promoter region and in front of the transcription complex. These sites might be accessible to complementary oligonucleotides, at least transiently. It has also been shown that triple-stranded structures can form in homopurine/homopyrimidine regions upon the association of an oligopyrimidine sequence through so-called Hoogsteen hydrogen bonding. Triple helices might interfere with either protein binding to, or RNA synthesis from, the target sequence. In addition, DNA-binding factors are known to modulate transcription efficiency. The specific association of oligonucleotides with such factors might result in either enhanced (if the factor were a repressor) or decreased (if the factor were an activator) RNA synthesis. All three possibilities of transcription interference have been investigated.

The single-stranded region of the transcription initiation complex can be targeted by oligonucleotides. Oligonucleotides complementary to the region from -8 to $+2$ of the T7 promoter A2 have been used to ascertain the size of the bubble created by the binding of the polymerase [Grachev et al., 1984]. It has also been reported that an oligonucleotide linked to an intercalating agent, complementary to the open complex formed by the *bla* gene and the *E. coli* RNA polymerase, blocked initiation of in vitro RNA synthesis in a specific way: This anti-*bla* oligomer had no effect on the transcription from the *lac* promoter [Hélène et al., 1985].

Recently, the possibility of targeting double-stranded regions of DNA has also been considered, which offers the potential advantage that there is only one site per gene compared with the many that must be hit if mRNA is targeted. The formation of triple helices was demonstrated a long time ago: binding of a thymine (T) to an adenine-thymine (A-T) pair or of a protonated cytosine (C^+) to a guanine-cytosine (G-C) pair via Hoogsteen hydrogen bonding results in the selective formation of T.A-T and C^+.G-C triplets [see Wells et al., 1988, for review] (see also Murray and Crockett, Fig. 3, this volume). Thus in the triple helix motif pyrimidine oligonucleotides bind in the major groove parallel to the purine strand of the Watson Crick double helix. The occurrence of such triple-stranded structures involving oligonucleotides has been demonstrated using oligopyrimidine sequences linked to various reagents: the association of the oligomer with a DNA duplex brought the reagent to the vicinity of the DNA bases, which introduced irreversible damage in both target strands. Crosslinking [Le Doan et al., 1987; Praseuth et al., 1988; Vlassov et al., 1988a], and cleaving groups [François et al., 1989a; Moser and Dervan, 1987; Perrouault et al., 1990; Povsic and Dervan, 1989] have been used to demonstrate the formation of the triplex. Up to now triple helices have been restricted to recognition of purine tracts by pyrimidine oligonucleotides, even if more recently the repertoire of base triplets has been formally extended to the formation of G.G-C, A.A-T [Letai et al., 1988], and G.T-A [Griffin and Dervan, 1989] triplets, although it should be noted that with such deviations from the homopyrimidine-homopurine pairing limitations on sequence composition exist. Normal triple helical structures form under ionic and temperature conditions compatible with the association of biological macromolecules. The stability of complexes can be extended to neutral pH by using pyrimidine residues substituted at position 5 with bromine (5-bromouracil) or a methyl group (5-methylcytosine) without alteration of their specificity for the homopurine target sequence [Povsic and Dervan, 1989].

Homopyrimidine oligomers, complementary to a DNA sequence, can therefore compete with DNA-binding proteins provided that the hybridization site overlaps (or is close to) the binding site. This has been demonstrated for restriction endonucleases, and for a transcription factor [François et al., 1989b; Hanvey et al., 1990; Maher et al., 1989; Strobel and Dervan, 1991]. Moreover the formation of a complex between a purine-rich 27-mer and a double-stranded sequence containing the *myc* promoter reduced the in vitro transcription of the target [Cooney et al., 1988]. A similar effect has been recently described in cultured HeLa cells [Postel et al., 1991].

The specific binding of transcription factors to regulatory sequences modulates the activity of the RNA polymerase complex. The introduction of a competitor sequence might trap such a protein, subsequently leading to a modification of the transcription yield. Of course, the competitor oligonucleotide should mimic the binding site of the regulatory factor and therefore should be provided as a double strand. Several attempts have been made to regulate gene expression at the transcription step through this approach. Harel-Bellan and coworkers failed to decrease the expression of the target gene using a 14-base pair heat shock element [Harel-Bellan et al., 1989]. A negative result has also been obtained with an oligomer mimicking the interferon consensus sequence (Gagnor and Lebleu, personal communication). There is no reason why such an approach should fail, if the protein–oligonucleotide complex forms, which has actually been demonstrated in the first case. Indeed, positive results have been reported using competing plasmid-born sequences. Finally, oligonucleotides that interact selectively with protein targets can be selected [Blackwell et al., 1990a,b; Ellington and Szostak, 1990; Tuerk and Gold, 1990] (see also Murray and Crockett, this volume).

III. WHICH ANTISENSE?

Conventional, unmodified oligonucleotides have weaknesses when used as regulatory agents in biological environments (cell-free systems, growth media, intact cells). In particular they are susceptible to degradation by DNases, and they are taken up by cells with a poor efficiency. Moreover, as the limiting parameter is the amount of bound antisense, oligonucleotides should have a high affinity for their target. Chemical modifications can be introduced into the molecule provided they do not impair the recognition of the complementary sequence and do not result in "toxic" effects. Taking into account these different criteria has led to the introduction of numerous chemical modifications in order to improve some of the properties of antisense oligonu-

TABLE II. Design of Chemically Modified Oligonucleotides*

Specificity	Target recognition	Length
		RNase H sensitivity
	Protein binding	Sequence
Stability	Degradation by nucleases	Phosphotriesters
		Alkylphosphonates
		α-Oligomers
		Phosphorothioates
		Phosphorodithioates
		Phosphoramidates
		2'-O-alkyl
RNase H induction	mRNA cleavage	Phosphodiester
		Phosphorothioate
Uptake	Membrane penetration	Length/charge
	Endocytosis	Hydrophobic groups
		Polycations
		Lipoproteins
		Liposomes
Affinity	Competition with proteins	Modified bases
		Modified backbone
		Intercalating agents
		Active groups
Toxicity	Nonspecific binding	
	Mutagenicity	
	Carcinogenicity	

*The criteria indicated in the first column are considered to take into account the problems listed in the second column. Improvement of antisense properties has been achieved by introduction of the modifications given in the last column (see text for details).

cleotides (Table II). Before describing the properties of these modified molecules, I will first consider the specificity of the effects induced by antisense oligomers.

A. Specificity

Discussion of this point is frequently confusing because people usually mean at the same time both the uniqueness of the target sequence and the restriction of the biological effects to the target gene (biological specificity). The uniqueness of the target is mathematically related to its length and to its sequence (not base composition) on the one hand and to the complexity of the genome on the other hand. This has been discussed in detail elsewhere [Hélène and Toulmé, 1989]. Suffice it to say here that a calculation taking into account the nonrandom distribution of dinucleotides in human genomic DNA led to a minimal length of 11–15 nucleotides, if one assumes that only 0.5% of the DNA is transcribed. The two figures are due to the fact that the human genome contains 60% A.T pairs, and they correspond to targets containing only G/Cs or A/Ts, respectively.

Of course, the biological specificity is also related to the length and to the sequence of the oligonucleotide, as the active oligonucleotide fraction is the one that is bound to the target, but at this point it should be realized that the optimal length (and sequence) for an antisense molecule will be a compromise resulting from opposing requirements. The uniqueness of the target will be guaranteed by lengthening the sequence. However, the biological specificity will primarily depend on the difference between free energies of formation of perfect and unperfect duplexes. This difference will be larger for short sequences than for long ones: The relative destabilizing contribution of a mismatch is inversely related to the oligonucleotide length. A single mismatch will prevent the formation of a 12 base pair duplex but will only slightly weaken a 30 base pair complex. In the latter case nonspecific biological effects will be observed [Verspieren et al., 1990]. It should also be added that longer oligomers can fold back, sequestering part of the molecules into an inactive form.

In addition, the biological specificity depends on parameters that are intrinsic to the system in which the antisense has to be used. The association between two complementary nucleic acid sequences depends on both ionic and temperature conditions. Although the ion concentration is not expected to vary widely from one cell to the other, translation takes place at 18°C in *Xenopus* oocytes and at 37°C in cultured eukaryotic cells. Therefore, a given oligonucleotide will bind more efficiently in the first case than in the second one. Moreover, translation inhibition by antisense oligonucleotides results from the interplay between numerous partners (see above): RNase H and unwinding activity have been shown to be involved in cell-free systems [Cazenave et al., 1987b; Minshull and Hunt, 1986; Shakin and Liebhaber, 1986]. This should also be taken into account as RNase H has been demonstrated to induce RNA cleavage at the level of mismatched duplexes [Cazenave et al., 1987b]. This means that the "ideal" antisense oligonucleotide will not be the same for investigations in different systems.

Although the above discussion concerns mainly unmodified oligonucleotides, it largely holds true for modified ones. Additional problems, most likely due to interactions with proteins, can be caused by chemical modifications. The nonspecific inhibition of reverse transcription by phosphorothioate [Matsukura et al., 1987] and α-oligomers [Bertrand et al., 1989] was ascribed to sequence-independent complexes formed by these modified oligonucleotides and reverse transcriptase. Similarly, we have observed a nonspecific inhibition of tran-

scription in S30 *E. coli* extracts by acridine-linked oligomers [Toulmé et al., 1986]. This was probably due to oligomer binding to the bacterial RNA polymerase. Incidentally, it should be noted that using a nonrelated oligonucleotide (e.g., sense, random) as a control does not guarantee that the biological effects induced by the antisense sequence are actually due to its nucleic acid target, as interactions with proteins might display both a sequence and a length dependence.

B. Nuclease Resistance and RNase H Sensitivity

Conventional oligonucleotides are substrates for nucleases. The degradation is rather limited in cell-free extracts, but oligomers are rapidly chopped up in HL-60 nuclear extracts [Furdon et al., 1989] and in serum-supplemented growth media. The half-life of an oligonucleotide has been reported to vary from minutes to several tens of hours [Cazenave et al., 1987a; Holt et al., 1988; Leonetti et al., 1988; Verspieren et al., 1987; Wickstrom, 1986] (see also Tidd, this volume; Wickstrom, this volume). The variation is due to several parameters: the oligonucleotide concentration used; the temperature of serum decomplementation, which results in variable nuclease inactivation; and the batch of serum. Heating the serum at 65°C for 30 minutes greatly reduced the nuclease content [Holt et al., 1988]. It seems that 3' exonuclease is the major activity present in fetal calf serum; consequently oligomers linked to a 3' substituent are degraded 10 times more slowly than normal ones [Stein et al., 1988a; Verspieren et al., 1987]. The degradation of oligonucleotides within the cell is also documented: In *Xenopus* oocytes the half-life for a 16-mer was determined to be about 10 minutes [Cazenave et al., 1987a]. In contrast, intact oligonucleotides have still been detected after a 6 hour incubation in rat pheochromocytoma cells [Teichman-Weinberg et al., 1988].

As indicated in Table II, the action of endonucleases can be overcome by various modifications of the molecule. Phosphodiester internucleoside linkages have been substituted with phosphotriester [Barrett et al., 1974; Miller et al., 1977], methylphosphonate [Blake et al., 1985b; Miller et al., 1981] (see also Tidd, this volume; Miller, this volume), phosphoramidate [Agrawal et al., 1988], phosphorothioate [Latimer et al., 1989; Stein et al., 1988b] (see also Matsukura, this volume), phosphorodithioate [Caruthers, 1989], or phosphoroselenoate [Mori et al., 1989], leading to compounds much less sensitive to DNases than the parent ones. Oligomers made up of α-isomers of nucleoside units also gave rise to nuclease-resistant molecules that hybridize to form α-β hybrids with the target in which the strands run in parallel, in contrast to the antiparallel orientation of β-β duplexes [Morvan et al., 1986; Thuong et al., 1987]. In *Xenopus* oocytes an α-16-mer was only 40% degraded after 8 hours [Cazenave et al., 1987a].

All these derivatives bind to their complementary sequence, generally with a reduced affinity compared with conventional molecules, but not all of them induce a decrease in protein synthesis. As discussed above, this is essentially related to their ability to elicit RNase H activity. From this point of view nuclease-resistant oligomers can be divided into two categories: 1) phosphorothioates and phosphoroselenoates, which induce RNA cleavage; and 2) phosphonates, phosphoramidates, and α-oligomers, which do not. Consequently, oligonucleotides of the second class do not arrest polypeptide chain elongation but they do block translation initiation [Boiziau et al., 1991b; Cazenave et al., 1989; Gagnor et al., 1987; Maher and Dolnick, 1988]. It should be noted that the relative susceptibility of oligonucleotide–RNA hybrids to RNase H also depends on the sequence: It has been reported that phosphorothioate oligo(dT) form better substrates than unmodified oligo(dT) when bound to poly(rA) [Stein et al., 1988b] but the reverse was true for an oligomer containing all four bases [Cazenave et al., 1989].

Because of their nuclease resistance and their ability to elicit RNase H activity, phosphorothioate analogs have been widely used in intact cells. Their advantage over unmodified oligonucleotides has been demonstrated by experiments carried out in frog oocytes with a 17-mer targeted to the coding region of the rabbit β-globin mRNA. A normal oligonucleotide was an efficient antisense molecule only if it were coinjected with the mRNA, whereas inhibition of β-globin synthesis was still observed when the injection of phosphorothioate antisense was carried out 6 hours prior to mRNA, indicating that intact molecules were still present [Cazenave et al., 1989]. Moreover, a catalytic antisense effect has been observed upon coinjection of mRNA and phosphorothioate 17-mer: 50% reduction of protein synthesis was achieved at an oligonucleotide concentration (4 nM) fourfold lower than that of the target. Similarly, as discussed in more detail by Matsukura (this volume), Matsukura et al. [1989] found that the expression of the HIV *rev* gene was significantly inhibited in chronically infected cells by a phosphorothioate oligomer complementary to the translation initiation region, but not by conventional phosphodiester oligonucleotides. This inhibition was still apparent after 28 days in culture.

However, phosphorothioate derivatives do not fulfill all the requirements listed in Table II. In particular, nonspecific association with proteins has been demonstrated to lead to nonspecific inhibition in the micromolar concentration range [Cazenave et al., 1989]. An alternative is constituted by oligonucleotides containing mixtures of modified and unmodified linkages: A block of natural nucleotides sandwiched between two blocks of resistant nucleotides (methylphosphonates or phosphoramidates) results in an oligomer that is protected from exonucleases, both in cell extracts and in *Xenopus* oocytes, while retaining its ability to induce the cleavage of the complementary RNA strand by RNase H [Agrawal and Goodchild, 1987; Dagle et al., 1990; Furdon et al., 1989; Quartin et al., 1989; Shibahara et al., 1987]. Such analogs with a long lifetime can be used at higher concentrations without inducing nonspecific toxic effects caused by the release of breakdown products [Tidd and Warenius, 1989]; their uses and potential advantages are discussed in detail by Tidd (this volume). Moreover these chimeras could give rise to improved specificity. An unmodified antisense oligomer, complementary to a region of the rabbit β-globin mRNA that displayed strong homology with the α-message, caused inhibition of synthesis of both globins in vitro. RNase H did not discriminate between the perfectly and partially matched hybrids (see also Minshull and Hunt, this volume). Using an antisense made of the insertion of a phosphodiester window, centered on the variant region, sandwiched between two blocks of methylphosphonate complementary to the conserved parts of the targets, restored some specificity (Boiziau, Shire, Blonski and Toulmé, unpublished results).

C. Uptake–Transport

Contrary to the prevailing view, oligonucleotides cross cellular membranes despite their high negative charge density, as indicated by the specific biological effects observed with intact cultured cells. However, limited reliable information is available regarding the quantitation of uptake and the intracellular location of antisense oligomers. First, ^{32}P end-labeled oligonucleotides have been used in most studies, and part of the cell-associated radiolabeled material might be breakdown products. Second, it is often difficult to discriminate between membrane-bound and internalized molecules.

It has been reported in several studies that the uptake of oligonucleotides by intact cultured cells reaches a plateau after a 1–4 hour incubation, at a level amounting to about 10% of the external concentration [Harel-Bellan et al., 1988; Loke et al., 1989; Yakubov et al., 1989; Zamecnik et al., 1986]. Intact unmodified oligomers have been extracted from cultured pheochromocytoma cells [Teichman-Weinberg et al., 1988], human leukemia T cells [Harel-Bellan et al., 1989], and Krebs ascite cells [Knorre et al., 1985] after several hours incubation. The mechanism by which oligonucleotides enter the cell is not clear yet. Two different studies identified an 80 kD protein that might mediate endocytosis of oligonucleotide

in HL-60 cells [Loke et al., 1989] and L929 cells [Yakubov et al., 1989]. This is a saturable, energy-dependent process. The enhancement of the oligonucleotide uptake by chloroquine (which prevents degradation of the endocytosed ligand and receptor) and its decrease by sodium azide (which inhibits internalization of endocytic vesicles) support the idea that at least part of the molecules are taken up through an endocytotic pathway [Neckers, 1989]. This has also been suggested for fluorescently tagged oligonucleotide–polylysine conjugates in L929 cells [Leonetti et al., 1990a]. But this process does not account for all the uptake, as neutral oligophosphonates do not compete with charged phosphodiester and phosphorothioate oligomers, suggesting a second mechanism that might be passive diffusion across the cell membrane [Miller et al., 1981].

Chemical modifications have been introduced to improve the uptake of antisense oligonucleotides by intact cells. Polycations and hydrophobic groups have been demonstrated to be efficient in a number of cases. Oligonucleotide–poly(L)lysine conjugates complementary to various regions of the VSV genome prevented the development of the virus in L929 cultured cells in the 10–100 nM concentration range, which is 1–2 orders of magnitude lower than other modified oligonucleotides [Degols et al., 1989; Lemaître et al., 1987]. However, some cell lines (HeLa or LM fibroblasts) were not protected against VSV infection by the conjugate, indicating that the efficiency of uptake may vary from cell to cell [Leonetti et al., 1988]. As discussed by Degols et al. (this volume), it has recently been demonstrated that polyanions like heparin potentiate the effects of such conjugates.

Oligomers linked to different hydrophobic groups have also been used. The internal concentration of cholesterol-linked oligonucleotides has been found to be higher than that obtained for unmodified ones [Boutorin et al., 1989]. A phosphorothioate linked to a cholesterol group, complementary to the *tat* gene of HIV, was more efficient at inhibiting viral development than both the homologous phosphodiester sequence and the nonconjugated phosphorothioate [Letsinger et al., 1989]. The expression of Ha-*ras* in T24 cells has been inhibited by an acridine-linked 9-mer conjugated to a dodecanol tail, whereas the parent compounds did not exhibit any inhibitory property [Saison et al., 1991]. However, in this case it has not been demonstrated that improved efficiency resulted from an increased uptake. We have determined that the uptake of a dodecanol-tailed 17-mer was only slightly increased in rabbit reticulocytes [Boiziau and Toulmé, 1992b]. Liposomes have also been used to deliver antisense molecules. A 15-mer complementary to the AUG region of c-*myc* that was inactive when added to HL-60 cells efficiently decreased c-*myc* protein synthesis when loaded into liposomes [Loke et al., 1988]. More recently, cell-specific delivery of antisense oligonucleotides has been achieved by liposomes associated with antibodies [Leonetti et al., 1990b].

It should be pointed out that any chemical modifications introduced into the oligonucleotide can perturb its uptake behavior: phosphorothioates that are trapped by membranes and taken up by cells more slowly than unmodified oligonucleotides [Neckers, 1989]. In contrast, the hydrophobic ring of the acridine derivative, used to enhance the binding of antisense oligonucleotides to their complementary sequence, improved their uptake by trypanosomes in culture [Hélène and Toulmé, 1989].

D. Affinity

Increased affinity of the oligonucleotide for its complementary sequence can be obtained by covalent linking to intercalating agents (Table II). If the linker is long enough, stacking interactions between the intercalator and base pairs of the oligonucleotide–RNA duplex will take place, thus providing the complex with additional energy for interaction. Indeed, an acridine derivative linked to the end of unmodified β- or α-oligomers led to duplexes of increased stability [Asseline et al., 1984; Thuong et al., 1987; Toulmé et al., 1986]. Such modified antisense oligomers have been used against dif-

ferent targets of either prokaryotic or eukaryotic origin. The expression of a T4 phage gene has been turned off in S30 *E. coli* extracts by oligomers linked to the acridine derivative, complementary to a region close to the ribosome attachment site [Toulmé et al., 1986]. Such modified 15- and 10-mers were more efficient antisense agents than homologous unmodified ones. Inhibition of rabbit β-globin synthesis has been observed, both in wheat germ extract and in injected frog oocytes, by an acridine-linked 11-mer spanning the AUG initiation codon [Cazenave et al., 1987b]. Interestingly, the unmodified oligonucleotide had no effect on the latter system. An octathymidylate, linked to the acridine derivative, inhibited the cytopathic effect of SV40 [Birg et al., 1990]. This was specific (noncomplementary acridine-linked oligonucleotides had no effect) and probably resulted from the association of the oligonucleotide with the A_8 sequence, which is close to the origin of replication of the viral DNA. However, it is not known whether this involves the formation of double- or triple-stranded structures.

We have also obtained a successful result with an oligonucleotide complementary to the 5' end of the so-called mini-exon sequence present on every message of the African trypanosome: An acridine-linked 9-mer was more efficient at inhibiting in vitro protein synthesis than an unmodified 12-mer [Verspieren et al., 1987]. Moreover, this modified 9-mer displayed trypanocidal properties when added to a culture of the insect stage of the parasite grown in a serum-free medium. This effect was selective, as neither noncomplementary acridine-linked oligonucleotides nor complementary unmodified oligomers exhibited any effect under the same conditions [Verspieren et al., 1987]. A similar approach was used against influenza virus, taking advantage of the presence of a conserved sequence at the end of viral RNAs. A complementary acridine-linked 7-mer prevented viral development in cultured cells. The specificity of the inhibition was demonstrated by the use of a different virus strain to which the oligomer was not complementary:

No inhibition was induced in this case [Zerial et al., 1987].

E. Active Oligonucleotides

The unmodified oligonucleotides and their analogs described thus far (e.g., nuclease resistant, linked to either intercalators or hydrophobic groups) are passive ones, which means that they can dissociate from the complementary sequence. The lifetime of such hybrids is finite, and therefore the inhibition they induce is transient. Coupling a reagent to the oligomer should cause irreversible damage to the target sequence, subsequently leading to a permanent block. Two classes of active compounds have been used, namely, crosslinking and cleaving reagents. Of course, agents of the second class offer the advantage of acting catalytically, whereas those belonging to the first one react stoichiometrically. These compounds and their biological applications have been extensively reviewed by Knorre et al. [1989].

In the first category, alkylating reagents were initially used more than 20 years ago to introduce sequence-specific modifications of nucleic acids [Belikova et al., 1967]. A chloroethyl derivative of benzylamine, which reacts with nucleophilic centers of nucleic acids (mainly N[7] of guanine and N[3] of adenine), has been most frequently used. Such compounds have been demonstrated to decrease selectively the phage M13 infectivity in JM103 *E. coli* cells to block the cytopathic effect of influenza and tick-borne encephalitis virus, and to prevent the synthesis of the immunoglobin light chain in mouse myeloma cells [Knorre et al., 1985, 1989]. Other crosslinked products leading to the inhibition of biological processes have been obtained with oligonucleotides bearing platinum derivatives [Chu and Orgel, 1989] and psoralens [Kean et al., 1988; Kulka et al., 1989]. In the latter case, the addition of the antisense to the target sequence is triggered by UV irradiation. This has allowed the control of the development of herpes virus in cultured Vero cells by a psoralen-linked 12-mer methylphosphonate analog complementary to the splice junction of immediate early

viral mRNA 4 because of the accumulation of unspliced RNA product [Kulka et al., 1989].

As far as translation is concerned, the location of the target sequence on the mRNA is less important for reactive oligonucleotides than for passive ones. Inhibition of the elongation step has been achieved by α-oligomers and methylphosphonate oligomers conjugated to either an alkylating group [Boiziau et al., 1991a] or to a photosensitizer [Kean et al., 1988], whose sequence was complementary to the coding region of the rabbit β-globin mRNA. Therefore, reactive antisense oligomers make RNase H activity superfluous.

Sequence-dependent cleavage of the RNA (or DNA) target can be achieved by antisense oligonucleotides linked to either nucleases or chemical reagents [see Hélène et al., 1989, for review]. The binding of the oligomer to its complementary sequence will bring the active group to the vicinity of the substrate. As a consequence, the cleaving reaction will be restricted to the area surrounding the oligonucleotide-binding site. The nonspecific staphylococcal nuclease has been given a sequence specificity by attachment to synthetic oligonucleotides [Corey and Schultz, 1987; Zuckermann et al., 1988; Zuckermann and Schultz, 1989]. Various chemical groups have been linked to the end of oligonucleotides in order to introduce strand breaks either directly (metal chelates and ellipticine) or indirectly (photosensitizers) following alkaline treatment. Fe^{2+}-EDTA, Cu^{+}-phenanthroline, and Fe^{2+}-porphyrin complexes generate hydroxyl radicals (OH·), which oxidize the sugar, subsequently leading to phosphodiester bond cleavage. These compounds introduce several cuts into the target sequence because of the diffusion of the active species. Interestingly their distribution is narrower for phenanthroline than for EDTA complexes. No biological application has been reported for such cleaving antisense oligonucleotides: This might in part be due to the low yield of cleavage (10%–15% for Fe^{2+} EDTA, about 60% for Cu^{+}-phenanthroline complexes). Indeed, a 10-mer linked to an Fe–EDTA complex did not reduce in vitro protein synthesis in S30 *E. coli* extracts any more efficiently than did the parent compound even though cleavage products derived from the target mRNA have been detected after an overnight incubation (Toulmé and Thuong, unpublished results).

Cleaving oligonucleotides (oligonucleotides linked to a nuclease) are inactivated as incubation proceeds because of the self-degradation of the chimera, ultimately leading to nonspecific cleavage. This can easily be overcome by the use of nuclease-resistant vector oligomers. Similarly, oligonucleotides carrying a chemical reagent can be self-inactivated following the reaction of the active species on the targeting oligomer.

Of course, nonspecific effects can be induced by oligonucleotides linked to active reagents because of their reaction with nontarget molecules, particularly with proteins. This has been demonstrated with oligomers carrying an alkylating group [Knorre et al., 1989; Yakubov et al., 1989].

Ribozymes constitute another interesting possibility for site-specific cleavage of RNA targets and are discussed elsewhere (see Murray and Crockett, Section II, this volume; Joyce, this volume; Koizumi and Ohtsuku, this volume). Short RNA molecules that display a sequence-dependent RNase activity [Uhlenbeck, 1987] have been derived from natural RNA species that undergo self-cleaving reactions [Cech, 1987; Forster and Symons, 1987]. Cleavage reactions by synthetic ribozymes have been described in vitro [Haseloff and Gerlach, 1988; Jeffries and Symons, 1989; Koizumi et al,. 1988]. However, their use in intact cells has not been widespread [for successful examples, see Cotten and Birnstiel, 1989; Scanlon et al., 1991; Sioud and Drlica, 1991]. Synthetic oligoribonucleotides can now be prepared, but the high nuclease sensitivity of natural RNA sequences will probably require the design of chemically modified RNA molecules.

IV. CONCLUSION

The utility of antisense oligonucleotides clearly appears in two different fields. Their value in molecular genetics has already been demonstrated [see Hélène and Toulmé, 1990, for review]. Their use in controlling the development of various pathogens—viruses and parasites—in culture and in controlling the expression of dysfunctional genes such as oncogenes (see Tidd, this volume; Wickstrom, this volume) indicates that they might constitute a new class of therapeutic agents even though a number of problems still remain to be solved before applications in human beings might be considered [Zon, 1989; Riordan and Martin, 1991].

ACKNOWLEDGMENTS

My work has been supported by the Institut National de la Santé et de la Recherche Médicale (INSERM), the Centre National de la Recherche Scientifique (CNRS), the World Health Organization (UNDP/World Bank/Special Programme for Research and Training in Tropical Diseases), and the Direction de la Recherche et des Études Techniques. I am grateful to Dr. D. Shire for critical reading of the manuscript.

V. REFERENCES

Agrawal S, Goodchild J (1987): Oligodeoxynucleoside methylphosphonates: Synthesis and enzymic degradation. Tetrahedron Lett 28:3539–3542.

Agrawal S, Goodchild J, Civeira MP, Thornton AH, Sarin PS, Zamecnik PC (1988): Oligodeoxynucleoside phosphoramidates and phosphorothioates as inhibitors of human immunodeficiency virus. Proc Natl Acad Sci USA 85:7079–7083.

Agrawal S, Ikeuchi T, Sun D, Sarin PS, Konopka A, Maizel J, Zamecnik PC (1989): Inhibition of human immunodeficiency virus in early infected and chronically infected cells by antisense oligodeoxynucleoside and their phosphorothioate analogues. Proc Natl Acad Sci USA 86:7790–7794.

Asseline U, Delarue M, Lancelot G, Toulmé F, Thuong NT, Montenay-Garestier T, Hélène C (1984): Nucleic acid–binding molecules with high affinity and base sequence specificity: Intercalating agents covalently linked to oligodeoxynucleotides. Proc Natl Acad Sci USA 81:3297–3301.

Barrett JC, Miller PS, Ts'o POP (1974): Inhibitory effect of complex formation with oligodeoxynucleotide ethyl phosphotriesters on transfer ribonucleic acid aminoacylation. Biochemistry 13:4897–4906.

Belikova AM, Zaritova VF, Grineva NI (1967): Synthesis of ribonucleosides and diribonucleoside phosphates containing 2-chloroethylamine and nitrogen mustard residues. Tetrahedron Lett 37:3557–3562.

Berget SM, Robberson BL (1986): U1, U2, and U4/U6 small nuclear ribonucleoproteins are required for in vitro splicing but not polyadenylation. Cell 46: 691–696.

Bertrand JR, Imbach JL, Paoletti C, Malvy C (1989): Comparative activity of α- and β-anomeric oligonucleotides on rabbit β-globin synthesis: Inhibitory effect of cap targeted α-oligonucleotides. Biochem Biophys Res Commun 164:311–318.

Birg F, Praseuth D, Zérial A, Thuong NT, Asseline U, Le Doan T, Hélène C (1990): Inhibition of simian virus 40 DNA replication in CV-1 cells by an oligodeoxynucleotide covalently linked to an intercalating agent. Nucleic Acids Res 18:2901–2908.

Black DL, Chabot B, Steitz JA (1985): U2 as well as U1 small nuclear ribonucleoproteins are involved in premessenger RNA splicing. Cell 42:737–750.

Blackwell TK, Kretzner L, Blackwood EM, Eisenman RN, Weintraub H (1990a): Sequence-specific DNA binding by the c-Myc protein. Science 250:1149–1151.

Blackwell TK, Weintraub H (1990b): Differences and similarities in DNA-binding preferences of MyoD and E2a protein complexes revealed by binding site selection. Science 250:1104–1110.

Blake KR, Murakami A, Miller PS (1985a): Inhibition of rabbit globin mRNA translation by sequence-specific oligodeoxyribonucleotides. Biochemistry 24:6132–6138.

Blake KR, Murakami A, Spitz SA, Glave SA, Reddy MP, Ts'o POP, Miller PS (1985b): Hybridization arrest of globin synthesis in rabbit reticulocyte lysate and cells by oligodeoxyribonucleoside methylphosphonates. Biochemistry 24:6139–6145.

Bloch E, Lavignon M, Bertrand JR, Pognan F, Morvan F, Malvy C, Rayner B, Imbach JL, Paoletti C (1988): α-Anomeric DNA:β-RNA hybrids as new synthetic inhibitors of Escherichia coli RNase-H, Drosophila embryo RNase-H and M-MLV reverse transcriptase. Gene 72:349–360.

Boiziau C, Boutorine AS, Loreau N, Verspieren P, Thuong NT, Toulmé JJ (1991a): Effect of antisense oligonucleotides linked to alkylating agents on in vitro translation of rabbit β-globin and Trypanosoma brucei mRNAs. Nucleosides Nucleotides 10:239–244.

Boiziau C, Kurfurst R, Cazenave C, Roig V, Thuong NT, Toulmé JJ (1991b): Inhibition of translation initiation by antisense oligonucleotides via an RNase-H independent mechanism. Nucleic Acids Res 19:1113–1119.

Boiziau C, Thuong NT, Toulmé JJ (1992a): Two mechanisms are responsible for the inhibition of reverse transcription by antisense oligonucleotides. Proc Natl Acad Sci USA (in press).

Boiziau C, Toulmé JJ (1992b): Modified oligonucleotides in rabbit reticulocytes: Uptake, stability and antisense properties. Biochimie (in press).

Bories D, Raynal MC, Solomon DH, Darzynkiewicz Z, Cayre YE (1989): Down-regulation of a serine protease, myeloblastin, causes growth arrest and differentiation of promyelocytic leukemia cells. Cell 59: 959–968.

Boutorin AS, Guskova LV, Ivanova EM, Kobetz ND, Zarytova VF, Ryte AS, Yurchenko LV, Vlassov VV (1989): Synthesis of alkylating oligonucleotide derivatives containing cholesterol or phenazidium residue at the 3′ termini and their interaction with DNA within mammalian cells. FEBS Lett 254:129–132.

Caruthers MH (1989): Synthesis of oligonucleotides and oligonucleotide analogues. In Cohen JS (ed): Oligodeoxynucleotides: Antisense Inhibitors of Gene Expression. London: Macmillan Press, pp 7–24.

Cazenave C, Chevrier M, Thuong NT, Hélène C (1987a): Rate of degradation of α- and β-oligodeoxynucleotides in Xenopus oocytes. Implications for anti-messenger strategies. Nucleic Acids Res 15:10507–10521.

Cazenave C, Loreau N, Thuong NT, Toulmé JJ, Hélène C (1987b): Enzymatic amplification of translation inhibition of rabbit β-globin mRNA mediated by antimessenger oligodeoxynucleotides covalently linked to intercalating agents. Nucleic Acids Res 15:4717–4736.

Cazenave C, Stein CA, Loreau N, Thuong NT, Neckers LM, Subasinghe C, Hélène C, Toulmé JJ (1989): Comparative inhibition of rabbit globin mRNA translation by modified antisense oligonucleotides. Nucleic Acids Res 17:4255–4273.

Cech TR (1987): The chemistry of self-splicing RNA and RNA enzymes. Science 236:1532–1539.

Chu BCF, Orgel LE (1989): Inhibition of DNA synthesis by crosslinking the template to platinum thiol-derivatives of complementary oligodeoxynucleotides. Nucleic Acids Res 17:4783–4798.

Cooney M, Czernuszewicz G, Postel EH, Flint S, Hogan ME (1988): Site-specific oligonucleotide binding represses transcription of the human c-myc gene in vitro. Science 241:456–459.

Corey DR, Schultz PG (1987): Generation of a hybrid sequence-specific single-stranded deoxyribonuclease. Sciennce 238:1401–1403.

Cotten M, Birnstiel ML (1989): Ribozyme mediated destruction of RNA in vivo. EMBO J 8:3861–3866.

Dagle JM, Walder JA, Weeks DL (1990): Targeted degradation of mRNA in Xenopus oocytes and embryos directed by modified oligonucleotides: Studies of An2 and cyclin in embryogenesis. Nucleic Acids Res 18:4751–4757.

Dash P, Lotan I, Knapp M, Kandel ER, Goelet P (1987): Selective elimination of mRNAs in vivo: Complementary oligodeoxynucleotides promote RNA degradation by an RNase-H-like activity. Proc Natl Acad Sci USA 84:7896–7900.

Degols G, Léonetti JP, Gagnor C, Lemaître M, Lebleu B (1989): Antiviral activity and possible mechanism of action of oligonucleotide–poly(L-lysine) conjugates targeted to the vesicular stomatitis virus mRNA and genomic RNA. Nucleic Acids Res 17:9341–9350.

Eckhardt H, Lührmann R (1979): Blocking of the initiation of protein biosynthesis by a pentanucleotide complementary to the 3' end of *Escherichia coli* 16 S rRNA. J Biol Chem 254:11185–11188.

Ellington AD, Szostak J (1990): In vitro selection of RNA molecules that bind specific ligands. Nature 346: 818–822.

Forster AC, Symons RH (1987): Self-cleavage of plus and minus RNAs of a virusoid and a structural model for the active sites. Cell 49:211–220.

François JC, Saison-Behmoaras T, Barbier C, Chassignol M, Thuong NT, Hélène C (1989a): Sequence-specific recognition and cleavage of duplex DNA via a triple-helix formation by oligonucleotides covalently linked to a phenantroline-copper chelate. Proc Natl Acad Sci USA 86:9702–9706.

François JC, Saison-Behmoaras T, Thuong NT, Hélène C (1989b): Inhibition of restriction endonuclease cleavage via triple-helix formation by homopyrimidine oligonucleotides. Biochemistry 28:9617–9619.

Furdon PJ, Dominski Z, Kole R (1989): RNase-H cleavage of RNA hybridized to oligonucleotides containing methylphosphonate, phosphorothioate and phosphodiester bonds. Nucleic Acids Res 17:9193–9204.

Gagnor C, Bertrand JR, Thenet S, Lemaître M, Morvan F, Rayner B, Malvy C, Lebleu B, Imbach JL, Paoletti C (1987): Alpha-DNA VI: Comparative study of α- and β-anomeric oligonucleotides in hybridization to mRNA and in cell-free translation inhibition. Nucleic Acids Res 15:10419–10436.

Gagnor C, Rayner B, Léonetti JP, Imbach JL, Lebleu B (1989): Alpha-DNA IX: Parallel annealing of α-anomeric oligodeoxyribonucleotides to natural mRNA is required for interference in RNase-H mediated hydrolysis and reverse transcription. Nucleic Acids Res 17:5107–5114.

Gewirtz AM, Anfossi G, Venturelli D, Valpreda S, Sims R, Calabretta B (1989): G_1/S transition in normal human T-lymphocytes requires the nuclear protein encoded by c-*myb*. Science 245:180–183.

Goodchild J (1990): Conjugates of oligonucleotides and modified oligonucleotides: A review of their synthesis and properties. Bioconjugate Chem 1: 165–187.

Grachev MA, Zaychikov EF, Ivanova EM, Komarova NI, Kutyavin IV, Sidelnikova NP, Frolova IP (1984): Oligonucleotides complementary to a promoter over the region $-8\ldots+2$ as transcription primers for *E. coli* RNA polymerase. Nucleic Acids Res 12:8509–8524.

Griffin LC, Dervan PB (1989): Recognition of thymine-adenine base pairs by guanine in a pyrmidine triple helix motif. Science 245:967–971.

Hanvey JC, Shimizu M, Wells RD (1990): Site-specific inhibition of EcoRI restriction/modification enzymes by a DNA triple helix. Nucleic Acids Res 18:157–161.

Harel-Bellan A, Brini A, Ferris DF, Robin P, Farrar WL (1989): In situ detection of a heat-shock regulatory element binding protein using a double short synthetic enhancer sequence. Nucleic Acids Res 17:4077–4087.

Harel-Bellan A, Durum S, Muegge K, Abbas AK, Farrar WL (1988): Specific inhibition of lymphokine biosynthesis and autocrine growth using antisense oligonucleotides in Th1 and Th2 helper T cell clones. J Exp Med 168:2309–2318.

Haseloff J, Gerlach WL (1988): Simple RNA enzymes with new and highly specific endoribonuclease activities. Nature 334:584–591.

Hastie ND, Held WA (1978): Analysis of mRNA populations by cDNA–mRNA hybrid-mediated inhibition of cell-free protein synthesis. Proc Natl Acad Sci USA 75:1217–1221.

Hélène C, Montenay-Garestier T, Saison T, Takasugi M, Toulmé JJ, Asseline U, Lancelot G, Maurizot JC, Toulmé F, Thuong NT (1985): Oligodeoxynucleotides covalently linked to intercalating agents: A new class of gene regulatory substances. Biochimie 67:777–783.

Hélène C, Thuong NT, Saison-Behmoaras T, François JC (1989): Sequence-specific artificial endonucleases. Trends Biotechnol 7:310–315.

Hélène C, Toulmé JJ (1989): Control of gene expression by oligodeoxynucleotides covalently linked to intercalating agents and nucleic acid cleaving reagents. In Cohen JS (ed): Oligodeoxynucleotides: Antisense Inhibitors of Gene Expression. London: Macmillan Press, pp 137–172.

Hélène C, Toulmé JJ (1990): Oligonucleotides as specific gene regulators. Biochim Biophys Acta 1049:99–125.

Holt JT, Redner RL, Nienhuis AW (1988): An oligomer complementary to c-*myc* mRNA inhibits proliferation of HL-60 promyelocytic cells and induces differentiation. Mol Cell Biol 8:963–973.

Inouye M (1988): Antisense RNA: its functions and applications in gene regulation—A review. Gene 72:25–34.

Jayaraman K, Mcparland K, Miller P, Ts'o POP (1981): Selective inhibition of *Escherichia coli* protein synthesis and growth by nonionic oligonucleotides complementary to the 3' end of 16S rRNA. Proc Natl Acad Sci USA 78:1537–1541.

Jeffries AC, Symons RH (1989): A catalytic 13-mer ribozyme. Nucleic Acids Res 17:1371–1377.

Jessus C, Cazenave C, Ozon R, Hélène C (1988): Specific inhibition of endogenous β-tubulin synthesis in *Xenopus* oocytes by anti-messenger oligodeoxynucleotides. Nucleic Acids Res 16:2225–2233.

Kean JM, Murakami A, Blake KR, Cushman CD, Miller PS (1988): Photochemical cross-linking of psoralen-derivatized oligonucleoside methylphosphonates to rabbit globin messenger RNA. Biochemistry 27: 9113–9121.

Kim SK, Wold BJ (1985): Stable reduction of thymidine kinase activity in cells expressing high levels of antisense RNA. Cell 42:129–138.

Knorre DG, Vlassov VV, Zarytova VF (1985): Reactive oligonucleotide derivatives and sequence-specific modification of nucleic acids. Biochimie 67:785–789.

Knorre DG, Vlassov VV, Zarytova VF (1989): Oligonucleotides linked to reactive groups. In Cohen JS (ed): Oligodeoxynucleotides: Antisense Inhibitors of Gene Expression. London: Macmillan Press, 12:173–196.

Koizumi M, Iwai S, Ohtsuka E (1988): Construction of a series of several self-cleaving RNA duplexes using synthetic 21-mers. FEBS Lett 228:228–230.

Krainer AR, Maniatis T (1985): Multiple factors including the small nuclear ribonucleoproteins U1 and U2 are necessary for pre-mRNA splicing in vitro. Cell 42:725–736.

Krämer A, Keller W, Appel B, Lührmann R (1984): The 5′ terminus of the RNA moiety of U1 small nuclear ribonucleoprotein particles is required for the splicing of messenger RNA precursors. Cell 38:299–307.

Kulka M, Smith CC, Aurelian L, Fishelevich R, Meade K, Miller P, Ts'o POP (1989): Site specificity of the inhibitory effects of oligo(nucleosidemethhylphosphonate)s complementary to the acceptor splice junction of herpes simplex virus type I immediate early mRNA 4. Proc Natl Acad Sci USA 86:6868–6872.

Latimer LJP, Hampel K, Lee JS (1989): Synthetic repeating sequence DNAs containing phosphorothioates: Nuclease sensitivity and triplex formation. Nucleic Acids Res 17:1549–1561.

Le Doan T, Perrouault L, Praseuth D, Habhoub N, Decout J-L, Thuong NT, Lhomme J, Hélène C (1987): Sequence-specific recognition, photocrosslinking and cleavage of the DNA double helix by an oligo-alpha-thymidylate covalently linked to an azidoproflavine derivative. Nucleic Acids Res 19:7749–7760.

Lemaître M, Bayard B, Lebleu B (1987): Specific antiviral activity of a poly(L-lysine) conjugated oligodeoxyribonucleotide sequence complementary to vesicular stomatitis virus N protein mRNA initiation site. Proc Natl Acad Sci USA 84:648–652.

Leonetti J-P, Degols G, Lebleu B (1990a): Biological activity of oligonucleotide-poly(L-lysine) conjugates: Mechanism of cell uptake. Bioconjugate Chem 1:149–153.

Leonetti J-P, Machy P, Degols G, Lebleu B, Leserman L (1990b): Antibody-targeted liposomes containing oligodeoxyribonucleotides complementary to viral RNA selectively inhibit viral replication. Proc Natl Acad Sci USA 87:2448–2451.

Leonetti JP, Mechti N, Degols G, Gagnor C, Lebleu B (1991): Intracellular distribution of microinjected antisense oligonucleotides. Proc Natl Acad Sci USA 88:2702–2706.

Leonetti JP, Rayner B, Lemaître M, Gagnor C, Milhaud PG, Imbach JL, Lebleu B (1988): Antiviral activity of conjugates between poly(L-lysine) and synthetic oligodeoxyribonucleotides. Gene 72:323–332.

Letai AG, Palladino MA, Fromm E, Rizzo V, Fresco JR (1988): Specificity in formation of triple-stranded nucleic acid helical complexes: Studies with agarose-linked polynucleotide affinity columns. Biochemistry 27:9108–9112.

Letsinger RL, Zhang G, Sun DK, Ikeuchi T, Sarin PS (1989): Cholesteryl-conjugated oligonucleotides: Synthesis, properties and activity as inhibitors of replication of human immunodeficiency virus in cell culture. Proc Natl Acad Sci USA 86:6553–6556.

Liebhaber SA, Cash FE, Shakin SH (1984): Translationally associated helix-destabilizing activity in rabbit reticulocyte lysate. J Biol Chem 259:15597–15602.

Loke SL, Stein C, Zhang X, Avigan M, Cohen JS, Neckers LM (1988): Delivery of c-myc antisense phosphorothioate oligodeoxynucleotides to hematopoietic cells in culture by liposome fusion: Specific reduction in c-myc protein expression correlates with inhibition of cell growth and DNA synthesis. Curr Top Microbiol Immunol 141:282–289.

Loke SL, Stein CA, Zhang XH, Mori K, Nakanishi M, Subasinghe C, Cohen JS, Neckers LM (1989): Characterization of oligonucleotide transport into living cells. Proc Natl Acad Sci USA 86:3474–3478.

Loreau N, Boiziau C, Verspieren P, Shire D, Toulmé JJ (1990): Blockage of AMV reverse transcriptase by antisense oligodeoxynucleotides. FEBS Lett 252:53–56.

Maher LJ III, Dolnick BJ (1988): Comparative hybrid arrest by tandem anti-sense oligodeoxynucleotides or oligodeoxynucleoside methylphosphonates in a cell-free system. Nucleic Acids Res 16:3341–3358.

Maher LJ III, Wold B, Dervan PB (1989): Inhibition of DNA binding proteins by oligonucleotide-directed triple helix formation. Science 245:725–730.

Majumdar C, Stein CA, Cohen JS, Broder S, Wilson SH (1989): Stepwise mechanism of HIV reverse transcriptase: Primer function of phosphorothioate oligodeoxynucleotide. Biochemistry 28:1340–1346.

Marcus-Sekura CJ (1988): Techniques for using antisense oligodeoxyribonucleotides to study gene expression. Anal Biochem 172:289–295.

Matsukura M, Shinokuza K, Zon G, Mitsuya H, Reitz M, Cohen JS, Broder S (1987): Phosphorothioate analogues of oligodeoxynucleotides: Inhibitors of replication and cytopathic effects of human immunodeficiency virus. Proc Natl Acad Sci USA 84:7706–7710.

Matsukura M, Zon G, Shinozuka K, Robert-Guroff M, Shimada T, Stein CA, Mitsuya H, Wong-Staal F, Cohen JS, Broder S (1989): Regulation of viral expression of human immunodeficiency virus in vitro by an antisense phosphorothioate oligodeoxynucleotide against rev (art/trs) in chronically infected cells. Proc Natl Acad Sci USA 86:4244–4248.

Miller PS, Braiterman LT, and Ts'o POP (1977): Effects of a trinucleotide ethyl phosphotriester, Gmp(Et)Gmp

(Et)U on mammalian cells in culture. Biochemistry 16:1988–1996.

Miller PS, McParland KB, Jayaraman K, Ts'o POP (1981): Biochemical and biological effects of nonionic nucleic acid methylphosphonates. Biochemistry 20: 1874–1880.

Minshull J, Hunt T (1986): The use of single-stranded DNA and RNase-H to promote quantitative "hybrid arrest of translation" of mRNA/DNA hybrids in reticulocyte lysate cell-free translations. Nucleic Acids Res 14:6433–6451.

Mori K, Boiziau C, Cazenave C, Matsukura M, Subasinghe C, Cohen JS, Broder S, Toulmé JJ, Stein CA (1989): Phosphoroselenoate oligodeoxynucleotides: Synthesis, physico-chemical characterization, antisense inhibitory properties and anti-HIV activity. Nucleic Acids Res 17:8207–8219.

Morvan F, Rayner B, Imbach JL (1990): α-Oligodeoxynucleotides (α-DNA): A new chimeric nucleic acid analog. In Setlow JK (ed): Genetic Engineering. New York: Plenum Press, vol 12, pp 37–52.

Morvan F, Rayner B, Imbach JL, Chang DK, Lown JW (1986): Alpha-DNA I. Synthesis, characterization by high field ^{1}H-NMR, and base-pairing properties of the unatural α-(d(CpCpTpTpCpC)) with its complement β-(d(GpGpApApGpGp)). Nucleic Acids Res 14:5019–5035.

Moser HE, Dervan PB (1987): Sequence-specific cleavage of double helical DNA by triple helix formation. Science 238:645–650.

Neckers LM (1989): Antisense oligodeoxynucleotides as a tool for studying cell regulation: Mechanism of uptake and application to the study of oncogene function. In Cohen JS (ed): Oligodeoxynucleotides: Antisense Inhibitors of Gene Expression. London: Macmillan Press, vol 12, pp. 211–231.

O'Keefe SJ, Wolfes H, Kiessling AA, Cooper GM (1989): Microinjection of antisense c-mos oligonucleotides prevents meiosis II in the maturing mouse egg. Proc Natl Acad Sci USA 86:7038–7042.

Paterson BM, Roberts BE, Kuff EL (1977): Structural gene identification and mapping by DNA–mRNA hybrid-arrested cell-free translation. Proc Natl Acad Sci USA 74:4370–4374.

Paules RS, Buccione R, Moschel RC, Vande Woude GF, Eppig JJ (1989): Mouse mos protooncogene product is present and functions during oogenesis. Proc Natl Acad Sci USA 86:5395–5399.

Perrouault L, Asseline U, Rivalle C, Thuong NT, Bisagni E, Giovanangelli C, Le Doan T, Hélène C (1990): Sequence-specific artificial photo-induced endonucleases based on triple helix-forming oligonucleotides. Nature 344:358–360.

Postel EH, Flint SJ, Kessler DJ, Hogan ME (1991): Evidence that a triplex-forming oligodeoxynucleotide binds to the c-myc promoter in HeLa cells, thereby reduc-

ing c-myc messenger RNA levels. Proc Natl Acad Sci USA 88:8227–8231.

Povsic TJ, Dervan PB (1989): Triple helix formation by oligonucleotide on DNA extended to the physiological pH range. J Am Chem Soc 111:3059–3061.

Praseuth D, Perrouault L, Le Doan T, Chassignol M, Thuong NT, Hélène C (1988): Sequence-specific binding and photocrosslinking of α and β oligodeoxynucleotides to the major groove of DNA via triple-helix formation. Proc Natl Acad Sci USA 85:1349–1353.

Quartin RS, Brakel CL, Wetmur JG (1989): Number and distribution of methylphosphonate linkages in oligodeoxynucleotides affect exo- and endonuclease sensitivity and ability to form RNase-H substrates. Nucleic Acids Res 17:7253–7262.

Riordan ML, Martin JC (1991): Oligonucleotide-based therapeutics. Nature 350:442–443.

Saison-Behmoaras T, Tocque B, Rey I, Chassignol M, Thuong NT, Hélène C (1991): Short modified oligonucleotides directed against Ha-ras point mutation induce selected cleavage of the messenger RNA and inhibit T26 cell proliferation. EMBO J 10: 1111–1118.

Sarin PS, Agrawal S, Civeira MP, Goodchild J, Ikeuchi T, Zamecnik PC (1988): Inhibition of acquired immunodeficiency syndrome virus by oligodeoxynucleoside methylphosphonates. Proc Natl Acad Sci USA 85: 7448–7451.

Scanlon KJ, Jiao L, Funato T, Wang W, Tone T, Rossi JJ, Kashani-Shabet M (1991): Ribozyme-mediated cleavage of c-fos mRNA reduces gene expression of DNA synthesis enzymes and methallothwonein. Proc Natl Acad Sci USA 88:10591–10595.

Shakin SH, Liebhaber SA (1986): Destabilization of messenger RNA/complementary DNA duplexes by the elongating 80 S ribosome. J Biol Chem 261: 16018–16025.

Shibahara S, Mukai S, Nishihara T, Inoue H, Ohtsuka E, Morisawa H (1987): Site directed cleavage of RNA. Nucleic Acids Res 15:4403–4415.

Shuttleworth J, Colman A (1988): Antisense oligonucleotide-directed cleavage of mRNA in Xenopus oocytes and eggs. EMBO J 7:427–434.

Simons RW (1988): Naturally occuring antisense RNA control—A brief review. Gene 72:35–44.

Simons RW, Kleckner N (1988): Biological regulation by antisense RNA in prokaryotes. Annu Rev Genet 22:567–600.

Sioud M, Drlica K (1991): Prevention of human immunodeficiency virus type 1 integrase expression in Escherichia coli by a ribozyme. Proc Natl Acad Sci USA 88:7303–7307.

Smith CC, Aurelian L, Reddy MP, Miller PS, Ts'o POP (1986): Antiviral effect of an oligo(nucleoside methylphosphonate) complementary to the splice junction of herpes simplex virus type 1 immediate early pre-mRNAs 4 and 5. Proc Natl Acad Sci USA 83: 2787–2791.

Stein CA, Cohen JS (1988): Oligodeoxynucleotides as inhibitors of gene expression: A review. Cancer Res 48:2659–2668.

Stein CA, Cohen JS (1989): Antisense compounds: Potential role in cancer therapy. In DeVita VT, Hellman S, Rosenberg SA (ed): Important Advances in Oncology. Philadelphia: Lippincott, pp 79–97.

Stein CA, Mori K, Loke SL, Subasinghe C, Shinozuka K, Cohen JS, Neckers LM (1988a): Phosphorothioate and normal oligodeoxyribonucleotides with 5′-linked acridine: Characterization and preliminary kinetics of cellular uptake. Gene 72:333–341.

Stein CA, Subasinghe C, Shinokuza K, Cohen JS (1988b): Physicochemical properties of phosphorothioates oligodeoxynucleotides. Nucleic Acids Res 16:3209–3221.

Stephenson ML, Zamecnik PC (1978): Inhibition of Rous sarcoma viral RNA translation by a specific oligodeoxyribonucleotide. Proc Natl Acad Sci USA 75:285–288.

Stevenson, M, Iversen, PL (1989): Inhibition of human immunodeficiency virus type 1-mediated cytopathic effects by poly(L-lysine)-conjugated antitsense oligodeoxyribonucleotides. J Gen Virol 70:2673–2682.

Strobel SA, Dervan PB (1991): Single-site enzymatic cleavage of yeast genomic DNA mediated by triple-helix formation. Nature 350:172–174.

Takayama KM, Inouye M (1990): Antisense RNA. Crit Rev Biochem Mol Biol 25:155–184.

Taniguchi T, Weissmann C (1978): Inhibition of Qβ RNA 70S ribosome initiation complex formation by an oligonucleotide complementary to the 3′ terminal region of E. coli 16S ribosomal RNA. Nature 275:770–772.

Teichman-Weinberg A, Littauer UZ, Ginzburg I (1988): The inhibition of neurite outgrowth in PC12 cells by tubulin antisense oligodeoxyribonucleotides. Gene 72:297–307.

Thuong NT, Asseline U, Roig V, Takasugi M, Hélène C (1987): Oligo(alpha-deoxynucleotides) covalently linked to intercalating agents: differential binding to ribo- and deoxyribopolynucleotides and stability towards nuclease digestion. Proc Natl Acad Sci USA 84:5129–5133.

Tidd DM, Warenius HM (1989): Partial protection of oncogene, antisense oligodeoxynucleotides against serum nuclease degradation using terminal methylphosphonate gropus. Br J Cancer 60:343–350.

Toulmé JJ, Hélène C (1988): Antimessenger oligodeoxynucleotides: an alternative to antisense RNA for artificial regulation of gene expression—A review. Gene 72:51–58.

Toulmé JJ, Krisch HM, Loreau N, Thuong NT, Hélène C (1986): Specific inhibition of mRNA translation by complementary oligonucleotides covalently linked to intercalating agents. Proc Natl Acad Sci USA 83:1227–1231.

Toulmé JJ, Verspieren P, Boiziau C, Loreau N, Cazenave C, Thuong NT (1990): Les oligonucléotides antisens: Outils de génétique moléculaire et agents thérapeutiques. Ann Parasitol Hum Comp 65:(Suppl I):11–14.

Trudel M, Dondon J, Grunberg-Manago M, Finelli J, Buckingham RH (1981): Effect of oligonucleotide AGAGGAGGU on protein synthesis in vitro. Biochimie 63:235–240.

Tuerk C, Gold L (1990): Systematic evolution of ligands by exponential enrichment: RNA ligands to bacteriophage T4 DNA polymerase. Science 249:505–510.

Uhlenbeck OC (1987): A small catalytic oligonucleotide. Nature 328:596–600.

Uhlmann E, Peyman A (1990: Antisense oligonucleotides: A new therapeutic principle. Chem Rev 90:544–579.

van der Krol AR, Mol JNM, Stuitje AR (1988a): Modulation of eukaryotic gene expression by complementary RNA or DNA sequences. Biotechniques 6:958–976.

van der Krol R, Mol JNM, Stuitje AR (1988b): Antisense genes in plants: An overview. Gene 72:45–50.

Verspieren P, Cornelissen AWCA, Thuong NT, Hélène C, Toulmé JJ (1987): An acridine-linked oligodeoxynucleotide targeted to the common 5′ end of trypanosome mRNAs kills cultured parasites. Gene 61:307–315.

Verspieren P, Loreau N, Thuong NT, Shire D, Toulmé JJ (1990): Effect of RNA secondary structure and modified bases on the in vitro inhibition of trypanosomatid protein synthesis by antisense oligodeoxynucleotides. Nucleic Acids Res 18:4711–4717.

Vlassov VV, Gaidamakov SA, Zarytova SA, Knorre DG, Levina AS, Nikonova AA, Podust LM, Fedorova OS (1988a): Sequence-specific chemical modification of double-stranded DNA with alkylating oligodeoxyribonucleotide derivatives. Gene 72:313–322.

Vlassov VV, Zarytova VF, Kutyavin IV, Mamaev SV (1988b): Sequence-specific chemical modification of a hybrid bacteriophage M13 single-stranded DNA by alkylating oligonucleotide derivatives. FEBS Lett 231:352–354.

Walder RY, Walder JA (1988): Role of RNase-H in hybrid-arrested translation by antisense oligonucleotides. Proc Natl Acad Sci USA 85:5011–5015.

Wells RD, Collier DA, Hanvey JC, Shimizu M, Wohlrab F (1988): The chemistry and biology of unusual DNA structures adopted by oligopurine-oligopyrimidine sequences. FASEB J 2:2939–2949.

Wickstrom E (1986): Oligodeoxynucleotide stability in subcellular extracts and culture media. J Biochem Biophys Methods 13:97–102.

Yakubov LE, Deeva EA, Zarytova VF, Ivanova EM, Ryte AS, Yurchenko LV, Vlassov VV (1989): Mechanism of oligonucleotide uptake by cells: Involvement of specific receptors? Proc Natl Acad Sci USA 86:6454–6458.

Zaia JA, Rossi JJ, Murakawa GJ, Spallone PA, Stephens DA, Kaplan BE, Eritja R, Wallace RB, Cantin EM (1988): Inhibition of human immunodeficiency virus by using an oligonucleoside methylphosphonate targeted to the *tat*-3 gene. J Virol 62:3914–3917.

Zamecnik PC, Goodchild J, Taguchi Y, Sarin PS (1986): Inhibition of replication and expression of human T-cell lymphotropic virus type III in cultured cells by exogenous synthetic oligonucleotides complementary to viral RNA. Proc Natl Acad Sci USA 83:4143–4146.

Zamecnik PC, Stephenson ML (1978): Inhibition of Rous sarcoma virus replication and cell transformation by a specific oligodeoxynucleotide. Proc Natl Acad Sci USA 75:280–284.

Zerial A, Thuong NT, Hélène C (1987): Selective inhibition of cytopathic effect of type A influenza viruses by oligodeoxynucleotides covalently linked to an intercalating agent. Nucleic Acids Res 15: 9909–9919.

Zon G (1988): Oligonucleotide analogues as potential chemothereapeutic agents. Pharmacol Res 5:539–549.

Zon G (1989): Pharmacological considerations. In Cohen JS (ed): Oligodeoxynucleotides: Antisense Inhibitors of Gene Expression. London: Macmillan Press, pp 233–247.

Zuckermann RN, Corey DR, Schultz PG (1988): Site-selective cleavage of RNA by a hybrid enzyme. J Am Chem Soc 110:1614–1615.

Zuckermann RN, Schultz PG (1989): Site-selective cleavage of structured RNA by a staphylococcal nuclease–DNA hybrid. Proc Natl Acad Sci USA 86: 1766–1770.

ABOUT THE AUTHOR

JEAN-JACQUES TOULMÉ is Director of the Laboratory of Molecular Biophysics at the University of Bordeaux, France. He received his ''Maitrise de Biochimie'' at the University of Orléans and his ''Doctorat des Sciences Physiques'' under Claude Hélène at the University Pierre et Marie Curie, Paris VI, in 1982. His early research in Paris was devoted to biophysical studies of nucleic acid–protein interactions. He contributed to the demonstration of photoreactivating and endonucleolytic properties of tryptophan-containing oligopeptides. Then he shifted to artificial regulation of gene expression. At present, Dr. Toulmé's main interest focuses on using synthetic oligonucleotides to block either mRNA translation or viral RNA reverse transcription. Part of his work is devoted to the control of protozoan parasite development by antisense oligomers.

Antisense RNA and DNA: 195–212
© 1992 Wiley-Liss, Inc.

Antisense Ablation of mRNA in Frog and Rabbit Cell-Free Systems

Jeremy Minshull and Tim Hunt

•

I. INTRODUCTION

In 1982, Evans et al. discovered proteins that were specified by maternal mRNA in sea urchin and clam embryos [Evans et al., 1983]. These proteins—the cyclins—showed unprecedented behavior during the mitotic cell cycle. They accumulated during interphase and were destroyed by proteolysis during a short window in mitosis that preceded the metaphase→anaphase transition, suggesting a role in the control of the cell cycle. The question immediately arose as to how to test this idea in organisms that "did not have genetics." Recently, antisense techniques have been used to provide "reverse genetic" analysis of genes that were characterized by molecular criteria rather than by classical genetics [Izant and Weintraub, 1984; Weintraub et al., 1985; Melton, 1985; Cazenave et al., 1986; Walder, 1988]. A potential solution to the problem of how to test the role of cyclins in the control of the cell cycle was thus to use microinjected antisense DNA or RNA to block cyclin synthesis. This required the isolation of cDNA clones corresponding to the cyclins, and here again the antisense approach seemed likely to be helpful. Antisense DNA was originally used by Paterson et al. [1977] to prevent translation of its cognate mRNA in cell-free systems in order to match viral genes with their translation products. Modification and development of their protocol allowed us to identify cDNA clones encoding cyclins.

We begin with a brief description of how cDNA and oligonucleotides can be used to ablate specifically the corresponding mRNA in a crude mixture of RNA. Translation of this mRNA in a cell-free system, such as the reticulocyte lysate, allows the encoded protein to be identified by its absence. This technique can be used directly in cDNA isolation and can also be very helpful in selecting oligonucleotides for use as probes or as primers for the polymerase chain reaction (PCR).

Cell-free translation systems perform no transcription, RNA processing, or transport; thus in these systems antisense molecules must work by blocking translation; as discussed by Toulmé (this volume), the other processes are potential targets for inhibition within living cells. Reliable hybrid arrest with antisense cDNA or oligonucleotides can be obtained in cell-free systems by digesting the hybrids with RNase H before assaying their translation [Minshull and Hunt, 1986]. We show here that addition of an oligonucleotide that is complementary to a sequence common to the coding regions of rabbit α- and β-globin quickly and specifically stops the translation of globin mRNA in crude reticulocyte lysate, provided that RNase H is added. The synthesis of proteins specified by other mRNA is unaffected. We have determined the effects of the length and concentration of oligonucleotide, concentration of RNase H, and temperature on the ablation of mRNA already engaged in translation. When we began these studies, there was very little information about these operationally important parameters.

We also investigated how the residual capped (5′) globin mRNA fragments were treated by the translation machinery. We found that ribosomes rapidly left the 3′ ends of mRNA that lacked termination codons and released their incomplete nascent chains as peptidyl tRNA. Perhaps unexpectedly, the released N-terminal globin fragment attached to tRNA is unstable, turning over with a half-life of about 10 minutes.

A major limitation of hybrid arrest revealed by these studies was that of semispecific mRNA degradation. As the concentrations of oligonucleotide or RNase H were raised, we saw that synthesis of proteins other than those targeted was inhibited. This highlighted a potential difficulty with the technique as a means of determining protein function: how to be certain that the observed effects are not due to elimination of another mRNA that contains a region complementary, or a close match, to the oligonucleotide used. It is fair to say that by the end of these studies, our initial enthusiasm for antisense approaches was tempered by the appreciation of these limitations; nonetheless, there are some situations in which careful use of the methodology can be extremely useful. In the light of this caveat, we describe how we have used antisense oligonucleotides to obtain information about the role of cyclin proteins in the induction of mitosis in a *Xenopus* cell-free system.

II. "HYBRID ARREST OF TRANSLATION" IS DUE TO RNase H

The original protocol in which viral genes were mapped by their ability to prevent the translation of their cognate mRNA worked with double-stranded (ds) DNA [Paterson et al., 1977; Miller et al., 1983]. High concentrations of formamide were used to favor the formation of RNA–DNA hybrids, and, after annealing, the nucleic acids were recovered by ethanol precipitation before testing their translatability. This somewhat lengthy and cumbersome procedure was not well suited for screening even a moderate number of clones. In 1982, Chandler showed that cDNA in a single-stranded (ss) vector formed DNA–RNA hybrids under conditions that were closer to those compatible with cell-free translation.

A curious feature of these pioneering studies was the observation that inhibition of mRNA translation by antisense DNA was variable (and sometimes nonexistent) in the reticulocyte lysate, whereas it worked reliably in the wheat germ translation assay [Paterson et al., 1977; Chandler, 1982; Haeuptle et al., 1986]. The variability of hybrid arrest in reticulocyte lysate

was thought to be due to an "unwinding activity" associated with translating 80S ribosomes [Liebhaber et al., 1984; Shakin and Liebhaber, 1986]. Similarly, in *Xenopus* embryos, antisense RNA was found to be displaced from its target mRNA by a dsRNA unwinding/modifying activity distinct from the one found in the reticulocyte lysate, which is discussed by Bass (this volume) [see also Harland and Weintraub, 1985; Melton, 1985; Bass and Weintraub, 1987; Rebagliati and Melton, 1987]. At about the same time, however, Kawasaki [1985] reported highly efficient inhibition of translation in *Xenopus* oocytes when he used antisense oligodeoxynucleotides.

It occured to us that the difference between the two observations in *Xenopus,* and between the wheat germ and reticulocyte translation systems, might lie in the activity of RNase H, the enzyme that degrades the RNA strand of a DNA–RNA duplex. In retrospect it is difficult to understand how this was overlooked, for in 1979 Donis-Keller had shown that short oligonucleotides could target specific hydrolysis of the RNA strand in a DNA–RNA hybrid duplex and this was utilized by several groups [Krainer and Maniatis, 1985; Krämer et al., 1984] to map the structures of snRNA.

We tested if RNase H was responsible for the hybrid arrest by reextracting RNAs that had been incubated with antisense DNA in the wheat germ extract. They were indeed cleaved in the region of duplex formation, presumably by an RNase H activity [Minshull and Hunt, 1986; Haeuptle et al., 1986]. The reticulocyte lysate had a much lower content of RNase H, and "hybrid arrest" could be obtained in this system either by digesting the hybrids with RNase H prior to translation or by adding the enzyme to the lysate [Minshull and Hunt, 1986]. We subsequently tested this method to screen a cDNA library made in the single-stranded vector M13 and were able to isolate clones for sea urchin cyclin, a very abundant mRNA whose specific loss of translation was easy to see even on one-dimensional gels [Pines and Hunt, 1987].

III. GLOBIN mRNA ABLATION IN THE RETICULOCYTE LYSATE

The view that the basis for the action of antisense oligonucleotides is to provide substrates for RNase H significantly alters one's understanding of the process and puts it on a somewhat more rational basis. Thus the mRNA is actually destroyed in the process, and its loss can be followed by standard procedures such as RNase protection mapping or Northern blotting as well as by loss of translation. Moreover, the rate and specificity of attack on a particular mRNA will depend on the rate of formation of DNA–RNA hybrids and the rate at which these are recognized and cut by RNase H. At the outset, we had very little feel for how well these requirements were met under physiological conditions, either in vivo or in translation-competent cell-free extracts, in which the ionic strength and pH were optimized for protein synthesis and could not be varied and in which the oligonucleotides could probably be regarded as having to compete for their targets with ribosomes and RNA binding proteins.

Since we were extremely familiar with the rabbit reticulocyte lysate translation system, we used it as a model in which to investigate the rates of annealing of oligonucleotides to mRNA and of RNase H scission under physiological conditions. The major translation products of the mRNA in rabbit reticulocyte lysate are the α- and β-chains of globin. Inspection of the sequences of the mRNA for these two proteins revealed a common sequence of 12 nucleotides (5'-GGCAAGAAGGUG-3'), located from nucleotides 214 to 225 in the α-globin mRNAs and from 246 to 257 in the mRNA for β-globin (Fig. 1). The 12-mer 5'-CACCTTCT-TGCC-3' is thus complementary to both globin mRNA and we tested whether it was long enough to form a hybrid with globin mRNA at 30°C, the optimal translation temperature.

A. Antisense Inhibition of Globin Synthesis in the Reticulocyte Lysate

1. A 12 residue oligonucleotide stops globin synthesis in the presence of RNase H. We began by testing the ability of the 12-mer to

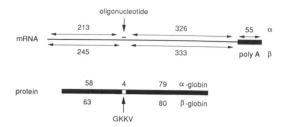

Fig. 1. *Map of globin mRNA and translation products. The position of complementarity to the 12-mer 5'-CACCTTCTTGCC-3', coding for the amino acids GKKV, is indicated in the mRNA for α- and β-globin. The positions of methionine codons are at 32 and 55, and the cysteines are at positions 104 and 112 in the α-and β-chain, respectively.*

prevent the translation of globin mRNA endogenous to the reticulocyte lysate under normal protein synthesis conditions. No inhibition of protein synthesis occurred when a 50-fold molar excess of the oligonucleotide over globin mRNA was added to reticulocyte lysate unsupplemented with RNase H and incubated for 1 hour with [^{35}S]methionine (Fig. 2). By contrast, when 25 U/ml of RNase H was added with the same concentration of oligonucleotide, complete inhibition of globin synthesis occurred rapidly. Full-length globin synthesis was reduced to less than 1% of control levels by 12 minutes, and a new polypeptide of about 6,500 Da appeared (see Fig. 2). This polypeptide represented the translation products of α- and β-globin mRNA cut at the position of the 12-mer. Synthesis of these truncated products diminished as time went by for reasons that are discussed in Section IV.

Oligonucleotide-induced inhibition of globin synthesis was specific, as shown by the continued synthesis of lipoxygenase, the strong 60 kD polypeptide made in reticulocytes that is the next most abundant translation product after globin [Thiele et al., 1982; Affara et al., 1985; Schewe et al., 1986]. Indeed, in some experiments lipoxygenase synthesis was actually stimulated (by up to twofold; data not shown).

Thus formation of DNA–RNA hybrids occurred rapidly under physiological conditions,

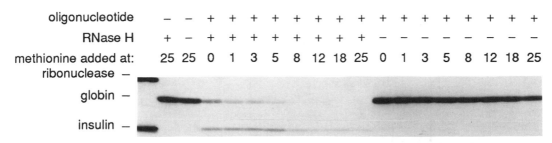

Fig. 2. *Hybrid arrest of globin mRNA translation causes synthesis of truncated polypeptides. Reticulocyte lysate was incubated with 25 U/ml RNase H alone, 10 μg/ml of the antiglobin mRNA oligonucleotide alone, or RNase H and oligonucleotide. [³⁵S]methionine (500 μCi/ml) was added at various times as indicated in minutes. Each sample was incubated for 30 minutes after the addition of label. Translation products were analyzed by electrophoresis on a 17.5% SDS polyacrylamide gel and by autoradiography. Positions of molecular weight markers are indicated in the left-hand lane. Reticulocyte lysates contain about 10 μg/ml globin mRNA [Jackson and Hunt, 1983], so 10 μg/ml of a 12-mer oligonucleotide is a 50-fold molar excess.*

and RNase H cut the mRNA despite competition from all the normal components of cytoplasm, including actively translating ribosomes and the mRNP proteins present in reticulocytes. This is in accord with findings in *Xenopus* oocytes, in which oligonucleotides delivered by microinjection have been used to ablate mRNA [Kawasaki, 1985; Dash et al., 1987; Shuttleworth and Colman, 1988]. It is noteworthy that no inhibition of protein synthesis occurred in the absence of added RNase H, even when high concentrations of oligonucleotides were added, as previously noted by others [Blake et al., 1985; Haeuptle et al., 1986; Maher and Dolnick, 1987; Walder and Walder, 1988]. Conversely, RNase H could be added to high concentrations (up to 4,000 U/ml of our homemade enzyme) without causing nonspecific inhibition of protein synthesis.

2. [³⁵S]cysteine or [³⁵S]methionine can be used to measure inhibition of globin synthesis. A single methionine residue is present in the truncated chains of rabbit globin specified by mRNA cut by the oligonucleotide (position 32 in α-globin and 55 in β-globin), so the translation products would carry the same amount of label whether full-length or truncated (see Fig. 1). In contrast, the single cysteine residues occur after the GKKV sequences (positions α104 and β112), so that [³⁵S]cysteine should not label the fragments. Cysteine labeling thus made it

possible to use trichloroacetic acid (TCA) precipitation to follow the decay of full-length globin synthesis during oligonucleotide-mediated scission of the mRNA.

We tested how rapidly incorporation into TCA-insoluble material shut off when either [³⁵S]cysteine or [³⁵S]methionine was used to label reactions with added oligonucleotide and RNase H. Surprisingly, there was little difference in the time of shutoff; incorporation of both amino acids stopped at about 8 minutes (data not shown), in good agreement with the data in Figure 2. This unexpected result suggested that there was little net methionine incorporation into the fragments after the mRNAs had been cut. As we show below, this is largely due to the instability of the truncated polypeptides. Operationally this meant that [³⁵S] methionine could be used to time the end of full-length globin synthesis in the experiments described below.

B. Factors Affecting Oligonucleotide-Mediated Cleavage of mRNA by RNase H

1. Globin synthesis shut-off time is inversely proportional to the concentrations of oligonucleotide and RNase H. The lysate system allowed precise measurements of the effect of oligonucleotide and RNase H concentrations on the rate of translational arrest. Figure 3 shows the kinetics of incorporation of [³⁵S]methionine

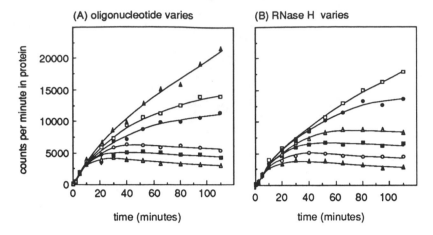

Fig. 3. *The effect of varying oligonucleotide and RNase H concentrations on hybrid arrest of translation. Reticulocyte lysate was incubated with [^{35}S]methionine, and protein synthesis was measured by TCA precipitation. The quality of the TCA precipitation data was greatly improved by adding the labeled amino acid at 200 μCi/ml and including 10 μM of the unlabeled amino acid.* **A:** *50 U/ml RNase H were present in all reactions and the concentration (μg/ml) of anti-globin oligonucleotide varied:* ▲ *none;* □, *0.31;* ●, *0.63;* ○, *1.25;* ■, *2.5;* △, *10.* **B:** *The oligonucleotide was present at 10 μg/ml and the RNase H concentration (expressed as U/ml) varied:* □, *none;* ●, *1.5;* △, *6.3;* ■, *12.5;* ○, *25;* ▲, *50.*

into protein when either oligonucleotide or RNase H concentrations were varied. In both cases, the time of shut-off varied inversely with the concentration. The main limitation in these experiments was how long globin synthesis continued in the controls; it was just possible to detect inhibition of synthesis when a twofold molar excess of oligonucleotide over mRNA was added (0.3 μg/ml). Reliable detection of inhibition was not possible at lower oligonucleotide concentrations, because, by the time an effect would be expected, globin synthesis in the control was too low and variable.

2. Effect of temperature on inhibition of globin synthesis. We wanted to use in vivo arrest of translation in the eggs and oocytes of cold-blooded animals such as frogs and sea urchins, which normally live at temperatures well below the 30°C optimal translation temperature of the reticulocyte lysate. Accordingly we examined the effect of temperature on the rate at which globin mRNA was cut. A 12-fold molar excess (2.5 μg/ml) of oligonucleotide caused globin synthesis to cease after 8 minutes at 37°C, 13 minutes at 30°C, and 39 minutes at 20°C in

the presence of 25 U/ml RNase H. Although globin synthesis inhibition took much longer at the lower temperature, it was just as complete. These data indicate that oligonucleotide-mediated hybrid arrest should work for experiments with cold-blooded organisms, in agreement with the findings of Dash et al. [1987] and Shuttleworth and Colman [1988].

3. Effect of oligonucleotide length and mismatch on inhibition of globin synthesis. It is difficult to guess what the melting temperature of the mRNA–oligonucleotide pair is likely to be in the complex environment of crude cytoplasm, and we were curious to know how long the matching stretch should be in order to obtain good arrest of translation. We tested equimolar amounts of an 8-mer, a 10-mer, and the 12-mer and found that they all gave arrest of globin synthesis, although the 8-mer worked much more slowly (Table I and data not shown).

We also synthesized 10-mers and 8-mers with sequence mismatches at either the sixth or the seventh base to determine the effect of mismatches on the ability of the oligonucleotide to arrest globin synthesis. Neither of two mis-

TABLE I. The Effect of Sequence Length and Mismatch on Hybrid Arrest*

Sequence	Length	Matching bases	Globin synthesis (%)
CAC CTT CTT GCC	12	12	17
C CTT CTT GCC	10	10	21
TT CTT GCC	8	8	45
C CTT CCT GCC	10	9	29
C CTT TTT GCC	10	9	82
TT CCT GCC	8	7	93
TT TTT GCC	8	7	93
CCA GTC CAC	9	2	100

*Reticulocyte lysate was incubated under standard translation conditions with RNase H at 50 U/ml and oligonucleotides at 50 μg/ml. [³⁵S]methionine was added at the beginning of the incubation, and samples were taken for TCA precipitation and scintillation counting after 60 minutes. As a control, each oligonucleotide was added to lysate without added RNase H. Percent globin synthesis data were obtained by dividing the counts found with RNase H by those without.

matched 8-mers showed significant arrest under our standard conditions with a 2 hour incubation, but, as Table I shows, one of the mismatched 10-mers gave a degree of arrest between those of the correct 10-mer and 8-mer sequences.

C. A Fraction of a Translating mRNA Population Is Inaccessible to RNase H Scission if Ribosome Movement Is Blocked.

Ribosomes covering the region of complementarity would be expected to protect mRNA from attack by oligonucleotides and RNase H. The fraction of messages protected from such digestion at any instant in time will depend on the spacing of ribosomes on the message. To see what fraction of globin mRNA is protected in this way, a reticulocyte lysate was incubated with emetine, a specific inhibitor that acts at the elongation stage to block ribosome movement [Vázquez, 1978]. Oligonucleotide and RNase H were added 5 minutes later. After 30 minutes at 30°C, RNA was extracted, denatured with glyoxal, and analyzed on a 1.4% agarose gel. The RNA was transferred to nitrocellulose and probed with a randomly primed cDNA clone for human β-globin. The agarose gel does not resolve the 5′ and 3′ fragments of β-globin mRNA, so the band shown in Figure 4 corresponding to the cleavage products should generate approximately the same intensity signal as the full-length globin mRNA band. A

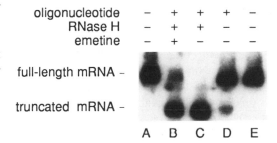

Fig. 4. *Ribosomes protect globin mRNA from attack by RNase H. Reticulocyte lysate was incubated with combinations of 25 μg/ml antiglobin oligonucleotide and 25 U/ml RNase H for 30 minutes. In one reaction (lane B), 20 μM emetine was added 5 minutes before the oligonucleotide. RNA was extracted from these reactions or from fresh lysate and analyzed by agarose gel electrophoresis, transferred to nitrocellulose, and probed with a [³²P] labeled human β-globin cDNA clone. Lane A shows RNA from fresh lysate, lanes B–E incubated 35 minutes at 30°C with added translation components, together with: lane B: oligonucleotide, RNase H, and emetine; lane C: oligonucleotide and RNase H; lane D: oligonucleotide; lane E: no additions.*

densitometer scan of a lighter exposure of the blot shown in Figure 4 indicated that 66% of the RNA was digested by RNase H and oligonucleotide in an emetine-arrested lysate (lane B) compared with complete cleavage when protein synthesis was taking place (lane C). Essentially all the globin mRNA was present on polysomes in this lysate at the start of the incubation (data not shown). Thus the cutting of the message observed in the presence of emetine is due to the accessibility of the 12 base sequence to the oligonucleotide and RNase H in 66% of translating β-globin messages at any instant.

The principle of using oligonucleotide and RNase H–mediated RNA ablation [Donios-Keller, 1979] has already seen use in probing snRNP and ribosome structure [Krämer et al., 1984; Krainer and Maniatis, 1985; Black et al., 1985; Tapprich and Hill, 1986]. The results shown here indicate that the method can also be used for mRNA. No doubt it could be used to explore other aspects of the translation machinery, for example, to probe for areas in the untranslated regions of mRNA that are protected by proteins.

The batch of reticulocyte lysate used in this experiment contained low RNase H activity, since incubation for 30 minutes with oligonucleotide and no added RNase H led to scission of about 15% of the β-globin mRNA (lane D). However, this level of RNase H was insufficient to cause noticeable effects on globin translation in the lysate during a 2 hour incubation (see Figs. 2 and 3). Lysates prepared according to different protocols probably vary in their RNase H content, since Walder and Walder [1988] found significantly higher levels of RNase H than we did.

IV. THE FATE OF TRUNCATED mRNA

A. The 5' and 3' Fragments of Globin mRNA Are Stable

Figure 4 suggested that the 5' and 3' portions of the cut message were relatively stable during a 30 minute incubation in the lysate, since the truncated mRNA band in lane C had the same intensity as the band of full-length mRNA from untreated lysate (lane E).

This result presented a puzzle. All the globin mRNA in a reticulocyte lysate was cut by RNase H in the presence of a suitable oligonucleotide, and by 10–15 minutes synthesis of globin has essentially stopped; but the capped

5' message fragment was stable and theoretically capable of directing the synthesis of methionine-labeled N-terminal globin polypeptides. Yet methionine incorporation stopped with almost the same kinetics as those of cysteine (see Section III.B). What is the origin of this apparent paradox?

It is widely believed that the lack of a termination codon on an mRNA slows or may even prevent release of the nascent polypeptide [Mueckler and Lodfish, 1986; Perara et al., 1986], though Haeuptle et al. [1986] showed in the wheat germ cell-free system that truncated mRNA led to the synthesis and release of peptidyl tRNA from the ribosomes. Thus, although sluggish recycling of ribosomes on the 5' fragments might account for the observed inhibition of methionine incorporation, we considered the alternative explanation, that the 6,500 Da N-terminal polypeptide fragment was synthesised at a rate comparable with that of full-length globin, but that it was rapidly degraded. The specificity and synchrony of the oligonucleotide/RNase H–mediated cleavage allowed us to explore the question of what happened when ribosomes reached the physical end of a message without meeting a termination codon.

B. Ribosomes Recycle as Rapidly on Truncated mRNA as on Intact Messages

To follow ribosome behavior after cleavage of the mRNA, we examined the polysome profiles. If ribosomes got stuck on reaching the ends of the 5' fragments, small, stable polysomes should persist well after globin synthesis stops. This was indeed what we found. Addition of RNase H and oligonucleotide to reticulocyte lysate resulted in a rapid change in the polysome profile. Within 9 minutes the normal globin profile (Fig. 5, left, zero time) was replaced by one in which 80S ribosomes dominated and the residual polysomes were smaller and less numerous than in the control, with dimers as the mode (Fig. 5, right, zero time). However, this pattern was more or less stably maintained for at least 60 minutes, with

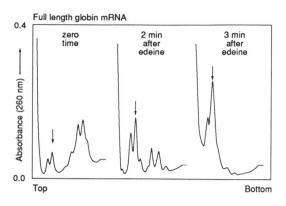

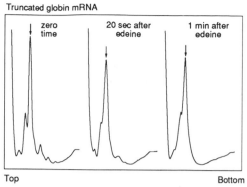

Fig. 5. *The polysomes containing truncated globin mRNA have a reduced modal size and are dynamically unstable. Reticulocyte lysate was incubated under standard translation conditions without (**left**) or with (**right**) 25 μg/ml antiglobin oligonucleotide and 25 U/ml RNase H. Edeine (10 μM final) was added at 30 minutes ("zero time"). The 40 μl reactions were stopped by dilution into 200 μl of ice-cold gradient buffer: 25 mM KCl,* *10 mM NaCl, 1.1 mM MgCl₂, 10 mM HEPES, pH 7.2, and 0.1 mM EDTA. Samples were analyzed on 5 ml sucrose density gradients (15–40% w/v sucrose in gradient buffer) by spinning in a Beckman SW50.1 ultracentrifuge rotor at 50,000 rpm for 45 minutes. Vertical arrows indicate the position of the 80S ribosomes.*

only a very gradual reduction in two, three, and four peak heights (not shown).

This finding was compatible with the "stuck-ribosome" model, but to test it we examined the dynamic stability of these small residual polysomes after the addition of edeine, a highly specific inhibitor of initiation of protein synthesis, which acts by preventing the binding of 60S ribosomes to the 40S/Met–tRNA$_f$/ mRNA complex [Vázquez, 1978]. Reticulocyte lysate was incubated with RNase H and oligonucleotide at 30°C for 30 minutes to cut all the globin mRNA. Edeine was then added and samples taken for analysis on sucrose density gradients. Figure 5 (right) shows that the polysomes, which were stable in the control without edeine from 9 to 30 minutes (data not shown), broke down within 1 minute after the addition of edeine (Fig. 5, right). This was significantly faster than the ribosomes on full-length mRNA in an equivalent experiment, which took about 4 minutes to run off in this experiment (Fig. 5, left). Thus the stable polysome profile seen with cut mRNA was not caused by ribosomes unable to leave mRNA without a termination codon, but must have

been due to constant rapid cycling of ribosomes on the shortened mRNA, presumably accompanied by synthesis of the truncated polypeptide. Thus the most likely cause of the lack of net polypeptide synthesis was instability of the truncated product.

C. The N-Terminal Truncated Globin Polypeptides are Unstable

To measure the stability of the truncated polypeptide, we performed a pulse–chase experiment. Lysate was treated with oligonucleotide and RNase H to cut the globin mRNA, [³⁵S] methionine was added to label nascent polypeptides, and emetine and unlabeled methionine were added 15 minutes later. Samples were taken at intervals after the addition of label and the TCA-precipitable counts measured. Figure 6 shows that [³⁵S]methionine incorporated into truncated polypeptide disappears with a half-life of about 10 minutes, whereas full-length globin is completely stable. The residual stable labeled protein at the end of the chase is probably due to the synthesis of other proteins in these lysates.

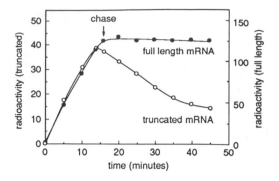

Fig. 6. *The N-terminal truncated globin peptides are unstable. Reticulocyte lysate was incubated under standard translation conditions without (filled circles) or with (open circles) 25 µg/ml antiglobin oligonucleotide and 25 U/ml RNase H. After 30 minutes [³⁵S] methionine was added, and 16 minutes later 1 mM unlabeled methionine and 0.1 mM emetine were added. Samples were taken at intervals thereafter for measuring TCA-precipitable radioactivity. Note the different radioactivity scales.*

D. Ribosomes Release Truncated Polypeptides as Peptidyl tRNA When They Reach the End of the mRNA

Data indicate that ribosomes cycle rapidly on the truncated mRNA and that their product is unstable. However, they do not show whether the ribosomes leaving the end of the mRNA continue to carry peptidyl tRNA or whether they release it. To distinguish between these possibilities, lysates containing truncated or full-length mRNA were labeled with [³⁵S] methionine for 32 minutes, at which time emetine was added to freeze the ribosomes on the mRNA. Samples were taken for analysis on sucrose gradients at the same time as, and 38 minutes after, adding the emetine. Fractions from the gradients were collected and their radioactivity analyzed by cetyltrimethylammonium bromide (CTAB) precipitation, a procedure that precipitates tRNA and peptidyl tRNA, but not completed proteins [Darnbrough et al., 1973]. The polysomes on the truncated mRNA contained only about one-fifth as many nascent chains as the controls, and no peak of radioactivity was associated with the 80S ribosomes

at either time point (Fig. 7). However, compared with the control, the incubation with truncated mRNA showed a large increase in the CTAB-precipitable counts at the top of the gradient. These extra counts, which presumably represent peptidyl tRNA, decayed with the same kinetics as the truncated polypeptide. We conclude that the polypeptide is released from the ribosomes as peptidyl tRNA. Similar results were obtained in the wheat germ system by Haeuptle et al. [1986].

Since the N termini of the released peptides are identical with normal α-and β-chains, it would appear that the N-terminal rule proposed by Bachmair et al. [1986] cannot account for their instability. Perhaps there is a way that peptides arising from hydrolysis of peptidyl tRNA can be marked, though it has long been known that the truncated globin peptides produced by puromycin action are very unstable [Etlinger and Goldberg, 1980]. The method described here provides relatively homogeneous substrates that could be used for further study of the problem.

V. ABLATION OF mRNA TO TEST PROTEIN FUNCTION

A. At High Levels of Oligonucleotide, Inhibition of Protein Synthesis Becomes Less Specific

Although oligonucleotides can give highly specific arrest of translation of particular mRNAs [Minshull and Hunt, 1986], the results in Table I suggested that inactivation of the "wrong" mRNA through fortuitous imperfect sequence matches might occur. Thus a 17-mer contains 10 possible octanucleotide sequences and a much larger number of 9-out-of-10 mismatched sequences that could cut any mRNA that happened to match one of them.

Until now, we had been studying a simple system whose predominant products were the α- and β-chains of globin, two relatively short and unusually abundant mRNA. To see if accidental sequence matches would be a problem in oligonucleotide-mediated hybrid arrest in cells making a more typically diverse range of

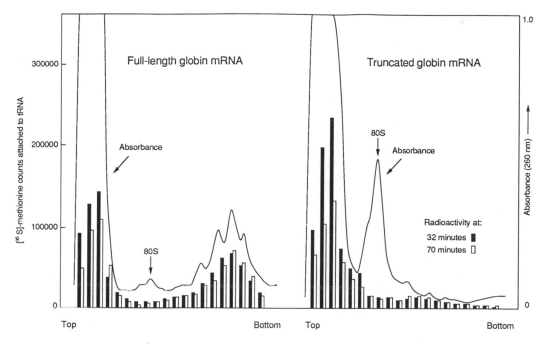

Fig. 7. *Nascent chains are released from truncated mRNAs as peptidyl tRNA. Reticulocyte lysate was incubated under standard translation conditions without* **(left)** *or with* **(right)** *25 μg/ml antiglobin oligonucleotide and 25 U/ml RNase H for 30 minutes before [³⁵S]methionine was added. Emetine (170 μM) was added 32 minutes after addition of the label, and samples (40 μl) were taken immediately after and 38 min-* *utes later for analysis by sucrose density gradient centrifugation as described in Figure 5. The gradients were collected into 0.25 ml fractions, which were counted after precipitation with 1 ml of 2% CTAB followed by 1 ml of 0.5 M NaOAc, pH 5.2, containing 0.5 mg/ml yeast RNA as carrier. The absorbance profiles of the polysomes did not change appreciably between the two time points.*

proteins, we turned to study maternal mRNA from early sea urchin and starfish embryos.

Increasing amounts of a 17-mer complementary to the small subunit of ribonucleotide reductase mRNA [Standart et al., 1985] were incubated with total starfish oocyte mRNA and RNase H for 30 minutes at 37°C and then translated in the reticulocyte lysate. At low doses of oligonucleotide (1.4 μg/ml), the synthesis of a single polypeptide of the expected size (about 44 kD) was specifically inhibited (Fig. 8; cf. lanes F and G). Further studies confirmed that this band corresponded to starfish ribonucleotide reductase; it bound to a monoclonal antibody (YL 1/2) that recognizes the small subunit of ribonucleotide reductase [Standart et al., 1986; N. Standart, personal communi-

cation]. However, as the dose of oligonucleotide was increased to a maximum of 330 μg/ml, at least two other polypeptides (indicated by arrows) showed reduced synthesis (Fig. 8). A similar effect was observed when the 17-mer was held constant at 70 μg/ml and the RNase H concentration was increased from 0 to 450 U/ml (data not shown). These high levels of oligonucleotide and RNase H presumably promote the scission of messages with partially matching sequences. We have found that "nonspecific" inhibition of protein synthesis can be a serious problem even at quite low oligonucleotide concentrations in experiments with *Xenopus* extracts (J. Minshull, J.J. Blow, and T. Hunt, unpublished experiments, and see below).

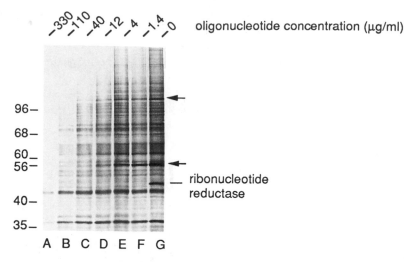

Fig. 8. *The specificity of hybrid arrest is reduced at high levels of oligonucleotide and RNase H. In all lanes, RNA from oocytes of the starfish* M. glacialis *was mixed with increasing concentrations of the oligonucleotide 5'-TGGATGTTCTCGATGGC-3' (anti-sea urchin ribonucleotide reductase) and digested with 450 U/ml RNase H in 100 mM KCl, 20 mM Tris-Cl, pH 7.5, 1 mM MgCl$_2$, 1 mM DTT, 50 μg/ml BSA. One microliter of the digestion mix was added to a nuclease-treated retic-* *ulocyte lysate with [^{35}S]methionine as label. Translation products were analyzed by electrophoresis on a 15% SDS polyacrylamide gel, followed by autoradiography. Lane* **A,** *330 μg/ml;* **B,** *110 μg/ml;* **C,** *40 μg/ml;* **D,** *12 μg/ml;* **E,** *4 μg/ml;* **F,** *1.4 μg/ml;* **G,** *no oligonucleotide. The positions of molecular weight markers are indicated in kD. Arrows indicate further bands that disappear at higher oligonucleotide concentrations.*

B. Extrapolating Hybrid Arrest Conditions From the Reticulocyte Lysate to a *Xenopus* Cell-Free System

We used the information given in Section III above to plan a rational hybrid arrest strategy to investigate the role of cyclin mRNA translation in mitosis in a cell-free extract made from *Xenopus* eggs. There are significant differences between the reticulocyte lysate and *Xenopus* egg in vitro systems that must be taken into account when extrapolating from one to the other.

1. Oligonucleotide selection. Oligonucleotides are very stable in the reticulocyte lysate, but degrade quite rapidly in *Xenopus* oocytes and eggs [Rebagliati and Melton, 1987; Woolf et al., 1990; J. Minshull and A. Colman, unpublished results). Thus, although in the reticulocyte lysate oligonucleotides only 10 or 12 residues long efficiently direct RNase H cleavage of mRNA, longer oligonucleotides appear to be

needed in the *Xenopus* system, probably because the extra length increases their lifetime and thus improves the chances of obtaining complete ablation of the target mRNA. We found that oligonucleotides containing 20–23 residues gave good results. Modified (phosphorothioate) oligonucleotides are more stable in *Xenopus* eggs and oocytes, so lower concentrations and shorter oligonucleotides can be used to acheive the same ablation [Baker et al., 1990; Dagle et al., 1990; Woolf et al., 1990].

Choosing the sequence of an oligonucleotide to use in antisense studies is a largely empirical matter. In our experience, some work well and others do not, and there is little obvious correlation between size, base composition, and effectiveness at promoting the degradation of a given mRNA. We have found that 1.3 μM of a 23-mer oligonucleotide (10 μg/ml) can be sufficient to produce total cleavage within 5 minutes of a translating mRNA present at 0.16 nM in *Xenopus* egg extracts

[Minshull et al., 1989]. Other oligonucleotides required up to 10–20-fold higher concentrations to achieve the same level of cutting [J. Minshull and T. Hunt, unpublished results; Baker et al., 1990]. There are no obvious reasons for these differences that can be found either in the sequence composition of the oligonucleotides or in the predicted secondary structures of the mRNA.

2. Measuring inhibition of translation. Before any conclusions can be drawn about the functional consequences of loss of a particular mRNA or its protein product, it is crucial to check the extent of the ablation. It was easy to assay the level of globin synthesis in the reticulocyte lysate since it was the major protein being synthesized. If the oligonucleotide is directed against a relatively rare mRNA in a complex mixture, however, looking at translation products may not be sufficiently sensitive to measure the degree of cleavage. An alternative and in most ways superior method is to measure the degree of scission of the target mRNA. RNase protection mapping is extremely sensitive and can distinguish between cut and full-length mRNA [Krieg and Melton, 1987; Minshull et al., 1989]. Recently, several laboratories have reported inhibition of cellular or viral replication by antisense oligonucleotides [e.g., Heikkila et al., 1987; Goodchild et al., 1988]. It would be interesting to see whether the targeted RNA were being cut in the expected locations in these situations. A potential difficulty about measuring mRNA destruction is of course that the targeted protein itself may be quite stable; destroying mRNA only prevents new protein synthesis. Moreover, in many situations, new synthesis of RNA can occur after the oligonucleotides have been destroyed. Neither of these considerations presented a problem in the early frog embryo, where no new transcription occurs between maturation and the mid-blastula transition and where many maternal mRNAs are recruited onto polyribosomes for the first time during maturation or after fertilization. Even so, there are still relatively few studies in which the use of antisense techniques have yielded reliable new information. One example was the demonstration that c-*mos*

translation was required for maturation; ablation of the c-*mos* mRNA blocked maturation, and injection of c-*mos* mRNA promoted maturation [Sagata et al., 1988]. We recently performed experiments in which all known cyclin mRNAs were ablated in stage VI *Xenopus* oocytes, and there was no inhibition of maturation from which we were able to conclude that no *new* cyclin synthesis was required for oocyte maturation; other studies had shown that a stockpile of maternal protein existed, which is probably adequate to supply the cyclins required for meiosis [Minshull et al., 1991].

C. Problems Associated With Using Antisense Oligonucleotides to Investigate Protein Function

While it is essential to demonstrate that an antisense oligonucleotide has prevented the translation of its cognate mRNA, this is not sufficient to conclude that any observed effects can be ascribed exclusively to lack of the targeted mRNA. In our opinion, inhibition of function by oligonucleotides needs to be interpreted very cautiously. We suggest that, logically speaking, only when it is possible to show that an mRNA is completely cut and the cell shows *no* effects (other than loss of synthesis of the particular protein) can one legitimately make the strong conclusion that the mRNA is *un*important. If the mRNA is cut and some process fails, it may be due to unsuspected side effects and not loss of the targeted mRNA or its translation product. Some major causes of side effects are described below.

1. Oligonucleotide specificity. A serious problem with the use of antisense oligonucleotides is that the specificity of mRNA cutting is reduced when the oligonucleotide concentration is increased (see Section V.A). Since RNase H requires as little as six base-paired nucleotides to cut the RNA, this is not surprising [Donis-Keller, 1979]. We have found this to be a problem in *Xenopus* egg extracts, where rapid and complete cutting of mRNA requires the use of 10–20 μg/ml oligonucleotide. The synthesis of other proteins in addition to those intended is clearly inhibited, with a tendency first to lose the longer proteins, as might be expected from considerations of target size [Minshull et al., 1989]. Even when no other specific protein synthesis changes can be seen, there is no guarantee that the synthesis of relatively minor proteins has not been affected.

Probably the most satisfactory way to deal with this problem is to use more than one oligonucleotide directed against the same mRNA. If all oligonucleotides produce the same phenotypic effect, this greatly strengthens the argument that the effects seen are specific and result from destruction of the presumed target [Sagata et al., 1988].

2. Nonspecific inhibition of cellular processes by oligonucleotides. The introduction into cells or a cell-free system of the relatively large amounts of oligonucleotides required to achieve rapid and complete arrest of translation can have nonspecific deleterious effects. These effects, such as general inhibition of protein synthesis [Minshull and Hunt, 1986] (Fig. 8), may give rise to phenotypes that could be mistaken for the genuine effect of ablating the target mRNA. Some, but not all, of this toxicity is inevitable: Oligonucleotides are degraded in cells to nucleotides, which if present in large excesses are themselves toxic [Woolf et al., 1990]. The inclusion of a control oligonucleotide is a helpful, but not always an adequate, way to assess nonspecific effects. This is because in our experience there is an appreciable difference in the amount of general inhibition seen with different oligonucleotide preparations, even those with the same sequence made on the same machine. Traditional purification by separation on polyacrylamide/urea gels is not normally very useful, since the exact size of the oligonucleotide is not critical; thus contamination with incomplete synthesis products will not be detrimental. Extracting the oligonucleotide with an organic solvent such as ether reduces the inhibitory effects at least as well as gel purification (which usually includes an organic extraction step in any case). More effective than ether extraction are the commercially available oligonucleotide purification systems. We found

that OPC columns from Applied Biosystems gave good results.

3. Surviving 5′ and 3′ mRNA cleavage products. In some cases the persistence of mRNA cleavage products may have to be borne in mind. This is significant in the reticulocyte lysate where cleaved mRNA fragments are stable and can continue to direct protein synthesis [Minshull and Hunt, 1986]. In other cases, such as *Xenopus* egg cell-free systems, cut mRNAs are rapidly degraded [Minshull et al., 1989], and in *Xenopus* oocytes mRNA fragment stability varies [Dash et al., 1987; Shuttleworth and Colman, 1988].

VI. CONCLUSIONS

The idea of using the antisense approach was originally suggested to us by the need to test the effects of loss of particular mRNAs in organisms like clams, sea urchins, and frogs in which genetics could never be used to create null mutations in the desired genes. Once we had realized the importance of RNAse H in the mode of action of antisense olinucleotides, we became enthusiastic advocates of the technique, as we discovered that it was a simple matter to ablate mRNA in cell-free extracts and in easily injected cells like frog oocytes. Almost all it took was the design and synthesis of oligonucleotides!

Our enthusiasm was relatively short lived, however, as we became increasingly aware of the problems of interpretation presented by the antisense approach. We suspect that relatively few people fully appreciate the logic that was made clear to us from experience. When we first ablated cyclin mRNA in frog oocytes, we expected to find that the oocytes would fail to mature. In our first experiments, this is precisely what happened, and we were very pleased. As we did more experiments, however, and did careful studies of the extent of mRNA scission, we found that doses of antisense oligonucleotides that gave 99% + cutting of particular cyclin mRNAs did not prevent oocyte maturation.

As described above, we now know that *Xenopus* oocytes can mature normally without any cyclin synthesis [Minshull et al., 1991]. This is a valid conclusion from such studies. All too often, however, the inhibition of cellular function by antisense oligonucleotides is interpreted as providing strong support for the hypothesis that loss of the targeted mRNA causes the observed result. This is a false conclusion. The only valid inference to be drawn is from the opposite result, when the targeted mRNA is destroyed and nothing happens. This means that the mRNA plays no part in the process. A nice example of this was reported at a recent *Xenopus* meeting by T. Pieler. He and his colleagues discovered that *Xenopus* embryos contained a class of mRNAs that encoded a family of zinc-finger proteins in *Xenopus* embryos. The mRNAs could be targeted for destruction en masse because they contained a highly conserved motif of an appropriate length for antisense ablation. When the investigators successfully cut a large fraction of these mRNA, they were surprised to find no developmental defect. They quickly realized the answer, however, which was that translation of these mRNAs occurred early in oogenesis. Loss of the mRNAs after fertilization was of no consesquence!

It should be recalled that antisense oligonucleotides sometimes stimulate the translation of mRNA, rather than inhibit it. We discovered this largely by accident during studies of masked maternal mRNA in clam embryos (so it is probably of somewhat limited and esoteric significance). Nevertheless, it is of some interest to consider the logic in this case. Clam oocytes contain abundant supplies of mRNA for cyclin A and the small subunit of ribonucleotide reductase, but do not synthesize significant amounts of these proteins until after fertilization. The messages are said to be "masked." We found that these mRNAs contain a short region that we call the "masking box," because when we added antisense RNA complementary to these regions, it led to tremendous stimulation of their translation in cell-free extracts of clam oocytes [Standart et al., 1990]! (We have not tried to see if this works in intact cells, partly because clam oocytes are difficult to microinject and partly because they

are highly seasonal.) We were also able to stimulate translation of ribonucleotide reductase mRNA by cutting off the 3' untranslated region of its mRNA with an antisense oligonucleotide. These experiments are not subject to our previous strictures vis à vis their interpretation, because the antisense RNA and DNA caused stimulation of synthesis, and highly specific stimulation at that.

There is a way around the problem of interpretation of the negative finding, that may be useful. It can be called the "rescue technique." Consider, for example, a current concern of ours, the case of the mRNA for Eg1 in *Xenopus* oocytes. This mRNA encodes a homolog of $p34^{cdc2}$ whose function is uncertain [Paris et al., 1991]. The Eg1 mRNA is not translated until meiotic maturation, and the levels of Eg1 protein are very low in nonmatured oocytes. It is possible to prevent the synthesis of Eg1 by microinjection of antisense oligonucleotides, and these oligonucleotides then are themselves destroyed by nucleases in these cells. If lack of Eg1 protein has an effect on the subsequent development of the *Xenopus* embryo, then this antisense treatment should cause developmental abnormalities. We can logically assign any such abnormalities to the specific loss of Eg1 mRNA if the observed defects can be rescued by injection of either Eg1 protein or Eg1 mRNA after all the original mRNA and the oligonucleotides have been destroyed. A particularly neat way to achieve this end is to target the oligonucleotides against the 5' untranslated region of the mRNA and rescue with a synthetic mRNA in which the Eg1 coding region has been placed downstream of a heterologous 5' untranslated region (and therefore not a target for the oligonucleotide), which is a simple matter. We are currently involved in experiments of this kind.

Thus we may end on an optimistic note. There are certainly times when antisense approaches are very useful. It is extremely important to perform appropriate controls and to bear in mind the everpresent spectre of nonspecific ablation of unknown RNA (quite apart from other as-yet-unknown nonspecific inhibitory effects) in drawing one's final conclusions.

VII. REFERENCES

Affara N, Fleming J, Goldfarb PS, Black E, Thiele B, Harrison PR (1985): Analysis of chromatin changes associated with the expression of globin and non-globin genes in cell hybrids between erythroid and other cells. Nucleic Acids Res 13:5629–5644.

Bachmair A, Finley D, Varshavsky A (1986): In vivo half-life of a protein is a function of its amino-terminal residue. Science 234:179–185.

Baker C, Holland D, Edge M, Colman A (1990): Effects of oligo sequence and chemistry on the efficiency of oligodeoxyribonucleotide-mediated mRNA cleavage. Nucleic Acids Res 18:3537–3543.

Bass BL, Weintraub H (1987): A developmentally regulated activity that unwinds RNA duplexes. Cell 48:607–613.

Black DL, Chabot B, Steitz JA (1985): U2 as well as U1 small nuclear ribonucleoproteins are involved in premessenger RNA splicing. Cell 42:737–750.

Blake KR, Murakami A, Miller PS (1985): Inhibition of rabbit globin mRNA translation by sequence-specific oligodeoxyribonculeotides. Biochemistry 24:6132–6138.

Cazenave C, Loreau N, Toulme J-J, Helene C (1986): Anti-messenger oligodeoxynucleotides: Specific inhibition of rabbit β-globin synthesis in wheat germ extracts and *Xenopus* oocytes. Biochimie 68:1065–1069.

Chandler PM (1982): The use of single-stranded phage DNAs in hybrid arrest and release translation. Anal Biochem 127:9–16.

Dagle JM, Walder JA, Weeks DL (1990): Targeted degradation of mRNA in *Xenopus* oocytes and embryos directed by modified oligonucleotides: Studies of An2 and cyclin in embryogenesis. Nucleic Acids Res 18:4751–4757.

Darnbrough C, Legon S, Hunt T, Jackson RJ (1973): Initiation of protein synthesis: Evidence for messenger RNA independent binding of methionyl-transfer RNA to the 40S ribosomal subunit. J Mol Biol 76:379–403.

Dash P, Lotan I, Knapp M, Kandel ER, Goelet P (1987): Selective elimination of mRNAs in vivo: Complementary oligodeoxynucleotides promote RNA degradation by an RNase H-like activity. Proc Natl Acad Sci USA 84:7896–7900.

Donis-Keller H (1979): Site specific enzymatic cleavage of RNA. Nucleic Acids Res 7:179–192.

Etlinger JD, Goldberg AL (1980): Control of protein degradation in reticulocytes and reticulocyte lysate extracts by hemin. J Biol Chem 255:4563–4568.

Evans T, Rosenthal ET, Youngblom J, Distel D, Hunt T (1983): Cyclin: A protein specified by maternal mRNA in sea urchin eggs that is destroyed at each cleavage division. Cell 33:389–396.

Goodchild J, Agrawal S, Civiera MP, Sarin PS, Sun D, Zamecnik P (1988): Inhibition of human immunodeficiency virus replication by antisense oligodeoxynucleotides. Proc Natl Acad Sci USA 85:5507–5511.

Haeuptle M-T, Frank R, Dobberstein B (1986): Translation arrest by oligodeoxynucleotides complementary to mRNA coding sequences yields polypeptides of predetermined length. Nucleic Acids Res 14:1427–1448.

Harland R, Weintraub H (1985): Translation of mRNA injected into *Xenopus* oocytes is specifically inhibited by antisense RNA. J Cell Biol 101:1094–1099.

Heikkila R, Schwab G, Wickstrom E, Loke SL, Pluznik DH, Watt R, Neckers LM (1987): A *c-myc* antisense oligodeoxynucleotide inhibits entry into S-phase but not progress from G_0 to G_1. Nature 328:445–449.

Izant JG, Weintraub H (1984): Inhibition of thymidine kinase gene expression by antisense RNA: A molecular approach to genetic analysis. Cell 36:1007–1015.

Jackson RJ, Hunt T (1983): The turnover of methionine in the Met-tRNA pool and the control of protein synthesis in reticulocyte lysates. FEBS Lett 143:301–305.

Kawasaki ES (1985): Quantitative hybridisation–arrest of mRNA in *Xenopus* oocytes using single-stranded complementary DNA or oligonucleotide probes. Nucleic Acids Res 13:4991–5004.

Krainer AR, Maniatis T (1985): Multiple factors including the small nuclear ribonucleoproteins U1 and U2 are necessary for pre-mRNA splicing in vitro. Cell 42:725–736.

Krämer A, Keller W, Appel B, Lührmann R (1984): The 5' terminus of the RNA moiety of U1 small nuclear ribonucleoprotein particles is required for the splicing of messenger RNA precursors. Cell 38:299–307.

Krieg PA, Melton DA (1987): In vitro RNA synthesis with SP6 RNA polymerase. Methods Enzymol 155:397–415.

Liebhaber SA, Cash FE, Shakin SH (1984): Translationally associated helix-destabilising activity in rabbit reticulocyte lysate. J Biol Chem 259:15597–15602.

Maher III LJ, Dolnick BJ (1987): Specific hybridization of dihydrofolate reductase mRNA in vitro using antisense RNA or anti-sense oligonucleotides. Arch Biochem Biophys 253:214–220.

Melton DA (1985): Injected anti-sense RNAs specifically block messenger RNA translation in vivo. Proc Natl Acad Sci USA 82:144–148.

Miller JS, Paterson BM, Ricciardi RP, Cohen L, Roberts BE (1983): Methods utilizing cell-free protein-synthesizing systems for the identification of recombinant DNA molecules. Methods Enzymol 101:650–674.

Minshull J, Hunt T (1986): The use of single-stranded DNA and RNase H to promote quantitative "hybrid arrest of translation" of mRNA/DNA hybrids in reticulocyte lysate cell-free translations. Nucleic Acids Res 14:6433–6451.

Minshull J, Blow JJ, Hunt T (1989): Translation of cyclin mRNA is necessary for extracts of activated *Xenopus* eggs to enter mitosis. Cell 56:947–956.

Minshull J, Murray A, Colman A, Hunt T (1991): *Xenopus* oocyte maturation does not require new cyclin synthesis. J. Cell Biol 114:767–772.

Mueckler M, Lodish HF (1986): The human glucose transporter can insert posttranslationally into microsomes. Cell 44:629–637.

Paris J, Le-Guellec R, Couturier A, Le-Guelle CK, Omilli F, Camonis J, MacNeill S, Philippe M (1991): Cloning by differential screening of a *Xenopus* cDNA coding for a protein highly homologous to cdc2. Proc Natl Acad Sci USA 88:1039–1043.

Paterson BM, Roberts BE, Kuff EL (1977): Structural gene identification and mapping by DNA.mRNA hybrid arrested cell-free translation. Proc Natl Acad Sci USA 74:4370–4374.

Perara E, Rothman RE, Lingappa VR (1986): Uncoupling translocation from translation: Implications for transport of proteins across membranes. Science 232:348–352.

Pines J, Hunt T (1987): Molecular cloning and characterization of the mRNA for cyclin from sea urchin eggs. EMBO J 6:2987–2995.

Rebagliati MR, Melton DA (1987): Antisense RNA injections in fertilized frog eggs reveal an RNA duplex unwinding activity. Cell 48:599–605.

Sagata N, Oskarsson M, Copeland T, Brumbaugh J, Vande Woude GF (1988): Function of *c-mos* proto-oncogene product in meiotic maturation in *Xenopus* oocytes. Nature 335:519–526.

Schewe T, Rapoport SM, Kühn H (1986): Enzymology and physiology of reticulocyte lipoxygenase: Comparison with other lipoxygenases. Adv Enzymol 58:191–272.

Shakin SH, Liebhaber SA (1986): Destabilization of messenger RNA/complementary DNA duplexes by the elongating 80S ribosome. J Biol Chem 261:16018–16025.

Shuttleworth J, Colman A (1988): Antisense oligonucleotide-directed cleavage of mRNA in *Xenopus* oocytes and eggs. EMBO J 7:427–434.

Standart NM, Bray SJ, George ELTH, and Ruderman J (1985): The small subunit of ribonucleotide reductase is encoded by one of the most abundant translationally regulated maternal RNAs in clam and sea urchin eggs. J Cell Biol 100:1968–1976.

Standart N, Dale M, Stewart E, Hunt T (1990): Maternal mRNA from clam oocytes can be specifically unmasked in vitro by antisense RNA complementary to the 3′ untranslated region. Genes Dev 4:2157–2168.

Standart N, Hunt T, Ruderman JV (1986): Differential accumulation of ribonucleotide reductase subunits in clam oocytes: The large subunit is stored as a polypeptide, the small subunit as untranslated mRNA. J Cell Biol 103:2129–2136.

Tapprich WE, Hill WE (1986): Involvement of bases 787–795 of *Escherichia coli* 16S ribosomal RNA in ribosomal subunit association. Proc Natl Acad Sci USA 83:556–560.

Thiele BJ, Andree H, Höhne M, Rapoport SM (1982): Lipoxygenase mRNA in rabbit reticulocytes. Eur J Biochem 129:133–141.

Vázquez D (1978): Inhibitors of Protein Biosynthesis. Molecular Biology and Biophysics, Vol 30. Berlin: Springer-Verlag.

Walder J (1988): Antisense DNA and RNA: Progress and prospects. Genes Dev 2:502–504.

Walder RY, Walder JA (1988): Role of RNase H in hybrid-arrested translation by antisense oligonucleotides. Proc Natl Acad Sci USA 85:5011–5015.

Weintraub H, Izant JG, Harland RM (1985): Anti sense RNA as a molecular tool for genetic analysis. Trends Genet 1:22–25.

Woolf TM, Jennings CGB, Rebagliati M, Melton DA (1990): The stability, toxicity and effectiveness of unmodified and phosphorothioate antisense oligodeoxynucleotides in *Xenopus* oocytes and embryos. Nucleic Acids Res 18:1763–1769.

ABOUT THE AUTHORS

JEREMY MINSHULL is a Postdoctoral Fellow in the Department of Physiology at the University of California, San Francisco. He received his B.A. from Cambridge University, where he went on to study for his Ph.D. in the laboratory of Dr. Tim Hunt. In Dr. Hunt's laboratory he demonstrated the role of RNase H in oligonucleotide-mediated "hybrid arrest of translation." He then used antisense techniques to study the control of the cell cycle in *Xenopus*, and showed that cyclin B synthesis is required for entry into mitosis. Dr. Minshull is currently working in Professor Andrew Murray's laboratory on feedback control of the cell cycle, and how spindle assembly regulates the exit from mitosis.

TIM HUNT is a Senior Scientist at the Clare Hall Laboratories of the Imperial Cancer Research Fund, where he heads the Cell Cycle Control Laboratory. He received his B.A. degree from the University of Cambridge, England, in 1964 and stayed on to do his Ph.D. under the supervision of Asher Kerner and Alan Munro in the Department of Biochemistry, studying the biosynthesis of hemoglobin in rabbit reticulocytes. He continued these studies with Irving London at the Albert Einstein College of Medicine, New York, working on the control of protein synthesis in the reticulocyte lysate cell-free system. Returning to Cambridge, he continued work on this system for many years with Richard Jackson, and was appointed University Lecturer in Biochemistry in 1981. From 1977 to 1985, Dr. Hunt taught at the Marine Biological Laboratory, Woods Hole, where he and his colleagues discovered cyclins in the summer of 1982. His interest in antisense techniques was stimulated by the need to eliminate cyclin mRNA from sea urchin eggs. In 1990 Dr. Hunt moved to the ICRF, where he continues to work on the protein kinases that control cell cycle transitions. He was elected a Fellow of the Royal Society in 1991.

Antisense RNA and DNA: 213–225
© 1992 Wiley-Liss, Inc.

Toward the Total Removal of Maternal mRNA From *Xenopus* Germ Cells With Antisense Oligodeoxynucleotides

Alan Colman

I. INTRODUCTION

The full grown *Xenopus* oocyte is a large cell (diameter ~1.2 mm) that, after its removal from the maternal ovary, can survive in vitro for long periods in simple salt media. The large size and hardiness of this cell have facilitated numerous studies involving the microinjection of a variety of macromolecules or organelles [Colman, 1984]. Indeed, it has long proved a popular choice of system for the molecular or cell biologist intent on examining some particular aspect of the control of expression of a partic-ular non-*Xenopus* gene. Microinjection offers two particular advantages: First, its use circumvents any difficulties posed by the passage of the conveyed material across the plasma membrane. Second, it allows the introduction of known and regulatable doses of injected materials into the cell. These uses of the oocyte as an ''in vivo'' test tube have overshadowed the fact that the oocyte is an extremely interesting cell in its own right. With the appropriate stimulus it changes into an unfertilized egg and, on fertilization, undergoes the full gamut of developmental changes that will lead to an adult frog. The molecular processes that underlie and generate these developmental changes have been the subject of continuing scrutiny.

A use of both genetics and experimental manipulation of embryos is essential for a full understanding of the embryology of any metazoan. However, the relative importance of these two approaches varies according to features of each organism and the way in which it develops. For example, the small genome size and short life cycle of invertebrates such as the fly *(Drosophila)* and the worm *(Caenorhabditis)* makes them more amenable to a genetic approach than a slowly growing organism with a large genome, such as the frog *(Xenopus)*. Conversely, the large size of *Xenopus* embryos makes them ideal for surgical techniques. Such techniques have demonstrated that interactions between the cells of a developing vertebrate are important in establishing the basic body plan. However, in only a few instances [e.g., retinoic acid in bird limb formation; Eichele, 1989] have the inducers (the molecules that convey the instructions) been identified. The unequivocal identification of inducers is one of many areas in vertebrate embryology where the absence of a conventional genetic approach is a handicap. This has led to much interest in alternative, "reverse" genetic strategies such as the overexpression of a specific gene [Dreiver and Nusslein-Volhard, 1988; McMahon and Moon, 1989], gene replacement [Rossant and Joyner, 1989], or antisense inhibition of gene action [for review, see Colman, 1990]. The antisense sequences can be RNA produced in embryos from introduced DNA templates (Giebelhaus et al., 1988), or RNA directly injected (Rosenberg et al., 1985); in these cases it is believed that the hybridization of the antisense strand interferes sterically with the processing of transcripts or their translation [McMahon and Moon, 1989] (see also Patel and Jacobs-Lorena, this volume; Levy et al., this volume). Alternatively, oligodeoxynucleotides can be injected into embryonic cells, where they mediate the cleavage of the hybridized RNA strand. I describe how *Xenopus* oocytes can be used to test and refine oligonucleotides for their ability to mediate the ablation of target mRNA and how, subsequently, the knowledge acquired is being employed in various strategies for interfering with the translation of mRNA thought to be important for early frog development. I begin with a description of germ cell and early embryonic development (Section II). This is followed by a summary of the current understanding of the molecular basis of some of the early developmental events (Section III). How oligonucleotides have been evaluated with *Xenopus* oocytes and how insight into the oocyte cell cycle has been obtained with these reagents are described in Section IV. The toxicity caused by the large doses of oligonucleotides needed, is also discussed in Section IV, thereby setting the scene for Section V in which the ways of reducing the amounts needed by oligonucleotide modification are reviewed. In Section VI are examined some of the more fundamental problems that affect the interpretation of oligonucleotide use, and finally, in Section VII are compared and contrasted the track records and future prospects of oligonucleotides and other antisense reagents. Hopefully this review not only will show how the *Xenopus* oocyte can provide a useful testing ground for studies of oligonucleotide-mediated RNA cleavage but also will present the oocyte, rather than the fertilized egg, as the optimal vehicle for the production of frog embryos where specific RNA sequences have been eliminated.

II. EARLY *XENOPUS* DEVELOPMENT

Approximately 6 weeks after metamorphosis in *Xenopus,* oogonia, the mitotic progenitors of the oocytes, undergo what is to be a highly prolonged meiotic division. Whether all the oogonia become meiotic or whether a continually regenerating population remains throughout the lifetime of the frog is unclear. During the meiotic phase, the oocyte grows considerably [Dumont, 1972], although the growth phases can be spasmodic. The major spurt in growth occurs after the oocytes develop vitellogenin receptors in their plasma membranes. This allows a rapid endocytic uptake of vitellogenin from the maternal blood into the oocyte; the vitellogenin is made in the maternal liver. As a result, the oocyte can increase in diameter from 0.6 to 1.2 mm within 28 days. This protein store will serve the nutritional needs of the embryo until the feeding tadpole stage, 7 days into development, and partly explains the ability of the oocyte to survive for so long in media containing only simple salts.

The full-grown oocyte is arrested at the prophase stage of the first meiotic division. On stimulation by progesterone (in vitro) or progesterone-like substance(s) (in vivo), meiosis proceeds, in a process known as *maturation,* as far as the second meiotic metaphase, when the cell is then called an *unfertilized egg*. After fertilization, a series of 12 rapid, synchronized, reductive cleavage divisions ensue, forming within 6 hours a hollow ball of about 2,000 cells known as a blastula [for detailed review, see Gerhardt, 1980]. During this period from fertilization to the blastula stage, there is negligible transcription of the zygotic genes and the protein synthesis that occurs is entirely dependent on the maternal mRNA i.e., the RNA that is accumulated during oogenesis. However as blastulation proceeds, a transition (midblastula transition) takes place and the zygotic genome is activated. Complex cell movements are initiated, and by the completion of gastrulation, 12 hours into development, the establishment of the basic body plan has occurred. During this crucial period between the blastula and gastrula stages, cells in different parts of the embryo receive instructions that predispose them to follow, subsequently, different developmental pathways. It is clear that some of these instructions are maternal in origin, and much effort has been expended in determining their nature.

III. IDENTIFICATION OF EARLY-ACTING GENES IN *XENOPUS* DEVELOPMENT

A. Current Status

The generation time in *Xenopus* is about 12–18 months. This, the large genome size (6 pg/somatic cell), and its pseudotetraploid nature make the generation and selection of useful genetic mutants a formidable task and one that has not met with much success. However, the large size of the embryo and its resilience to manipulation have given the frog a venerable place in the history of embryology this century. We understand a lot about the movements and regional specification of cells in the early embryo, and in the last 20 years the crucial role of intercellular communication in the process of specifying cell fates has been identified.

In embryonic terms, the earliest example of such instructive communication is mesoderm induction. In this process, signalling molecule(s) produced by cells in vegetal regions of the embryo that belong to the endodermal lineage (precursor to gut, lungs, and so forth) induce overlying marginal zone cells to form mesoderm (precursor to muscle, blood, kidney, and so forth). This process can be simulated by exposing ectodermal cells, which would otherwise form epidermis, to purified growth factors, such as transforming growth factors, basic fibroblast growth factors, and various activins [Slack et al., 1987; Smith, 1987; Kimelman and Kirschner, 1987; Thomsen et al., 1990]. Much recent effort is going into identifying the endogenous mesodermal inducing factor(s); however, these efforts are frustrated by the poor genetics of the amphibian. Consequently attempts have been made to find other means of identifying unambiguously the contributions of the molecules responsible. One such method involves the overproduction of a suspected protein after the injection of synthetic RNA into the embryo. Harvey and Melton [1988] used this approach for a homeobox-containing gene, *Xhox* 1A.

B. Overexpression

The *Xenopus Xhox* 1A gene is first expressed during gastrulation [Harvey and Melton, 1988]. Subsequently the major site of expression is in the somitic mesoderm. The injection of synthetic *Xhox* 1A mRNA into a cleaving embryo leads to embryos with a specific disturbance in somite morphogenesis, leading the authors to conclude that this gene is important for somite patterning and that this patterning is disrupted by its overexpression.

C. Ectopic Expression

A second and often overlapping approach is to express the suspected protein in an ectopic site. The first use of this technique as a developmental tool in *Xenopus* was made by Kintner [1988]. He injected N-CAM mRNA into *Xenopus* embryos and showed that the ectopic N-CAM expression in the somites disrupted their organization, whereas its overexpression at its normal site, the developing neural tube, had no effect. A similar strategy was used by Ruiz i Altaba and Melton [1989] to implicate the *Xhox* 3 gene in anterioposterior patterning.

D. Antisense RNA Inhibition of Gene Expression

A final method that has been used in *Xenopus* is the inhibition of gene expression using antisense. This can involve the use of antisense DNA or RNA. Antisense DNA is discussed in Section IV.

Early experiments by Melton [1985] and Harland and Weintraub [1985] using microinjected *Xenopus* oocytes demonstrated that antisense RNAs can work by interfering with mRNA translation. Both studies were performed with antisense RNAs directed against non-*Xenopus* transcripts, as were experiments by Sumikawa and Miledi [1988]. While antisense RNA-mediated inhibition of the translation of endogenous mRNAs has also been achieved in oocytes [Wormington, 1986], several failed attempts in fertilized eggs have been reported.

On investigating the failure of injected antisense RNA to inhibit expression from specific mRNA in *Xenopus* embryos, two groups [Rebagliati and Melton, 1987; Bass and Weintraub, 1987] independently discovered that injected RNA duplexes were unwound by the unwinding/modifying activity present in the cytoplasm of the embryonic cells (which is discussed in detail by Bass, this volume). The activity is segregated in the nucleus of the oocyte, is released into the cytoplasm on maturation, and finally returns to a nuclear location at the

gastrula stage of development. The duplex is destabilized as the enzyme modifies as many as 50% of adenine (A) residues to inosine (I) [Bass and Weintraub, 1988]. Since inosine base pairs with guanosine, the inevitable consequence of this modification is the corruption of translational reading frames in any coding region involved in the duplex structure.

Recently Kimelman and Kirschner [1989] found a naturally occurring antisense RNA in *Xenopus* oocytes that is complementary to a 900 bp region of basic fibroblast growth factor (bFGF) mRNA, which is also present. (Surprisingly this ''antisense'' RNA has an open reading frame encoding a 25 kD protein.) It appears that all the bFGF mRNA is in hybrid form, since on maturation of the oocyte all the bFGF mRNA molecules were shown to become modified by the unwindase that is released into the cytoplasm by the events of maturation. This unwinding/modifying activity is now believed to be of widespread occurrence in eukaryotic cells [Wagner and Nishikura, 1988] (see also Bass, this volume). Ironically it would now seem that the failure to detect stable hybrids in the original *Xenopus* experiments cannot explain the lack of inhibition seen, since any A-I modifications should lead to truncated and/or inactive proteins.

A more likely explanation for the failure is that no hybrids form in the first place. All the antisense RNA used in these experiments are quite long (>100 bases), so it is possible that secondary structure in target or antisense RNA prevented their union, since, as described below, small oligonucleotides do not appear to have the same difficulties of access. Quite possibly, these difficulties are only manifested in the early embryonic stages of *Xenopus,* where the target mRNAs are not newly synthesized or in transit as they are in the other systems discussed in this book, but instead are in some sort of long-term store. This view is supported by the success using an antisense RNA approach for later embryonic stages [Giebelhaus et al., 1988], where sense and antisense RNA transcription were designed to occur concurrently, and by success in using relatively long antisense

RNA in other systems (see the first eight chapters of this volume).

IV. APPLICATIONS OF ANTISENSE OLIGODEOXYNUCLEOTIDES

A. In Cultured Cells

The first successful demonstration of an antisense strategy in vivo was the inhibition of Rous sarcoma virus proliferation in cell culture by the addition of synthetic oligonucleotides to the culture medium [Zamecnik and Stephenson, 1978]; these oligonucleotides contained an unmodified phosphodiester backbone (n-oligos). Subsequent studies (reviewed by Toulmé, this volume, and discussed in greater detail elsewhere in this volume) have shown that cultured cells can to varying extents be protected from infection from vesicular stomatitis virus [Agris et al., 1986] (see also Degols et al., this volume), retroviruses such as human immunodeficiency virus [Matsukura et al., 1987] (see also To, this volume; Matsukura, this volume), influenza virus [Zerial et al., 1987] (see also Agrawal and Leiter, this volume), and herpes simplex virus [Smith et al., 1986]. Many examples also exist of the regulation of gene expression in cultured cells by antisense oligonucleotides (see Toulmé, this volume; Tidd, this volume; Miller, this volume; Deglos et al., this volume; To, this volume; Matsukura, this volume; Agrawal and Leiter, this volume; Wickstrom, this volume).

B. In *Xenopus* Oocytes

This alternative approach of using antisense DNA (rather than antisense RNA) was first pioneered in *Xenopus* by Kawasaki [1985], who showed that translation from injected RNA could be inhibited by injection of short complementary n-oligos. Subsesquently, several other groups have shown that a variety of mRNA, both endogenous (calmodulin [Dash et al., 1987], histone [Shuttleworth and Colman, 1988], heat shock [Shuttleworth and Colman, 1988], β-tubulin [Jessus et al., 1988a,b], cyclins [Minshull et al., 1991], Vg1 [Shuttleworth and Colman, 1988; Woolf et al., 1990])

and exogenous (interleukin 2 [Kawasaki, 1985], globin [Cazenave et al., 1986, 1987a; 1989; Izant and Sardelli, 1988], and strychnine-sensitive glycine receptors [Akagi et al., 1989]) are specifically cleaved by n-oligos, although the degree of cleavage varies in an unpredictable way from oligonucleotide to oligonucleotide between different mRNAs, and even on the same mRNA [Shuttleworth et al., 1988]. I would point out that the techniques involved in oligonucleotide microinjection into oocytes are routine, and detailed protocols have been published by Colman [1984].

1. Studies with anti-H4 oligonucleotides. We have studied the susceptibility of one species of endogenous oocyte mRNA, histone H4, to unmodified phosphodiester oligonucleotides (n-oligos) in some detail. A comparison of a nested family of anti-H4 n-oligos comprising 6, 8, 10, 12, 14, 16, 18, 20, 25, 30, 35, or 40 bases demonstrated that the minimum length required in oocytes was 10 bases and that maximal cleavage was obtained with a 12-mer [Shuttleworth et al., 1988; Baker et al., 1990]. The observed ablation of mRNA has all the hallmarks of an RNAse H-mediated reaction of the type that has been demonstrated to provide the basis for hybrid-arrested translation in cell-free translation systems, as discussed by Minshull and Hunt (this volume). This family of n-oligos were designed around an anti-H4, 20-mer oligonucleotide called H4-1 (see Fig. 1). We were able to show that, with this oligonucleotide, up to 95% of H4 mRNA could be

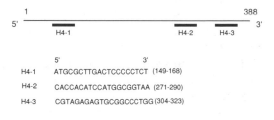

Fig. 1. *Antisense n-oligos directed against* Xenopus *H4 mRNA. The sequences of three 20-mer antisense n-oligodeoxynucleotides against the 388b H4 mRNA are shown, with the positions of complementarity given in parentheses.*

destroyed at an oligonucleotide/mRNA molar ratio of approximately 1,500; further ablation was not achieved with larger amounts of oligonucleotide, whether this was introduced in one dose or in successive doses, 3 hours apart. The failure of sequential injections (into diametrically opposed oocyte regions) to increase ablation has demonstrated that the survival of the residual 5% H4 mRNA is not caused by a combination of slow n-oligo diffusion and/or instability, which is known to be very short in oocytes ($t_{1/2} < 20$ minutes) [Cazenave et al., 1987a; Woolf et al., 1990]. Since deproteinized oocyte H4 mRNA does not show this behavior in an in vitro assay, we have conjectured that the resistant mRNA is protected in vivo by association with protein.

We have also investigated the relative efficacy of different anti-H4 n-oligos on H4 mRNA both in oocytes and in an in vitro assay using exogenous RNAse H [Baker et al., 1990]. Three oligonucleotides were used: H4-1, H4-2, and H4-3 (Fig. 1). The amounts of ablation achieved with saturating doses of each oligo were 95%, 30%, and 10% for oligonucleotides H4-1, H4-3, and H4-2, respectively. Interestingly this pattern of cleavage efficiency, with H4-1 > H4-3 > H4-2, is preserved in a completely in vitro analysis. Prior annealing in vitro results in all H4 mRNA being cleaved by each oligo. Since calculations [according to the equation of Meinkoth and Wahl, 1984] on the hypothetical stability of these oligos indicates that the oligos should form hybrids with dissociation temperatures (T_d) of 64°C, H4-1; 62°C, H4-2; and 68°C, H4-3, ranking these oligos in order of increasing T_d would not lead us to predict their performance correctly. We believe that the feature limiting the effectiveness of oligos in vitro (and probably in vivo) is the secondary structure of the target RNA. Unfortunately, as we have stated before, computer predictions from the known H4 mRNA sequence give an unreliable guide to the behavior of our oligos. If we assume that single-stranded regions are more accessible, then treatment of the H4 sequence using the algorithms of Zuker and Stiegler [1981] would give

a ranking of H4-3 > H4-1 > H4-2. Furthermore, with the same algorithms on a synthetic H4 RNA only 88% homologous (at the nucleotide level) to native H4 mRNA, a strikingly different structure was predicted (data not shown). Yet when this synthetic RNA was challenged with H4-1 and H4-2, both in vivo and in vitro, H4-1 was much better in both situations. We conclude both from our work and from that of others that the a priori prediction of the optimal oligonucleotide sequence is presently problematical, and we would advocate the use of the cell-free assay described by Baker et al. [1990] as a useful preliminary step for selecting the best oligonucleotide for use in oocytes.

2. Effects of oligonucleotides on the oocyte cell cycle. The ability of phosphodiester oligonucleotides to destroy maternal mRNAs in oocytes has been recently exploited to investigate the molecular basis of maturation of oocytes into unfertilized eggs. A major regulator of the process is maturation-promoting factor (MPF). Recently MPF has been shown to contain a heterodimer consisting of cyclin B and $p34^{cdc2}$, the *Xenopus* homolog of the yeast cell cycle control protein cdc2 [see Gautier et al., 1990]. $p34^{cdc2}$ is a serine/threonine kinase and in its inactive state contains phosphorylated serine, threonine, and tyrosine residues. Activation is associated with dephosphorylation of tyrosine. Although many details of maturation are poorly understood, it is clear that a cascade of phosphorylation and dephosphorylation reactions occur. Oligo-mediated ablation studies have recently provided insight into some of the steps involved. Sagata et al. [1988] found that oligo-mediated depletion of the mRNA encoding the $p39^{c-mos}$ protooncogene product resulted in inhibition of maturation. Sagata et al. [1989] also found that injection of this RNA caused maturation even in the absence of hormonal stimulation. These results were interpreted as demonstrating the involvement of $p39^{c-mos}$ protein in the activation of MPF.

We have described above how difficult it can be to remove totally certain maternal mRNA. Fortunately, interesting information can be gleaned from experiments in which only partial cleavage has been obtained. Recently we cloned the cDNA, $p40^{MO15}$, corresponding to an oocyte mRNA encoding a putative cdc2-related protein kinase [Shuttleworth et al., 1990]. When oocytes were injected with a complementary oligo (oligo 15-1), up to 80% of the MO15 mRNA was cleaved. A control oligo (H4-1) had no effect. When the kinetics of maturation were monitored using the assay of germinal vesicle (oocyte nucleus) breakdown, it was found that MO15 RNA depletion caused maturation to occur earlier. This effect of RNA depletion could be reversed if, following oligonucleotide-mediated depletion, MO15 synthetic RNA was injected into the same oocytes. We conclude that $p40^{MO15}$ is involved in negatively regulating meiosis in *Xenopus* oocytes.

C. Toward Removing Maternal mRNA

Attempts to interfere with early developmental processes using injected oligonucleotides have had a patchy success. Shuttleworth and Colman [1988] found that phosphodiester oligonucleotides were effective in ablating specific mRNA after injection into one cell embryos; however, this ablation was achieved at the expense of developmental toxicity and nonspecificity (i.e., nontargeted mRNAs were partially ablated). Consequently they adopted a different strategy: Oocytes were injected with the n-oligos and then left to recover for several hours during which the oligonucleotides were degraded. The oocytes were then matured in vitro, color-coded with vital stains, and transferred to the body cavity of a laying female frog. This passage through a foster female is essential for the ability of the matured oocytes to be fertilized by added sperm [Holwill et al., 1987]. We have used this strategy to remove the maternal mRNA Vg1 from *Xenopus* embryos. This mRNA, which is localized to a vegetal position in oocytes and embryos, has homology to TGF-β and may be involved in mesoderm induction [Weeks and Melton, 1987]. However, the amount of the n-oligos necessary to remove the maternal Vg1 mRNA proved toxic to subsequent development for unknown reasons, although we have speculated that the expansion of the

deoxynucleotide pool as a result of oligonucleotide degradation might be a contributory factor. These toxic effects might be oligonucleotide-sequence specific, since Kloc et al. [1989] showed recently that they could deplete embryos of the maternal sequence Xlgv7 and still obtain normal development. Likewise El-Baradi et al. [1991] have used n-oligos to remove mRNA encoding a family of zinc-finger–containing, putative transcription factors from embryos. Again no new phenotype was obtained, and this work revealed a further shortcoming of the antisense strategy as applied to *Xenopus* development; namely, that since the maternal pool of transcription factors was so large, mRNA depletion caused no noticeable reduction in protein levels. However, Wylie and Heasman (personal communication) recently observed that oligonucleotide-mediated ablation of oocyte vimentin mRNA resulted in embryos showing regional-specific defects after gastrulation. It remains to be seen in this case what the effect, if any, mRNA depletion has on embryonic vimentin levels.

V. USE OF MODIFIED OLIGONUCLEOTIDES

As discussed by Toulmé (this volume), one immediate problem that was recognized early on in oligonucleotide studies with cultured cells was that unmodified oligonucleotides were often [but not always; see Holt et al., 1988] degraded by nucleases present in the culture medium [Wickstrom, 1986], thus necessitating their addition at concentrations up to 100 μM. This factor, combined with a prevalent belief that the negative charge on unmodified, phosphodiester (n-oligo)–based oligonucleotides compromised penetration through the plasma membrane, quickly led to a search for effective derivatives that would be more stable in vivo and more lipid soluble. The first generation of such derivatives include oligonucleotides with methyl phosphonate [P-CH$_3$; see detailed review by Miller (this volume)] or phosphorothioate (S-oligo) backbones. Oligomers of the unnatural α-anomer of deoxynucleoside have also been used, and some have been tested in *Xenopus* [Cazenave et al., 1989; Boiziau et al., 1991]. Subsequent development has focused on the chemical modification of oligonucleotides with a view to increasing their reactivity (for detailed review, see Cohen, 1989] (see also Toulmé, this volume). Such modified structures include the crosslinking reagent psoralen [Kean et al., 1989], alkylating reagents [Knorre et al., 1985], metal complexes that cleave both RNA and DNA [Boutorin et al., 1984], acridine rings that increase the affinity of oligonucleotides for RNA [Toulmé et al., 1986; Cazenave et al., 1987a,b] and various adducts designed to encourage membrane penetration as discussed by Degols et al. (this volume), such as polylysine [Lemaitre et al., 1987] and cholesterol. Some of these modified reagents have also been applied to *Xenopus* oocytes and embryos. Cazenave et al. [1987b] were able to show that the introduction of intercalating moieties at the 3' end of n-oligos resulted in greater hybrid stability; unfortunately these reagents had an affinity for the mass of yolk particles inside oocytes. Baker et al. [1990] and Woolf et al. [1990] have found that fully derivatized, phosphorothioate oligonucleotides (S-oligos) are considerably more stable than the identical n-oligo sequences in oocytes and embryos. Because of this prolonged stability, these oligonucleotides are effective at much lower concentrations. Unfortunately, despite the reduced quantities need, the doses of S-oligo necessary for complete mRNA ablation were still toxic to development.

More recently, Dagle et al. [1990] reported the successful use of a new type of oligonucleotide analog in their attempts to remove the maternal mRNA An2 and cyclin B2 from *Xenopus* embryos. They showed that the substitution of several (optimally four) phosphoramidates for phosphodiester linkages at the 5' and 3' ends of n-oligos conferred much greater nuclease resistance to the oligonucleotides. However, the retention of a phosphodiester core still allowed the oligonucleotide to mediate RNAse H activity. These oligonucleotides worked very efficiently, with 10 ng of modified oligonucle-otide causing 30 times more ablation than 10 ng of unmodified oligonucleotide. Moreover, the use of oligonucleotides against both mRNAs resulted in the appearance of abnormal embryos. Finally, Dagle et al. [1990] tested terminally modified oligonucleotides where phosphorothioate linkages replaced the phosphoramidate linkages. Although these oligonucleotides also had increased stability to nucleases, they worked more slowly. (For a further discussion of the use of such chimeric oligomers in mammalian cells, see Tidd, this volume.)

VI. SPECIFICITY OF OLIGONUCLEOTIDE ACTION

A major consideration in the design and use of any antisense reagent is the specificity of its action. This in turn will be a function of the specificity of hybridization. This will be enhanced if the target sequence is unique within the cell. Computations on the likelihood of a random sequence of given length reccurring more than once in the RNA complement of a cell are necessarily inaccurate, but figures of 11–15 (Hélène, personal communication) or 12 [Miller and Tso, 1987] consecutive nucleotides as defining a unique sequence have been quoted, even though ''random'' is an uncomfortable concept in this area [Agrawal et al., 1990]. Longer oligonucleotides will form more stable hybrids; however, increasing length brings with it potential self-foldback problems and increasing permutations of shorter stretches of consecutive nucleotides that might facilitate hybridization to other RNA, especially as relatively short regions of duplex can serve as RNAse H templates (a 4-mer in vitro [Donis-Keller, 1979] and a 10-mer in oocytes [Shuttleworth et al., 1988]). Surprisingly cells can show remarkable discrimination between perfect and mismatched hybrids. Wang et al. [1985] synthesized 64 oligonucleotides (14-mers) based on a short tumor necrosis factor (TNF) protein fragment and coinjected pools of these oligonucleotides in *Xenopus* oocytes, along with a poly(A)$^+$ preparation that contained TNF mRNA. Oligonucleotide pools capable of inhibiting TNF translation were further subdivided until one oligonucleotide, with the sequence GCTACA-GGCTTGTC, was identified and used in the successsful screening for a TNF cDNA. In this process of oligonucleotide elimination, GCT-ACAGGCTTGTC was clearly distinguished from GCCACAGGCTTGTC. Unfortunately very few details were given of the effects of the various oligonucleotides on the TNF assay, so that the quantitative consequences of mismatch recognition were not revealed.

In another study also based on a biological assay, Holt et al. [1988] found that small degrees of mismatch could completely compromise oligonucleotide action when the mismatched base pairs were appropriately sited. They found that a two base mismatch (at positions 6 and 10 in the oligonucleotide) between a 15-mer oligonucleotide and c-*myc* mRNA was sufficient to prevent the elongation of HL-60 cell doubling times, which occurred when the perfectly matched oligonucleotide was used. Neither of these studies rule out some inhibition from regions of partial homology, and the in vitro experiments of Minshull and Hunt (this volume) support the conclusion that mismatched oligos can give rise to RNA cleavage. A further indication that oligonucleotides can cleave mRNA at secondary sites was reported by Cazenave et al. [1987b], who showed that in vitro treatment of rabbit β-globin mRNA with a 17-mer that had perfect complementarity to one part of the mRNA sequence generated two fragments in a ratio of about 4:1. Although the minor fragment could have been a postcleavage degradation product, its size was consistent with oligonucleotide-mediated cleavage at a site in the mRNA to which the oligonucleotide had partial (13/17 with 7 contiguous base pairs) complementarity.

The short size of oligonucleotides and the irreversible nature of the cleavage reaction that certain oligonucleotides (n-oligos and S-oligos) mediate by the action of RNase H make the issue of specificity particularly acute. We are forced to the conclusion that however judicious the choice of oligonucleotide sequence is, it is impossible to exclude the gratuitous cleav-

age of unknown mRNAs through partial hybrid formation, and this could complicate interpretation. In addition, very high oligonucleotide concentrations seem to cause completely nonspecific RNA cleavage [Cazenave et al., 1987b], and in some circumstances oligonucleotides can directly inhibit the activity of specific enzymes in a process not involving hybridization [Matsukura et al., 1987] (see also Matsukura, this volume). Finally, Smith et al. [1990] reported that certain oligonucleotides have an intrinsic toxicity in oocytes that, although sequence specific, cannot be attributed to the complementarity of that sequence with a known oocyte RNA.

In view of all these potential problems, the only rigorous way of definitively attributing a function to an mRNA by these methods would be to show that the effects of oligonucleotide ablation are reversed by the reintroduction of fresh target mRNA. Failing this, evidence should be produced showing that more than one oligonucleotide directed against the same target mRNA is capable of the same biological effect, since it is most unlikely that oligonucleotides of radically different sequence would have similar secondary targets [e.g., Minshull et al., 1989].

A final note of caution that has emerged particularly from *Xenopus* studies is that it is sometimes easy to generate apparently specific developmental lesions by the antisense approach. This point, made forcefully by Rebagliati and Melton [1987], probably reflects the acute sensitivity of certain developmental processes to the slightest perturbation and highlights the need for carefully designed controls. As stated above, the ultimate control involves mRNA replacement therapy. Unfortunately this is rarely done and is often not technically possible. Consequently, a combination of factors—the nonspecific toxicity, the interdependence of different regions of the embryo, the variable levels of antisense inhibition—all conspire to make the early *Xenopus* embryo a particularly troublesome target for this technology.

VII. CONCLUSIONS AND FUTURE PROSPECTS

Three types of antisense reagent have been used in oocytes: oligonucleotides, antisense RNAs, and ribozymes (the latter two classes are covered elsewhere in this volume). Each has been demonstrated to work, although complete target inactivation has proved the exception rather than the rule [for a further review, see Colman, 1990]. That this approach works even inefficiently is remarkable. It is one thing to anneal complementary sequences in vitro as in hybrid-arrested translation and quite another to expect similar effects with nucleoprotein complexes in vivo. Oligonucleotides offer the advantage that target mRNAs can be irreversibly cleaved; however, they are unstable, and often toxic, and selection of the most effective target sequence is still empirical.

Antisense RNA and ribozymes share with oligonucleotides the empiricism of target choice while lacking the toxicity and offer the great advantage of constitutive or inducible production in the cell or tissue of choice, after integration of their genes into the genome of the host cell or organism. In the context of removing maternal mRNA from oocytes, this would be useful in instances in which pools of maternally accumulated protein obscure the effect of mRNA depletion on G_0 development, since these pools could be eliminated in the oocytes of the G_0 females. Unfortunately integration of injected genes occurs extremely rarely in Xenopus. In addition, because it does not mediate RNA cleavage, antisense RNA does not have the catalytic potential of the other reagents, though it is generally more stable than either of the others. Nevertheless, Cotten et al. [1989], in a direct in vitro comparison of all three reagents directed against the same region of a histone pre-mRNA, found the antisense RNA to be the most effective inhibitor of RNA splicing.

To date, ribozymes have also proved disappointing in vivo. If one considers the irreversible cleavage brought about by ribozyme action, together with their low toxicity and potential "immortality" through integration, these reagents would seem to offer the best of all worlds. However, although ribozymes have been demonstrated to work to varying degrees in bacteria [Chuat and Galibert, 1989], in monkey and human cells [Cameron and Jennings, 1989; Sarver et al., 1990], and in Xenopus oocytes [Cotten and Birnstiel, 1989; Saxena and Ackerman, 1990], thus far they have been found to work very inefficiently in vivo.

It would also seem, then, that oligonucleotides still have a place in antisense strategies with Xenopus. In conclusion, I suggest a new oligonucleotide-related strategy that may have specific potential for interfering with the expression from Xenopus maternal mRNA. This involves the synthesis, in vitro, of antisense oligoribonucleotides (ORN) by the methods of Milligan and Uhlenbeck [1989]. These ORN could be expected to access target mRNA with the same facility as oligonucleotides. However, they would be more stable and less toxic than oligonucleotides and, if injected into matured oocytes or embryos, could cause the translational corruption of their target RNAs through the mediation of the unwindase activity described earlier. This in turn could lead to biologically inactive proteins. We are presently evaluating this strategy.

VIII. REFERENCES

Agrawal S, Mayrand S, Zamecnik PC (1990): Site-specific excision from RNA by RNAse H and mixed-phosphate-backbone oligodeoxynucleotides. Proc Natl Acad Sci USA 87:1401–1405.

Agris C, Blake K, Miller P, Reddy P, T'so P (1986): Inhibition of vesicular stomatitis virus protein synthesis by sequence-specific oligodeoxyribonucleoside methyl phosphonates. Biochemistry 25:6268–6275.

Akagi H, Patton DE, Miledi R (1989): Discrimination of heterogenous mRNA encoding strychnine-sensitive glycine receptors in Xenopus oocytes by antisense oligonucletides. Proc Natl Acad Sci USA 86:8103–8107.

Baker C, Holland D, Edge M, Colman A (1990): Effects of oligo sequence and chemistry on the efficiency of oligodeoxynucleotide-mediated cleavage. Nucleic Acids Res 18:3537–3543.

Bass B, Weintraub H (1987): A developmentally regulated activity that unwinds RNA duplexes. Cell 48:607–613.

Bass B, Weintraub H (1988): An unwinding activity that

covalently modifies its double-stranded RNA substrate. Cell 55:1089–1098.

Boiziau C, Kurfurst R, Cazenave C, Roig V, Thuong NT, Toulmé J-J (1991): Inhibition of translation initiation by antisense oligonucleotides via an RNase-H independent mechanism. Nucleic Acids Res 19:1113–1119.

Boutorin AS, Vlassov VV, Kazakov SA, Kutiavin IV, Podyminogin MA (1984): Complementary addressed reagents carrying EDTA-Fe(II):groups for directed cleavage of single-stranded nucleic acids. FEBS Lett 172:43–46.

Cameron FH, Jennings PA (1989): Specific gene suppression by engineered ribozymes in monkey cells. Proc Natl Acad Sci USA 86:9139–9143.

Cazenave C, Loreau N, Toulmé J-J, Hélène C (1986): Antimessenger oligodeoxynucleotides: Specific inhibition of rabbit β-globin synthesis in wheat germ extracts and *Xenopus* oocytes. Biochimie 68:1063–1069.

Cazenave C, Chevrier M, Thuong N, Hélène C (1987a): Rate of degradation of α and β-oligodeoxynucleotides in *Xenopus* oocytes: Implications for anti-messenger strategies. Nucleic Acids Res 15:10507–10521.

Cazenave C, Loreau N, Thuong N, Toulmé J-J, Hélène C (1987b): Enzymatic amplification of translation inhibition of rabbit β-globin mRNA mediated by antimessenger oligodeoxynucleotides covalently linked to intercalating agents. Nucleic Acids Res 15:4717–4736.

Cazenave C, Stein CA, Loreau N, Thuong NT, Neckers LM, Subasinghe C, Hélène C, Cohen JS, Toulmé J-J (1989): Comparative inhibition of rabbit globin mRNA translation by modified antisense oligodeoxynucleotides. Nucleic Acids Res 17:4255–4273.

Chuat JC, Galibert F (1989). Can ribozymes be used to regulate procaryote gene expression? Biochem Biophys Res Commun 162:1025–1029.

Cohen J (ed) (1989): Oligodeoxyribonucleotides. Antisense Inhibitors of Gene Expression. Topics in Molecular and Structural Biology, 12. London: Macmillan.

Colman A (1984): Translation of eucaryotic messenger RNA in *Xenopus* oocytes. In Hames BD, Higgins SJ (eds): Transcription and Translation—A Practical Approach. Oxford: IRL, pp 271–302.

Colman A (1990): Antisense strategies in cell and developmental biology. J Cell Sci 97:399–409.

Cotten M, Birnstiel ML (1989): Ribozyme mediated destruction of RNA in vivo. EMBO J 8:3861–3866.

Cotten M, Schaffner G, Birnstiel ML (1989): Ribozyme, antisense RNA, and antisense DNA inhibition of U7 small nuclear ribonucleoprotein-mediated histone pre-mRNA processing in vitro. Mol Cell Biol 9:4479–4487.

Dagle J, Walder J, Weeks D (1990): Targeted destruction of mRNA in *Xenopus* oocytes and embryos directed by modified oligonucleotides: Studies of An2 and cyclin in embryogenesis. Nucleic Acid Res 18:4751–4757.

Dash P, Lotan, Knapp M, Kandel E, Goelet P (1987): Selective elimination of mRNAs in vivo: Complementary oligonucleotides promote RNA degradation by an

RNAse H–like activity. Proc Natl Acad Sci USA 84:7896–7900.

Donis-Keller H (1979): Site-specific cleavage of RNA. Nucleic Acids Res 7:179–192.

Driever W, Nusslein-Volhard C (1988): The bicoid protein determines position in the *Drosophila* embryo in a concentration dependent manner. Cell 54:95–104.

Dumont JN (1972): Oogenesis in *Xenopus laevis* (Daudin). I. Stages of oocyte development in laboratory maintained animals. J Morphol 136:153–180.

Eichele G (1989): Retinoids and vertebrate limb pattern formation. Trends Genet 5:246–250.

El-Baradi T, Bouwmeester T, Giltay R, Peiler T (1991) The maternal store of zinc finger protein encoding mRNAs in fully grown oocytes is not required for early embryogenesis. EMBO J 10:1407–1413.

Gautier J, Minshull J, Kohka M, Glotzer M, Hunt T, Maller J (1990): Cyclin is a component of maturation promoting factor from *Xenopus* eggs. Cell 60:487–494.

Gerhardt J (1980): Mechanisms regulating pattern formation in the amphibian egg and early embryo. In Godberger RF (ed): Biological Regulation and Development, Vol 2, Molecular Organization and Cell Function. New York: Plenum Press, pp 133–316.

Giebelhaus DH, Eib DW, Moon RT (1988): Antisense RNA inhibits expression of membrane skeleton protein 4.1 during embryonic development of *Xenopus*. Cell 53:601–615.

Harland R, Weintraub H (1985): Translation of mRNA injected into *Xenopus* oocytes is specifically inhibited by antisense RNA. J Cell Biol 101:1094–1099.

Harvey RP, Melton DA (1988): Microinjection of synthetic *Xhox*-1A homeobox mRNA disrupts somite formation in developing *Xenopus* embryos. Cell 53:687–697.

Holt JT, Redner RL, Nienhius AW (1988): An oligomer complementary to c-*myc* mRNA inhibits proliferation of HL-60 cells and induces differentiation. Mol Cell Biol 8.963–973.

Holwill S, Heasman J, Crawley C, Wylie C (1987): Axis and germ line deficiencies caused by UV irradiation of *Xenopus* oocytes cultured in vitro. Development 100:735–743.

Izant JG, Sardelli AD (1988): Anti-sense suppression of RNA maturation. J Cell Biol 107:102a.

Jessus C, Cazenave C, Ozon R, Hélène C, (1988a): Specific inhibition of endogenous β-tubulin synthesis in *Xenopus* oocytes by anti-messenger oligodeoxyribonucleotides. Nucleic Acids Res 16:2225–2233.

Jessus C, Chevrier M, Ozon R, Hélène C, Cazenave C (1988b): Specific inhibition of β-tubulin synthesis in *Xenopus* oocytes using anti-sense oligodeoxyribonucleotides. Gene 72:311–312.

Kawasaki ES (1985): Quantitative hybridization-arrest of mRNA in *Xenopus* oocytes using single-stranded complementary DNA or oligonucleotide probes. Nucleic Acids Res 13:4991–5004.

Kean JM, Murakami A, Blake K, Cushman C, Miller P (1989): Photochemical cross-linking of psoralin-derivatised oligonucleoside methylphosphonate to rabbit globin messenger RNA. Biochemistry 27: 9113–9121.

Kimelman D, Kirschner M (1987): Synergistic induction of mesoderm by FGF and TGF-β and the identification of an mRNA coding for FGF in the early *Xenopus* embryo. Cell 51:869–877.

Kimelman D, Kirschner M (1989): An antisense mRNA directs the covalent modification of the transcript encoding fibroblast growth factor in *Xenopus* oocytes. Cell 59:687–696.

Kintner C (1988): Effects of altered expression of the neural cell adhesion molecule, N-CAM, on early neuronal development in *Xenopus*. Neuron 1:545–555.

Kloc M, Miller M, Carraco A, Eastman E, Etkin L (1989): The maternal store of the xlgv 7 mRNA in full grown oocytes is not required for normal development in *Xenopus*. Development 107:899–907.

Knorre DG, Vlassov VV, Zarytova VF (1985): Reactive oligonucleotide derivatives and sequence-specific modification of nucleic acids. Biochemie 67:785–789.

Lemaitre M, Bayard B, Lebleu B (1987): Specific antiviral activity of poly (L-lysine)-conjugated oligodeoxyribonucleotide sequence complementary to vesicular stomatitis virus N protein mRNA intiation site. Proc Natl Acad Sci USA 84:648–652.

Matsukura M, Shinozuka K, Zon G, Mitsuya H, Reitz M, Cohen J, Broder S (1987): Phosphorothioate analogs of oligodeoxynucleotides: Inhibitors of replication and cytopathic effects of human immunodeficiency virus. Proc Natl Acad Sci USA 84:7706–7710.

McMahon A, Moon R (1989): Ectopic expression of the proto-oncogene *int-1* in *Xenopus* embryos leads to duplication of the embryonic axis. Cell 58:1075–1084.

Meinkoth J, Wahl G (1984): Hybridisation of nucleic acids immobilised on solid supports. Anal Biochem 138: 267–284.

Melton DA (1985): Injected antisense RNAs specifically block messenger translation in vivo. Proc Natl Acad Sci USA 82:144–148.

Miller P, Tso P (1987): A new approach to chemotherapy based on molecular biology and nucleic acid chemistry: Matagen (masking tape for gene expression). Anti-Cancer Drug Design 2:117–128.

Milligan J, Uhlenbeck O (1989): Synthesis of small RNAs using T7 RNA polymerase. Methods Enzymol 180:51–61.

Minshull J, Blow J, Hunt T (1989): Translation of cyclin mRNA is necessary for extracts of activated *Xenopus* eggs to enter meiosis. Cell 56:947–956.

Minshull J, Murray A, Colman A, Hunt T (1991): *Xenopus* oocyte maturation does not require new cyclin synthesis. J Cell Biol 114:767–772.

Rebagliati M, Melton DA (1987): Antisense RNA injections in fertilised frog eggs reveal an RNA duplex unwinding activity. Cell 48:599–605.

Rosenberg UB, Preiss A, Seifert E, Jackle H, Knipple DC (1985): Production of phenocopies by *Krüppel* antisense RNA injection into *Xenopus* embryos. Nature 313:703–706.

Rossant J, Joyner A (1989): Towards a molecular genetic analysis of mammalian development. Trends Genet 5:277–282.

Roy L, Singh B, Gautier J, Arlinghaus R, Nordeen S, Maller J (1990): The cyclin B2 component of MPF is a substrate for the c-mos[xe] proto-oncogene product. Cell 61:825–831.

Ruiz i Altaba A, Melton DA (1989): Involvement of the *Xenopus* homeobox gene, *Xhox* 3, in pattern formation along the anterior-posterior axis. Cell 57:317–326.

Sagata N, Daar I, Oskarson M, Showalter SD, Vande Woude GF (1989): The product of the c-*mos* proto-oncogene as a candidate initiator for oocyte maturation. Science 245:643–645.

Sagata N, Oskarson M, Copeland T, Brumbaugh J, Vande Woude GF (1988): Function of c-*mos* proto-oncogene product in meiotic maturation in *Xenopus* oocytes. Nature 335:519–525.

Sarver N, Cantin EM, Chang PS, Zaia JA, Ladne PA, Stephens DA, Rossi JJ (1990): Ribozymes as potential anti-HIV-1 therapeutic agents. Science 247:1222–1225.

Saxena SK, Ackerman EJ (1990): Ribozymes correctly cleave a model substrate and endogenous RNA in vivo. J Biol Chem 265:17106–17109.

Shuttleworth J, Colman A (1988): Antisense oligonucleotide-directed cleavage of mRNA in *Xenopus* oocytes and eggs. EMBO J 7:427–434.

Shuttleworth J, Matthews G, Dale L, Baker C, Colman A (1988): Antisense oligodeoxynucleotide-directed cleavage of maternal mRNA in *Xenopus* oocytes and embryos. Gene 72:267–275.

Shuttleworth J, Godfrey R, Colman A (1990): p40mo15, a cdc2-related protein kinase involved in negative regulation of meiotic maturation in *Xenopus* oocytes. EMBO J 9:3233–3240.

Slack J, Darlington B, Heath J, Godsave S (1987): Mesoderm induction in early *Xenopus* embryos by heparin-binding growth factors. Nature 326:197–200.

Smith JC (1987): A mesoderm inducing factor is produced by a *Xenopus* cell line. Development 99:3–14.

Smith C, Aurelian L, Reddy P, Miller P, T'so P (1986): Antiviral effect of an oligo (nucleotide methyl phosphonate) complementary to the splice junction of herpes simplex virus type 1 immediate early pre-mRNAs 4 and 5. Proc Natl Acad Sci USA 83:724–726.

Smith R, Bement W, Dersch M, Dworkin-Rastl E, Dworkin M, Capco D (1990): Nonspecific effects of oligodeoxynucleotide injection in *Xenopus* oocytes: A reevaluation of previous D7 mRNA ablation experiments. Development 110:769–780.

Sumikawa K, Miledi R (1988): Repression of nicotinic

acetylcholine receptor expression by antisense RNAs and an oligonucleotide. Proc Natl Acad Sci USA 85:1302–1306.

Thomsen G, Woolf T, Whitman M, Sokol S, Vaughan J, Vale W, Melton DA (1990): Activins are expressed early in *Xenopus* embryogenesis and can induce axial mesoderm and anterior structures. Cell 63:485–493.

Toulmé J-J, Krisch HM, Loreau N, Thuong NT, Hélène C (1986): Specific inhibition of mRNA translation by complementary oligodeoxynucleotides covalently linked to intercalating agents. Proc Natl Acad Sci USA 83:1227–1231.

Wagner RW, Nishikura K (1988): Cell cycle expression of RNA duplex unwinding activity in cells. Mol Cell Biol 8:770–777.

Wang AM, Creasy A, Ladner MB, Lin LS Strickler J, van Arsdell JN, Yamamoto R, Mark DF (1985): Molecular cloning of the complementary DNA for human tumor necrosis factor. Science 228:149–154.

Weeks D, Melton DA (1987): A maternal mRNA localised to the vegetal hemisphere in *Xenopus* eggs codes for a growth factor related to TGF β. Cell 51:861–867.

Wickstrom E (1986): Oligonucleotide stability in subcellular extracts and culture media. J Biochem Biophys Methods 13:97–102.

Woolf T, Jennings C, Rebagliati M, Melton DA (1990): The stability, toxicity and effectiveness of unmodified and phosphorothioate antisense oligodeoxynucleotides in *Xenopus* oocytes and embryos. Nucleic Acids Res 18:1763–1769.

Wormington M (1986): Stable repression of ribosomal protein L1 synthesis in *Xenopus* oocytes by microinjection of antisense RNA. Proc Natl Acad Sci USA 83:8639–8643.

Zamecnik PC, Stephenson ML (1978): Inhibition of Rous sarcoma virus replication and cell transformation by a specific oligodeoxynucleotide. Proc Natl Acad Sci USA 75:280–284.

Zerial A, Thuong NT, Hélène C (1987): Selective inhibition of the cytopathic effect of type A influenza viruses by oligodeoxynucleotide covalently linked to an intercalating agent. Nucleic Acids Res 15:9909–9919.

Zuker M, Stiegler P (1981): Optimal computer folding of large RNA sequences using thermodynamics and auxillary information. Nucleic Acids Res 9:133–148.

ABOUT THE AUTHOR

ALAN COLMAN is Professor of Biochemistry at the University of Birmingham in the United Kingdom, where he teaches developmental and cell biology courses. After receiving his B.A. from Oxford University in 1971, he received his Ph.D. under John Gurdon at the Laboratory of Molecular Biology in Cambridge. The thesis work involved the specific transcription of DNAs injected into the oocytes and eggs of the frog, *Xenopus laevis*. Dr. Colman then took a lectureship at Warwick University where he used the frog oocyte as an in vivo test tube to identify targeting signals in a variety of eukaryotic proteins. Dr. Colman moved to Birmingham in 1987 and is currently investigating how proteins of the Wint 1 gene superfamily cause mesoderm induction in early frog development. His research papers have appeared in such journals as *Nature, Cell,* the *Journal of Cell Biology* and *EMBO Journal*. He was elected to membership of the European Molecular Biology Organisation in 1989.

Antisense RNA and DNA: 227–240
© 1992 Wiley-Liss, Inc.

Anticancer Drug Design Using Modified Antisense Oligonucleotides

David M. Tidd

I. INTRODUCTION

A major goal of cancer chemotherapy research has been to identify qualitative biochemical differences between malignant cells and normal cells that might be exploited in the development of tumor-selective clinical treatments. Most current therapies are selective only in so far as they affect dividing cells, and consequently they are dose limited by their toxicity to normal proliferating tissues. However, it is now clear that cancer is a disease of genetic damage that results in the inappropriate expression or altered function of protein products involved in signal transduction pathways controlling cell proliferation. The normal cellular genes, or protooncogenes, coding for such proteins are limited in number, and only specific types of genetic lesions can convert these into the "activated" oncogenes of transformed cells.

Thus the long-sought biochemical differences between malignant cells and normal cells may only be the presence in the former of specific damaged or activated oncogenes. If indeed this is the case, then it may be reasoned that inhibiting expression of the appropriate oncogenes in a particular tumor cell should force the cell to become less transformed and might even trigger entry into a terminal differentiation pathway. There is already some evidence to support this hypothesis, suggesting that the possibility exists for development of an entirely novel approach to cancer therapy that would be both nontoxic to the patient and, for the first time, truly tumor specific.

It is too early, however, to speculate about the ultimate clinical form of such a therapy, since the general validity of the concept remains to be established in a variety of model transformed cell systems in vitro. In addition, fully transformed cells may have accumulated genetic damage that superceded early events in malignant progression [Toksoz et al., 1987; Mulder et al., 1989] and hence the choice of oncogenes to be targeted is another important consideration.

Antisense RNA technology has been applied to the problem of inhibiting the expression of activated oncogenes. However, antisense oligonucleotides could potentially offer several advantages in that this approach is considerably less laborious, since large amounts of material may be readily synthesized chemically as a variety of short sequences chosen to have optimum sensitivity to short mismatches between activated oncogenes and the normal parent protooncogenes from which they are derived. It is also possible for large numbers of cells to be treated exogenously. In addition, there is always the possibility, however remote, that new drugs for treatment of cancer may be developed in which antisense oligonucleotides provide the targeting specificity [Miller and Ts'o, 1987].

For these reasons, our group has recently become involved in an attempt to apply the antisense oligonucleotide and oligonucleotide analog technology, pioneered by Zamecnik [Zamecnik and Stephenson, 1978; Stephenson and Zamecnik, 1978; Zamecnik et al., 1986] and Miller and Ts'o [1987], to the problem of inhibiting oncogene expression. It became clear at an early stage in our work that, although these molecules are undoubtedly able to inhibit cell-free translation and gene expression in certain model intact cell systems [e.g., McManaway et al., 1990], they are not generally applicable as routine tools for use with cultures of transformed cells. Therefore we have undertaken a study to try and understand why antisense oligonucleotides did not work in our systems and how we might develop a more universal type of oligonucleotide antisense effector. Our work, so far, has involved normal oligodeoxynucleotides, the nonionic methylphosphonate oligodeoxynucleotide analogs (Fig. 1, $X = CH_3$) developed by Miller and Ts'o [1987] and reviewed in detail by Miller (this volume), and chimeric molecules incorporating both structures. Table I lists some major factors affecting the biological activity of antisense oligonucleotides and compares phosphodiester and methylphosphonodiester oligonucleotides in terms of their capacity to fulfill the implicit requirements for optimal activity. Our results and those of others are discussed in terms of the points listed in Table I. It should be stressed that, although there is considerable optimism

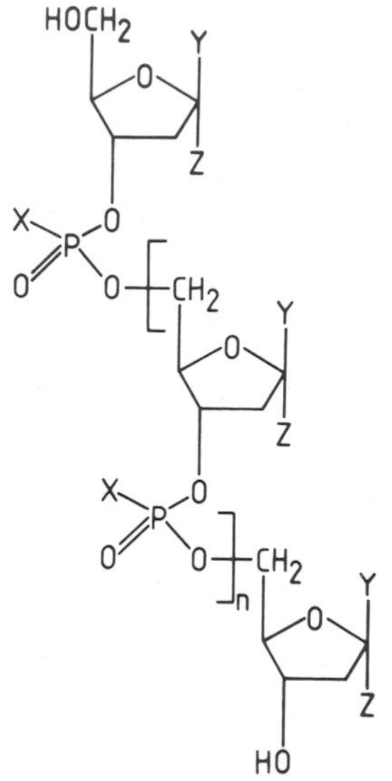

Fig. 1. *General structures of normal oligodeoxyribonucleotides and methylphosphonate oligodeoxyribonucleotide analogs.*

X	Y	Z	
O–	A,G,C, or T	H	Normal phosphodiester oligodeoxyribonucleotide
CH_3	A,G,C, or T	H	Methylphosphonate oligodeoxyribonucleotide analog

(Reproduced from Tidd, 1990, with permission of the publisher.)

TABLE I. Factors Affecting Biological Activity of Antisense Oligodeoxynucleotides

	Normal phosphodiester	Methylphosphonate analogs
Biological stability	−	+ + +
Nontoxicity	±	+ +
Cell uptake	+	+ +
Hybridization efficiency and stability	+ + +	+
Ribonuclease H activation	+ + +	−

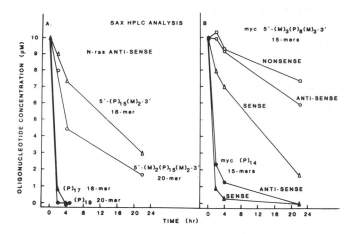

Fig. 2. *Persistence of intact phosphodiester and methylphosphonodiester/phosphodiester chimeric oligodeoxynucleotides at 37°C in McCoy's 5A medium containing 15% heat-inactivated fetal calf serum. SAX HPLC analyses. P in parentheses with a subscript numeral indicates the number of consecutive phosphodiester linkages and, similarly, M, the number of consecutive nuclease-resistant methylphosphonodiester linkages (see Fig. 1).* **A:** *The sequence of the human N-ras antisense 18-mer was CAG TTT GTA CTC AGT CAT, and the 20-mer sequence was ACC AGT TTG TAC TCA GTC AT.* **B:** *The sequences of the human myc 15-mers were antisense, AAC GTT GAG GGG CAT; nonsense (a random permutation of the bases present in the antisense oligodeoxynucleotide), GTA CGG TAA CGG GAT; sense, ATG CCC CTC AAC GTT. Terminal methylphosphonate linkages at the 3' end of the oligodeoxynucleotides alone were sufficient to afford considerable protection against degradation (A), demonstrating that the predominant nuclease in fetal calf serum is a 3'-phosphodiesterase. (Reproduced from Tidd, 1990, with permission of the publisher.)*

about the potential development of new cancer chemotherapeutic agents that incorporate antisense oligonucleotide analog structures as sequence recognition elements, ultimately these may serve only in in vitro systems to identify specific oncogenes as targets for chemotherapeutic attack. The pharmacological and financial considerations discussed by Zon [1990] may well mean that other types of drugs would be required for clinical modulation of such targets.

II. BIOLOGICAL STABILITY OF OLIGONUCLEOTIDES

The phenotypic effects on transformed cells of inhibiting oncogene expression will only be observed if the inhibition is maintained long enough to allow preexisting levels of oncogene protein to decay through turnover within the cells. Consequently antisense oligonucleotides should be sufficiently stable within cell cultures such that effective concentrations are maintained

for the appropriate length of time. Normal phosphodiester oligodeoxynucleotides are deficient in this requirement, since they are susceptible to rapid degradation by nucleases both within cells and in the serum component of cell culture media (Fig. 2). On the other hand, Smith et al. [1986] have reported that nonionic methylphosphonate oligonucleotide analogs are resistant to nuclease hydrolysis, and this result was confirmed in our own work when no detectable breakdown of N-*ras* sequence methylphosphonate 9-mers or reduction in their concentration occurred in cell cultures during incubations of up to 48 hours at 37°C [Tidd et al., 1988]. To determine the relative contributions of endonuclease, 3'-phosphodiesterase, and 5'-phosphodiesterase activities to degradation of oligodeoxynucleotides by fetal calf serum, chimeric structures were prepared in which either the 3' or both the 3' and 5' ends of the molecules were protected from exonuclease attack by two consecutive methylphosphonodiester linkages [Tidd and Warenius, 1989]. Two analog linkages were incorporated, since venom phosphodiesterase, at least, was reported to be able to bypass one terminal methylphosphonodiester and cleave the phosphodiester bond two base residues removed from the 3'-OH terminus [Miller et al., 1980], whereas phosphodiester oligodeoxynucleotides with two methylphosphonate linkages at each end were resistant to degradation by purified exonucleases [Agrawal and Goodchild, 1987].

Two chimeric oligodeoxynucleotides representing an N-*ras* antisense sequence were used (Fig. 2A). One, $(P)_{15}$ $(M)_2$, an 18-mer with two methylphosphonodiester linkages at the 3' end only, was designed to resist 3'-phosphodiesterase attack while permitting degradation by 5'-phosphodiesterase and endonucleases. The other, $(M)_2$ $(P)_{15}$ $(M)_2$, a 20-mer with two methylphosphonodiester linkages at both ends of the molecule, permitted an evaluation of the contribution of endonuclease-mediated breakdown alone. The core phosphodiester 16-mer sequence was the same in both oligodeoxynucleotides. The enhanced lifetime of both chimeric structures in tissue culture medium relative to their all-phosphodiester counterparts (Fig. 2A) demonstrated that 3'-phosphodiesterase is the major nuclease in fetal calf serum responsible for oligodeoxynucleotide breakdown. The data for $(M)_2$ $(P)_{15}$ $(M)_2$ showed that endonuclease was by no means insignificant, but 5'-phosphodiesterase activity could not be detected since $(P)_{15}$ $(M)_2$ was even less readily degraded than was $(M)_2$ $(P)_{15}$ $(M)_2$. Terminal methylphosphonodiester linkages were also shown to endow considerable protection to *myc* sequence oligodeoxynucleotides when incubated in cell culture medium containing 15% fetal calf serum (Fig. 2B). Similar proportionate survival of intact methylphosphonodiester/phosphodiester chimeras in cell culture medium was observed whether the initial oligodeoxynucleotide concentration was 10 or 100 μM [Tidd and Warenius, 1989]. However, the serum endonuclease activity responsible for degradation of the chimeric oligonucleotides appeared to decay somewhat during incubation

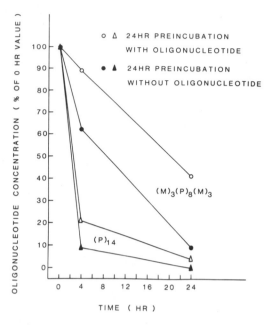

Fig. 3. *Stability of phosphodiester and methylphosphonodiester/phosphodiester chimeric* myc *sense oligodeoxynucleotide 15-mers (10 μM) in McCoy's 5A medium containing 15% heat-inactivated fetal calf serum at 37°C following preincubation (37°C, 24 hours) of the medium with or without oligonucleotide (10 μM). See also Figure 2.*

with, but not without, these molecules, and consequently further increases in overall persistence of intact chimeric oligonucleotide were achieved, e.g., by adding the total amount in two portions separated by an interval of 24 hours (Fig. 3).

III. NONSPECIFIC TOXICITY

It is essential that oligonucleotides should be completely nontoxic to cells over the requisite concentration range if valid conclusions are to be reached about the phenotypic effects of inhibiting oncogene expression. Normal phosphodiester oligodeoxynucleotides may not be toxic in themselves. However, it is apparent that high concentrations of 2'-deoxynucleosides released through degradation of these molecules may have profound biological effects.

For example, in the absence of protective levels of adenosine deaminase, 2′-deoxyadenosine is toxic, while thymidine in high concentration blocks entry into S-phase, causing cells to accumulate at the G_1–S boundary, and, indeed, this action has been exploited in order to synchronize cell populations for passage through the cell cycle [Doida and Okada, 1967].

In our hands a *myc* antisense 15-mer oligodeoxynucleotide complementary to codons 1–5 of the human c-*myc* gene [Wickstrom et al., 1988; Holt et al., 1988] failed to affect the levels of Myc protein in HL-60 cells at a concentration of 25 μM. However, partial inhibition of cell proliferation was observed in cultures exposed to the antisense oligodeoxynucleotide, a nonsense random permutation of the antisense sequence, and a *myc* sense 15-mer oligodeoxynucleotide (data not shown). We ascribed these effects to breakdown products released through rapid degradation of the oligodeoxynucleotides (Fig. 2B) by nucleases present in the serum component of the cell culture medium [Tidd and Warenius, 1989]. Nuclease-resistant methylphosphonate oligonucleotide analogs were nontoxic at comparable concentrations [Smith et al., 1986; Tidd et al., 1988], and *myc* nonsense and sense methylphosphonodiester/phosphodiester chimeric 15-mer oligonucleotides, with three methylphosphonodiester linkages at each end of the molecules (Fig. 2B), were less growth inhibitory to HL-60 cells than the all-phosphodiester oligodeoxynucleotides. Apparent sequence-specific inhibition of HL-60 cell proliferation and Myc protein synthesis were observed with the antisense chimeric oligonucleotide (data not shown). However, the latter effects were not consistently reproducible, and we concluded that more detailed biochemical investigations were required to define the reasons for this variability.

IV. CELL UPTAKE

The demonstration of antisense effects of normal phosphodiester oligodeoxynucleotides on intact cells suggests that despite their polyanionic nature they are able to gain access to the interior of at least some types of cell. At the same time it is evident that the cell membrane represents a barrier that will limit the efficacy of oncogene antisense oligonucleotides and that potentially their activity might be enhanced by facilitating intracellular delivery. There have been few in-depth studies of actual cell uptake, and most investigations have utilized oligodeoxynucleotides terminally tagged with ^{32}P- or ^{35}S-phosphate or with acridine [Harel-Bellan et al., 1988b; Wickstrom et al., 1988; Holt et al., 1988; Becker et al., 1989; Zamecnik et al., 1986; Stein et al., 1988; Loke et al., 1988; Loke et al., 1989; Vlassov et al., 1986; Boutorin et al., 1989; Yakubov et al., 1989]. Several of these reports failed to address the possibility that extracellular degradation preceeded intracellular accumulation of the label. In general, the results of these experiments suggested that intact cells accumulated greater amounts of oligodeoxynucleotides than did fixed cells over the course of several hours by the action of receptor-mediated endocytosis. However, the intracellular concentration was always less than that in the medium, and, when measured, cell-associated oligodeoxynucleotides were largely intact at early times but became progressively degraded. Degols et al. (this volume) discuss how the enhanced cell delivery of oligodeoxynucleotides may be achieved by linking them to poly-L-lysine [see also Leonetti et al., 1988; Westermann et al., 1989; Stevenson and Iversen, 1989; Degols et al., 1989; Lemaitre et al., 1987] or by their encapsulation in liposomes [see also Loke et al., 1988]. Alternative approaches of linkage to cholesterol [Boutorin et al., 1989] or acridine [Hélène and Toulmé, 1989] have also been suggested.

In contrast to normal phosphodiester oligodeoxynucleotides, the more lipophilic, uncharged methylphosphonate oligonucleotide analogs have been reported to enter cells slowly by simple diffusion until the intracellular and extracellular concentrations are essentially equilibrated [Marcus-Sekura et al., 1987; Miller et al., 1981]. It is possible that the methylphosphonate sections of chimeric methylphosphonodiester/phosphodiester oligodeoxynucleotides may enhance uptake of the molecules when compared with their all-phosphodiester counterparts. This possibility is currently under investigation.

V. HYBRIDIZATION EFFICIENCY AND STABILITY

It is essential that oncogene antisense oligonucleotides should recognize and bind tightly to their target sequences in mRNA at 37°C, the temperature of incubation of mammalian cell cultures. Consequently, the temperature for dissociation of half the hybrids formed between the oligonucleotide and its complementary sequence in vitro, the melting temperature (T_m), should be well in excess of this figure under physiological salt conditions, although the actual T_m within the cellular milieu, could it be measured, would probably differ from that recorded in vitro, being dependent on both nucleic acid concentrations and environment. For short oligodeoxynucleotides, under a given set of conditions, the hybrid T_m increases with increasing chain length, and, generally speaking, normal phosphodiester 15–20-mers with approximately 50% G + C content at micromolar concentrations form hybrids with T_ms between 50° and 70°C in physiological salt solutions (Fig. 4).

In addition to this fundamental requirement, Zon [1988] has pointed out several other factors that should be taken into consideration when selecting the base sequence of an antisense oligonucleotide. In particular, external and internal self-complementarity should be avoided since this may lead to bimolecular self-association or intramolecular hairpin formation, which could reduce the efficiency of hybridization with the target sequence. Also, the oligonucleotide should not contain runs of five or more purines, as this may result in the formation of secondary structure through base stacking interactions, as exemplified by the *myc* antisense 15-mer studied by Wickstrom (this volume) and others [Heikkila et al., 1987; Harel-Bellan et al., 1988b; Wickstrom et al., 1988, 1989; Holt et al., 1988; Tidd and Warenius, 1989]. Consideration of the secondary

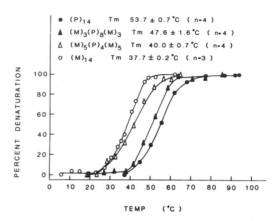

Fig. 4. *Melting curves of hybrids formed between a human* myc *antisense oligodeoxynucleotide 15-mer sequence with varying degrees of substitution of terminal phosphodiester linkages by methylphosphonate and a chemically synthesised* myc *sense RNA 27-mer (see also Fig. 2). The sequence of the RNA 27-mer was CCU CCC GCG ACG AUG CCC CUC AAC GUU. Percent denaturation was calculated from the observed hyperchromic shift at 260 nm for each temperature point divided by the maximum hyperchromicity measured upon complete dissociation of the hybrids. The concentrations of oligodeoxynucleotide and oligoribonucleotide were both 2 µM in 0.1 M sodium chloride, 0.01 M sodium cacodylate buffer, pH 7. (Reproduced from Tidd, 1990, with permission of the publisher.)*

structure of the target mRNA may also be important, since open loop regions are likely to hybridize more readily with an antisense oligonucleotide than are those regions involved in intramolecular hydrogen bonding [Wickstrom et al., 1988].

The validity of the oncogene antisense oligonucleotide experiment rests on the assumption that the target base sequence is unique to the oncogene mRNA. The number of bases required to define a unique sequence in the human genome may be calculated statistically as at least 15–19 by assuming a random distribution of the total base content [Leonetti et al., 1988; Hélène and Toulmé, 1989; Zon, 1989; van der Krol et al., 1988; Cazenave et al., 1987]. However, since only a small fraction of the genome is transcribed, a sequence of at least 11–15 bases is likely to be unique to an mRNA, where the lower figure is for oligodeoxynucleotides containing only G and C, and the higher figure is for those comprised only of A and T [Hélène and Toulmé, 1989]. In the past, oligodeoxribonucleotides of 15–20 bases were considered of sufficient length both to define a unique sequence in mRNA and to form hybrids of sufficient stability for antisense effects to be elicited, assuming that cell uptake and nucleolytic degradation were not limiting factors. However, long oligonucleotides may be partially complementary to sequences in nontargeted mRNAs, and nonspecific effects could result from interactions at these sites. Consequently, a case has been made for using shorter antisense oligonucleotides on the grounds that these would be less likely to induce nonspecific effects through partial complementarity to other mRNAs, would probably be taken up by cells more readily than longer sequences, and would exhibit more favorable intracellular hybridization kinetics [Hélène and Toulmé, 1989; Zon, 1989; Cazenave et al., 1987]. On the other hand, short oligonucleotides may require modifications to enhance hybrid stability, e.g., by linking an intercalating agent [Cazenave et al., 1987; Zerial et al., 1987; Bazile et al., 1989; Gautier et al., 1987; Thuong et al., 1987; Toulmé et al., 1986] if antisense effects are to be induced, and on the basis of the original statistical considerations they may be entirely complementary to sequences in nontargeted mRNAs.

The poor biological stability and cell uptake properties of normal phosphodiester oligodeoxynucleotides led to the search for nuclease-resistant structural analogs, with enhanced ability to cross cell membranes but that still retain the hybridization characteristics of the parent molecule. The nonionic methylphosphonate oligodeoxynucleotide analogs were shown to be nuclease resistant, to enter cells by simple diffusion, and to hybridize with complementary nucleic acids [Miller and Ts'o, 1987; Miller et al., 1981]. However, the stability of methylphosphonate–phosphodiester nucleic acid hybrids and the efficiency of hybridization are poor in comparison with normal

phosphodiester oligodeoxynucleotides [Miller and Ts'o, 1987; Maher and Dolnick, 1988; Tidd et al., 1988]. The experimental results shown in Figure 4 demonstrated that progressive replacement of the terminal phosphodiester linkages by methylphosphonodiesters in a *myc* antisense 15-mer [Heikkila et al., 1987] led to proportionate decreases in the T_m of hybrids formed with a complementary RNA 27-mer. In addition to the lower melting temperatures of methylphosphonate hybrids relative to their phosphodiester counterparts, it was apparent that even at temperatures well below the T_m the dissociation constant for methylphosphonate hybrids was high, such that in 1:1 molar mixtures of oligonucleotide analog and target sequence only a small proportion of the molecules were present as duplexes. This was shown by the experiment depicted in Figures 5 and 6, in which hybrids were selectively removed from solution by binding the negatively charged N-*ras* sense oligodeoxynucleotide 20-mer target sequence to the weak anion exchanger DEAE Sephadex, while the uncharged nonhybridized methylphosphonate antisense 9-mer molecules remained in solution. The methylphosphonate oligonucleotide was subsequently recovered from the resin-bound hybrids by thermal dissociation and was shown to represent only about 10% of the total methylphosphonate in the original hybridization mixture (Fig. 5). This result was in rough agreement with the data obtained from mixing curve experiments [Tidd et al., 1988]. The affinity-purified N-*ras* antisense 9-mer was rehybridized to the sense phosphodiester 20-mer and the melting curve compared with that of the parent preparation (Fig. 6, top), when a 5°C increase in T_m was observed. However, the hyperchromic shift upon complete dissociation of hybrids formed by the affinity-purified material was small and similar to that of the parent preparation, indicating that even at low temperature (15°–20°C) only a small proportion of antisense molecules previously bound by the target 20-mer were present as duplexes when remixed with the sense sequence. In addition, when sense phosphodiester 20-mer was added back to the

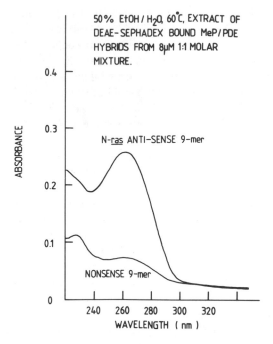

50% EtOH / H₂O, 60°C, EXTRACT OF DEAE-SEPHADEX BOUND MeP/PDE HYBRIDS FROM 8µM 1:1 MOLAR MIXTURE.

N-ras ANTI-SENSE 9-mer

NONSENSE 9-mer

Fig. 5. *Isolation of human N-ras antisense methylphosphonate oligodeoxynucleotide analog 9-mer molecules hybridized at 20°C in solution to a normal phosphodiester N-ras sense oligodeoxynucleotide 20-mer sequence. Shown is the UV spectrum of the supernatant obtained following binding of all sense 20-mer and hybrids by addition of the weak anion exchanger DEAE Sephadex, removal of the resin, and thermal dissociation of methylphosphonate oligonucleotide from the resin bound hybrids into solution. The sequences were antisense oligonucleotide analog, CTC AGT CAT; nonsense analog control, CAC GAT TCT; normal phosphodiester 20-mer sense target, ATG ACT GAG TAC AAA CTG GT. (Reproduced from Tidd, 1990, with permission of the publisher.)*

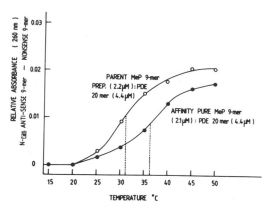

PARENT MeP 9-mer PREP. (2.2µM):PDE 20 mer (4.4µM)

AFFINITY PURE MeP 9-mer (2.1µM): PDE 20 mer (4.4µM)

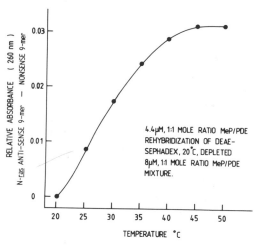

4.4µM, 1:1 MOLE RATIO MeP/PDE REHYBRIDIZATION OF DEAE-SEPHADEX, 20°C, DEPLETED 8µM, 1:1 MOLE RATIO MeP/PDE MIXTURE.

Fig. 6. *Melting curves of hybrids formed between the human N-ras antisense methylphosphonate oligodeoxynucleotide analog 9-mer (parent, affinity-purified, and affinity-depleted preparations) and the normal phosphodiester N-ras sense oligodeoxynucleotide 20-mer (see Fig. 5).* **Top:** *○, Parent methylphosphonate preparation; ●, rehybridized methylphosphonate previously isolated by hybridization to the target 20-mer sequence (see Fig. 5).* **Bottom:** *Hybridization of methylphosphonate remaining following removal of the target 20-mer and associated hybridized analog oligonucleotide in the experiment shown in Figure 5. (Reproduced from Tidd, 1990, with permission of the publisher.)*

antisense methylphosphonate 9-mer solution remaining after removal of hybrids in the experiment shown in Figure 5, equilibrium was reestablished, with a small proportion of duplexes again being formed, which gave a melting curve similar to that of the original preparation, except that the T_m was somewhat reduced (Fig. 6, bottom). These results confirm that, if significant antisense effects are to be achieved with methylphosphonate oligonucleotide analogs, vast excesses over target sequence will be necessary to convert most of the latter into hybrids

[Miller and Ts'o, 1987] and this is indeed found to be the case [Maher and Dolnick, 1988; Yu et al., 1989].

The higher T_m of hybrids formed by the affinity-purified N-*ras* antisense methylphosphonate 9-mer (Fig. 6, top) may be explained on the basis of the chirality of the methylphosphonodiester linkage, since synthesis of the oligonucleotide analog was not stereospecific and the original product was a mixture of 256 (2^n, where n = 8 internucleoside linkages) stereoisomers. Presumably, affinity purification selected for those molecules with a lower proportion of linkages in the helix-destabilizing pseudoaxial conformation [Durand et al., 1989; Bower et al., 1987]. The high dissociation constant of methylphosphonate oligonucleotide hybrids was observed even with affinity-purified material at well below the T_m. This may be related to stereoelectronic effects of replacing oxygen by methyl, which disturbs the conformation of the $P-O_3$ and the $P-O_5$ linkages and tends to counter adoption of a helical geometry over an extended chain length [Bower et al., 1987; Moody et al., 1989]. On the other hand, the exonuclease-resistant chimeric methylphosphonodiester/phosphodiester *myc* antisense 15-mer, $([M]_3 [P]_8 [M]_3$, with three methylphosphonate linkages at each end of the molecule, formed hybrids with a T_m only 6°C lower than the all-phosphodiester 15-mer (Fig. 4), and the absorbance data (not shown) indicated that the efficiency of hybridization was the same for both. This confirmed that the helical configuration is tolerated over short runs of methylphosphonate linkages [Moody et al., 1989]. A further desirable property of the methylphosphonodiester/phosphodiester chimeras is that they are readily soluble in water, whereas the all-methylphosphonate *myc* antisense 15-mer (Fig. 4) was very sparingly soluble.

VI. MECHANISMS OF ACTION

As discussed by Minshull and Hunt (this volume) and by Colman (this volume), work with cell-free protein synthesizing systems and microinjected oocytes has demonstrated that, when inhibition of translation of added mRNA was achieved with antisense oligodeoxynucleotides targeted at translated sequences downstream of the initiation codon region, the effects were the result of degradation of the mRNA by ribonuclease H at the site of hybridization with the oligodeoxynucleotide [Haeuptle et al., 1986; Walder and Walder, 1988; Cazenave et al., 1989; Jessus et al., 1988; Shuttleworth et al., 1988; Shuttleworth and Colman, 1988; O'Keefe et al., 1989; Harel-Bellan et al., 1988a; Donis-Keller, 1979; Hausen and Stein, 1970; Wyatt and Walker, 1989; Furdon et al., 1989]. Once fully assembled, the ribosomal complex is able to destabilize secondary structure locally as it moves along the message, and, in the absence of ribonuclease H, antisense oligodeoxyribonucleotides complementary to downstream translated sequences in mRNA were unable to affect protein synthesis [Blake et al., 1985a; Sankar et al., 1989; Teichman-Weinberg et al., 1988; Liebhaber et al., 1984]. Ribonuclease H may also contribute to antisense effects of oligodeoxyribonucleotides directed at the initiation codon region and 5' upstream untranslated sequences in mRNA and at splice sites in pre-mRNA. However, there is some evidence that inhibition of initiation complex formation or splicing reactions may be achieved by targeting these sites in the absence of any involvement of the enzyme [Blake et al., 1985a; Maher and Dolnick, 1988; Walder and Walder, 1988; Sankar et al., 1989; Boiziau et al., 1991].

Methylphosphonate oligonucleotide analogs do not activate ribonuclease H–mediated cleavage of RNA [Maher and Dolnick, 1988; Furdon et al., 1989; Quartin et al., 1989], and yet sequence-specific inhibition of protein synthesis has been reported using high concentrations of these molecules, generally when upstream and initiation codon regions or splice sites were the targets [Maher and Dolnick, 1988; Marcus-Sekura et al., 1987; Smith et al., 1986; Brown et al., 1989; Blake et al., 1985b; Miller et al., 1985; Sarin et al., 1988]. In view of the foregoing discussion, it is surprising that antisense methylphosphonate 8-mers spanning codon 12 of the c-Ha-*ras* gene were able to inhibit Ras protein synthesis in an in vitro system, although high concentrations and preannealing to the mRNA were required [Yu et al., 1989]. The N-*ras* antisense methylphosphonate 9-mer (see Figs. 5 and 6), complementary to the first three codons of the human N-*ras* gene, failed to affect dexamethasone-induced accumulation of p21^{N-ras} in T15 cells, a line of NIH-3T3 cells transfected with a murine mammary tumor virus promoter–human N-*ras* gene construct [Tidd et al., 1988].

It is assumed that ribonuclease H may contribute to the effects of antisense oligodeoxynucleotides in intact mammalian cells, although as yet this has not been demonstrated directly. However, Harel-Bellan et al. [1988a] reported decreases in intracellular concentrations of target mRNA following treatment of lymphocytes with lymphokine antisense oligodeoxynucleotides, which they suggested could result from a ribonuclease H–like activity. It is important that the extent of the participation of ribonuclease H in the action of antisense oligodeoxynucleotides in intact cells be determined. If indeed the enzyme is a significant factor, then methylphosphonate oligonucleotides would be at a disadvantage when compared with normal phosphodiester oligodeoxynucleotides, since hybrids of RNA with the analog molecules are not substrates. However, chimeric methylphosphonodiester/phosphodiester oligodeoxynucleotides are endowed with the ability to direct ribonuclease H cleavage of RNA by virtue of their phosphodiester component (Fig. 7) while being afforded a degree of protection against their own degradation by nucleases through incorporation of the methylphosphonates [Tidd and Warenius, 1989; Furdon et al., 1989; Quartin et al., 1989]. Ribonuclease H cleavage of RNA may be restricted to a single base site through the use of chimeric oligonucleotides comprised of nonactivating analog sections and a short phosphodiester oligodeoxynucleotide region to direct the enzyme [Inoue et al., 1987; Shibahara et al., 1987]. Assuming that ribonuclease H is involved in the generation

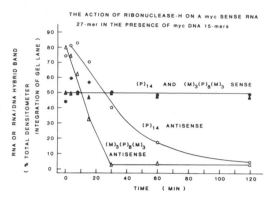

Fig. 7. *Degradation of a chemically synthesized human* myc *sense sequence RNA 27-mer by ribonuclease H following hybridization with all-phosphodiester and methylphosphonodiester/phosphodiester chimeric* myc *antisense oligodeoxynucleotide 15-mers (see Figs. 2 and 4). Oligonucleotide mixtures were incubated at 37°C with ribonuclease H from* Escherichia coli. *Samples of the reaction mixtures were analyzed by gel electrophoresis; gels were stained with ethidium bromide and photographed under UV illumination. Band intensities were quantitated by densitometric measurements on the photographic negatives. (Reproduced from Tidd, 1990, with permission of the publisher.)*

of antisense effects in intact cells, it is conceivable that chimeric oligonucleotides could be designed to activate cleavage of mRNA carrying a point mutation, without affecting the normal message. In this way it might be possible, for example, to investigate the effects of inhibiting expression of an activated *ras* oncogene without disturbing the function of the normal *ras* protooncogene [Shen et al., 1987].

VII. CONCLUSION

Neither normal phosphodiester oligodeoxynucleotides nor methylphosphonate analogs exhibit all the properties of an ideal antisense effector structure (Table I). Phosphodiester oligodeoxynucleotides hybridize efficiently and with good stability to complementary sequences and activate ribonuclease H cleavage of RNA, but they exhibit poor biological stability and cell uptake properties. In contrast, nonionic methylphosphonates are biologically stable and enter cells by simple diffusion, but they hybridize poorly to complementary nucleic acid sequences, fail to direct ribonuclease H cleavage of RNA, and become progressively less soluble with increasing chain length above nine bases.

Certain advantages can be achieved by combining both structures in chimeric molecules in which the methylphosphonates occupy the terminal positions. Such chimeric molecules are readily soluble, exonuclease but not endonuclease resistant, and retain the capacity to activate ribonuclease H cleavage of mRNA. In addition, the methylphosphonate sections may well enhance cell uptake, although this remains to be demonstrated.

Methylphosphonate/phosphodiester chimeras are easily prepared on standard commercial DNA synthesizers, without alteration of the synthesis cycle program, by merely replacing phosphoramidite bottles by methylphosphonamidites (currently available commercially from Glen Research Corporation, Sterling, VA) and vice versa as appropriate.

It may well be that oligonucleotide analogs linked to chemically reactive groups or chimeric molecules incorporating the RNA active site of ribozymes will become the antisense oligonucleotide structures of the future. However, for the present the immediate priority must be for more detailed investigations of the uptake and biochemical interactions of oligonucleotides in intact cells, since mere demonstration of an antisense effect provides little information on how the structures might be improved to make them more universally applicable as tools to inhibit oncogene expression in any desired tumor cell type.

ACKNOWLEDGMENTS

I thank Andrea Reynolds for typing the manuscript, Anita Chapman and Rosalind White for preparing photographs of the figures, and Carl Goodwin for laboratory assistance. The support of the Cancer Research Campaign and the Cancer and Polio Research Fund is gratefully acknowledged.

VIII. REFERENCES

Agrawal S, Goodchild J (1987): Oligodeoxynucleoside methylphosphonates: Synthesis and enzymic degradation. Tetrahedon Lett 28:3539–3542.

Bazile D, Gautier C, Rayner B, Imbach J-L, Paoletti C, Paoletti J (1989): α-DNA X: α and β tetrathymidylates covalently linked to oxazolopyridocarbazolium (OPC): comparative stabilization of oligo β-[dT]:oligo β-[dA] and oligo α-[dT]:oligo β-[dA] duplexes by the intercalating agent. Nucleic Acids Res 17:7749–7759.

Becker D, Meier CB, Herlyn M (1989): Proliferation of human malignant melanomas is inhibited by antisense oligodeoxynucleotides targeted against basic fibroblast growth factor. EMBO J 8:3685–3691.

Blake KR, Murakami A, Miller PS (1985a): Inhibition of rabbit globin mRNA translation by sequence-specific oligodeoxyribonucleotides. Biochemistry 24:6132–6138.

Blake KR, Murakami A, Spitz SA, Glave SA, Reddy MP, Ts'o POP, Miller PS (1985b): Hybridization arrest of globin synthesis in rabbit reticulocyte lysates and cells by oligodeoxyribonucleoside methylphosphonates. Biochemistry 24:6139–6145.

Boiziau C, Kurfurst R, Cazenave C, Roig V, Thuong NT, Toulmé J-J (1991): Inhibition of translation initiation by antisense oligonucleotides via an RNase-H independent mechanism. Nucleic Acids Res 19:1113–1119.

Boutorin AS, Gus'kova LV, Ivanova EM, Kobetz ND, Zarytova VF, Ryte AS, Yurchenko LV, Vlassov VV (1989): Synthesis of alkylating oligonucleotide derivatives containing cholesterol or phenazinium residues at their 3'-terminus and their interaction with DNA within mammalian cells. FEBS Lett 254:129–132.

Bower M, Summers MF, Powell C, Shinozuka K, Regan JB, Zon G, Wilson WD (1987): Oligodeoxyribonucleoside methylphosphonates. NMR and UV spectroscopic studies of R_p-R_p and S_p-S_p methylphosphonate (Me) modified duplexes of (d[GGAATTCC])₂. Nucleic Acids Res 15:4915–4930.

Brown D, Yu Z, Miller P, Blake K, Wei C, Kung H-F, Black RJ, Ts'o POP, Chang EH (1989): Modulation of *ras* expression by anti-sense, nonionic deoxynucleotide analogs. Oncogene Res 4:243–252.

Cazenave C, Loreau N, Thuong NT, Toulmé J-J, Hélène C (1987): Enzymatic amplification of translation inhibition of rabbit β-globin mRNA mediated by antimessenger oligodeoxynucleotides covalently linked to intercalating agents. Nucleic Acids Res 15:4717–4736.

Cazenave C, Stein CA, Loreau N, Thuong NT, Neckers LM, Subasinghe C, Hélène C, Cohen JS, Toulmé J-J (1989): Comparative inhibition of rabbit globin mRNA translation by modified antisense oligodeoxynucleotides. Nucleic Acids Res 17:4255–4273.

Degols G, Leonetti J-P, Gagnor C, Lemaitre M, Lebleu B (1989): Antiviral activity and possible mechanisms of action of oligonucleotides-poly (L-lysine) conjugates targeted to vesicular stomatitis virus mRNA and genomic RNA. Nucleic Acids Res 17:9341–9350.

Doida Y, Okada S (1967): Synchronization of L5178Y cells by successive treatment with excess thymidine and colcemid. Exp Cell Res 48:540–548.

Donis-Keller H (1979): Site specific enzymatic cleavage of RNA. Nucleic Acids Res 7:179–192.

Durand M, Maurizot JC, Asseline U, Barbier C, Thuong NT, Hélène C (1989): Oligothymidylates covalently linked to an acridine derivative and with modified phosphodiester backbone: Circular dichroism studies of their interactions with complementary sequences. Nucleic Acids Res 17:1823–1837.

Furdon PJ, Dominski Z, Kole R (1989): RNase H cleavage of RNA hybridized to oligonucleotides containing methylphosphonate, phosphorothioate and phosphodiester bonds. Nucleic Acids Res 17:9193–9204.

Gautier C, Morvan F, Rayner B, Huynh-Dinh T, Igolen J, Imbach J-L, Paoletti C, Paoletti J (1987): α-DNA IV: α-anomeric and β-anomeric tetrathymidylates covalently linked to intercalating oxazolopyridocarbazole. Synthesis, physicochemical properties and poly (rA) binding. Nucleic Acids Res 15:6625–6641.

Haeuptle M-T, Frank R, Dobberstein B (1986): Translation arrest by oligodeoxynucleotides complementary to mRNA coding sequences yields polypeptides of predetermined length. Nucleic Acids Res 14:1427–1448.

Harel-Bellan A, Durum S, Muegge K, Abbas AK, Farrar WL (1988a): Specific inhibition of lymphokine biosynthesis and autocrine growth using antisense oligonucleotides in Th1 and Th2 helper T cell clones. J Exp Med 168:2309–2318.

Harel-Bellan A, Ferris DK, Vinocour M, Holt JT, Farrar WL (1988b): Specific inhibition of c-myc protein biosynthesis using an antisense synthetic deoxyoligonucleotide in human T lymphocytes. J Immunol 140:2431–2435.

Hausen P, Stein H (1970): Ribonuclease H. An enzyme degrading the RNA moiety of DNA–RNA hybrids. Eur J Biochem 14:278–283.

Heikkila R, Schwab G, Wickstrom E, Loke SL, Pluznik DH, Watt R, Neckers LM (1987): A c-myc antisense oligodeoxynucleotide inhibits entry into S phase but not progress from G_0 to G_1. Nature 328:445–449.

Hélène C, Toulmé J-J (1989): Control of gene expression by oligodeoxynucleotides covalently linked to intercalating agents and nucleic acid-cleaving reagents. In Cohen JS (ed): Oligodeoxynucleotides: Antisense Inhibitors of Gene Expression. London: Macmillan Press Ltd, pp 137–172.

Holt JT, Redner RL, Nienhuis AW (1988): An oligomer complementary to c-myc mRNA inhibits proliferation of HL-60 promyelocytic cells and induces differentiation. Mol Cell Biol 8:963–973.

Inoue H, Hayase Y, Iwai S, Ohtsuka E (1987): Sequence-dependent hydrolysis of RNA using modified oligo-

nucleotide splints and RNase H. FEBS Lett 215:327–330.

Jessus C, Cazenave C, Ozon R, Hélène C (1988): Specific inhibition of endogenous β-tubulin synthesis in Xenopus oocytes by anti-messenger oligodeoxynucleotides. Nucleic Acids Res 16:2225–2233.

Krol AR van der, Mol JNM, Stuitje AR (1988): Modulation of eukaryotic gene expression by complementary RNA or DNA sequences. BioTechniques 6:958–976.

Lemaitre M, Bayard B, Lebleu B (1987): Specific antiviral activity of a poly(L-lysine)–conjugated oligodeoxyribonucleotide sequence complementary to vesicular stomatitis virus N protein mRNA initiation site. Proc Natl Acad Sci USA 84:648–652.

Leonetti JP, Rayner B, Lemaitre M, Gagnor C, Milhaud PG, Imbach J-L, Lebleu B (1988): Antiviral activity of conjugates between poly(L-lysine) and synthetic oligodeoxyribonucleoside. Gene 72:323–332.

Liebhaber SA, Cash FE, Shakin SH (1984): Translationally associated helix-destabilizing activity in rabbit reticulocyte lysate. J Biol Chem 259:15597–15602.

Loke SL, Stein C, Zhang X, Avigan M, Cohen J, Neckers LM (1988): Delivery of c-myc antisense phosphorothioate oligodeoxynucleotides to hematopoietic cells in culture by liposome fusion: Specific reduction in c-myc protein expression correlates with inhibition of cell growth and DNA synthesis. Curr Top Microbiol Immunol 141:282–289.

Loke SL, Stein CA, Zhang XH, Mori K, Nakanishi M, Subasinghe C, Cohen JS, Neckers LM (1989): Characterization of oligonucleotide transport into living cells. Proc Natl Acad Sci USA 86:3474–3478.

Maher LJ III, Dolnick BJ (1988): Comparative hybrid arrest by tandem antisense oligodeoxyribonucleotides or oligodeoxyribonucleoside methylphosphonates in a cell-free system. Nucleic Acids Res 16:3341–3358.

Marcus-Sekura CJ, Woerner AM, Shinozuka K, Zon G, Quinnan GV Jr (1987): Comparative inhibition of chloramphenicol acetyltransferase gene expression by antisense oligonucleotide analogues having alkyl phosphotriester, methylphosphonate and phosphorothioate linkages. Nucleic Acids Res 15:5749–5763.

McManaway ME, Neckers LM, Loke SL, Al-Nasser AA, Redner RL, Shiramizu BT, Goldschmidts WL, Huber BE, Bhatia K, Magrath IT (1990): Tumour-specific inhibition of lymphoma growth by an antisense oligodeoxynucleotide. Lancet 335:808–811.

Miller PS, Agris CH, Aurelian L, Blake KR, Murakami A, Reddy MP, Spitz SA, Ts'o POP (1985): Control of ribonucleic acid function by oligonucleoside methylphosphonates. Biochimie 67:769–776.

Miller PS, Dreon N, Pulford SM, McParland KB (1980): Oligothymidylate analogues having stereoregular, alternating methylphosphonate/phosphodiester backbones. J Biol Chem 255:9659–9665.

Miller PS, McParland KB, Jayaraman K, Ts'o POP (1981):

Biochemical and biological effects of nonionic nucleic acid methylphosphonates. Biochemistry 20:1874–1880.

Miller PS, Ts'o POP (1987): A new approach to chemotherapy based on molecular biology and nucleic acid chemistry: Matagen (masking tape for gene expression). Anti-Cancer Drug Design 2:117–128.

Moody HM, van Genderen MHP, Koole LH, Kocken HJM, Meijer EM, Buck HM (1989): Regiospecific inhibition of DNA duplication by antisense phosphate-methylated oligodeoxynucleotides. Nucleic Acids Res 17:4769–4782.

Mulder MP, Keijzer W, Verkerk A, Boot AJM, Prins MEF, Splinter TAW, Bos JL (1989): Activated *ras* genes in human seminoma: Evidence for tumour heterogeneity. Oncogene 4:1345–1351.

O'Keefe SJ, Wolfes H, Kiessling AA, Cooper GM (1989): Microinjection of antisense c-*mos* oligonucleotides prevents meiosis II in the maturing mouse egg. Proc Natl Acad Sci USA 86:7038–7042.

Quartin RS, Brakel CL, Wetmur JG (1989): Number and distribution of methylphosphonate linkages in oligodeoxynucleotides affect exo- and endonuclease sensitivity and ability to form RNase H substrates. Nucleic Acids Res 17:7253–7262.

Sankar S, Cheah K-C, Porter AG (1989): Antisense oligonucleotide inhibition of encephalomyocarditis virus RNA translation. Eur J Biochem 184:39–45.

Sarin PS, Agrawal S, Civeira MP, Goodchild J, Ikeuchi T, Zamecnik PC (1988): Inhibition of acquired immunodeficiency syndrome virus by oligodeoxynucleoside methylphosphonates. Proc Natl Acad Sci USA 85:7448–7451.

Shen WPV, Aldrich TH, Venta-Perez G, Franza BR Jr, Furth ME (1987): Expression of normal and mutant *ras* proteins in human acute leukemia. Oncogene 1:157–165.

Shibahara S, Mukai S, Nishihara T, Inoue H, Ohtsuka E, Morisawa H (1987): Site directed cleavage of RNA. Nucleic Acids Res 15:4403–4415.

Shuttleworth J, Colman A (1988): Antisense oligonucleotide-directed cleavage of mRNA in *Xenopus* oocytes and eggs. EMBO J 7:427–434.

Shuttleworth J, Matthews G, Dale L, Baker C, Colman A (1988): Antisense oligodeoxynucleotide-directed cleavage of maternal mRNA in *Xenopus* oocytes and embryos. Gene 72:267–275.

Smith CC, Aurelian L, Reddy MP, Miller PS, Ts'o POP (1986): Antiviral effect of an oligo (nucleoside methylphosphonate) complementary to the splice junction of herpes simplex virus type 1 immediate early pre-mRNAs 4 and 5. Proc Natl Acad Sci USA 83:2787–2791.

Stein CA, Mori K, Loke SL, Subasinghe C, Shinozuka K, Cohen JS, Neckers LM (1988): Phosphorothioate and normal oligodeoxyribonucleotides with 5'-linked acridine: Characterization and preliminary kinetics of cellular uptake. Gene 72:333–341.

Stephenson ML, Zamecnik PC (1978): Inhibition of Rous sarcoma viral RNA translation by a specific oligodeoxyribonucleotide. Proc Natl Acad Sci USA 75:285–288.

Stevenson M, Iversen PL (1989): Inhibition of human immunodeficiency virus type 1–mediated cytopathic effects by poly(L-lysine)-conjugated synthetic antisense oligodeoxyribonucleotides. J Gen Virol 70:2673–2682.

Teichman-Weinberg A, Littauer UZ, Ginzburg I (1988): The inhibition of neurite outgrowth in PC12 cells by tubulin antisense oligodeoxyribonucleotides. Gene 72:297–307.

Thuong NT, Asseline U, Roig V, Takasugi M, Hélène C (1987): Oligo (α-deoxynucleotide)s covalently linked to intercalating agents: Differential binding to ribo- and deoxyribopolynucleotides and stability towards nuclease digestion. Proc Natl Acad Sci USA 84:5129–5133.

Tidd DM (1990): A potential role for antisense oligonucleotide analogues in the development of oncogene targeted cancer chemotherapy. Anticancer Res 10:1169–1182.

Tidd DM, Hawley P, Warenius HM, Gibson I (1988): Evaluation of N-*ras* oncogene anti-sense, sense and nonsense sequence methylphosphonate oligonucleotide analogues. Anti-Cancer Drug Design 3:117–127.

Tidd DM, Warenius HM (1989): Partial protection of oncogene, anti-sense oligodeoxynucleotides against serum nuclease degradation using terminal methylphosphonate groups. Br J Cancer 60:343–350.

Toksoz D, Farr CJ, Marshall CJ (1987): *ras* gene activation in a minor proportion of the blast population in acute myeloid leukaemia. Oncogene 1:409–413.

Toulmé JJ, Krisch HM, Loreau N, Thuong NT, Hélène C (1986): Specific inhibition of mRNA translation by complementary oligonucleotides covalently linked to intercalating agents. Proc Natl Acad Sci USA 83:1227–1231.

Vlassov VV, Godovikov AA, Kobetz ND, Ryte AS, Yurchenko LV, Bukrinskaya AG (1986): Nucleotide and oligonucleotide derivatives as enzyme and nucleic acid targeted irreversible inhibitors. Biochemical aspects. Adv Enzyme Regul 24:301–322.

Walder RY, Walder JA (1988): Role of RNase H in hybrid-arrested translation by antisense oligonucleotides. Proc Natl Acad Sci USA 85:5011–5015.

Westermann P, Gross B, Hoinkis G (1989): Inhibition of expression of SV40 virus large T-antigen by antisense oligodeoxyribonucleotides. Biomed Biochim Acta 48:85–93.

Wickstrom EL, Bacon TA, Gonzalez A, Freeman DL, Lyman GH, Wickstrom E (1988): Human promyelocytic leukemia HL-60 cell proliferation and c-*myc* protein expression are inhibited by an antisense pentadecadeoxynucleotide targeted against c-*myc* mRNA. Proc Natl Acad Sci USA 85:1028–1032.

Wickstrom EL, Bacon TA, Gonzalez A, Lyman GH, Wickstrom E (1989): Anti-c-*myc* DNA increases differentiation and decreases colony formation by HL-60 cells. In Vitro Cell Dev Biol 24:297–302.

Wyatt JR, Walker GT (1989): Deoxynucleotide-containing oligoribonucleotide duplexes: Stability and susceptibility to RNase V_1 and RNase H. Nucleic Acids Res 17:7833–7842.

Yakubov LA, Deeva EA, Zarytova VF, Ivanova EM, Ryte AS, Yurchenko LV, Vlassov VV (1989): Mechanism of oligonucleotide uptake by cells: Involvement of specific receptors? Proc Natl Acad Sci USA 86: 6454–6458.

Yu Z, Chen D, Black RJ, Blake K, Ts'o POP, Miller P, Chang EH (1989): Sequence specific inhibition of in vitro translation of mutated or normal ras p21. J Exp Pathol 4:97–108.

Zamecnik PC, Goodchild J, Taguchi Y, Sarin PS (1986): Inhibition of replication and expression of human T-cell lymphotropic virus type III in cultured cells by exogenous synthetic oligonucleotides complementary to viral RNA. Proc Natl Acad Sci USA 83:4143–4146.

Zamecnik PC, Stephenson ML (1978): Inhibition of Rous sarcoma virus replication and cell transformation by a specific oligodeoxynucleotide. Proc Natl Acad Sci USA 75:280–284.

Zerial A, Thuong NT, Hélène C (1987): Selective inhibition of the cytopathic effect of type A influenza viruses by oligodeoxynucleotides covalently linked to an intercalating agent. Nucleic Acids Res 15: 9909–9919.

Zon G (1988): Oligonucleotide analogues as potential chemotherapeutic agents. Pharm Res 5:539–549.

Zon G (1989): Oligonucleotide analogues as potential chemotherapeutic agents. In Martin JC (ed): Nucleotide Analogues as Antiviral Agents. ACS Symposium Series, Vol 401. Washington, DC: American Chemical Society, pp 170–184.

Zon G (1990): Pharmaceutical considerations for oligonucleotide drugs: General points and comments on phosphorothioates. Proc Am Assoc Cancer Res 31:487–488.

ABOUT THE AUTHOR

DAVID M. TIDD is Cancer Research Campaign Research Lecturer within the Department of Biochemistry at the University of Liverpool, where he heads a small research group, supported by Programme Grants from the CRC and the Cancer and Polio Research Fund, to investigate the potential application of antisense oligonucleotides in cancer chemotherapy. After receiving his B.Sc. in pure chemistry from Nottingham University, he pursued his interests in biological aspects of the subject in a conversion course in biochemistry at the Imperial College of Science and Technology, University of London, followed by a Ph.D. under Alan Paterson at the University of Alberta, Canada, where he concentrated upon biochemical mechanisms of action of the antineoplastic 6-thiopurines. Dr. Tidd undertook postdoctoral research first in Oxford and then at the University of Southern California Comprehensive Cancer Center, studying experimental cancer chemotherapy within an environment of clinical collaboration, in the laboratories of Roger Berry and Thomas Hall, respectively. Dr. Tidd's current research interest is in exploiting the recent gains in the understanding of malignancy at the molecular genetic level for the development of tumor-specific therapy based upon an antisense approach. He has published recently on this subject in *Anti-Cancer Drug Design*, the *British Journal of Cancer*, and *Anticancer Research*.

Antisense RNA and DNA: 241–253
© 1992 Wiley-Liss, Inc.

Antisense Oligonucleoside Methylphosphonates

Paul S. Miller

I. INTRODUCTION

A wide variety of oligonucleotide analogs have been synthesized, and organic chemists continue to devise methods to prepare new analogs. These analogs are interesting in their own right because studies on their conformation and their interactions with proteins and nucleic acids can give new insights into the factors that affect nucleic acid structure and function. The analogs also have properties that make them attractive candidates for use as antisense agents in both cell culture and animals.

Most of the analogs synthesized to date result from modification of the sugar phosphate backbone. From the standpoint of use as antisense reagents, these modifications have mainly focused on changes designed to enhance the uptake of the oligomers and to increase their resistance to hydrolysis by cellular nucleases. This has resulted in the synthesis of oligomers with modified phosphate linkages, which gives a nonionic sugar phosphate backbone. A number of these nonionic oligonucleotide analogs have been studied, including oligonucleotide alkylphosphotriesters, oligonucleoside alkylphosphoramidates, and oligonucleoside methylphosphonates.

In this review I focus on the antisense

oligonucleoside methylphosphonates. First the structure and general properties of the nonionic oligonucleotide analogs are described. This is followed by a brief description of the methods that have been used to synthesize the oligomers, with particular emphasis on the solid support method using phosphonamidite synthons.

Oligonucleoside methylphosphonates form hydrogen-bonded duplexes with complementary nucleic acids, a property that is crucial to their ability to function as antisense reagents. The physical properties of these analogs, the properties and stabilities of duplexes formed by these analogs, and the effect of chirality of the methylphosphonate linkage on duplex stability are discussed. Oligonucleoside methylphosphonates have also been derivatized with functional groups such as psoralen and ethylenediaminetetracetate. The syntheses and properties of these derivatized methylphosphonates and their interactions with complementary nucleic acids are described.

The biochemical properties of oligonucleoside methylphosphonates, including their interaction with proteins and polymerizing enzymes and their uptake by cells in culture, are discussed. That discussion is then followed by a description of the antisense activities of the oligomers in bacterial cells and in mammalian cells in culture. The information in this section is organized according to the specific type of target, initiation codon, coding region, and splice junction region, for the antisense oligomer. Finally, a brief description of preliminary pharmacokinetic studies, that have been carried out in mice is presented.

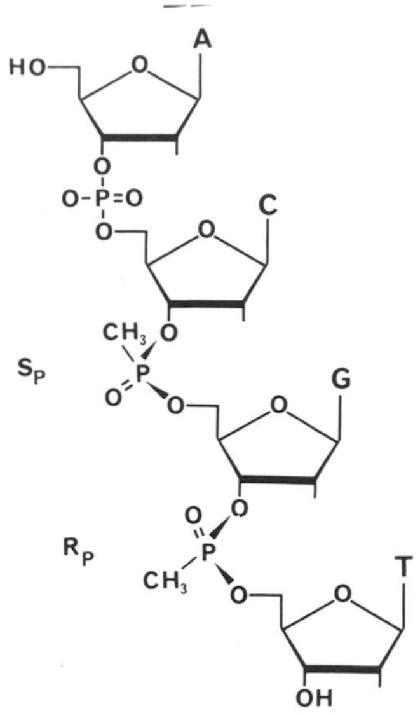

II. OLIGONUCLEOSIDE METHYLPHOSPHONATES

A. General Structure and Properties of Oligonucleoside Methylphosphonates

Oligodeoxyribonucleoside methylphosphonates contain a nonionic ($3'$-$5'$) linked internucleotide methylphosphonate bond in place of the naturally occurring phosphodiester internucleotide bond. The general structure of these analogs is shown in Figure 1. The electroneu-

Fig. 1. *General structure of oligo-2'-deoxyribonucleoside methylphosphonates. A normal phosphodiester bond (**top**) and the two chiral forms S_p and R_p of the internucleotide methylphosphonate bond (**bottom**) are shown.*

tral methylphosphonate group is isoteric with the phosphate group. Because the tetrahedral phosphorous of the methylphosphonate group is bonded to four different substituents, it is chiral and can exist in either an R_p or S_p form, with the result that each oligomer consists of a mixture of 2^n diastereoisomers, where n is the number of methylphosphonate linkages in the oligomer. The bases of the methylphosphonate oligomer are not modified, and therefore the methylphosphonate oligomers are capable of forming normal Watson Crick base pairs with complementary bases of single-stranded (ss) DNA and RNA. As described below, these analogs are particularly attractive as antisense reagents because of their nuclease resistance and their ability to be taken up intact by mammalian cells.

B. Synthesis of Oligonucleoside Methylphosphonates

Oligonucleoside methylphosphonates can be prepared by a variety of procedures. The syn-

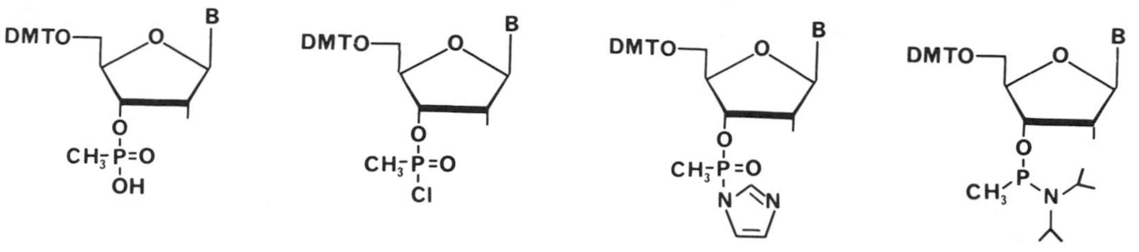

Fig. 2. *Synthons used in the chemical synthesis of oligonucleoside methylphosphonates. 1, 5'-O-dimethoxytrityldeoxynucleoside-3'-O-methylphosphonic acid; 2, 5'-O-dimethoxytrityldeoxynucleoside-3'-O-methyl-phosphonic chloride; 3, imidazolide; 4, 5'-O-dimethoxytrityldeoxynucleoside-3'-O-N, N-diisopropylaminomethylphosphonamidite.*

thetic intermediates, or synthons, that have been used are shown in Figure 2. The oligomers can be synthesized in solution by joining a 5'-*O*-dimethoxytrityldeoxynucleoside-3'-*O*-methylphosphonic acid (Fig. 2, No. 1) with a suitably protected nucleoside using coupling reagents such as dicyclohexylcarbodiimide, mesitylenesulfonylchloride, or mesitylenesulfonylnitrotriazole [Miller et al., 1979, 1983b; Jayaraman et al., 1981]. Oligomers containing up to 12 nucleoside units have been prepared with the latter coupling reagent [Miller et al., 1983b]. A variety of deoxynucleoside-3'-*O*-methylphosphonic acid derivatives have also been used as synthons to prepare oligonucleoside methylphosphonates both in solution and on insoluble polymer supports. These derivatives include 5'-*O*-dimethoxytrityldeoxynucleoside-3'-*O*-methylphosphonic chlorides (Fig. 2, No. 2) and imidazolides (Fig. 2, No. 3) [Agarwal and Riftina, 1979; Miller et al., 1983a, 1986].

The phosphorous atom in the synthons described above is in the five oxidation state. These synthons are thus analogous to those in the phosphotriester method, which has been widely used to prepare both oligodeoxyribo- and oligoribonucleotides. The rate and yield of the coupling reactions can be considerably increased by using synthons containing phosphorous derivatives in the three oxidation state. Thus 5'-*O*-dimethoxytrityldeoxynucleoside-

3'-*O*-N, N-diisopropylaminomethylphosphonamidite synthons (Fig. 2, No. 4) have been used with great success to prepare oligonucleoside methylphosphonates on controlled pore glass polymer supports [Sinha et al., 1983; Dorman et al., 1984; Jager and Engels, 1984; Stec et al., 1985; Agrawal and Goodchild, 1987; Zon, 1987]. The coupling reaction is carried out in the presence of tetrazole, which acts as an acid catalyst for the reaction. The reaction is complete within 3 minutes at room temperature and proceeds in 97% or greater yield. Following the coupling reaction, the resulting internucleotide methylphosphonate linkage is quantitatively oxidized to the methylphosphonate linkage by aqueous iodine treatment. This method has been used to prepare oligomers up to 21 nucleosides in length and is readily adapted to use in existing commercial DNA synthesizers with minimal changes in the synthetic programs. This method does, however, require the use of rather large excesses of monomer in order to drive the reactions to completion, and this factor could limit its utility for the preparation of large amounts of oligomers.

The methylphosphonate linkage is sensitive to hydrolysis under strong basic conditions. Thus the usual oligonucleotide deprotection procedure, which employs concentrated ammonium hydroxide to remove the oligomer from the support and to remove base protecting groups, cannot be used with methylphosphonate oligomers. Instead, the protecting groups are readily removed from these oligomers by treatment with hydrazine hydrate in acetic acid–pyridine buffer followed by treatment with ethylenediamine in 95% ethanol at room temperature [Miller et al., 1983b].

It is advantageous to synthesize methylphosphonate oligomers with the following backbone arrangement: d-NpNpNp . . . NpN, where p is a phosphodiester linkage and *p* is a methylphosphonate linkage. This backbone arrangement increases the solubility of the oligomers in aqueous solution and allows the oligomers to be purified readily by affinity chromatography on DEAE-cellulose anion exchange columns [Miller et al., 1983b, 1986] and by reversed phase HPLC. The 5'-terminal phosphodiester linkage also enables the oligomer to be phosphorylated by polynucleotide kinase. Thus it is possible to prepare 5'-[^{32}P]end-labeled oligomers. These oligomers can be readily characterized by polyacrylamide gel electrophoresis and by chemical sequencing procedures [Murakami et al., 1985].

C. Physical Properties of Oligonucleoside Methylphosphonates and Their Interactions With Complementary Nucleic Acids

The structure of the methylphosphonate group is very similar to that of the phosphodiester group as shown by X-ray crystallographic structure analysis of the S_p isomer of d-A*p*T [Chacko et al., 1983]. The conformations of the two diastereoisomers of dinucleoside methylphosphonates, d-N*p*N, have been studied in solution by circular dichroism and nuclear magnetic resonance spectroscopy [Miller et al., 1979; Kan et al., 1980]. The stacking interactions and the sugar conformation of these dimers are very similar to those of the corresponding dinucleoside monophosphates d-NpN. For dimers in which strong stacking interactions occur, it is possible to use the nuclear overhauser effect to assign the absolute configurations of the phosphonate methyl groups. In the case of d-A*p*A, the S_p diastereoisomer has a conformation that is virtually identical to that of d-ApA, whereas the R_p isomer shows

less base–base stacking interactions [Kan et al., 1980].

Oligodeoxyribonucleoside methylphosphonates form stable complexes with complementary nucleic acids. For example, d-A*p*A forms triple-stranded complexes with poly(U) and poly(dT), having a base stoichiometry of 2U(T):1A [Miller et al., 1979]. The melting temperature (T_m) of the d-A*p*A/poly(U) triplex, as determined by measuring the change in ultraviolet absorbance as a function of temperature, is dependent on the configuration of the methylphosphonate group. The T_m of the triplex containing the S_p isomer of d-A*p*A is 15.4°C, whereas that of the R_p isomer is 19.8°C. The difference in T_m was attributed to the necessity of overcoming the greater stacking interaction present in the S_p isomer prior to triplex formation.

The T_m of the R_p- and S_p-d-A*p*A/poly(U) triplexes are both higher than that of the d-ApA/poly(U) triplex, which is 7.0°C. The same trend is observed with longer oligo(A) methylphosphonates. Thus the T_m of the d-A*p*A*p*A/poly(U) triplex is 43°C, whereas that of the d-ApApA/poly(U) triplex is 32°C [Miller et al., 1981]. The increased T_m of the methylphosphonate oligomer triplexes relative to those of the oligodeoxyribonucleotide triplexes was attributed to the reduced charge repulsion between the nonionic methylphosphonate backbone and the phosphodiester backbone of the poly(U).

The position and configuration of a single methylphosphonate linkage affected the T_m of a self-complementary octadeoxyribonucleotide, d-(GGAATTCC)$_2$ [Bower et al., 1987]. Duplexes containing methylphosphonate groups with the R_p–R_p configuration generally had T_m that were similar to those of the unmodified duplex. These T_m were 5°–11°C higher than those of duplexes containing S_p–S_p methylphosphonate linkages. It appeared that the greatest effect of configuration on T_m occurred when the methylphosphonate linkage was positioned at the center of the oligomer. Proton NMR nuclear overhauser effect (NOE) measurements indicated that the S_p-methyl group interacts with the H-3′ on the adjacent ribose residue in the duplex. This steric interaction could contribute, in part, to the lower stability of the duplex.

The configuration of a single methylphosphonate linkage was also found to affect the ability of the self-complementary hexanucleotide d-(CG)$_3$ to undergo a B to Z conformational transition at high salt concentrations [Callahan et al., 1986]. An oligomer containing an R$_p$ methylphosphonate linkage as the first 5′-internucleotide bond underwent this transition to the same extent as the unmodified oligomer, as measured by circular dichroism spectroscopy. In contrast, the S$_p$ isomer did not undergo the B–Z transition. These results suggested that the methyl group of the S$_p$ isomer disrupted a water bridge between the N-2 amino group of the guanine and the oxygen of the 3′-phosphate group of deoxyguanylic acid, an interaction that is required for the maintenance of the Z conformation.

More dramatic effects of methylphosphonate configuration on T$_m$ were observed in complexes formed between oligothymidylates containing stereoregular alternating methylphosphonate–phosphodiester backbones and poly (dA) or poly(A) [Miller et al., 1980]. Two oligomers having the structure of d-Tp(TpTp)$_4$T, in which the configuration of each methylphosphonate linkage was either R$_p$ or S$_p$ throughout the backbone, were studied. The absolute configurations of the methylphosphonate linkages were not determined. One of the oligomers formed a duplex with poly(dA) having a T$_m$ of 32.4°C, whereas the other oligomer formed a triplex having a T$_m$ of 2°C. The reason for the difference in stoichiometry and the large difference in stabilities is not understood but may be due in part to the unusual conformations that are assumed by poly(dT)–poly (dA) duplexes.

Longer oligonucleoside methylphosphonates that contain only methylphosphonate backbones or backbones with a single phosphodiester group at the 5′-terminus of the oligomer form stable duplexes with complementary ssDNA oligonucleotide "targets" [Froehler et al., 1988; Sarin

et al., 1988; Lin et al., 1989]. The T$_m$ of these oligonucleoside methylphosphonate–target duplexes are similar to those of duplexes formed by the corresponding oligodeoxyribonucleotide and target DNA. The breadth of the optical transition curve is similar to that of oligonucleotide–target duplexes. This suggests that the various diastereomeric forms of the methylphosphonate oligomers each form duplexes having similar stabilities. The T$_m$ of methylphosphonate–target duplexes are essentially independent of the ionic strength of the medium, whereas the T$_m$ of oligodeoxyribonucleotide–target duplexes decrease with decreasing ionic strength [Miller et al., 1980; Quartin and Wetmur, 1989]. As in the case of the triple-stranded complexes described above, this effect is ascribed to the reduced intermolecular charge repulsion between the nonionic backbone of the methylphosphonate oligomer and the negatively charged phosphodiester backbone of the target.

D. Derivatives of Oligodeoxyribonucleoside Methylphosphonates

Derivatives of oligodeoxyribonucleoside methylphosphonates have been prepared with 4′-N-(aminoethyl)aminomethyl-4,5′,8-trimethylpsoralen ([ae]AMT) and 2-(aminoethyl)-3-carboxamidopsoralen ([ae]CP), and their interactions with ssDNA and mRNA have been studied [Lee et al., 1988a,b; Kean et al., 1988; Miller et al., 1988; Bhan and Miller, 1990]. As shown in Figure 3, the psoralen, which is a photoreactive crosslinking group, is linked to the 5′-end of the oligomer via a nuclease-resistant phosphoramidate bond. The derivatized oligomers are capable of forming normal hydrogen-bonded duplexes with complementary nucleic acids. When the oligomer is bound to its target, the psoralen is positioned such that it can crosslink with a pyrimidine residue that is located one nucleotide 3′ to the last base pair formed between the 5′-nucleotide of the methylphosphonate oligomer and complementary base of the target (n + 1 crosslinking position). Crosslinking is initiated by irradiation with 365 nm light, a wavelength that is not absorbed by the nucleic acid bases. The pre-

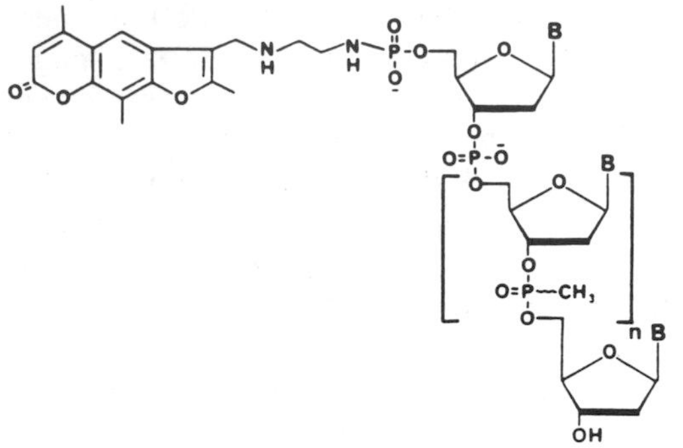

Fig. 3. *Oligonucleoside methylphosphonate derivatized with 4'-N-(aminoethyl)aminomethyl-4,5',8-trimethylpsoralen.*

ferred crosslinking site for (ae)AMT-derivatized oligomers appears to be a pyrmidine residue in the $n+1$ position, whereas oligomers derivatized with (ae)CP efficiently crosslink to pyrimidines located in the last base pair (n) position as well as to pyrimidines located in the $n+1$ position [Bhan and Miller, 1990].

Psoralen-derivatized methylphosphonate oligomers can crosslink to the extent of 95% or greater with ssDNA targets. The crosslinking reaction is dependent on the fidelity of base pairing between the oligomer and the target. Single base pair mismatches significantly reduce the extent of crosslinking, and extensive noncomplementarity eliminates crosslinking entirely, even at 0°C. The crosslinking reaction dramatically decreases at temperatures that correspond to the T_m of the oligomer–target duplex.

Psoralen-derivatized oligomers can also crosslink with complementary regions of globin mRNA [Kean et al., 1988; Miller et al., 1988]. Oligomers complementary to rabbit α- or β-globin mRNA crosslink specifically with their complementary mRNA when mixtures of the oligomer and mRNA are irradiated at 365 nm. The extent of crosslinking appears to be dependent on the secondary structure of the mRNA. Oligomers that are complementary to putative single-stranded regions in the mRNA crosslink more extensively than those oligomers whose binding sites are in stem regions of the mRNA.

In addition, the extent of crosslinking is approximately 15-fold greater when the psoralen is targeted to a U residue as opposed to a C residue in the mRNA.

Oligonucleoside methylphosphonates have been derivatized with other crosslinking functional groups such as moieties containing 2-chloroethylamino alkylating groups (such as the 4-[*N*-2-chloroethyl-*N*-methylamino]benzyl group). Like psoralen-derivatized oligomers, these oligomers crosslink with target nucleic acids in a site specific manner [Amirkhanov and Zarytova, 1989].

Methylphosphonate oligomers have also been derivatized with an Fe^{II}–ethylenediamine tetracetate complex [Lin et al., 1989]. This EDTA–Fe^{II} complex generates hydroxyl radicals in the presence of reducing agents such as dithiothreitol. The hydroxyl radicals cause degradation and strand scission of the sugar phosphate backbone of the target. When the EDTA–Fe^{II} derivatized oligomer binds to its target, the hydroxyl radicals generated in the vicinity of the 5' end of the oligomer cleave the target in the region of the 5' end of the oligomer binding site.

E. Biochemical Properties of Oligonucleoside Methylphosphonates

The methylphosphonate internucleotide bond is resistant to hydrolysis by purified exo- and endonucleases and by the nucleases found in bovine and human serum. Oligothymidylates having alternating methylphosphonate–phosphodiester linkages are slowly hydrolyzed at the phosphodiester bond by snake venom phosphodiesterase and by micrococcal nuclease [Miller et al., 1980]. However, the phosphodiester linkages of these oligomers are completely resistant to cleavage by spleen phosphodiesterase and S_1 endonuclease.

In addition to being resistant to nuclease degradation, methylphosphonate linkages in oligodeoxyribonucleotides can prevent the RNA strand of hybrids formed between the oligomer and RNA from being cleaved by ribonuclease H. Thus a duplex formed between RNA and an oligomer that contains only methylphosphonate linkages is resistant to hydrolysis by RNase H. However, when some of the methylphosphonate linkages are replaced by three or more contiguous diester linkages, the duplex formed between this oligomer and RNA is cleaved by the enzyme [Quartin et al., 1989; Furdon et al., 1989].

Neither isomer of the decathymidylates containing alternating methylphosphonate–phosphodiester linkages served as a primer for *Escherichia coli* DNA polymerase I or for calf thymus DNA polymerase [Miller et al., 1982]. It appears that the methylphosphonate linkage at the 3' end of the primer prevents the polymerase from productively interacting with the primer template complex. However, addition of two or more nucleotides containing phosphodiester linkages to the 3' end of the oligomer resulted in priming activity. This activity increased as the length of the phosphodiester "tail" increased. In contrast to their behavior with DNA polymerases, oligonucleoside methylphosphonates did serve as primers for avian myeloblastosis virus reverse transcriptase [Murakami et al., 1985; Furdon et al., 1989]. The efficiency of priming was approximately one-tenth that of a normal oligodeoxyribonucleotide primer of the same sequence.

It appears that the methylphosphonate linkage can affect the interaction of polymerizing enzymes and possibly other proteins that act upon nucleic acid substrates. This suggests that the methylphosphonate group may be used as a probe to monitor such interactions. This was directly demonstrated by studying the interaction of methylphosphonate modified synthetic *lac* operator DNA duplexes with *lac* repressor [Noble et al., 1984]. Introduction of single S_p or R_p methylphosphonate linkages near the middle of either strand of the 21-mer operator duplex had essentially no effect on binding of repressor as measured by a filter binding assay. When the methylphosphonate group was positioned three bases from the 5' end of the duplex, equilibrium dissociation constants for the two diastereomers were 100- and 10,000-fold less than that observed with the unmodified duplex. Thus it appears that the interaction of the repres-

sor with the operator duplex is extremely sensitive to the configuration of the methylphosphonate group, although the absolute configurations of these groups were not known.

Oligodeoxyribonucleoside methylphosphonates of the type d-NpNp . . . NpN or d-NpN-pNp . . . NpN are taken up intact by mammalian cells in culture. The kinetics of uptake by transformed Syrian hamster fibroblasts of oligomers with chain lengths from three to nine are essentially the same. After 2 hours, the oligomer concentration inside the cells appears to be approximately the same as the concentration of the oligomer in the cell culture medium, consistent with the entry of the oligomer into cells by diffusion [Miller et al., 1981]. Oligomers 18 and 21 nucleotides in length have been reported to be taken up in a linear manner respectively by CV-1 cells and daunorubicin-resistant K562/III cell cultures [Marcus-Sekura et al., 1987; Vasanthakumar and Ahmed, 1989] (see also Tidd, this volume).

In contrast to their behavior in mammalian cells, methylphosphonate oligomers longer than four nucleoside units are not taken up by *E. coli* [Jayaraman et al., 1981]. It appears that the bacterial cell wall prevents oligomers above a certain molecular weight from gaining access to the interior of the cell. However, bacterial spheroplasts and mutants of *E. coli* that do not have an intact cell wall readily take up the methylphosphonate oligomers.

III. ANTISENSE ACTIVITIES OF OLIGONUCLEOSIDE METHYLPHOSPHONATES

A. Studies With Bacteria and Cell Systems

Antisense oligonucleoside methylphosphonates can inhibit protein synthesis in both bacterial and mammalian cell-free systems and in cells in culture in a sequence-specific manner [see Miller, 1991, for review]. In addition, oligomers have been designed that are capable of selectively inhibiting virus protein synthesis, virus replication, and other viral functions in virus-infected cells. Oligonucleoside methylphosphonates have been targeted against the functional regions of bacterial ribosomal RNA (rRNA) and mammalian mRNA as well as against splice junction regions of precursor mRNA.

1. Oligomers complementary to bacterial rRNA. The heptamer d-[ApGpGp)$_2$T], which is complementary to the Shine-Dalgarno ribosomal docking sequence of *E. coli* 16S ribosomal RNA, inhibited translation of bacterial mRNA in cell-free extracts derived from *E. coli* [Jayaraman et al., 1981]. This oligomer prevents the ribosome from binding to bacterial mRNA. As predicted, the oligomer had no effect on translation of globin mRNA in a rabbit reticulocyte cell-free extract. The heptamer was also an effective inhibitor of protein synthesis and cell growth in a mutant of *E. coli* that contained an altered cell wall and that allowed the oligomer to be taken up by the cell. The oligomer had no inhibitory effect on mammalian cellular protein synthesis or growth.

2. Oligomers complementary to translated regions of mRNA. Translation of rabbit α- and β-globin mRNA in rabbit reticulocyte lysates is inhibited by methylphosphonate oligomers 8–12 nucleosides in length, which are complementary to the 5′ end, the initiation codon, and the coding regions of the mRNA [Blake et al., 1985]. Inhibition ranged from 7% to 76% at an oligomer concentration of 100 μM. Both oligomer chain length and the presence of mRNA secondary structure at the oligomer binding site appeared to affect the extent of inhibition. Thus some of the oligomers were more effective inhibitors after they were preannealed with the mRNA.

These antisense methylphosphonate oligomers also inhibited globin synthesis in rabbit reticulocytes. Interestingly those oligomers that were effective inhibitors of protein synthesis in vitro and did not require preannealing were also effective in the reticulocytes, whereas those oligomers that required preannealing in vitro did not exhibit inhibitory effects in the reticulocytes. This observation suggested that the secondary structure of the globin mRNA was the same in vitro and in the cells.

Translation of globin mRNA in a rabbit reticulocyte lysate is specifically inhibited by psoralen-derivatized methylphosphonate oligomers (12-mers) that specifically crosslink to either α- or β-globin mRNA [Kean et al., 1988]. An α-specific oligomer inhibited translation of α-globin mRNA by 43%, whereas a β-specific 12-mer inhibited translation of β-globin mRNA 67% at a concentration of 5 μM. This represents a 20–30-fold increase in the efficiency of inhibition by the psoralen-derivatized methylphosphonate oligomers compared with the underivatized oligomers of similar sequence that inhibit in the concentration range 100–150 μM. A noncomplementary oligomer that did not crosslink with either mRNA did not inhibit translation. This result shows that irradiation of the mRNA in the presence of the psoralen-derivatized oligomers does not lead to nonspecific photoinduced damage of the mRNA.

Nonanucleoside (9-mer) methylphosphonates targeted against the initiation codon regions of vesicular stomatitis virus (VSV) N, NS, and G protein mRNA inhibited cell-free translation of VSV mRNA [Agris et al., 1986]. For example, d-ApApCpApGpApCpApT, which is complementary to N protein mRNA, specifically inhibited N protein synthesis by 36% in a rabbit reticulocyte lysate at an oligomer concentration of 100 μM. The VSV antisense oligomers also inhibited VSV protein synthesis in virus-infected mouse L929 cells. Syntheses of all five viral proteins were inhibited 42%–99% by the oligomers at a concentration of 150 μM [Agris et al., 1986]. As a consequence of their inhibitory effects on protein synthesis, treatment of VSV-infected cells with the oligomers also resulted in 1.0–1.5 log reductions of virus titers. Although the oligomers inhibited synthesis of all five VSV proteins, they had no measurable effect on protein synthesis in uninfected mouse L929 cells. Furthermore, the oligomers did not inhibit the rate of growth or colony forming ability of L929 cells. These results show that methylphosphonate oligomers as short as octamers can be used to inhibit selectively viral nucleic acid expression in cell culture.

Methylphosphonate oligomers of varying chain lengths complementary to the initiation codon region of chloramphenicol acetyltransferase mRNA were found to inhibit translation in transformed cells [Marcus-Sekura et al., 1987]. Oligomers 9, 12, 15, and 21 nucleotides in length were found to give 8%–65% inhibition at a concentration of 30 μM. The greatest inhibition was achieved by the 12-mer, 60%, and the 15-mer, 65%, respectively, whereas the 21-mer gave only 40% inhibition. This result suggested that the 21-mer either was not taken up as well by the cells as was the 15-mer or that the mRNA binding site of the 21-mer was not completely single stranded.

A dodecanucleoside methylphosphonate targeted against the first 11 nucleotides beginning at the initiation codon of Balb-*ras*-oncogene mRNA, almost completely inhibited synthesis of p21 protein encoded by the mRNA in a rabbit reticulocyte system at an oligomer concentration of 100 μM [Blake et al., 1985]. This oligomer, at a concentration of 50 μM, also inhibited Ha-*ras* p21 synthesis by 90% in *ras*-transformed NIH3T3 cells. Oligomers complementary to this region but containing one or two base pair mismatches were less effective as inhibitors in a cell-free translation system.

Octanucleoside methylphosphonates, complementary to the regions of human c-Ha-*ras* (Ras-O) or activated c-Ha-*ras*-T24 mRNA (Ras-1) which encode the twelfth amino acid, were tested as inhibitors of p21 synthesis in a rabbit reticulocyte translation system [Brown et al., 1989; Yu et al., 1989]. The c-Ha-*ras*-T24 mRNA contains a single G to U transition, which results from a point mutation in the twelfth codon of the mRNA. The Ras-O oligomer inhibited translation of c-Ha-*ras* mRNA approximately 50% at a concentration of 100 μM, whereas 200 μM Ras-I was required to achieve the same level of inhibition. In contrast, 50 μM Ras-I inhibited translation of c-Ha-*ras*-T24 32%, whereas 50 μM Ras-O had no inhibitory effect. These results suggest that it may be possible to use oligonucleoside methylphosphonates to study the role of normal and mutated oncogenes in the initiation and maintenance of cell transformation.

Recent results have shown that an oligonucleoside methylphosphonate 15 nucleotides in length whose sequence is complementary to the initiation codon region of a drug resistance gene (*mdr1*) mRNA inhibits synthesis of P-glycoprotein in K562/III cells in culture [Vasanthakumar and Ahmed, 1989]. Under the conditions of the experiment, 30 μM oligomer completely inhibited synthesis of the protein. As a consequence of this inhibition, the K562/III cells were 85–119-fold more sensitive to the cytotoxic effects of daunorubicin after the cells had been treated with the oligomer for 72 hours.

As mentioned in Section II, hybrids between RNA and methylphosphonate oligomers are not a substrate for RNaseH, the enzyme that is believed to be responsible for the antisense activity of oligodeoxyribonucleotides and oligonucleotide phosphorothioates directed to downstream portions of mRNAs (see Minshull and Hunt, this volume). However, as reviewed by Toulmé (this volume), RNase-H–independent inhibition by oligomers directed at the 5′ portion of an mRNA has been observed. It seems most likely therefore that the methylphosphonate oligomers inhibit translation by physically blocking the ribosome from interacting with the mRNA or by preventing its progress along the mRNA. The much greater effectiveness of crosslinking oligomers would then be explained, since in such cases the "unwinding activity" that appears to be associated with translating 80S ribosomes (see Minshull and Hunt, this volume, for discussion) would not be able to unwind the crosslinked hybrid.

3. Oligomers complementary to splice junction regions of precursor mRNA. Oligonucleoside methylphosphonates complementary to the splice sites of virus precursor mRNAs are effective inhibitors of viral function and replication in cultured cells. For example, a nonanucleoside methylphosphonate, d-ApA*p*T*p*A*p*C*p*C*p*T*p*C-*p*A, which is complementary to the donor splice site of SV40 large T-antigen precursor mRNA, inhibited large T-antigen synthesis in SV40-infected African green monkey kidney cells [Miller et al., 1985]. The extent of inhibition was 19% and the oligomer had no detectable inhibitory effect on cellular protein synthesis.

An octanucleoside methylphosphonate, d-TpCpCpTpCpCpTpG, which is complementary to the acceptor splice site of herpes simplex virus type-1 (HSV-1) immediate early mRNA 4 and 5, specifically inhibited HSV-1 protein synthesis by 90% and DNA synthesis by 70%–75% in infected Vero cells when added to the culture medium at a concentration of 250 μM [Smith et al., 1986]. The oligomer (100 μM) also inhibited HSV-1 growth approximately 85% and growth of HSV-2 approximately 21%. The sequences of the precursor mRNA 4 and 5 splice acceptor sites of HSV-1 and HSV-2 are similar but not identical, and this may account for the low level of inhibition of HSV-2 by the oligomer. The oligomer had no inhibitory effect on uninfected Vero cell protein synthesis, DNA synthesis, cell growth, or colony-forming ability.

An increase in the extent and specificity of inhibition was observed when the longer (12-mer) methylphosphonate d-TpTpCpCpTpCp-CpTpGpCpGpG was used, which is also complementary to the splice site of HSV-1 immediate early mRNA 4 and 5 [Kulka et al., 1989]. This oligomer inhibited HSV-1 replication 98% at 100 μM but had no inhibitory effect on HSV-2 replication. The oligomer was also shown to inhibit the formation of spliced immediate early mRNA 4. This result suggests that the oligomer is able to enter the nucleus and interact with its targeted splice site on the virus precursor mRNA.

A psoralen derivative of the dodecamer d-(ae)AMTpTpTpCpCpTpCpCpTpGpCp-GpG inhibited virus growth 90%–98% when cells that were treated with 5 μM oligomer were irradiated 1–3 hours after infection with the virus. Decreasing levels of inhibition from (80% to 15%) were observed when irradiation was carried out 6–12 hours postinfection. The dependence of the inhibitory effect upon the time of irradiation during the virus replication cycle is consistent with the temporal role of the immediate early proteins in activating early and late viral genes. The results suggest that psoralen-

derivatized methylphosphonates could be used to probe gene expression as a function of the viral replication or cell cycle.

Oligonucleoside methylphosphonates targeted to the acceptor splice site of the tat-3 gene of HIV have inhibitory effects on virus expression in cultured cells. An octanucleoside methylphosphonate inhibited both syncytial cell formation and reverse transcriptase activity 94% and 92%, respectively, at 100 μM oligomer concentration, when the oligomer was added 1 hour prior to infection [Zaia et al., 1988]. The antisense octamer also inhibited HIV RNA synthesis approximately 98% 7 days after virus infection, and significant inhibition could be observed for up to 9 days after treatment with oligomer. Only moderate inhibition, 12% and 23%, was produced by the sense oligomer, and no inhibition was observed with the HSV-1–specific oligomer d-TpCpCpTpCpCpTpG. The sense oligomer inhibited RNA synthesis 88%, whereas the HSV-1–specific oligomer had no inhibitory effect. None of the oligomers appeared to be toxic to noninfected cells, as determined by trypan blue dye exclusion and incorporation of [^{3}H]thymidine.

A 20-mer complementary to the splice acceptor site of the tat-3 gene was found to inhibit syncytia formation and p24 synthesis 98% and 100%, respectively, when added to the cell culture medium at a concentration of 3 μM at the time of infection [Sarin et al., 1988]. The corresponding oligodeoxyribonucleotide gave only 33% and 39% inhibition at the same concentration, whereas a shorter (15-mer) methylphosphonate oligomer complementary to the same site gave only 22% and 28% inhibition.

B. Studies in Animals

Little is known about the effect of oligonucleoside methylphosphonates in animals. The distribution and fate of the tritium-labeled oligomer d-Tp[^{3}H]TpCpCpTpCpCpTpGpCpGpG in mice has been studied [Chen et al., 1990]. When injected into the tail vein, the oligomer was found in most of the organs with the exception of the brain. The highest levels were found in the kidney, lung, and liver. The oligomer was rapidly cleared from the mouse, with most of the label being excreted in the urine. The half-life for distribution of the oligomer was 6 minutes, whereas the half-life for its elimination was 17 minutes. Radioactivity recovered from the urine and from various tissues was examined by reversed phase high-pressure liquid chromatography. Two products were found, intact oligomer and d-[^{3}H]TpCpCpTpCpCpT-pGpCpGpG, which arose as a result of cleavage of the terminal phosphodiester bond. No cleavage of the methylphosphonate linkages was observed in this assay.

IV. CONCLUSION

The studies described above demonstrate that oligonucleoside methylphosphonates are effective and specific inhibitors of mRNA function and expression in mammalian cells. Thus oligonucleoside methylphosphonates will be useful probes for studying the expression of cellular and viral genes in cell culture. Advances in developing derivatized oligomers, such as the psoralen-derivatized methylphosphonate oligomers, will lead to even greater specificity and will provide tools for unique mechanistic studies on gene expression.

The nuclease resistance and sequence-specific antisense activity of these oligomers suggest that they and/or their derivatives could be developed for use as highly selective antiviral and chemotherapeutic agents. For this goal to be realized, it will be necessary to develop synthetic methods capable of producing large quantities of these compounds. This will enable extensive testing in cell culture and in animals and will lead to a better understanding of the pharmacology and mechanisms of action of these novel nucleic acid analogs.

V. REFERENCES

Agarwal KL, Riftina F (1979): Synthesis and enzymatic properties of deoxyribooligonucleotides containing methyl and phenylphosphonate linkages. Nucleic Acids Res 6:3009–3024.

Agrawal S, Goodchild J (1987): Oligodeoxynucleoside methylphosphonates: Synthesis and enzymic degradation. Tetrahedron Lett 28:3539–3542.

Agris CH, Blake KR, Miller PS, Reddy MP, Ts'o POP (1986): Inhibition of vesicular stomatitis virus protein synthesis and infection by sequence-specific oligodeoxyribonucleoside methylphosphonates. Biochemistry 25:6268–6275.

Amirkhanov N, Zarytova F (1989): Reactive oligonucleotides bearing methylphosphonate groups. III. Affinity modification of nucleic acid target by 4-(N-2-chloroethyl-N-methylamino)benzyl-3'- and 5'-phosphoamide derivatives having stereoregular methylphosphonate residues. Bioorg Khim 15:379–385.

Bhan P, Miller PS (1990): Photo-cross-linking of psoralen derivatized oligonucleoside methylphosphonates to single stranded DNA. Bioconjugate Chem 1:82–88.

Blake KR, Murakami A, Spitz SA, Glave SA, Reddy MP, Ts'o POP, Miller PS (1985): Hybridization arrest of globin synthesis in a rabbit reticulocyte lysate and in rabbit reticulocytes by oligodeoxyribonucleoside methylphosphonates. Biochemistry 24:6139–6145.

Bower M, Summers MF, Powell C, Shinozuka K, Regan JB, Zon G, Wilson WD (1987): Oligodeoxyribonucleoside methylphosphonates. NMR and UV spectroscopic studies of R_p–R_p and S_p–S_p methylphosphonate modified duplexes of [d(GGAATTCC)]$_2$. Nucleic Acids Res 15:4915–4930.

Brown D, Yu Z, Miller P, Blake K, Wei C, Kung H-F, Black R, Ts'o P, Chang E (1989): Modulation of *ras* expression by anti-sense, nonionic deoxyoligonucleotide analogs. Oncogene Res 4:243–252.

Callahan L, Han F-S, Watt W, Duchamp D, Kezdy FJ, Agarwal K (1986): B- to Z-DNA transition probed by oligonucleotides containing methylphosphonates. Proc Natl Acad Sci USA 83:1617–1621.

Chacko KK, Lindner K, Saenger W, Miller PS (1983): Molecular structure of deoxyadenylyl-3'-methylphosphonate-5'-thymidine dihydrate, (d-ApT·2H$_2$O), a dinucleoside monophosphate with neutral phosphodiester backbone. An X-ray crystal study. Nucleic Acids Res 11:2801–2814.

Chen T-K, Miller PS, Ts'o POP, Colvin OM (1990): Disposition and metabolism of oligodeoxynucleoside methylphosphonate following a single IV injection in mice. Drug Metab Dispos 18:815–818.

Dorman MA, Noble SA, McBride MJ, Caruthers MH (1984): Synthesis of oligonucleotides and oligodeoxynucleotide analogs using phosphoramidite intermediates. Tetrahedron 49:95–102.

Froehler B, Ng P, Matteucci M: Phosphoramidate analogues of DNA (1988): Synthesis and thermal stability of heteroduplexes. Nucleic Acids Res 16:4831–4839.

Furdon P, Dominski Z, Kole R (1989): RNase H cleavage of RNA hybridized to oligonucleotides containing methylphosphonate, phosphorothioate and phosphodiester bonds. Nucleic Acids Res 17:9193–9204.

Jager A, Engels J (1984): Synthesis of deoxynucleoside methylphosphonates via a phosphonamidite approach. Tetrahedron Lett 25:1437–1440.

Jayaraman K, McParland K, Miller P, Ts'o POP (1981): Selective inhibition of *Escherichia coli* protein synthesis and growth by nonionic oligonucleotides complementary to the 3′ end of 16s rRNA. Proc Natl Acad Sci USA 78:1537–1542.

Kan LS, Cheng DM, Miller PS, Yano J, Ts'o POP (1980): Proton nuclear magnetic resonance studies on dideoxyribonucleoside methylphosphonates. Biochemistry 19:2122–2132.

Kean JM, Murakami A, Blake KR, Cushman CD, Miller PS (1988): Photochemical cross-linking of psoralen-derivatized oligonucleoside methylphosphonates to rabbit globin messenger RNA. Biochemistry 27:9113–9121.

Kulka M, Smith C, Aurelian L, Fishelevich R, Meade K, Miller P, Ts'o P (1989): Site specificity of the inhibitory effects of oligo(nucleoside methylphosphonates) complementary to the acceptor splice junction of herpes simplex virus type 1 immediate early mRNA 4. Proc Natl Acad Sci USA 86:6868–6872.

Lee BL, Blake KR, Miller PS (1988a): Interaction of psoralen-derivatized oligodeoxyribonucleoside methylphosphonates with synthetic DNA containing a promoter for T7 RNA polymerase. Nucleic Acids Res 16:10681–10697.

Lin S-B, Blake KR, Miller PS, Ts'o POP (1989): Use of EDTA derivatization to characterize interactions between oligodeoxyribonucleoside methylphosphonates and nucleic acids. Biochemistry 28:1054–1061.

Lee BL, Murakami A, Blake KR, Lin SB, Miller PS (1988b): Interaction of psoralen-derivatized oligodeoxyribonucleoside methylphosphonates with single-stranded DNA. Biochemistry 27:3197–3203.

Marcus-Sekura CJ, Woerner AM, Shinozuka K, Zon G, Quinnan QV, Jr (1987): Comparative inhibition of chloramphenicol acetyltransferase gene expression by antisense oligonucleotide analogues having alkyl phosphotriester, methylphosphonate and phosphorothioate linkages. Nucleic Acids Res 15:5749–5763.

Miller PS (1991): Oligonucleoside methylphosphonates as antisense reagents. BioTechnology 9:358–362.

Miller PS, Agris CH, Aurelian L, Blake KR, Murakami A, Reddy MP, Spitz SA, Ts'o POP (1985): Control of ribonucleic acid function by oligonucleoside methylphosphonates. Biochimie 67:769–776.

Miller PS, Agris CH, Blandin M, Murakami A, Reddy MP, Spitz SA, Ts'o POP (1983a): Use of methylphosphonic dichloride for the synthesis of oligonucleoside methylphosphonates. Nucleic Acids Res 11:5189–5204.

Miller PS, Agris CH, Murakami A, Reddy MP, Spitz SA, Ts'o POP (1983b): Preparation of oligodeoxyribonucleoside methylphosphonates on a polystyrene support. Nucleic Acids Res 11:6225–6242.

Miller PS, Annan ND, McParland KB, Pulford SM (1982): Oligothymidylate analogues having stereoregular, alternating methylphosphonate/phosphodiester backbones as primers for DNA polymerase. Biochemistry 21:2507–2512.

Miller PS, Blake KR, Cushman CD, Kean JM, Lee BL, Lin S-B, Murakami A (1988): Antisense oligonucleoside methylphosphonates: Interaction of psoralen-derivatized oligomers with RNA and DNA. Nucleic Acids Res Symp Ser 20:113–114.

Miller PS, Dreon N, Pulford SM, McParland KB (1980): Oligothymidylate analogues having stereoregular, alternating methylphosphonate/phosphodiester backbones. Synthesis and physical studies. J Biol Chem 235:9659–9665.

Miller PS, McParland KB, Jayaraman K, Ts'o POP (1981): Biochemical and biological effects of nonionic nucleic acid methylphosphonates. Biochemistry 20:1874–1880.

Miller PS, Reddy MP, Murakami A, Blake KR, Lin S-B, Agris CH (1986): Solid-phase syntheses of oligodeoxyribonucleoside methylphosphonates. Biochemistry 25:5092–5097.

Miller PS, Yano J, Yano E, Carroll C, Jayaraman K, Ts'o POP (1979): Nonionic nucleic acid analogues: Synthesis and characterization of dideoxyribonucleoside methylphosphonates. Biochemistry 18:5134–5142.

Murakami A, Blake KR, Miller PS (1985): Characterization of sequence-specific oligodeoxyribonucleoside methylphosphonates and their interaction with rabbit globin mRNA. Biochemistry 24:4041–4046.

Noble SA, Fisher EF, Caruthers MH (1984): Methylphosphonates as probes of protein–nucleic acid interactions. Nucleic Acids Res 12:3387–3404.

Quartin R, Brakel C, Wetmur J (1989): Number and distribution of methylphosphonate linkages in oligodeoxynucleotides affect exo- and endonuclease sensitivity and ability to form RNase H substrates. Nucleic Acids Res 17:7253–7262.

Quartin RS, Wetmur JG (1989): Effect of ionic strength on the hybridization of oligodeoxynucleotides with reduced charge due to methylphosphonate linkages to unmodified oligodeoxynucleotides containing the complementary sequence. Biochemistry 28:1040–1047.

Sarin PS, Agrawal S, Civeira MP, Goodchild J, Ikeuchi T, Zamecnik TC (1988): Inhibition of acquired immunodeficiency syndrome virus by oligodeoxynucleoside methylphosphonates. Proc Natl Acad Sci USA 85:7448–7451.

Sinha ND, Grossbuchhaus V, Koster H (1983): A new synthesis of oligodeoxynucleoside methylphosphonates on controlled pore glass support using phosphite approach. Tetrahedron Lett 24:877–880.

Smith CC, Aurelian L, Reddy MP, Miller PS, Ts'o POP (1986): Antiviral effect of an oligo(nucleoside methylphosphonate) complementary to the splice junction of herpes simplex virus type 1 immediate early premRNAs 4 and 5. Proc Natl Acad Sci USA 83:2787–2791.

Stec WJ, Zon G, Egan W, Byrd RA, Phillips LR, Gallo KA (1985): Solid-phase synthesis, separation, and stereochemical aspects of *p*-chiral methane- and 4,4′-dimethoxytriphenylmethanephosphonate analogues of oligodeoxyribonucleotides. J Org Chem 50:3908–3913.

Vasanthakumar G, Ahmed NK (1989): Modulation of drug resistance in a daunorubicin resistant subline with oligonucleoside methylphosphonates. Cancer Comm 1:225–232.

Yu Z, Chen D, Black RJ, Blake K, Ts'o POP, Miller P, Chang E (1989): Sequence specific inhibition of in vitro translation of mutated or normal *ras* p21. J Exp Pathol 4:97–108.

Zaia JA, Rossi JJ, Murakawa GJ, Spallone PA, Stephens DA, Kaplan RE, Eritja R, Wallace RB, Cantin EM (1988): Inhibition of human immunodeficiency virus by using an oligonucleoside methylphosphonate targeted to the TAT-3 gene. J Virol 62:3914–3917.

Zon G (1987): Synthesis of backbone-modified DNA analogues for biological applications. J Protein Chem 6:131–145.

ABOUT THE AUTHOR

PAUL S. MILLER is Professor of Biochemistry in the School of Hygiene and Public Health at the Johns Hopkins University, where he teaches bioorganic chemistry. He received his B.S. in chemistry from the State University of New York at Buffalo in 1965 and his Ph.D. in chemistry under the direction of Dr. Robert Letsinger at Northwestern University in 1969. While at Northwestern, he carried out studies on the chemical synthesis of oligonucleotides. Dr. Miller then carried out postdoctoral research on the physical properties of oligonucleotide analogs at the Johns Hopkins University in collaboration with Dr. Paul O.P. Ts'o. Dr. Miller's current research involves the synthesis of antisense oligonucleotide analogs and studies on their physical and biochemical properties. This research has been described in a number of journals including *Biochemistry*, *Nucleic Acids Research*, and *Proceedings of the National Academy of Sciences USA*. Dr. Miller also serves as associate editor of *Bioconjugate Chemistry*.

Antisense RNA and DNA: 255–265
© 1992 Wiley-Liss, Inc.

Transmembrane Passage and Cell Targeting of Antiviral Synthetic Oligonucleotides

**Geneviève Degols, Patrick Machy, Jean-Paul Leonetti,
Lee Leserman, and Bernard Lebleu**

I. INTRODUCTION

Gene expression involves specific interactions between nucleic acids or between nucleic acids and proteins; single- or double-stranded nucleic acid fragments of appropriate sequence and length should therefore allow interference with the expression of a given gene. This so-called antisense approach has received increasing attention during the past few years from both academic and industrial research groups. Although any gene could in principle be used as a target, particular emphasis has been devoted to the control of oncogenes and viral genes, since such approaches might ultimately provide a novel form of highly specific chemotherapy.

Unfortunately many problems have yet to be solved; these include the scale-up of the synthesis of oligonucleotides by cost-effective methodologies, the metabolic stability of oligonucleotides in biological fluids and in cells, and the transmembrane passage of oligonucleotides and their distribution within various cell compartments. In addition, a better knowledge of their intracellular targets and of the rules involved in efficient oligonucleotide target recognition is necessary (even if base recognition through the well-known Watson Crick rules is the basis). Appreciable progress in these various points has already been made since the pioneering work of Zamecknik, Ts'O, Vlassov and their colleagues in the early 1980s [van der Krol et al., 1988] (see also Toulmé, this volume). Our own interest has mainly been geared toward the development of tools that allow protection of oligonucleotides against serum nuclease degradation, toward an efficient transfer across the plasma membrane, and toward the possible targeting of oligonucleotides to cells (or tissues) expressing specific determinants at their surface.

Instead of being administered in cell culture in their free form, oligonucleotides of the appropriate specificity have been covalently linked to poly(L-lysine) [Lemaitre et al., 1987] (with or without sulfated polyanions) or encapsulated in antibody-targeted liposomes [Leonetti et al., 1990]. An antiviral activity specific for a given virus (vesicular stomatitis virus [VSV] in our experiments) or a cell growth inhibitory activity can be conferred on the same cell line (L929 murine fibroblasts for instance), depending on the oligonucleotide sequence. A review of our experimental approaches and some recent data are presented.

II. RESULTS

A. Oligomer Choice and Experimental Design Rationale

Murine L929 fibroblasts growing as monolayers were used throughout, since either an antiviral activity in VSV-infected confluent cells or an antiproliferative activity in uninfected exponentially growing cells could be generated using the same experimental system. Moreover, growth of these cells in serum-supplemented or completely synthetic (serum replaced by Opti-MEM) media allows evaluation of the influence of serum components (and in particular of nucleases) on the biological activity of free or vector-associated oligonucleotides.

Natural or 3'-modified oligonucleotides have been used in order to minimize problems potentially associated with alternative chemistries. Several interesting modifications of oligonucleotides have been described either in the phosphodiester [Stec et al., 1984] (e.g., methylphosphonates [see Miller, this volume, for review], phosphorothioates) or the sugar configuration [Morvan et al., 1986] (e.g., α-anomers). These modifications increase the resistance of oligonucleotides to nuclease degradation as desired; however, they are not devoid of adverse effects. For example, phosphorothioates hybridize to their target RNAs with somewhat lower melting temperatures (T_m) [Stein

et al., 1988]; and hybrids generated from an interaction of methylphosphonates or α-anomeric oligonucleotides with their RNA targets are not recognized by RNAse H, a potentially important factor in antisense oligonucleotide action [Maher and Dolnick., 1988; Gagnor et al., 1989]. In addition, toxic or mutagenic properties of such oligonucleotides are conceivable, and this possibility needs to be carefully evaluated to avoid adverse effects in vivo.

The choice of 15-mer oligonucleotides was dictated by the following considerations: First, a 12–14-mer sequence has been calculated as being represented only once statistically in the total mRNA of higher eukaryotes. Second, it is not advantageous to increase the oligonucleotide size much above this value, since this increases the cost as well as the tolerance to mismatches. Oligonucleotides overlapping the AUG initiation codon of the c-*myc* protooncogene mRNA or of the VSV N (nucleocapsid) protein mRNA were synthesized in the phosphodiester series with a 3'-ribosylated adenosine end (Fig. 1); this latter modification was strictly required only for poly(L-lysine) conjugation but was used throughout for the sake of homogeneity. Oligonucleotides bearing several mismatches or random oligonucleotides with the same base composition were generally used as controls; moreover, as might be expected, the anti c-*myc* oligonucleotide turned out to be an excellent negative control for antiviral experiments and anti-VSV oligonucleotides for the antiproliferative ones. The oligonucleotide sequences used, their target sequences around the oligonucleotide-binding site, and oligonucleotide nomenclature are described in Table I.

B. Biological Activity of Poly(L-Lysine)-Conjugated Oligonucleotides as Compared With Nonconjugated Oligonucleotides

Nonconjugated oligonucleotides with the appropriate sequence exhibit a specific biological activity when incubated with cells growing continuously in (our experiments), or transiently transferred to, a culture medium partially or totally devoid of serum nucleases (this is generally achieved by omitting serum at the time

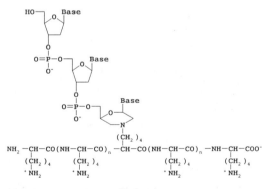

of experimentation, by heat decomplementation of the serum, or by prior screening of serum lots for low nuclease content). Although not always clarified in some papers, this point is critical when using natural oligonucleotides, as illustrated in Figure 2A in our antiviral model, from which it is clear that serum free medium is needed to see an effect. In contrast, covalent conjugation with poly(L-lysine) (see Fig. 1 for a general outline of the procedure), but not the simple mixing of poly(L-lysine) and 3'-modified oligonucleotides, makes oligonucleotides biologically active without any manipulation of the culture medium. This conjugation does not lead to any loss of specificity, as demonstrated in the few controls shown here and in our previously published material.

Aside from these considerations, poly(L-lysine)-conjugated oligonucleotides are 10–50 times more active than unconjugated ones, as illustrated under strictly comparable conditions in Figure 2A,B (see the abcissae scales): Similar general conclusion can be drawn when comparing our own data to those in the published literature [Stevenson and Iversen, 1989], although differences in target, experimental design, and so forth, render the comparative evaluation more difficult.

The mechanism(s) through which poly(L-lysine) conjugation potentiates the biological activity of antisense oligonucleotides is not completely understood. Poly(L-lysine) serves as a transmembrane carrier of conjugated small molecular weight material, as demonstrated by Ryser and Shen [1978] for methotrexate and by ourselves for antisense oligonucleotides [Lemaitre et al., 1987; Stevenson and Iversen, 1989]; the conjugate is taken up by a nonspecific receptor-mediated endocytic pathway, and the conjugated material is eventually released somewhere in a cell acidic compartment through

Fig. 1. *Outline of antisense oligonucleotide–poly(L-lysine) conjugate synthesis. 15-mer oligonucleotides of the appropriate specificity were synthesized on a Biosearch Cyclone automatic DNA synthesizer using phosphoramidite chemistry: 3'-ribosylated oligodeoxyribonucleotides were synthesized by the same procedure except for the use of a ribosylated CPG support, as described previously [Leonetti et al., 1988]. Oligonucleotides were purified by conventional HPLC chromatographic procedures. Ribosylated oligodeoxy-*

ribonucleotides were covalently coupled to lysine ε-amino groups (in a molar ratio of 0.5–1.0 per polypeptide chain) of 14,000 MW poly(L-lysine) as described previously. In some cases conjugates were supplemented with various polyanions to form ternary complexes at the concentrations indicated.

TABLE I. Partial Nucleotide Sequences of VSV N Protein mRNA and Human c-*myc* mRNA and Location of the Complementary Synthetic Oligos

VSV N protein mRNA translation initiation site:
m^{7}GpppACAGUAAUCAAA<u>AUG</u>UCUGU– – – – – –3′

Oligo VSV: TGTCATTAGTTTTAC– – – – –5′

Human c-*myc* mRNA translation-initiation site:
5′ – – – CCGCGACG<u>A</u><u>UG</u>CCCCUCAACGUUAG– – – – –3′

Oligo c-*myc*: CTACGGGGAGTTGCA– – – – –5′

AUG codons used for translation initiation in N protein mRNA and c-*myc* mRNA are underlined.

proteolytic degradation of the polypeptide moiety. Experiments involving inhibitors of the endocytic pathway, such as poly(D-lysine) conjugates or fluorochrome-linked oligonucleotides, are in line with this scheme; how and where the oligonucleotide escapes from the endocytic compartment to reach its cytoplasmic (and/or nuclear) target is a matter of conjecture. In addition, modifications of the 3′ end of oligonucleotides render them more resistant to phosphodiesterase degradation, which is generally recognized as a major factor of their catabolism by serum nucleases [Verspieren et al., 1987]. Other factors may be involved, including steric hindrance by the poly(L-lysine) moiety that may inhibit various nucleases, and a contribution of positively charged lysine residues to hybridization of the oligonucleotide to its target, if any such residues remain associated with the oligonucleotides after passage through the endocytic compartment.

C. Polyanions Potentiate the Biological Activity of Antisense Oligonucleotides

Complexing of poly(L-lysine) with polyanions such as heparin has been shown to reduce its cytotoxicity at high concentration in in vitro experiments [Morgan et al., 1988]. Sulfated polyanions are now also being extensively explored as potential antiviral agents active against enveloped viruses such as herpes virus (HSV-1, HSV-2), AIDS-associated viruses (HIV-1, HIV-2) [Ito et al., 1987], or VSV [Gonzales et al., 1987]. We took advantage of the low antiviral activity of polyanions on VSV in L929 cells. This allows an evaluation of the antiviral effect of heparin-complexed or free oligo-poly(L-lysine) conjugates. Heparin addition clearly appears to be beneficial in this approach (Fig. 3), since 1) it does not alter the specificity of oligo-poly(L-lysine) conjugates, at least in this model (additional control experiments with unrelated oligos, not shown, have been performed), while 2) it potentiates the sequence-specific antiviral activity of oligo-poly(L-lysine) conjugates by lowering the threshold active dose and allowing the maintenance of a significant antiviral state over longer periods of time (Fig. 3).

Moreover, heparin, as well as other polyanions (e.g., alginate), dramatically reduce the nonspecific cytotoxicity of poly(L-lysine), which has been observed by us and by others when administered at higher concentrations (2–10 μM) to some cell lines, for example, lymphocytes [Morgan et al., 1988]. Taken together these data might lead to interesting new information applicable to antiviral chemotherapy. Indeed, it is anticipated that complexes between appropriate oligonucleotide-poly(L-lysine) conjugates and some of the very efficient sulfated polyanions that have been selected for the absence of adverse anticoagulant properties might be very efficient tools against HIV infection; they would indeed potentiate the sequence-specific antiviral activity of oligonucleotide poly(L-lysine) conjugates in chronically infected cells and interfere with the interaction of virus and uninfected cells.

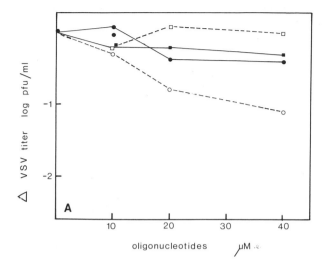

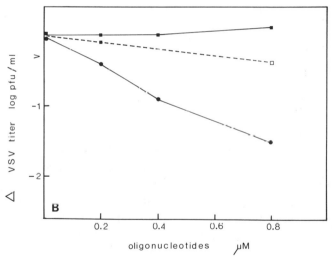

Fig. 2. *Antiviral activity of free (**A**) and poly(L-lysine) conjugated (**B**) antisense oligonucleotides. A: L929 cells were grown in minimal essential medium supplemented with 10% (v/v) fetal calf serum (●, ■). In some cases culture medium was eliminated and replaced by a completely synthetic serum-free (Opti-MEM, Gibco) medium 4 hours before the addition of oligonucleotides (○, □). Free oligonucleotides complementary to VSV N protein mRNA (○, ●) or c-myc mRNA (□, ■) were added to the culture medium at the indicated concentrations 2 hours before virus infection. B: L929 cells were grown in minimal essential medium supplemented with 10% (v/v) fetal calf serum. Poly(L-lysine)-conjugated oligonucleotides complementary to VSV N protein mRNA (●) or c-myc (■) were added to the culture medium at the indicated concentrations 2 hours before virus infection; a mixture of VSV oligo and poly(L-lysine) was used as a control (○). Virus output was monitored by quantification of its cytopathic effects, as described previously; results are expressed as a reduction in virus titer in logarithmic units in oligonucleotide treated samples as compared with untreated ones.*

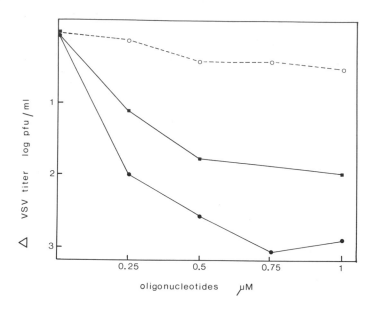

Fig. 3. *Potentiation of the antiviral activity of antisense oligonucleotide-poly(L-lysine) conjugates by polyanions. Experiments were carried out essentially as described in Figure 2B. The VSV oligo was conjugated to poly(L-lysine) and tested for its antiviral activity at* the indicated concentrations alone (■) or in the presence of 100 μg/ml heparin (●). A mixture of poly(L-lysine) and heparin at the same concentrations was used as a control (○).

D. In Vitro Cell Targeting of Antisense Oligonucleotides Through Liposome Encapsulation

Encapsulation of various drugs, such as immunomodulators and antineoplastic, antifungal, or antiviral agents, into lipid vesicles bearing cell recognition determinants is the subject of intense research, since this approach appears to constitute the elusive "magic bullet" as originally formulated by Ehrlich. Experience has recently been accumulated by our group in the successful in vitro targeting of oligo- and polynucleotide material such as (2'–5')(A)n [Bayard et al., 1985], an antiviral oligonucleotide, and the double-stranded RNA inducers (e.g., poly(rI)n, poly(rC)n, and related compounds) [Milhaud et al., 1989) of interferons and other cytokines. A similar approach, using antibody-targeted liposomes containing antisense RNA to the HIV *env* region, has been shown to be successful in inhibiting *tat* gene expression by 90% and gp160 production by 100% when applied at the same time as the virus infection [Renneisen et al., 1990].

We have made use of small, unilamellar vesicles whose phosphatidylethanolamine moiety has been covalently coupled to *Staphylococcus aureus* protein A, which can be bound to an immunoglobulin Fc fragment recognition polypeptide [Leserman et al., 1980] (see Fig. 4 for a general outline). This provides a versatile drug carrier, since by adding the cognate monoclonal antibody, liposomes could in principle be targeted to any cell surface determinant; direct coupling of antibodies to liposomes is also possible. The strict requirement for the presence of the relevant antibody to allow delivery of the liposome contents to a given cell has been described in detail [Machy et al., 1982]; the use of small, unilamellar liposomes is advantageous in providing stable material and allowing cell internalization through receptor-mediated endocytosis.

As shown in Figure 5 and discussed in ear-

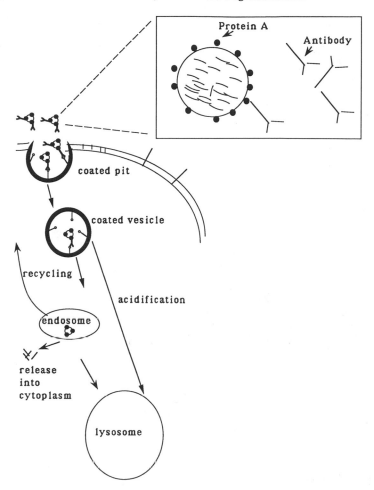

Fig. 4. *General outline of antibody-targeted–mediated delivery of antisense oligonucleotides. Liposomes, composed of 65% dipalmitoylphosphatidylcholine, 34% cholesterol, and 1% N-succinimidyl-3-(2-pyridyldithio) propionate–modified phosphatidylethanolamine, are prepared in the presence of an aqueous solution of* oligomer (15 mg/ml) and then covalently coupled to S. aureus protein A via the phosphatidylethanolamine moiety [Leserman et al., 1980]. Liposomes are associated with the cells via an antibody directed to a surface determinant of these cells and are internalized by an endocytic pathway.

lier publications [Leonetti et al., 1990], such antisense oligonucleotide-loaded liposomes lead to a dose-dependent reduction in VSV titer with a threshold value well below 1 μM. As for oligonucleotide poly(L-lysine) conjugates, the antisense effect remains sequence specific (Fig. 5; and control experiments with nonrelevant sequences, data not shown). In addition, the development of antiviral activity is strictly dependent on the recognition of a particular cell surface determinant, the H-2K molecule of the major histocompatibility antigen complex in our experimental model (Fig. 5) [Leonetti et al., 1990]. This allows a selective delivery of the antisense oligonucleotide to the corresponding cell type, at least in vitro. In addition and as expected, liposome encapsulation fully protects the antisense oligonucleotide from degradation by serum nucleases [Leonetti et al., 1990].

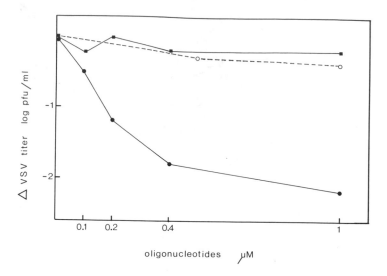

Fig. 5. *Antiviral activity of liposome encapsulated antisense oligonucleotides. 3'-Ribosylated VSV (○, ●) or myc (■) oligos were encapsulated into liposomes as described in Figure 4 and added to L929 cells at the indicated concentrations with an antibody specific for the expressed H-2K molecule (●) or with an antibody directed at a nonexpressed HLA determinant (○), as a control. Cells were infected 4 hours later. Antiviral activity is expressed as described in Figure 2.*

These points will probably turn out to be of considerable importance when performing in vivo experiments, since it is hardly conceivable that an entire organism may be loaded with oligonucleotides; manufacturing cost, potential toxicity problems, degradation by nucleases, and pharmacokinetics (a large part of antisense oligonucleotides injected into mice are rapidly eliminated in urine in the preliminary data reported thus far) constitute additional factors in favor of the use of drug delivery systems. Whether the actual solution will make use of our particular formulation still relies on much additional work. Drawbacks of this approach include the poor (around 3%) efficiency of oligonucleotide encapsulation, the fact that liposomes do not appear to escape from the vascular compartment, and difficulties encountered with in vivo targeting; we and others are actively searching for solutions to these problems. Liposomes have a propensity to be taken up by the reticuloendothelial compartment; this might, however, be of interest for the treatment of AIDS, since macrophages now appear to be a major virus reservoir in this disease.

III. SUMMARY

Synthetic oligodeoxynucleotides offer several possible mechanisms for the artificial regulation of specific cellular and viral genes, as reviewed by Toulmé (this volume). These include blocking of pre-mRNA splicing or mRNA translation, sequence-specific activation of RNase H, triple helix formation, or interference with protein binding. Problems related to applications of antisense technology include the metabolic stability of synthetic oligonucleotides, their transmembrane passage and cell targeting, their intracellular distribution, and their interaction with specific intracellular DNA, RNA, or protein targets. We have demonstrated that the chemical conjugation of 15-mer oligodeoxyribonucleotides to poly(L-lysine) or their encapsulation into antibody-targeted liposomes allows their internalization in active form in several cell lines in vitro. Specific targeting toward cell surface determinants has been obtained with liposomes. These approaches permit sequence-specific inhibition of protein synthesis at doses below 1 μM, that is, within a concentration range appreciably lower than previously used with oligonucleotides that have not been associated with drug delivery moieties. We discuss how the blockage of viral (i.e., VSV) and cellular (i.e., c-*myc* protooncogene) gene expression has been obtained; studies using various other targets, as well as the mechanisms of oligonucleotide internalization and action are currently being explored. We also discuss the prospects for in vivo uses of this technology.

ACKNOWLEDGMENTS

Work cited from the authors' laboratories was funded by grants from CNRS, INSERM, Association pour le Developpement de la Recherche sur le Cancer, and Agence Nationale de Recherche sur le SIDA.

IV. REFERENCES

Bayard B, Leserman L, Bisbal C, Lebleu B (1985): Antiviral activity in L1210 cells of liposome-encapsulated (2′-5′)oligo(adenylate) analogues. Eur J Biochem 151:319–325.

Gagnor C, Rayner B, Leonetti JP, Imbach JL, Lebleu B (1989): Parallel annealing of anomeric oligodeoxyribonucleotides to natural mRNA is required for interference in RNAse H mediated hydrolysis and reverse transcription. Nucleic Acids Res 17:5107–5113.

Gonzales ME, Alarcon B, Carrasco L (1987): Polysaccharides as antiviral agents: Antiviral activity of carrageenan. Antimicrob Agents and Chemother 31:1388–1393.

Ito M, Baba M, Sato A, Pauwels R, De Clerq E, Shigeta S (1987): Inhibitory effect of dextran sulfate and heparin on the replication of human immunodeficiency virus (HIV) in vitro. Antiviral Res 7:361–367.

Lemaitre M, Bayard B, Lebleu B (1987): Specific antiviral activity of a poly(L-lysine)-conjugated oligodeoxyribonucleotide sequence complementary to vesicular

stomatitis virus N protein mRNA initiation site. Proc Natl Acad Sci USA 84:648–652.

Leonetti JP, Machy P, Degols G, Lebleu B, Leserman L (1990): Antibody-targeted liposomes containing oligodeoxyribonucleotides complementary to viral RNA selectively inhibit viral replication. Proc Natl Acad Sci USA 87:2448–2451.

Leonetti JP, Rayner B, Lemaitre M, Gagnor C, Milhaud PG, Imbach JL, Lebleu B (1988): Antiviral activity of conjugates between poly(L-lysine) and synthetic oligodeoxyribonucleotides. Gene 72:323–332.

Leserman L, Barbet J. Kourilsky FM, Weinstein JN (1980): Targeting to cells of fluorescent liposomes covalently coupled with monoclonal antibody or protein A. Nature 288:602–604.

Machy P, Pierres M, Barbet J, Leserman L (1982): Drug transfer into lymphoblasts mediated by liposomes bound to distinct sites on H-2 encoded I-A, I-E and K molecules. J Immunol 129:2098–2102.

Maher III LJ, Dolnick BJ (1988): Comparative hybrid arrest by tandem antisense oligodeoxyribonucleotides or oligodeoxyribonucleoside methylphosphonates in a cell-free system. Nucleic Acids Res 16:3341–3357.

Milhaud PG, Machy P, Lebleu B, Leserman L (1989): Antibody targeted liposomes containing poly(rI)·poly(rC) exert a specific antiviral and toxic effect on cells primed with interferons α, β or γ. Biochim Biophys Acta 987:15–20.

Morgan DML, Clover J, Pearson JD (1988): Effects of synthetic polycations on leucine incorporation, lactate dehydrogenase release and morphology of human umbilical vein endothelial cells. J Cell Sci 91:231–238.

Morvan F, Rayner B, Imbach JL, Chang DK, Lown JW (1986): α-DNA I. Synthesis, characterization by high field ¹H-NMR, and base-pairing properties of the unnatural hexadeoxyribonucleotide α-d(CpCpTpTpCpCp) with its complement β-d(GpGpApApGpGp). Nucl Acids Res 14:5019–5035.

Renneisen K, Leserman L, Matthes E, Schröder HC, Müller WEG (1990): Inhibition of expression of human immunodeficiency virus-1 in vitro by antibody-targeted liposomes containing antisense RNA to the *env* region. J Biol Chem 265:16337–16342.

Ryser HP, Shen WC (1978): Conjugation of methotrexate to poly(L-lysine) increases drug transport and overcomes drug resistance in cultured cells. Proc Natl Acad Sci USA 75:3867–3870.

Stec WJ, Zon G, Egan W, Stec B (1984): Automated solid phase synthesis, separation, and stereochemistry of phosphorothioate analogues of oligodeoxyribonucleotides. J Am Chem Soc 106:6077–6079.

Stein CA, Subasinghe C, Shinozuka K, Cohen JS (1988): Physicochemical properties of phosphorothioate oligodeoxynucleotides. Nucleic Acids Res 16:3209–3221.

Stevenson M, Iversen PL (1989) Inhibition of human immunodeficiency virus type I mediated cytopathic effects by poly(L-lysine)-conjugated synthetic antisense oligodeoxyribonucleotides. J Gen Virol 70: 2673–2682.

van der Krol AR, Mol JNM, Stuitje AR (1988): Modulation of eukaryotic gene expression by complementary RNA and DNA sequences. BioTechniques 6:958–975.

Verspieren P, Cornelissen AWCA, Thuong NT, Hélène C, Toulmé J-J (1987) An acridine-linked oligodeoxynucleotide targeted to the common 5′ end of the trypanosome mRNAs kills cultured parasites. Gene 61:307–315.

ABOUT THE AUTHORS

GENEVIÈVE DEGOLS is a scientist in the Laboratoire de Biochimie des Proteins at the University Montpellier II Sciences et Techniques de Languedoc. She was the recipient of a fellowship from the Association pour le Développement de la Recherche sur le Cancer. Her research papers have appeared in such journals as the *Proceedings of the National Academy of Sciences USA, Nucleic Acids Research,* and *Bioconjugate Chemistry.*

PATRICK MACHY is a scientist (Chargé de Recherche) of the French Centre National de la Recherche Scientifique (C.N.R.S) at the Centre d'Immunologie INSERM–CNRS de Marseille-Luminy in Marseille. He received his Ph.D. at the University of Aix-Marseille II in 1984 under Lee Leserman, working on the targeting of liposomes to cell surface determinants via monoclonal antibodies and the internalization of the liposomes' contents into target cells in culture. He then pursued postdoctoral research at SmithKline & French laboratories in Philadelphia in the laboratory of Zdenka Jonak, studying targeted liposomes for gene delivery and endocytosis of cell surface molecules with Alemseged Truneh and Sylvia Hoffstein. Patrick Machy's current research involves the role of major histocompatibility complex-encoded class I molecules in T lymphocyte activation. His research papers have been published in *Nature*, the *Proceedings of the National Academy of Sciences USA*, the *Journal of Immunology, Biochimica et Biophysica Acta*, and *EMBO Journal*. He received an award from the French national cancer research campaign in 1984, a Bronze Medal of the C.N.R.S. in 1988, and the prize of "Young Scientist" in 1990.

JEAN-PAUL LEONETTI is a scientist at the Centre National de la Recherche Scientific. He received his Ph.D. under Bernard Lebleu, working on the targeting of antisense oligonucleotides on cultured cells, using liposomes and polypeptides. More recently he worked on the cellular location and the microinjection of these oligomers. His research papers have appeared in such journals as the *Proceedings of the National Academy of Sciences USA, Nucleic Acids Research,* and *Bioconjugates Chemistry.*

LEE LESERMAN has been head of the group "Cell Biology of the Immune System" at the Centre d'Immunologie de Marseille-Luminy since 1981. He received an A.B. at Beloit College and the M.D. and Ph.D. degrees in 1973 at the University of Chicago, where he studied cellular immunology with Donald Rowley. After clinical training at Duke University, he spent three years in the laboratory of William Terry in the Immunology Branch of the National Cancer Institute, where he studied liposome–cell interactions in collaboration with John Weinstein. He began working at the Centre d'Immunologie in 1979 as a postdoctoral fellow in the laboratory of François Kourilsky. In 1985 he was an American Cancer Society–Eleanor Roosevelt–International Cancer Fellow in the Genetics Department at Stanford University School of Medicine. He now holds the rank of Director of Research at the National Scientific Research Center (C.N.R.S.). He is on the editorial boards of *Journal of Biopharmaceutical Science* and *Critical Reviews in Therapeutic Drug Carrier Systems*. His current research interests include studies of antigen presentation and immunoregulation and use of liposomes for delivery of genes and oligonucleotides. His papers have appeared in *Nature*, the *Proceedings of the National Academy of Sciences USA*, *EMBO Journal*, the *Journal of Biological Chemistry*, and the *Journal of Immunology*. He has received awards from the Ligue Nationale Française contre le Cancer and the International Union Against Cancer.

BERNARD LEBLEU is Professor of Biochemistry at the University Montpellier II Sciences et Techniques du Languedoc in France, with teaching duties in both biochemistry and molecular biology. He received his Licence in Chemistry from the University of Brussels in 1966. He continued his studies there under Professors Hubert Chantrenne and Arsène Burny, earning a Ph.D. in biochemistry for his work on the isolation and characterization of globin mRNA from mammalian reticulocytes. Dr. Lebleu's postdoctoral research, with Professor Michel Revel at the Weizmann Institute of Sciences and with Professor Peter Lengyel at Yale University, focused on mechanisms of translation control in mammalian cells and the molecular basis of interferon action. His current research examines the molecular basis of interferon action and the control of viral and cellular gene expression by synthetic oligonucleotides. Bernard Lebleu has over 100 published articles in journals such as the *Proceedings of the National Academy of Sciences USA*, *Nucleic Acids Research*, and *Biochemistry*.

Antisense RNA and DNA: 267–284
© 1992 Wiley-Liss, Inc.

Antisense Control of Retrovirus Replication and Gene Expression

Richard Y. To

I. INTRODUCTION

Retroviruses are characterized by a unique pathway of replication. These viruses all contain plus-strand RNAs in their genomes and replicate through a reverse transcription step in their life cycles in which the single-stranded genomic RNAs are converted into double-stranded viral DNAs (proviruses). These proviruses can be covalently integrated into the chromosomes of their host cells (Fig. 1). Transcription of integrated viral genes and translation of the resultant viral mRNAs lead to synthesis of the structural proteins required for viral maturation. Infection by retroviruses generally does not lead to cytopathic effects in host cells (with the exception of some isolates), but instead establishes the chronic production of progeny viruses that bud from the cell surface

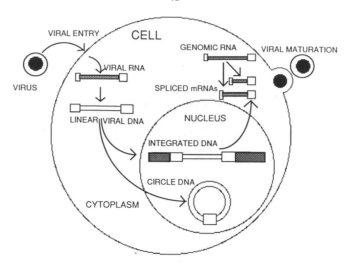

Fig. 1. *Diagrammatic representation of the retroviral life cycle. Open boxes at the ends of viral DNA and RNA structures are the long terminal repeat regions of the genome. Shaded areas flanking the integrated DNA represent host chromosomal DNA. Arrows denote synthesis or replication pathways. Translation of the viral mRNA results in viral proteins that assemble at the cell membrane and package the viral genomic RNA into mature virions.*

[Weller et al., 1980; Fenyo et al., 1988; Goff, 1989; Hanafusa, 1989]. However, the expression of transforming viral oncogenes or the expression of virus-induced cellular protoon-cogenes may elicit oncogenic transformation of the host cells and subsequently lead to tumorigenesis in the host organism. Such host organisms include a wide spectrum of animal species, ranging from insects to mammals [Teich, 1982]. Infections by field isolates of retroviruses are therefore legitimate health concerns for the agricultural industry as well as for humans. For example, the avian sarcoma-leukosis viruses (ASV-ALV) can spread oncogenic diseases in commercial flocks of chickens and is a nuisance to the poultry industry [Wang and Hanafusa, 1988]. The cattle industry has problems with diseases caused by bovine leukemia virus (BLV) [Burny et al., 1984]. Feline leukemia virus (FeLV) also poses a veterinary concern in domestic cats [Hardy, 1990]. Fatal diseases in humans such as acquired immunodeficiency syndromes (AIDS) and human T-cell lymphoma are purportedly caused by the human immunodeficiency virus (HIV) [Braun et al., 1990] and by the human T-cell lympho-tropic virus (HTLV), respectively [Manns and Blattner, 1990]. These clinical concerns accentuate the urgent need for measures that can either prevent or suppress retroviral infections.

Because of the fact that retroviruses integrate into and subsequently replicate along with their host genomes, they are particularly difficult to eradicate by the usual routes of drug therapy, without detrimental effects on the host [Mitsuya and Broder, 1989]. However, since the sequences of both the viral genomic RNAs and mRNAs are of the same sense, these RNAs should potentially provide ample targets for antisense inhibition during some stages of the viral replication cycle. Therefore the use of antisense agents targeted to specific regions of the viral genome has been suggested as a promising feasible approach for inhibiting viral replication [Stephenson and Zamecnik, 1978; Marimanat, 1985; Tellier and Weber, 1985].

The purpose of this discussion is to review current efforts toward inhibiting retroviral infection by specifically targeting viral RNAs with antisense technologies. A brief review of the retroviral structures and life cycle in Section II help to conceive the different strategies being used at various potential sites of antisense inhibition. In Section III, problems encountered with the use of antisense techniques against retroviruses are discussed. The different strategies are categorized according to the type of antisense agents, namely, synthetic oligomers, expression vectors, and retrovectors, that are currently being explored. Alternative strategies are also mentioned in Section IV.

II. VIRAL STRUCTURE AND REPLICATION
A. Viral Particle and Genome Structure

A typical retrovirus contains two copies of the genomic RNA that are packaged in a protein core together with the RNA-dependent DNA polymerase (reverse transcriptase). This core particle is encapsulated by a glycoprotein envelope to make up the virion. The viral genome of a replication-competent retrovirus contains distinct regions that code for these structures. The major regions are designated as *gag* (for core proteins), *pol* (for reverse transcriptase), and *env* (for envelope glycoproteins). At both ends of the genome are unique sequences termed *U5* (for the 5' end) and *U3* (for the 3' end) that contain regulation signals for the expression of viral genes. At the 5' terminus of U5 and at the 3' terminus of U3 are two identical and directly repeated sequences, called *R*, that are important for viral DNA synthesis. Immediately downstream from U5 is a short sequence that is the primer binding site (PBS) for a tRNA primer that hybridizes to the genome RNA during packaging and serves as the primer for the start of minus-strand DNA synthesis. Further downstream in the genome (Fig. 2), and into the *gag* region, is a sequence (ψ) important for signalling the packaging of genomic RNA into viral particles through an as-yet unclear mechanism [Eglitis and Anderson, 1988]. At the other end of the genome and immediately upstream of the U3 region is a stretch of 10–12 purine bases called the *polypurine tract* (PPT), which serves as the essential primer for the start of plus-strand DNA

synthesis during reverse transcription [Sorge and Hughes, 1982; Bizub et al., 1984; Sanchez-Pescador et al., 1985]. Other genes, such as viral oncogenes or HIV functional genes, may also make up part of the genome structure of different viruses. Both genomic RNAs and all viral messenger RNAs (mRNA) are capped at the 5' end and polyadenylated at the 3' end. A more detailed description can be obtained from the review by John Coffin [1982].

B. Viral Replication

1. Viral entry and reverse transcription of viral genomic RNA. Figure 1 diagrams the life cycle of a retrovirus. Infection starts when a virus attaches itself to the cell surface by the affinity between the viral envelope and a cell surface receptor. Entry into the cell probably involves phagocytosis by the host cell. Once inside the cell, the genomic RNA is released into the cytoplasm together with the reverse transcriptase in a ribonucleoprotein complex. Reverse transcription starts at the 3' end of the tRNA primer, which is packaged in the viral particle and is already attached to the PBS of the RNA.

A current model for the synthesis of viral DNA [Watson et al., 1987] is presented in Figure 2. A short stretch of minus-strand DNA is synthesized in the 5' to 3' direction by using the R-U5 region of the viral RNA as template. Synthesis stops when the 5' end of the RNA template is copied, and this stretch of DNA is termed the *minus-strand strong stop DNA*. At this point, the RNase H moiety of the reverse transcriptase (which digests RNAs in RNA–DNA hybrids) removes the RNA sequences that are in duplex formation with the newly synthesized minus-strand strong stop DNA (step 1 in Fig. 2). This leaves an overhanging single-stranded R region of the strong stop DNA that is able to make a "first jump" and hybridize to the complementary R region at the 3' end of the plus RNA template (or to the 3' end of another RNA molecule; step 2, Fig. 2) [Lobel and Goff, 1985]. After this step, minus-strand DNA extends along the full length of the viral genome and eventually displaces the tRNA

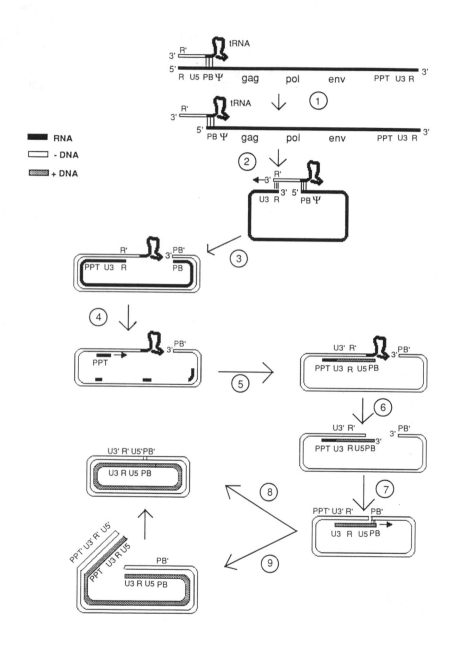

from the PBS at the 5' end of the RNA template (step 3, Fig. 2).

2. Plus-strand viral DNA synthesis. Soon after the minus-strand DNA has been synthesized over a region of the viral RNA, RNase H removes most of the RNA template within the RNA–DNA hybrid. But the removal is not complete, and short stretches of intact RNA may remain hybridized to various regions of the minus-strand DNA (step 4, Fig. 2). One of these undigested RNA stretches includes the PPT immediately adjacent to the U3 region of the RNA [Champoux et al., 1984; Resnick et al., 1984]. The 3' cleavage site for this particular RNA stretch is highly specific and is between the last (3') base of the PPT and the first (5') base of U3. The accuracy of this cleavage is important for generating the proper

---

Fig. 2. *A model of viral DNA synthesis from viral genomic RNA. At the top of the diagram is the viral RNA with tRNA primer attached to primer binding site (PB), and the minus-strand strong stop DNA has already been synthesized by reverse transcriptase. Regions of the genome are designated as R (direct repeat), U5 (5' unique), PB (primer binding), ψ (packaging signal), gag (encodes core proteins), env (envelope glycoproteins), PPT (polypurine tract), and U3 (3' unique). 5' and 3' ends of the strands are as marked. tRNA is represented by a solid loop. Regions involved in initial hybridizations are denoted by three short vertical lines. 5' to 3' DNA polymerization directions are indicated by small arrows in front of the 3' ends. Step 1: removal of RNA in duplex with minus-strand strong stop DNA by RNaseH. Step 2: strong stop DNA makes the "first jump" to hybridize with the R region at the 3' end of the viral RNA. Step 3: extension of minus-strand DNA along the length of the RNA and displacing of tRNA from the PB of the RNA. Step 4: removal of RNA to generate the PPT primer for plus-strand DNA. Step 5: synthesis of plus-strand strong stop DNA. Step 6: removal of tRNA by RNaseH. Step 7: "second jump" by plus-strand DNA to hybridize with PB at 3' end of minus-strand DNA. Elongation of the plus-strand DNA along the minus-strand template leads to steps 8 and 9. Step 8: formation of single-LTR circle by ligation to fill in the gaps. Step 9: formation of double-LTR linear DNA by strand-displacement extensions of both plus and minus DNA strands. Double-LTR circle can be formed by ligation of the ends of the linear molecule. An alternative pathway for single-LTR circle formation is the loss of one LTR from double-LTR circle by recombination.*

primer for plus-strand DNA synthesis [Finston and Champoux, 1984; Rattray and Champoux, 1987]. The precise positioning of the cleavage at this juncture depends on both the upstream RNA sequence (and maybe the downstream sequence as well) and the source of RNase H [Luo et al., 1990]. Although the specific 5' cleavage site for the PPT has not been detected yet, it has been shown that mutations in the genomic sequence around this region result in replication-defective viruses, probably because of incorrect formation of the primer for plus-strand DNA synthesis. Other RNA stretches may also serve as primres for plus-strand DNA, but the DNAs initiated from these nonspecific primers apparently do not lead to proper DNA intermediates suitable for integration. With the proper PPT primer, plus-strand DNA synthesis can be initiated by the DNA polymerase function of the reverse transcriptase, using the U3–R–U5 of the minus-strand DNA as template.

Synthesis of plus-strand DNA continues along the full length of the minus-strand strong stop DNA and stops only when the PBS region of the still covalently attached tRNA has also been copied (step 5, Fig. 2). This stretch of DNA is called the *plus-strand strong stop DNA* [Mitra et al., 1979]. RNase H then digests away the tRNA primer from the newly synthesized plus-strand DNA, which results in a 3'-protruding PBS region (step 6, Fig. 2). This sticky-end from the plus-strand strong stop DNA then makes a "second jump" (step 7, Fig. 2) and hybridizes to the complementary PBS region at the 3' end of the minus-strand DNA. This transplanted stretch of plus-strand strong stop DNA then continues its polymerization along the minus-strand DNA template for the completion of a double-stranded viral DNA replication intermediate.

3. Formation of unintegrated viral DNA. The formation of a closed circular DNA molecule with a single long terminal repeat (LTR) sequence (which includes U3–R–U5) requires ligation between the ends of each strand of the duplex DNA as depicted in step 8 of Figure 2. However, synthesis by strand displacement (step

9, Fig. 2) at the 3′ end of both plus and minus strands of DNA will result in a linear double-stranded DNA structure with one LTR at each end of the molecule. A third but minor form of the viral DNA has two LTR joined end to end to form a closed circular structure [Olsen and Swanstrom, 1985]. This third form of unintegrated viral DNA can presumably be generated by blunt end intramolecular ligation between the two LTR ends of the linear DNA molecules. An alternative pathway for the formation of single-LTR circle DNA may depend on the loss of one LTR from the double-LTR circle by recombination (as indicated in Fig. 2 by an arrow above the linear DNA produced by step 9).

4. Integration of viral DNA. The unintegrated forms of intermediate viral DNA can be detected in the cytoplasm of infected cells as early as 6–10 hours postinfection and usually can be detected for up to 10 days postinfection, depending on virus and host. Circular forms are usually found only within the cell nuclei. Although these intermediate DNA molecules contain the full genomic sequences of the virus, they cannot efficiently serve as templates for gene expression [Wang and Hanafusa, 1988; Stevenson et al., 1990; Robinson and Zinkus, 1990]. Mutant viruses defective in DNA integration exhibit very low levels of transient expression of viral proteins, and progeny virus production is minimal. Therefore integration of viral DNA into the host genome is a crucial step in the replication cycle for the virus to establish a successful infection. Earlier observations had suggested that the circular forms are the immediate precursors to integrated proviruses [Panganiban and Temin, 1984]. However, more recent studies have indicated that the linear forms are as effective in integration and may even be the preferred integration precursors [Brown et al., 1987; Ellis and Bernstein, 1989; Lobel and Goff, 1985].

Evidence has come from in vitro experiments in which the integrase function (IN) of the reverse transcriptase can efficiently integrate linear DNA into target DNA [Duyk et al., 1985; Katz et al., 1990]. Integrase is believed to rec-

ognize about six to eight base pairs (depending on the source of integrase) of inverted and imperfectly repeated sequences at both ends of the linear viral DNA. Two pyrimidine bases are removed by IN from each 3′ end of the double-stranded linear molecule early in the formation of the proviruses [Fujiwara and Craigie, 1988; Katzman et al., 1989; Roth et al., 1988]. At the site of integration in the target chromosomal DNA, IN makes a four to six base (again depending on the source of IN) staggered cut leaving 5′ overhanging ends on both sides [Vink et al., 1990]. These 5′ ends in host DNA are then covalently joined to the 3′ recessed ends in the viral DNA, probably by a novel mechanism [Fujiwara and Mizuuchi, 1988; Thode et al., 1990; Craigie et al., 1990]. Final filling in of the gaps by polymerase results in an integrated provirus that has lost two bases at each end of its linear form and is flanked by short and directly repeated cellular sequences [see Grandgenett and Mumm, 1990, for review].

5. Transcription from integrated DNA proviruses. Full-length genomic RNA is transcribed by the host's RNA polymerase from the integrated provirus. This RNA is capped at the 5′ end and polyadenylated at the 3′ end as a normal mRNA and serves as mRNA for the *gag–pol* precursor polypeptides. Subgenomic RNAs serving as mRNAs for the *env* gene and other viral genes are independently modified from the full-length RNA by splicing. RNA splicing requires the use of a splice donor site near the 5′ end of the *gag* region and splice acceptor sites 5′ to the individual genes. All resulting viral RNA molecules contain 5′ capped U5 and polyadenylated U3 regions for effective translation by the peptide synthesis machineries provided by the host cell. It is still unclear as to how the relative quantities of various species of viral RNA molecules can be regulated in the cell. It has been suggested that some *cis*-acting control elements upstream of individual splice acceptor sites may regulate the splicing events [Katz and Skalka, 1990]. The function of these elements may ensure that enough full-length RNA molecules remain unspliced to be packaged into mature virions.

6. Maturation. Translation of the mRNAs into polyproteins also depends on host machinery. Cleavage of these polyproteins into individual viral proteins and other posttranslational modifications probably occur at the same time as virion assembly, which takes place near the cell membrane. The ψ sequence near the *gag* region of the genomic RNA provides a signal for the preferential incorporation of full-length RNA molecules into the protein core. Two genomic RNA molecules are usually hydrogen bonded at their 5' ends and are copackaged into each core particle. This nucleoprotein particle is then encapsulated by the envelope glycoproteins together with some cellular membrane materials at the cell surface. The mature virion is finally released from the cell by a budding process and hence completes the life cycle.

One aspect of viral glycoprotein synthesis needs to be pointed out. Soon after infection, the newly synthesized envelope glycoproteins usually bind up all the surface receptors for that specific envelope at the surface of the infected cell. Superinfection of the same cell by another virus carrying the same glycoprotein envelope will be inhibited by the absence of receptors; this phenomenon is termed *subgroup interference* [Vogt and Ishizaki, 1966]. It has been suggested that the cytopathic effect seen in some HIV-1–infected cells may be due to insufficient synthesis of envelope glycoprotein, which then fails to prevent massive secondary infections [Robinson and Zinkus, 1990].

III. STRATEGIES FOR INHIBITING VIRAL INFECTION

A. Factors to Be Considered in Antisense Design

The aim of applying antisense techniques for the inhibition of viral infection is either to suppress the expression of integrated provirus in chronically infected cells or to prevent the virus from establishing itself in uninfected cells. To accomplish this goal, the obvious route is to disrupt the viral replication cycle, as mentioned earlier. One way of achieving this is to transfect antisense oligodeoxynucleotides into host cells. Each of these oligomers can be designed to be complementary to a specific region of the viral genome. Another method is to transfect cells with recombinant vectors that can transcribe antiviral sequences into antisense RNAs. A third technique is to use retrovectors that can infect target cells and express antisense RNAs. Each method is discussed further.

Antisense techniques for the blocking of retroviruses encounter the usual obstacles in antisense applications. Such difficulties include the inability of the cell to absorb antisense agents at a high enough concentration to exert an antisense effect, the susceptibility of antisense molecules to enzymatic degradation, nonspecific binding of these agents to essential cellular proteins [van der Krol et al., 1988], and the difficulties in delivering these molecules to the target RNAs. In addition, antisense inhibition of retroviruses presents a few more problems that warrant special attention. First, viral RNA may contribute up to 10% of the total polyadenylated RNA in infected cells [Varmus and Swanstrom, 1984]. An efficient antisense system will have to introduce into the target cell an amount of stable antisense molecules sufficient to overcome this level of expression without overwhelming the normal functions of the host cell by their sheer quantity [Izant and Weintraub, 1985]. Second, for maximum therapeutic effect in vivo, both the antisense molecules and the retrovirus should have the same target cells. Obviously, the antisense effect will be greatly diminished if the antisense molecules have difficulty in reaching the same cells as the retrovirus, and nonspecific uptake of antisense molecules by cells other than the target cells will reduce the quantity of these molecules available to elicit the antisense effect in the target cells, as well as provoking possible nonspecific side effects. Third, for the immunization of uninfected cells against retroviral infection, it is clear that transient expression of antisense agents will not provide any long-term protection for the target cells. The antisense molecules must be able to exist in these cells for an extended period of their life span. Retrovectors have the advantage of being able to express inserted genes as efficiently as wild-type retroviruses, and, with the correct envelope, retrovectors should have the same host range as the target virus. The antisense gene can then be integrated into host genomes by the integration mechanism of the retrovector, and the antisense effect should be permanent. Unfortunately this ability to integrate also poses a potential disadvantage in the use of retrovectors. It is known that LTR sequences from integrated proviruses can cause downstream promotion of cellular genes [Isfort et al., 1987] and may result in undesirable manifestation of certain genes in the host cell. Therefore the full potential of antisense techniques may still depend on the perfection of the delivery system. The use of each of these techniques in inhibiting retroviruses is discussed in the following sections.

B. The Use of Synthetic Oligodeoxynucleotides and In Vitro Synthesized RNA

1. Unmodified oligodeoxynucleotides. In 1978, Stephenson and Zamecnik reported the first evidence that a synthetic oligodeoxynucleotide could be used to block the production of Rous sarcoma virus (RSV) in chicken embryo fibroblasts (CEF). The oligomer was complementary to a 13-ribonucleotide segment of the R regions of the RSV genome. When a culture of CEF was flooded with this oligomer, the cells exhibited inhibition of RSV replication. This inhibition effect could be enhanced by increasing the molar ratio of blocking molecules to the viral RNAs. By varying the amount of oligomers added to the culture relative to the viral multiplicity of infection (MOI) of the CEF, they were able to obtain a dosage response from the cells with respect to viral inhibition effects.

Following this initial experiment, other synthetic oligodeoxynucleotides have been tried as antisense agents against retroviral replications in vitro. Zamecnik et al. [1986] reported using oligomers directed against the PBS as

well as splice sites of HIV. Human T-cell lines exposed to these oligomers exhibited suppression of viral replication (assayed by both reverse transcriptase activity and viral protein expression) by as much as 95%. Goodchild and colleagues [1988] designed oligodeoxynucleotides complementary to 20 target sites within the HIV genomic RNA, with the aim of arresting viral transcription or translation. By assaying for syncytia formation and viral protein (p24) expression, they found that targets in the R region and splice sites were the most susceptible sequences for viral inhibition. However, one of the side effects with these oligodeoxynucleotides is their toxicity to the host cells, which therefore limits their general application. The relatively inefficient absorption of these synthetic oligodeoxynucleotides by cells and their susceptibility to enzymatic degradation inside the host cells are also common problems.

2. Modified oligodeoxynucleotides. Because of the toxic side effects and other difficulties encountered in the use of synthetic oligomers, efforts were made to modify these agents to circumvent the problems. One such modified oligomer was a potent 28-mer phosphorothioate analog of oligodeoxycytidine reported by Matsukura et al. [1987] and discussed in detail by Matsukura (this volume). When this 28-mer was introduced into human T cells, HIV infection could be completely inhibited at the integration level. But since this modified oligodeoxycytidine, S-d(C)$_{28}$, did not appear to be specifically complementary to any viral genomic sequence, it has been suggested that this compound may have acted by binding directly to the viral reverse transcriptase, thereby preventing the synthesis of viral DNA [Majumdar et al., 1989]. Besides this potent compound (S-d[C]$_{28}$), antisense phosphorothioate oligodeoxynucleotides against the initiation sequences of the HIV *rev* gene have also been tested in chronically infected cells [Matsukura et al., 1989]. Those results showed greater than 95% reduction of HIV viral genomic RNA in the cells and are also discussed by Matsukura (this volume).

At the same time, oligodeoxynucleoside phosphoramidates and phosphorothioates were also studied by Agrawal et al. [1988]. These compounds were designed to be antisense to the splice donor and splice acceptor sites in the HIV genomic RNA in order to prevent mRNA formation. HIV-1 expression (assayed by such criteria as cell viability and syncitia formation, as well as viral protein syntheses) in human cell lines was inhibited over 90%. When these antisense agents were applied to cell lines chronically infected with HIV, the inhibition effects were less potent and of a more transient nature than when the agents were added to early infected cells [Agrawal et al., 1989]. Other modified oligomers have also been used as antisense agents. For example, an oligonucleoside methylphosphonate targeted at the HIV *tat*-3 gene was reported to exhibit a transient inhibitory effect against HIV as measured by reverse transcriptase activity [Zaia et al., 1988]. An unfortunate side effect in the use of short oligonucleotides is their relatively high affinity for nonspecific binding to other cellular proteins. To minimize toxic effects to the cells, the dosage of these oligomers must therefore be limited.

3. In vitro synthesized RNA targeted in liposomes. An interesting new approach for targeting antisense oligomers to cells has recently been applied to the control of HIV by Renneisen et al. [1990], who showed that antisense RNA, synthesized in vitro with T7 or SP6 RNA polymerase and encapsulated in liposomes, displayed an anti-HIV-1 effect in cells in vitro. The liposomes were targeted to T cells by antibodies to the T-cell receptor molecule CD3 and almost completely inhibited HIV-1 production when antisense RNA to the viral *env* region covering a part of exon II of the HIV-1 *tat* gene was used. The anti-HIV effect was dependent on the targeting antibody, and there was no effect when control sense RNA was used; nor was antisense RNA to the *pol* region successful. (It is interesting to note that Sullenger et al. [1990] also failed to obtain inhibition when they targeted the *pol* region with antisense RNA expressed from a replication-defective retro-

vector, as discussed in Section III. D.1, below.) In the cells in which viral replication was inhibited, *tat* gene expression was reduced by 90% and gp160 production by 100%. Since all major HIV RNA species were found (although at a reduced level), the authors came to the conclusion that in this case the antisense *env* RNA inhibited viral protein production primarily at the level of translation. This type of approach has also been used to target antisense oligodeoxynucleotides to cells as described by Degols et al. (this volume) and may offer the potential for developing in vivo drug delivery systems for antisense agents.

C. The Use of Expression Vectors

1. Cotransfection of antisense genes with proviral DNA. The alternative to using synthetic oligonucleotides is to introduce into cells specific sequences that can be transcribed into antisense RNAs targeted at specific regions of the retroviral genome. In 1985, Chang and Stoltzfus reported inhibition of RSV by using antisense sequences transfected into cells. In that study, they used a Bryan high-titer RSV (BH-RSV) as the target virus. BH-RSV has extensive deletions in the *env* region and is therefore defective in replication. When a plasmid containing this virus was transfected into quail cells, no viral progeny could be obtained. However, this defective condition could be alleviated by the coexpression of an *env* gene that was introduced by a separate expression vector into the same cell. Thus, when quail cells were cotransfected with both BH-RSV and *env*-containing plasmids, infectious and transforming virus could be rescued by such a complementation. When a third vector that could express anti-*env* RNA was cotransfected together with the above two recombinant plasmids, the resultant RSV titer was decreased by up to 80%. The data implied that the sense *env* RNA could be blocked by the anti-*env* RNA and resulted in a decrease in the cell of envelope glycoproteins available for rescue of the *env*-defective BH-RSV.

In addition to the above result, Chang and Stoltzfus [1987] also found that vectors that expressed anti-*src* RNA were able to suppress the rescue of BH-RSV by over 80% in a similar test system. Moreover, these anti-*src* sequences could even inhibit production of a replication-competent PragueA strain of RSV (PrA-RSV) in culture. In contrast to the general opinion that smaller oligomers are more efficient antisense molecules than larger ones, they found that the plasmid containing a 1.7 kb anti-*src* sequence was more efficient in blocking BH-RSV than plasmids with shorter anti-*src* sequences. Apparently, the sense–antisense interaction in the *src* region is sufficient to suppress viral replication even though this region is not essential in the life cycle of a normal replication-competent retrovirus. These data suggest that the mechanism employed in antisense inhibition of viral replication may involve more than simply arresting translation by the blocking of mRNAs.

2. Transfection of preinfected cells. Subsequent to the results reported by Chang and Stoltzfus, we reported our own findings on blocking viral replication in cells that were already expressing antisense sequences prior to viral infection [To et al., 1986]. We chose to target a region of the viral genome that is not normally required for viral replication. The original intent of our experiment was to inhibit oncogenic retroviruses by using antisense sequences complementary to the oncogene regions of the viral genomes. However, since the antisense oncogene sequences might also have an adverse antisense effect on the normal expression of cellular protooncogenes, it is possible that the interpretation of such results may be askew. To circumvent this problem, we used a model system containing a bacterial-neomycin-resistant gene Neo^r, which has no counterpart in either the cell or wild-type retrovirus. A replication competent retrovector carrying the anti-Neo^r sequence served as the model target virus. The blocking molecules in host cells were sense Neo^R RNA expressed from a transfected Neo^r-containing expression vector. The target viral constructs were all derived from the retrovector RCAS, which was modified from a Schmidt-Ruben A strain of Rous sarcoma virus (SRA-RSV) by Steve Hughes and colleagues [1987]. Essentially, the *v-src* region in SRA-RSV has been replaced by the Neo^r gene in reverse orientation with respect to the rest of the viral genome. The constructed plasmid was transfected into CEF to produce the resultant progeny virus αN10. The antisense Neo^r sequences in αN10 could be transcribed into both viral genomic RNA and mRNA. A similarly constructed control virus N10 has the Neo^r gene inserted in the retrovector in a sense orientation.

Inhibition attempts were carried out in a chemically immortalized quail cell line QT35 [Moscovici et al., 1977]. A DNA vector containing the Neo^r gene, flanked by viral LTR, was transfected into QT35 cells. Expression of the Neo^r gene rendered the cells resistant to G418. A G418-resistant cell clone, A1, was expanded for use as target cells. In this system, the sense Neo^r RNA in A1 cells served as blocking molecules that were complementary to the antisense Neo^r region of the target viral genome. Experimental results showed that while αN10 can replicate as efficiently in normal QT35 cells as in CEF, αN10 infection of G418-resistant A1 cells did not lead to the production of progeny virus (Fig. 3). Therefore, in the target cell, resistance to αN10 infection appeared to be correlated with resistance to G418. To show that this viral resistance was not an inherent property of the clonal A1 cells, we infected the same cells with the control virus N10, which contained the sense Neo^r sequence. Replication of N10 was not inhibited. These results (also presented in Fig. 3) showed that when Neo^r sequences in both the viral RNA and cellular RNA were in the same orientation, the target virus was not blocked by the host cells. Therefore, the inhibition effect of sense Neo^r RNA could only be demonstrated against virus carrying a complementary sequence.

However, this inhibition of αN10 replication by G418-resistant QT35 cells was not without limitations. We found that viral inhibition occurred only in those QT35 clones that synthesize Neo^r-containing RNA at a level greater than 0.003% of total cellular RNA. Clones

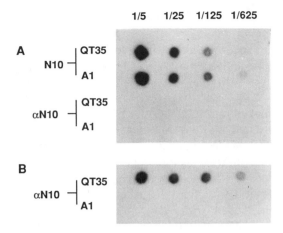

Fig. 3. *Sense Neo^r virus (N10) and antisense Neo^r virus (αN10) were used to infect a quail cell line, QT35, and a G418-resistant derivative clone, A1, expressing sense Neo^r transcripts. Viral RNA extracted from virions in the tissue culture supernatant were serially diluted as indicated at the top, dotted onto membranes, and hybridized to single-stranded radioactive probes specific for the sense Neo^r sequence (**A**) or specific for antisense Neo^r sequence (**B**). N10 virus infected both QT35 and A1. αN10 replicated in the control line QT35, but failed to replicate in A1.*

expressing lower levels of *Neo^r* RNA were not resistant to αN10 replication. These data established a lower limit for the quantitative requirement of antisense molecules. Furthermore, from our data, the amount of *Neo^r* RNA that gave viral inhibition did not appear to be in sufficient excess to suppress the quantity of viral RNA generated in a normal infection. We assumed therefore that the inhibition step might have occurred before viral RNA synthesis and that the inhibition mechanism might involve more than simply hybridization of antisense molecules to the viral RNAs.

To investigate this further, we carried out analysis of unintegrated viral DNA intermediates, which revealed that both αN10 and N10 could enter G418-resistant A1 cells and could even carry out reverse transcription. Unintegrated viral DNA molecules could be recovered from detergent-disrupted cells that had been exposed to αN10 for 10–48 hours. The extracted DNA was fractionated to separate

unintegrated DNA from chromosomal DNA. *Neo^r*-containing unintegrated viral DNA was recovered at similar levels from both αN10 and N10 infections and migrated in agarose gel electrophoresis at positions equivalent to linear viral DNA molecules (Fig. 4). However, our previous data [To et al., 1986] had shown that no αN10 proviral integration could be found in the chromosomes of cells that blocked viral production, whereas N10 provirus could be easily detected in genomes of the same cells. However, integrated αN10 proviruses could be detected in control cells not expressing any *Neo^r* sequences. Taken together, these results implied that viral inhibition probably occurred as a result of the inability of the viral DNA intermediates to integrate into the chromosomes of cells inhibited because they express the antiviral RNA, and that viral replication was interrupted prior to viral DNA integration into the host genome. The key to unravelling the mechanism of antisense inhibition on viral replication in our particular system may therefore depend on the understanding of how integration is prevented.

D. The Use of Retrovectors

1. Replication-defective retrovectors. The third method for introducing antisense molecules into target cells is the use of retrovectors. von Ruden and Gilboa [1989] used a *Neo^r*-containing nonreplicating retrovector to carry sequences either antisense to the 5′ end of the HTLV-1 genome or antisense to the *tax* gene. These recombinant constructs were used to transfect packaging cell lines. Successfully transfected cells were selected by G418 resistance. Packaging cell lines contain all the viral machinery to produce viral particles but cannot package their own viral genomes because of defects in the packaging signal (ψ) region. The genomes of the recombinant retrovectors did contain wild-type packaging signals and LTR but no other viral structural genes. As a result, the genomic RNA of these retrovectors could be incorporated into viral particles capable of undergoing only one round of infection in a fresh host cell. With the use of proper pack-

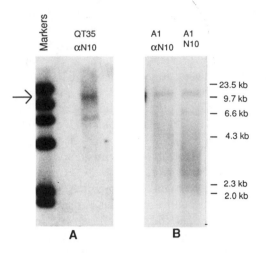

Fig. 4. *Unintegrated viral DNA from infected cells were fractionated by the HIRT method and electrophoresed on an agarose gel. The DNA bands were transferred to membranes and hybridized to Neo^r-specific probes.* **A:** *DNA extracted from QT35 cells infected by the αN10 virus.* **B:** *DNA extracted from the G418-resistant A1 cells exposed to αN10 or N10 virus, as indicated. The sizes of the marker bands are given in the right-hand column in kilobase pairs (kb). Position of linear unintegrated viral DNA is indicated by an arrow at about 10 kb.*

aging cells, subgroup interference would be avoided. Such particles containing the antisense sequences were used for a first round of infection in mononuclear cells from the umbilical cord blood of human infants. The vectors were allowed to integrate and express antisense RNA in the cells. The cells were then challenged with HTLV-1 in the presence of exogenous T-cell growth factor IL-2 to support maximal T-lymphocyte proliferation. In control cells without antisense molecules, HTLV-1 induced rapid proliferation of primary T lymphocytes within 5 days. In contrast, in cells preinfected with the proper antisense viruses and selected for G418 resistance, the cell numbers declined, and no proliferation was observed for more than 2 weeks. After 2 weeks, blast cells started to reappear, presumably because of a small number of target cells escaping the antisense effect. Nevertheless, the feasibility of using retrovectors to express antisense se-

quences for viral inhibition was successfully demonstrated.

The feasibility of using retrovectors to introduce antisense sequences was again demonstrated by the work of Rhodes and James [1990]. Various sequences complementary to the central portion of the HIV genome were inserted into a retrovector, including regions responsible for splicing and translation initiation. Jurkat cells transduced with these retrovectors were able to express anti-HIV RNA. Their data showed that when these cells were challenged with HIV, antisense sequences encompassing the initiation codon of the HIV transactivator (*tat*) gene exhibited the greatest inhibition effect of over 70%. However, the effect was only transient. They speculated that this transiency is due to the few noninhibitory cells that escaped antisense blocking. These "escaped" cells might have synthesized enough *tat* protein that could be transported through the culture media to the antisense-inhibited cells. The *tat* products then activated the LTR in these inhibited cells, and the result was that the *tat*-activated viral transcripts were too abundant relative to the quantity of antisense RNA present. If these workers are correct, then the major problem in using antisense agents still rests in inadequate blocking molecules.

To overcome this problem of insufficient antisense RNA, Sullenger and colleagues [1990] devised a system to amplify the synthesis of antisense RNA in target cells by exploiting RNA polymerase III. In mammalian cells, tRNA molecules transcribed by polymerase III are generally about 100-fold more abundant than polyadenylated RNA on a molar basis. Therefore a tRNA/polymerase III system coupled to antisense viral sequences could express high levels of antisense molecules. They fused a mutant human tRNA$_i^{met}$ gene to Moloney murine leukemia virus (M-MuLV) sequences to generate chimeric antisense–tRNA genes. A retrovector was constructed that contained an indicator *Neo^r* gene in addition to sequences antisense to the *gag* and *pol* genes of M-MuLV. The tRNA gene was inserted into the U3 region of the 3' LTR of the retrovector. After propagation in packaging cell lines and G418 selection, as described previously, high titer antisense–tRNA chimeric retrovectors were used to infect NIH 3T3 mouse cells to establish high levels of transcription of antisense RNA. Antisense M-MuLV tRNA were produced by these cells, corresponding to 15%–25% of the cell mRNA content on a molar basis. This should be compared with the normal abundance of M-MuLV RNA following infection, which are also extremely abundant and comprise 1%–5% of polyadenylated RNA. The antisense RNA-expressing cells were then challenged with M-MuLV. The infection, at an MOI of 2, reportedly yielded inhibition of virus replication of up to 97% with the antisense construct corresponding to the gag gene, but less than twofold in cells expressing antisense RNA corresponding to the *pol* gene, when compared with control cells. They also reported that the levels of both unspliced and spliced M-MuLV RNA were not affected in antisense-containing cells, but that pulse–chase experiments indicate an 8–10-fold reduction in the amount of *gag* gene product synthesized, suggesting that antisense control is mediated at the level of translation. The reduction in virus replication was therefore caused by a reduction in the release of viral particles, suggesting that the antisense inhibition was mediated at a late stage in the viral life cycle rather than at the proviral integration level.

2. Replication-competent retrovectors. In Paul Neiman's laboratory, we have also tried using retrovectors to introduce antisense molecules into target cells. Shane C. Booth attempted to inhibit a Prague C strain of Rous sarcoma virus (PrC-RSV) from replicating in, and transforming, CEF cells. RNA molecules expressed from an anti-*src* sequence were used as blocking molecules. A truncated sequence (Δsrc) of the viral transforming gene *v-src* was inserted in a reverse orientation into the replication-competent subgroup A retrovector RCAS. The resulting virus, RCASαsrc, was used to infect a bulk population of CEF that expressed the anti-*src* RNA without any apparent detrimental effect to the cells. These infected CEF cells were passaged

TABLE I. PrC-Rouse Sarcoma Virus Production (FFU/ml)*

MOI	CEFΔS	CEF/αS
0.1	1.8×10^5	$<1 \times 10^2$
1.0	3.2×10^5	$<1 \times 10^2$
10.0	2.1×10^5	1.3×10^3

*PrC-RSV superinfections in CEF already infected with sense but truncated *src* virus (CEFΔS) and in CEF already infected with antisense *src* virus (CEFαS). MOI, multiplicity of infection for input PrC-RSV. Progency PrC-RSV titer from each superinfected culture was determined by focus-forming assays on quail cells and is presented as focus-forming units (FFU) per milliliter of tissue culture supernatant.

in culture for 2 weeks to allow for viral spread in the cell population, since the cells could not be selected for expression of the antisense *v-src* sequence. Evidence of viral infection was established by assaying for anti-*src* RNA extracted from virions purified from culture supernatant. Infected cells were exposed to the PrC-RSV, which is a subgroup C transforming virus. Since PrC-RSV and RCASαsrc belong to different subgroups, subgroup interference would not have occurred to prevent superinfection. The inhibition of PrC-RSV was measured by the titers of progeny PRC-RSV produced by superinfected cells. The culture fluids were serially diluted and viral titers were assayed on normal CEF as focus-forming units per unit volume of culture supernatant. The results are presented in Table I. The titers of progeny PrC-RSV in a population of cells harboring the anti-*src* sequences showed a two to three log decrease as compared with progeny viral titers from a population of control CEF. The control CEF cells were infected with a control virus containing the truncated *v-src* (ΔS) sequence in the sense orientation prior to challenge by PrC-RSV. Since the population of target cells could not be selected for expression of anti-*src* sequences, a few cells in the culture might not be expressing sufficient amounts of anti-*src* RNA and consequently might allow the PrC-RSV to escape blockage. Furthermore, since PrC-RSV is a transforming virus, these few cells would eventually proliferate and be responsible for producing the low titers of progeny PrC-RSV observed, even in a population of cells infected with antisense virus. This result supported the observations made by Chang and Stoltzfus [1985] on the inhibiting effect of anti-*src* molecules on PrA-RSV as described earlier. It also demonstrated that oncogenic viruses can be inhibited by antioncogene RNA, which is not complementary to any of the viral replicative genes. This again suggested that a mechanism of antisense inhibition other than arrest of viral transcription or translation is responsible.

IV. CONCLUSION

It is quite apparent that the question is no longer whether antisense agents can inhibit retroviral replication, but how effectively can this technology be applied to an in vivo situation. In the interest of advancing these techniques, three major considerations must be discussed.

First, the effectiveness of antisense agents is the most important consideration. Antisense agents that do not completely block viral replication may still be applicable against cytotoxic viruses such as HIV, since the infected cells will eventually be destroyed and thus rid the virus from remaining blocked cells and the host. But for transforming retroviruses such as HTLV-1, a less than 100% viral blockage may not be of clinical pertinence, since the few cells that are able to support the virus will eventually be transformed and proliferate. In such cases, antisense agents designed to prevent viral integration, when viral RNAs are still at relatively low levels, will probably be more effective than agents designed to suppress viral transcription and translation.

The second consideration is the method of delivering the antisense agents into the host. All the reported results demonstrate that viral inhibition effect is more profound when the antisense agents are introduced into cells prior to viral infection. Therefore it appears that the ideal method of antisense inhibition is to immunize target cells with antisense agents. However, given the toxic side effects of oligodeoxynucleotides, the difficulty in administering expression vectors in vivo, and the possible downstream promotional effect of retrovectors, all the current methods of delivering antisense molecules into hosts have undesirable side effects. Modified oligomers that can be stably channelled to target cells in sufficient quantities without toxic effects need to be explored. Improved retrovectors [Miller and Buttimore, 1986; Bordingnon et al., 1989] that minimize the nonspecific promotional effects of LTR may be a promising method of expressing antisense molecules in hosts.

Third, the amount of antisense oligodeoxynucleotides or expressed antisense RNAs in the target cell must be maintained at a high level relative to the target viral RNA. This point is especially important in suppressing virus replication in cells that are chronically infected. Novel approaches such as using chimeric antisense–tRNA as reported by Sullenger et al. [1990] may have to be explored in order to fulfill this criterion.

It appears that, in order to meet the challenges of viral inhibition, we need to focus our attention both on the mechanism of the antisense effect and on achieving a better understanding of the details of retroviral replication. Antisense agents apparently inhibit viral replication by various different, and unclear, mechanisms. Each mechanism depends on the nature of the blocking molecule and its target sequence. Undoubtedly, proposed mechanisms would include transcription and translation arrest of essential viral gene expression, the inhibition of maturation of virions by blocking genomic viral RNA from being packaged, and the interruption of reverse transcription by interfering with the RNA template. The current suggestions on the mechanisms of antisense action, such as the RNase H removal of target RNA [Walder and Walder, 1988], the modification of target sequences by unwinding/modifying activity [Bass et al., 1989; Wagner and Nishikura, 1988] (see also Bass, this volume), or the activation of interferons [SenGupta and

Silverman, 1989] as reviewed by Stein and Cohen [1988], will no doubt play key roles in viral inhibition. But we still need to target our explorations on other steps of the viral replication cycle. For example, in our laboratory, we are currently investigating the mechanism of how antisense sequences can antagonize the synthesis of unintegrated viral DNA. The working hypothesis is that by blocking the PPT region of the viral genome, both the RNase H and IN activities of the virus will be ravaged to the point of being unable to process the viral DNA molecules into integrable forms. Our hope is to prevent the virus from ever integrating into the host genome. With better understanding of the viral replication process, I have no doubt that better targets can be found for applying antisense technologies against retroviruses.

V. REFERENCES

Agrawal S, Goodchild J, Civeira MP, Thornton AH, Sarin PS, Zamecnik PC (1988): Oligodeoxynucleoside phosphoramidates and phosphorothioates as inhibitors of human immunodeficiency virus. Proc Natl Acad Sci USA 85:7079–7083.

Agrawal S, Ikeuchi T, Sun D, Sarin PS, Konopka A, Maizel J, Zamecnik P (1989): Inhibition of human immunodeficiency virus in early infected and chronically infected cells by antisense oligodeoxynucleotides and their phosphorothioate analogues. Proc Natl Acad Sci USA 86:7790–7794.

Bass BL, Weintraub H, Catlaneo R, Billeter MA (1989): Biased hypermutation of viral RNA genomes could be due to unwinding/modification of double-stranded RNA. Cell 56:331.

Bizub D, Katz RA, Skalka AM (1984): Nucleotide sequence of noncoding regions in Rous-associated virus 2: Comparisons delineate conserved regions important in replication and oncogenesis. J Virol 49:557–565.

Bordingnon C, Yu S, Smith CA, Hantzopoulos P, Ungers GE, Keever CA, O'Reilly RJ, Gilboa E (1989): Retroviral vector-mediated high-efficiency expression of adenosine deaminase (ADA) in hematopoietic long-term cultures of ADA-deficient marrow cells. Proc Natl Acad Sci USA 86:6748–6752.

Braun MM, Heyward WL, Curran JW (1990): The global epidemiology of HIV infection and AIDS. Annu Rev Microbiol 44:555–577.

Brown PO, Browerman B, Varmus HE, Bishop JM (1987): Correct integration of retroviral DNA in vitro. Cell 49:347–356.

Burny A, Bruck C, Cleuter Y, Dekegel D, Deschamps J, Ghysdael J, Gilden RV, Kettmann R, Marbaix G, Mammerickx M, Portetelle D (1984): Leukaemogenesis by bovine leukemia virus. In Goldman JM, Jarrett O (eds): Mechanism of Viral Leukaemogenesis. London: Churchill Livingstone, pp 229–260.

Champoux JJ, Gilboa EE, Baltimore D (1984): Mechanism of RNA primer removal by RNase H activity of AMV Reverse transcriptase. J Virol 49:686–691.

Chang L-J, Stoltzfus M (1985): Gene expression from both intronless and intron-containing Rous sarcoma virus clones is specifically inhibited by anti-sense RNA. Mol Cell Biol 5:2341–2348.

Chang L-J, Stoltzfus CM (1987): Inhibition of Rous-sarcoma virus replication by antisense RNA. J Virol 61:921–924.

Coffin J (1982): Structures of the retroviral genome. In Weiss R, Teich N, Varmus H, Coffin J (eds): RNA Tumor Viruses. Cold Spring Harbor, NY: Cold Spring Harbor Laboratory Press, pp 261–268.

Craigie R, Fujiwara T, Bushman F (1990): The IN protein of Moloney murine leukemia virus processes the viral DNA ends and accomplishes their integration in vitro. Cell 62:829–837.

Duyk G, Longiaru M, Cobrinik D, Kowal R, de Haseth P, Skalka AM, Leis J (1985): Circles with 2 tandem long terminal repeats are specifically cleaved by *pol* gene-associated endonuclease from avian sarcoma and leukosis viruses: Nucleotide sequences required for site-specific cleavage. J Virol 56:589–599.

Eglitis MA, Anderson WF (1988): Retroviral vectors for introduction of genes into mammalian cells. BioTechniques 6:608–614.

Ellis J, Bernstein A (1989): Retrovirus vectors containing an internal attachment site: Evidence that circles are not intermediates to murine retrovirus integration. J Virol 63:2844–2846.

Fenyo EM, Morfeldt-Manson L, Chiodi F, Lind BM, von Gegerfelt A, Albert J, Olausson E, Asjo B (1988): Distinct replicative and cytopathic characteristics of human immunodeficiency virus isolates. J Virol 62:4414–4419.

Finston WI, Champoux JJ (1984): RNA-primed initiation of MoMuLV plus-strands by reverse transcriptase in vitro. J Virol 51:26–33.

Fujiwara T, Craigie R (1988): Integration of mini-retroviral DNA: A cell-free reaction for biochemical analysis of retroviral integration. Proc Natl Acad Sci USA 86:3065–3069.

Fujiwara T, Mizuuchi K (1988): Retroviral DNA integration: Structure of an integration intermediate. Cell 54:497–504.

Goff SD (1989): Genetics of replication of Moloney murine leukemia virus. In Hanafusa H, Pinter A, Pullman ME (eds): Retroviruses and Disease. San Diego: Academic Press, pp 1–19.

Goodchild J, Agrawal S, Civeira MP, Sarin PS, Sun D,

Zamecnik P (1988): Inhibition of human immunodeficiency virus replication by antisense oligodeoxynucleotides. Proc Natl Acad Sci USA 85:5507–5511.

Grandgenett DP, Mumm SR (1990): Unraveling retrovirus integration. Cell 60:3–4.

Hanafusa H (1989): Transformation by Rous sarcoma virus. In Hanafusa H, Pinter A, Pullman ME (eds): Retroviruses and Disease. San Diego: Academic Press, pp 40–56.

Hardy Jr WD (1990): Biology of Feline retroviruses. In Gallo RC, Wong-Staal F (eds): Retrovirus Biology and Human Disease. Basel: Marcel Dekker, pp 33–86.

Hughes S, Greenhouse JJ, Petropoulos CJ, Sutrave P (1987): Adaptor plasmids simplify the insertion of foreign DNA into helper-independent retroviral vectors. J Virol 61: 3004–3012.

Isfort R, Witter RL, Kung H-J (1987): C-*myc* activation in an unusual retrovirus-induced avian T-lymphoma resembling Marek's disease: Proviral insertion 5' of exon one enhances the expression of an intron promoter. Oncogene Res 2:81–94.

Izant JG, Weintraub H (1985): Constitutive and conditional suppression of exogenous and endogenous genes by anti-sense RNA. Science 229:345–352.

Katz RA, Merkel G, Kulkosky J, Leis J, Skalka AM (1990): The avian retroviral IN protein is both necessary and sufficient for integrative recombination in vitro. Cell 63:87–95.

Katz RA, Skalka AM (1990): Control of retroviral RNA splicing through maintenance of suboptimal processing signals. Mol Cell Biol 10:696–704.

Katzman M, Katz RA, Skalka AM, Leis JZ (1989): The avian retroviral integration protein cleaves the terminal sequences of linear viral DNA at the in vivo sites of integration. J Virol 63:5319–5327.

Lobel LI, Goff SP (1985): Reverse transcription of retroviral genome: Mutations in terminal repeat sequences. J Virol 53:449–455.

Lobel LI, Murphy JE, Goff SP (1989): The palindromic LTR-LTR junction of MoMuLV is not an efficient substrate for proviral integration. J Virol 63:2629–2637.

Luo G, Sharmeen L, Taylor J (1990): Specificities involved in the initiation of retroviral plus-strand DNA. J Virol 64:592–597.

Majumdar C, Stein CA, Cohen JS, Broder S, Wilson SH (1989): Stepwise mechanism of HIV reverse transcriptase: Primer function of phosphorothioate oligodeoxynucleotide. Biochemistry 28:1340–1346.

Manns A, Blattner W (1990): Epidemiology of adult T-cell leukemia/lymphoma and the acquired immunodeficiency syndrome. In Gallo RC, Wong-Staal F (eds): Retrovirus Biology and Human Disease. Basel: Marcel Dekker, pp 209–239.

Mariman ECM (1985): New strategies for AIDS therapy and prophylaxis (Letter). Nature 318:414.

Matsukura M, Shinozuka K, Zon G, Mitzuya H, Reitz M, Cohen JS, Broder S (1987): Phosphorothioate analogs of oligodeoxynucleotides: Inhibitions of replication and cytopathic effects of human immunodeficiency virus. Proc Natl Acad Sci USA 84:7706–7710.

Matsukura M, Zon G, Shinozuka K, Robert-Guroff M, Shimada T, Stein CA, Mitsuya H, Wong-Staal F, Cohen JS, Broder S (1989): Regulation of viral expression of human immunodeficiency virus in vitro by an antisense phosphorothioate oligodeoxynucleotide against *rev* (*art/trs*) in chronically infected cells. Proc Natl Acad Sci USA 86:4244–4248.

Miller AD, Buttimore C (1986): Redesign of retrovirus packaging cell lines to avoid recombination leading to helper virus production. Mol Cell Biol 6:2895–2902.

Miller DA, Rosman GJ (1989): Improved retroviral vectors for gene transfer and expression. BioTechniques 7:980–990.

Mitra SW, Goff SP, Gilboa E, Baltimore D (1979): Synthesis of a 600-nucleotide long plus-strand DNA by virions of Moloney murine leukemia virus. Proc Natl Acad Sci USA 76:4355–4359.

Mitsuya H, Broder S (1989): Second generation antiviral therapy against human immunodeficiency virus (HIV). In Groopman JE, Golde DW, Evans CH (eds): Mechanisms of Action and Therapeutic Applications of Biologicals in Cancer and Immune Deficiency Disorders. New York: Alan R Liss, pp 343–359.

Moscovici C, Moscovici MG, Jiminez H, Lai MMC, Hayman MI, Vogt PK (1977): Continuous tissue culture cell lines derived from chemically induced tumors of Japanese quail. Cell 11:95–103.

Olsen JC, Swanstrom R (1985): A new pathway in the generation of defective retrovirus DNA. J Virol 56:779–789.

Panganiban AT, Temin HM (1984): Circles with two tandem LTRs are precursors to integrated retrovirus DNA. Cell 36:673–679.

Ratner L, Haseltine W, Patarca R, Livak KJ, Starcich B, Joseph SF, Doran ER, Rafalski JA, Whitehorn EA, Baumeister K, Ivanoff L, Petteway SR Jr, Pearson ML, Lautenberger JA, Papas TS, Ghrayeb J, Chang NT, Gallo RC, Wong-Staal F (1985): Complete nucleotide sequence of the AIDS-virus HTLV-III. Nature 313: 277–283.

Rattray AJ, Champoux JJ (1987): The role of Moloney murine leukemia virus RNase H activity in the formation of plus-strand primer. J Virol 61:2843–2851.

Renneisen K, Leserman L, Matthes E, Schroder HC, Muller WE (1990): Inhibition of expression of human immunodeficiency virus-1 in vitro by antibody-targeted liposomes containing antisense RNA to the *env* region. J Biol Chem 265:16337–16342.

Resnick R, Omer CA, Faras AJ (1984): Involvement of retrovirus reverse transcriptase-associated RNase H in the initiation of strong-stop (+) DNA synthesis and the generation of the long terminal repeat. J Virol 51:813–821.

Rhodes A, James W (1990): Inhibition of human immunodeficiency virus replication cell culture by endogenously synthesized antisense RNA. J Gen Virol 71:1965–1974.

Robinson HL, Zinkus DM (1990): Accumulation of human immunodeficiency virus type-1 DNA in T cells: Result of multiple infection events. J Virol 64: 4836–4841.

Roth MJ, Schwartzberg PL, Goff SP (1988): Structure of the termini of DNA intermediates in the integration of retroviral DNA dependency on IN function and terminal DNA sequence. Cell 58:47–54.

Sanchez-Pescador R, Power MD, Barr PJ, Steimer KS, Stempien MM, Brown-Shimer SL, Gee WW, Renard A, Randolph A, Levy JA, Luciw PA (1985): Nucleotide sequence and expression of an AIDs-associated retrovirus (ARV-2). Science 227:484–49.

SenGupta DN, Silverman RH (1989): Activation of an interferon-regulated dsRNA-dependent enzymes by human immunodeficiency virus-1 leader RNA. Nucleic Acids Res 17:969–978.

Sorge J, Hughes SH (1982): Polypurine tract adjacent to the U3 region of the Rous sarcoma virus genome provides a *cis*-acting function. J Virol 43:482–488.

Stein CA, Cohen JS (1988): Oligodeoxynucleotides as inhibitors of gene expression: A review. Cancer Res 48:2659–2668.

Stephenson ML, Zamecnik PC (1978): Inhibition of Rous sarcoma viral RNA translation by a specific oligodeoxyribonucleotide. Proc Natl Acad Sci USA 75: 285–288.

Stevenson M, Haggerty S, Lamonica C, Meier C, Welch S-K, Wasiak A (1990): Integration is not necessary for expression of human immunodeficiency virus type 1 protein product. J Virol 64:2421–2425.

Sullenger BA, Lee TC, Smith CA, Ungers GE, Gilboa E (1990): Expression of chimeric tRNA-antisense transcript renders NIH3T3 cells highly resistant to Moloney murine leukemia virus replication. Mol Cell Biol 10:6512–6523.

Teich N (1982): Taxonomy of retroviruses. In Weiss R, Teich N, Varmus H, Coffin J (eds): RNA Tumor Viruses. Cold Spring Harbor, NY: Cold Spring Harbor Laboratory Press, pp 25–207.

Tellier R, Weber JM (1985): New strategies for AIDS therapy and prophylaxis (Letter). Nature 318:414.

Thode S, Schafer A, Pfeiffer P, Vielmetter W (1990): A novel pathway of DNA end to end joining. Cell 60:921–928.

To RY, Booth SC, Neiman PE (1986): Inhibition of retroviral replication by antisense RNA. Mol Cell Biol 6:4758–4762.

van der Krol AR, Mol JNM, Stuitje AR (1988): Modulation of eukaryotic gene expression by complementary RNA or DNA sequences. BioTechniques 6:958–976.

Varmus H, Swanstrom L (1984): Replication of retroviruses. In Weiss R, Teich N, Varmus H, Coffin J (eds): RNA Tumor Viruses, 2nd ed. Cold Spring Harbor, NY: Cold Spring Harbor Laboratory Press, pp 369–512.

Vink C, van Gent D, Plasterk RH (1990): Integration of human immunodeficiency virus type1 and 2 DNA in vitro by cytoplasmic extracts of Moloney murine leukemia virus-infected NIH3T3 cells. J Virol 64: 5219–5222.

Vogt PK, Ishizaki R (1966): Patterns of viral interference in the avian leukosis and sarcoma complex. Virology 30:368–374.

von Ruden T, Gilboa E (1989): Inhibition of human T-cell leukemia virus type I replication in primary human T cells that express antisense RNA. J Virol 63:677–682.

Wagner RW, Nishikura K (1988): Cell cycle expression of RNA duplex unwindase activity in mammalian cells. Mol Cell Biol 8:770–777.

Walder RY, Walder JA (1988): Role of RNaseH in hybrid-arrested translation by antisense oligonucleotides. Proc Natl Acad Sci USA 85:5011–5015.

Wang L-H, Hanafusa H (1988): Avian sarcoma viruses. Virus Res 9:159–203.

Watson JD, Hopkins NH, Steiz JA, Weiner AM (1987): The extraordinary diversity of eukaryotic viruses. In Gillen JR (ed): Molecular Biology of the Gene II, 4th ed. Menlo Park, CA: Benjamin/Cummings, pp 898–959.

Weller SA, Joy AE, Temin HM (1980): Correlation between cell killing and massive second-round superinfection by members of some subgroups of avian leukosis virus. J Virol 33:494–506.

Zaia JA, Rossi JJ, Murakawa GJ, Spallone PA, Stephens DA, Kaplan BE, Eritja R, Wallace RB, Cantin EM (1988): Inhibition of human immunodeficiency virus by using an oligonucleotide methylphosphonate targeted to the *tat*-3 gene. J Virol 62:3914–3917.

Zamecnik P, Goodchild J, Taguchi Y, Sarin PS (1986): Inhibition of replication and expression of human T-cell lymphotropic virus type III in cultured cells by exogenous synthetic oligonucleotides complementary to viral RNA. Proc Natl Acad Sci USA 83: 4143–4146.

ABOUT THE AUTHOR

RICHARD Y. TO is Staff Scientist in the Basic Science Division at Fred Hutchinson Cancer Research Center in Seattle, and is also a faculty staff member in the Microbiology Department at the University of Washington. He received B.S. degrees in both zoology and microbiology from the University of Oklahoma, and moved on to graduate studies on retroviruses at Oregon State University under George Beaudreau in the Department of Biochemistry and Biophysics, where he researched on the cellular counterparts of the *myb* oncogene from avian myeloblastosis virus in a variety of avian species. After receiving his Ph.D. in 1982, he joined Dr. Paul E. Neiman at the Fred Hutchinson Cancer Research Center as a Leukemia Society of America fellow. His works relevant to antisense technology have been published in *Molecular and Cellular Biology* and *Blood*. Besides working on inhibiting retrovirus replication with antisense techniques, Dr. To is also studying the phenomenon of senescence and programmed cell death in chicken fibroblasts and bursal stem cells. He is also a collaborator in the gene therapy research ongoing at the Fred Hutchinson Cancer Center.

Antisense RNA and DNA: 285–304
© 1992 Wiley-Liss, Inc.

Regulation of HIV Gene Expression by Antisense Oligonucleotides

Makoto Matsukura

I. INTRODUCTION

It has been recognized for some time that antisense oligodeoxynucleotides complementary to mRNA could theoretically inhibit the translation of mRNA to give an encoded protein. This process has been called *translation arrest*. However, the results reported to prove such a phenomenon [Izant and Weintraub, 1985; Stephenson and Zamecnik, 1978; Zamecnik and Stephenson, 1978] were in need of further experimentation. In particular, assay methods were not properly developed for specific detection of the inhibitory activity of an antisense oligomer directed against the translation of a target mRNA.

In experiments using viruses, almost all of the experiments were conducted in a de novo infection state [Agris et al., 1986; Gupta, 1987; Lemaitre et al., 1987; Sarin et al., 1988; Shibahara et al., 1989; Smith et al., 1986; Stephenson and Zamecnik, 1978; Wickstrom et al., 1986; Zaia et al., 1988; Zamecnik et al., 1986; Zamecnik and Stephenson, 1978; Zerial et al., 1987]. Although in some cases there was pretreatment of cells with virus for several hours prior to giving the antisense treatment, there were no examples using chronically or persistently infected, virus-producing cells. This complicates the interpretation of the results obtained in previous studies. Therefore, we employed two different assay systems, namely a cytopathic effect inhibition assay and a viral gene expression inhibition assay (see sections III and IV).

In the previous reports cited above, not as many negative control sequences as antisense sequences were tested in order to demonstrate sequence specificity. Particularly in antisense experiments dealing with oncogenes, inhibitory effects on the growth of target cells were often used to assess the inhibitory activities of antisense oligomers directed against oncogene expression. This presumes that cell growth retardation is a result of inhibition of oncogene expression by the antisense oligomer. However, retardation of cell growth could also be due to toxicity of the oligomer, and in fact we found

that different sequences could have different profiles of toxicity, apparently independently of their intended antisense activity. Therefore, it is very important to use various types of negative control oligomers. Furthermore, the production of proteins encoded by the oncogenes, which was used to assess the activity in some experiments [Wickstrom et al., 1988], could also be inhibited as a consequence of cell toxicity of the oligomers. We therefore decided to have various types of control sequences to demonstrate clearly the sequence specification of the anti-HIV activity of an oligomer. We also sought to establish a proper assay system to prove the antisense activity of the oligomers. In this chapter, I describe the anti-HIV activities of phosphorothioate oligomers and point out some of the problems that still need to be solved.

II. MATERIALS

A. Synthesis of Unmodified, Normal Oligodeoxynucleotides and Phosphorothioate Analogs

Unmodified oligomers (Fig. 1) were synthesized by the conventional phosphoramidite method. The phosphorothioate oligomers (Fig. 1) were synthesized by either stepwise or single-step sulfurization methods using an Applied Biosystems Model 380B DNA Synthesizer that employed manufacturer-supplied reagents and solvents, except for the sulfurization step and the pre- and post-sulfurization wash steps, which respectively used freshly prepared solution A and solution B: solution A = 2.5 g sulfur (Aldrich, > 99.999%), 23.7 ml carbon disulfide (Aldrich, HPLC grade), 23.7 ml pyridine (Aldrich, highest purity, anhydrous), and 2.5 ml triethylamine (Aldrich, highest purity); solution B = 50 ml carbon disulfide and 50 ml pyridine.

1. Stepwise sulfurization. The details of this method have been published elsewhere [Matsukura et al., 1988] as an improved version of the procedure originally reported by Stec et al. [1984, 1985]. This stepwise sulfurization offers

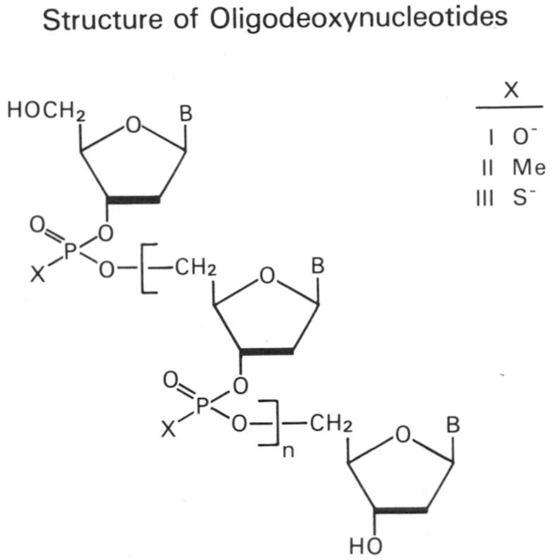
Structure of Oligodeoxynucleotides
HOCH2
O
B
X
I O-
II Me
III S-
CH2
HO
n

Fig. 1 *General molecular structure of normal (I), methylphosphonate (II), and phosphorothioate oligomers (III) (n-oligo, M-oligo, and S-oligo, respectively). B is adenine, guanine, cytosine, or thymine. [Reproduced from Matsukura et al., 1987, with permission of the publisher.]*

an advantage in cases in which there is a mixture of unmodified and/or other modifications in the phosphate linkages in the oligomers (P=O, P=S, P-Me, and so forth).

2. Single-step sulfurization. This method of synthesis is called the *H (hydrogen)-phosphonate method*. The details have also been reported elsewhere [Matsukura et al., 1988]. Since the sulfurization step is performed only once following the automated synthesis on the DNA synthesizer, convenient radiolabeling of the oligomer by ^{35}S is made possible by this method [Stein et al., 1990].

B. Sequence of the Target Regions in the HIV Genome

The *rev* region (Fig. 2) was employed for the following reasons: 1) Although HIV has a high mutation rate in its genetic sequence, the *rev* sequence is very well conserved among HIV isolates (Table I). 2) Since HIV cannot efficiently replicate without *rev* [Feinberg et al., 1986; Sodroski et al., 1986], inhibition of Rev protein synthesis by antisense oligomers against *rev* would profoundly affect HIV replication. In addition to the sequence above, we also targeted the intiation site of *gag*, the frameshift region between *gag* and *pol* (*g/p*), and unspliced *tat* (tat_{unsp}) and spliced *tat* (tat_{spl}) sequences (Table II).

C. Control Oligomers Used

To demonstrate clearly the sequence specificity of antisense activity, we synthesized and tested various types of oligomer controls relative to *rev* (Table II). Controls included the sense sequence, a random sequence (having the same net base composition as the antisense), and homo-oligomer and N^3-methylthymidine containing antisense sequences. N^3-methylthymidines (N-Me-Thd [Kyogoku et al., 1967] in an antisense sequence are known to block hybridization to the target mRNA carrying the sense sequence. In fact, there is no measurable melting temperature-Tm (midpoint of transition) between an N-Me-Thd containing antisense oligomer and a sense sequence oligomer (G. Zon, personal communication). Oligomers used were phosphorothioate oligomers (S-oligos) and unmodified "normal oligomers" (n-oligos; Fig. 1).

III. SEQUENCE NONSPECIFIC ANTI-HIV ACTIVITY

A. Methods

1. Cytopathic effect inhibition assay. The target cells ATH8 are a $CD4^+$ T-cell line immortalized by infection of HTLV-1 (human T-lymphotropic virus type 1) and is profoundly sensitive to the cytopathic effect of HIV [Mitsuya and Broder, 1986; Mitsuya et al., 1984, 1987]. ATH8 cells are kept in culture in complete medium (RPMI 1640 supplemented with 15% fetal calf serum, 4 mM L-glutamine, 50 nM 2-mercaptoethanol, and 50 units of penicillin and 50 μg of streptomycin per ml) in the presence of 20 units of recombinant interleukin-2 [IL-2, Amgen Biologicals, Thousand Oaks, CA) per milliliter and 15% (v/v) of conventional IL-2 (Advanced Biologicals, Silver Spring, MD).

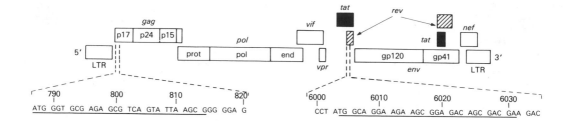

Antisense   α-*rev*: 5'-TCG TCG CTG TCT CCG CTT CTT CCT GCC A-3'

            N-Me-α-*rev*: 5'-TCG T*CG CTG TCT* CCG CT*T CTT CCT* GCC A-3'
                          (* denotes ³N-methyl-thymidine)

            α-*gag*: 5'-CGC TTA ATA CTG ACG CTC TCG CAC CCA T-3'

Sense   : 5'-TGG CAG GAA GAA GCG GAG ACA GCG ACG A-3'

Random   : 5'-TCG TCT TGT CCC GTC ATC GTT GCC CCT C-3'

Homo-oligomer dC₂₈: 5'-(dC)₂₈-3'

Fig. 2. *Genomic structure of HIV-1. The structural genes are* gag, pol, *and* env, *and the regulatory genes are* vif, tat, rev, *and* nef, *respectively. [Reproduced from Matsukura et al., 1989, with permission of the publisher.]*

TABLE I.   Conserved Sequence of HIV-1 *rev* Initiation Site Among Viral Isolates

HIV clones	Target sequences of *rev*	No. of different bases
BH10	ATG GCA GGA AGA AGC GGA GAC AGC GAC GA	Prototype
HXB2	ATG GCA GGA AGA AGC GGA GAC AGC GAC GA	0
HXB3	ATG GCA GGA AGA AGC GGA GAC AGC GAC GA	0
BH102	ATG GCA GGA AGA AGC GGA GAC AGC GAC GA	0
BH5	ATG GCA GGA AGA AGC GGA GAC AGC GAC GA	0
PV22	ATG GCA GGA AGA AGC GGA GAC AGC GAC GA	0
BRU	ATG GCA GGA AGA AGC GGA GAC AGC GAC GA	0
SF2	ATG GCA GGA AGA AGC GGA GAC AGC GAC GA	0
CDC42	ATG GCA GGA AGA AGC GGA GAC AGC GAC GA	0
RF	ATG GCA GGA AGA AG<u>A</u> GGA GAC AGC GAC GA	1*
MN	ATG GCA GGA AGA AGC GGA GAC AGC GAC GA	0
SC	ATG GCA GGA AGA AGC GGA GAC AGC GA<u>A</u> GA	1*
MAL	ATG GCA GGA AGA AGC GGA GAC AGC GAC GA	0
ELI	ATG GCA GGA AGA AGC GGA GAC AGC GAC GA	0
Z6	ATG GCA GGA AGA AGC GGA GAC AGC GAC GA	0
NL43	ATG GCA GGA AGA AGC GGA GAC AG<u>A</u> GAC GA	1*

*3.6%.

TABLE II. Sequences of Oligomers Used

1. Cytopathic effect inhibition assay using uninfected cells

ODN-1: 5'-TCG TCG CTG TCT CC-3',	antisense
ODN-2: 5'-GGA GAC AGC GAC GA-3',	sense
ODN-3: 5'-CAT AGG AGA TGC CT-3',	antisense
ODN-4: 5'-CTG GTT CGT CTC CC-3',	random
dC_n: 5'-$(dC)_n$-3',	homo-oligomer
dA_n: 5'-$(dA)_n$-3',	homo-oligomer

2. Viral expression inhibition assay using chronically infected cells

Anti-*rev*$_{28}$:	5'-TCG TCG CTG TCT CCG CTT CTT CCT GCC A-3'
N-Me-anti-*rev*$_{28}$:	5'-TCG T*CG CTG TCT* CCG CT*T CTT CCT* GCC A-3
Anti-*rev*$_{27}$:	5'-TCG TCG CTG TCT CCG CTT CTT CCT GCC-3'
Sense-*rev*:	5'-TGG CAG GAA GAA GCG GAG ACA GCG ACG A-3'
Random-*rev*:	5'-TCG TCT TGT CCC GTC ATC GTT GCC CCT C-3'
Anti-*gag*:	5'-CGC TTA ATA CTG ACG CTC TCG CAC CCA T-3'
Anti-*tat*$_{(unspl)}$:	5'-GTC GAC ACC CAA TTC TGA AAA TGG ATA A-3'
Anti-*tat*$_{(spl)}$:	5'-GTC GAC ACC CAA TTC AGT CGC CGC CCC T-3'
Anti-*G/P*-1:	5'-TCT TCC CTA AAA AAT TAG CCT GTC TCT C-3'
Anti-*G/P*-2:	5'-CCT GGC CTT CCC TTG TAG GAA GGC CAG A-3'
Anti-*G/P*-3:	5'-TGG CTC TGG TCT GCT CTG AAG AAA ATT C-3'
dC_{28}:	5'-CCC CCC CCC CCC CCC CCC CCC CCC CCC C-3'

Asterisks denote ^3N-methylthymidine.

The assay procedure has been published in detail (Mitsuya et al., 1987; Matsukura et al., 1987] (see also Fig. 3). Briefly, the target cells (2×10^5 ATH8 cells per tube) are preexposed to 1 µg/ml Polybrene (although this is not necessary) for 30 minutes, pelleted, exposed to purified HIV virus (HIV/III$_B$, 500 virus particles per cell) for 60 minutes, and resuspended in 2 ml of complete medium in the presence or absence of various concentrations of oligomers. On day 7 or later, the number of viable cells are counted in a hemocytometer under the microscope by the trypan blue dye exclusion method. Nonvirus-treated control samples were similarly cultured, except for exposure to the virus.

2. Sequences of oligomers tested. The sequences of oligomers tested in the cytopathic effect inhibition assay are shown in Table II. They consist of 14-mers that included antisense, sense, and random sequences as well as homo-oligomers. To assess length effects, 5-, 14-, 18-, 21-, and 28-mers of deoxycytidine oligomer were tested. The effects of base composition were evaluated with homo-oligomers of deoxycytidine, deoxyadenosine, and deoxythymidine in 5-, 14-, and 28-mers and are also evaluated by using oligomers that have different ratios of guanosine (G) and cytidine (C) deoxynucleotides (G + C% in the sequence).

B. Results

1. Sequence nonspecific anti-HIV activity. Fig. 4 shows the results of anti-HIV activity relative to the nuclease sensitivity and sequence specificity of oligomers. Without a protective agent, most of the ATH8 cells were killed by the viral cytopathic effect on day 7. As expected, unmodified antisense compounds (n-ODN-1, -2) showed no detectable anti-HIV activity in the assay we used. In contrast, the corresponding phosphorothioate oligomers (S-ODN-1, -2) showed potent anti-HIV activity against the cytopathic effect of HIV. However, it was unexpectedly found that even phosphorothioate oligomers with sense and random sequences, and a 14-mer homo-oligomer of deoxycytidine (S-dC$_{14}$) exhibited similar anti-HIV activity in the cytopathic effect inhibition assay system. These data strongly

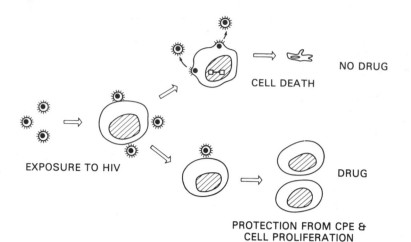

Fig. 3. *Scheme of a cytopathic effect inhibition assay that uses de novo infection with HIV. While, without protective agents, cells sensitive to the cytopathic effect of HIV are dying, in the presence of potent antiviral*

drugs cells can escape from the cytopathic effect and proliferate. Anti-HIV activity can be quantitated by counting the number of viable cells by the trypan blue dye exclusion method.

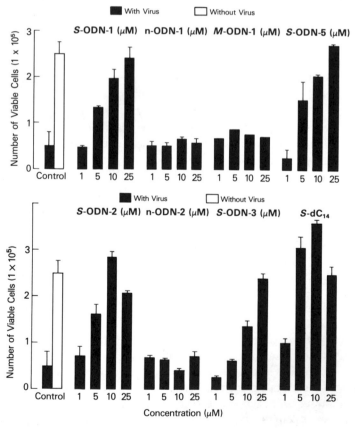

Fig. 4. *Phosphorothioate oligomers inhibit the cytopathic effect of HIV in ATH8 cells. Open columns represent the cultures without exposure to the virus, and closed columns represent the cultures exposed to the virus in the presence or absence of oligomers. S-*

ODN-1, -2, -3, -5, and S-dC₁₄ exhibit potent anti-HIV activity in the de novo infection assay. Neither n-ODN-1, -2, nor M-ODN-1 showed anti-HIV activity under the conditions used.

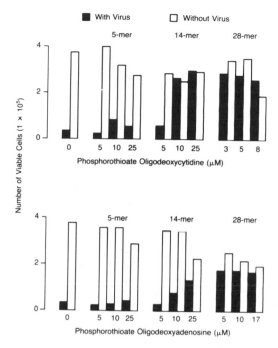

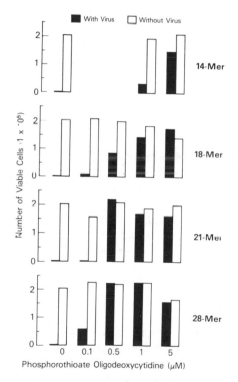

Fig. 5. *Comparison of anti-HIV activity of three lengths of S-dC and S-dA. Filled columns represent virus-exposed cells, and open columns represent nonvirus-exposed cells. The inhibitory effects of $S-dC_n$ are greater and more persistent ($S-dC_{28}$ was still active on day 28, but $S-dA_{28}$ had lost its activity), than those of $S-dA_n$ for the 14-mer and 28-mer. [Reproduced from Matsukura et al., 1987, with permission of the publisher.]*

suggest that the protective effects of phosphorothioate oligomers against de novo infection of HIV and the viral cytopathic effect do not require a specific sequence, which we therefore refer to as *sequence nonspecific anti-HIV activity*. Similar results using an indirect immunofluorescent assay were obtained in a de novo infection with H9 cells, which are resistant against the viral cytopathic effect (Matsukura, unpublished data).

2. Base composition effect on anti-HIV activity. There seemed to be some differences among the phosphorothioate oligomers initially tested, possibly because of different base compositions. Therefore we tested other homo-oligomers, namely, deoxythymidine and deoxyadenosine phosphorothioates ($S-dT_n$ and $S-dA_n$) in addition to $S-dC_n$. As shown in

Figure 5, deoxycytidine phosphorothioate oligomer was significantly more active than deoxyadenosine or deoxythymidine (data not shown) phosphorothioate oligomers. Persistence of the anti-HIV activity was also better with the deoxycytidine phosphorothioate oligomer than with the deoxyadenosine or deoxythymidine phosphorothioate oligomers (the anti-HIV activity of $S-dA_n$ and $S-dT_n$ were lost on day 14 after exposure to HIV, while the activity of $S-dC_n$ was still apparent even after 30 days; Matsukura, unpublished data). Furthermore, phosphorothioate oligomers that had a wide variety of $G+C$ content were tested in the cytopathic effect inhibition assay. Base composition seemed to be a more significant factor in shorter oligomers [Stein et al. 1989].

Fig. 6. *Detailed comparison of anti-HIV activity between the 14-mer and 28-mer of $S-dC_n$. Filled columns represent virus-exposed cells, and open columns represent nonvirus-exposed cells. There is an obvious length effect in sequence-nonspecific anti-HIV activity even with an increase of several nucleotide lengths as short as three nucleotides (see the data for the 18-mer and 21-mer). [Reproduced from Matsukura et al., 1987, with permission of the publisher.]*

3. Length effect on anti-HIV activity.

The 14-, 18-, 21-, and 28-mers of deoxycytidine phosphorothioate oligomers (S-dC$_{14}$, -dC$_{18}$, -dC$_{21}$, and -dC$_{28}$) were simultaneously tested. The results in Figure 6 show a clear length dependency in their anti-HIV activities. It is worthwhile noting that 0.5 μM of S-dC$_{28}$ showed more activity than 1 μM of S-dC$_{14}$, even though both have the same number of monomeric nucleoside residues (dCs), suggesting a real length effect. In other words, the oligomers probably functioned in their full length but not much in degraded, shorter lengths.

4. Combination of S-oligo and 2′,3′-dideoxynucleoside (ddN).

A possible enhancement of the anti-HIV activity of ddNs by S-oligos was tested with a combination of 2′,3′-dideoxyadenosine (ddA) and S-dC$_{14}$ in the cytopathic effect inhibition assay using ATH8 cells. As shown in Figures 7 and 8, 2 μM ddA and 5 μM S-dC$_{14}$ each exhibited a very marginal effect; however, the combination of both almost nullified the cytopathic effect of HIV, thus suggesting a synergistic enhancement of anti-HIV activity. The mechanisms of such synergism is not established at this time, but could well result from a difference in the mechanisms of action of ddA and S-oligos.

5. Mechanism of sequence nonspecific anti-HIV activity.

The results obtained in the cytopathic effect inhibition assay were confounding because sequence-specific inhibition of virus at the level of viral expression was originally expected. Later, as described below, it was found that phosphorothioate oligomers do exhibit a sequence-specific inhibitory effect on HIV expression in a different assay system, and therefore unique mechanisms other than translation arrest were postulated for this sequence nonspecific antiviral activity. To see the effect on viral DNA synthesis after HIV infection, the synthesis of viral DNA was studied by using Southern blot analysis and found to be completely inhibited by S-oligo in de novo infection (Fig. 9). This indicated that the intervention of the viral life cycle by S-oligo is prior to and/or at the stage of reverse transcription in the HIV

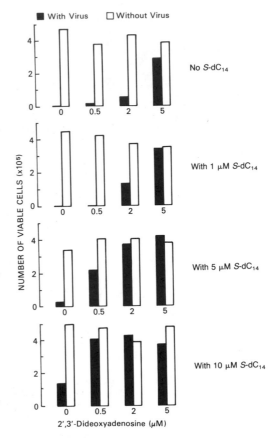

Fig. 7. *Synergistic enhancement of sequence-nonspecific anti-HIV activity of 2′,3′ dideoxyadenosine (ddA) with S-dC$_{14}$. Since this experiment involved a more potent viral inoculum for a longer duration than in the other experiments reported here, 5 μM S-dC$_{14}$ or 2 μM ddA alone exhibits very marginal protection compared with the data in Figures 4 and 6. In combination with 2 μM of ddA, 5 μM S-dC$_{14}$ exhibits a 100% protective effect. [Reproduced from Matsukura et al., 1987, with permission of the publisher.]*

life cycle. It is also found that S-oligos can inhibit syncytium formation by the virus (H. Mitsuya, personal communication). Since S-oligos do not significantly block the binding of HIV to CD4$^+$ H9 cells at concentrations at which it is effective as an anti-HIV compound [Matsukura et al., 1987], we now postulate that S-oligos inhibit viral replication, at least in part, by inhibiting the fusion step between the virus and target cells that follows

SYNERGISTIC ANTI-HTLV-III/LAV ACTIVITY OF DIDEOXYADENOSINE AND OLIGODEOXYCYTIDINE PHOSPHOROTHIOATE

NO DRUG

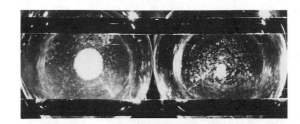

2μM DIDEOXYADENOSINE

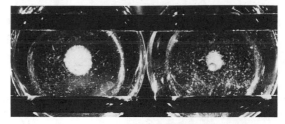

5μM OLIGODEOXYCYTIDINE PHOSPHOROTHIOATE

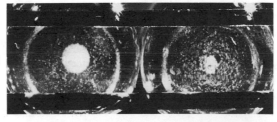

2μM DIDEOXYADENOSINE + 5μM OLIGODEOXYCYTIDINE PHOSPHOROTHIOATE

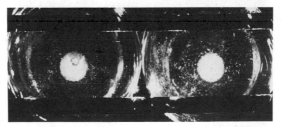

NO VIRUS   HTLV-III

Fig. 8. *The demonstration of synergistic enhancement of anti-HIV activity of ddA with S-dC$_{14}$. Round pellets are viable ATH8 cells at the bottom of culture tubes. In the presence of the virus, all cultures except for the combination treatment with 2 μM ddA and 5 μM S-dC$_{14}$ show very small pellets, which indicate complete cell death in the culture (see Fig. 7 for viable cell numbers).*

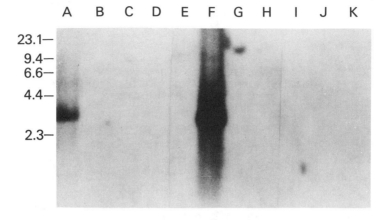

Fig. 9 *S-dC$_{28}$ inhibits de novo HIV DNA synthesis in ATH8 cells exposed to the virus. On day 4 (lanes **A–E**) and day 7 (lanes **F–J**) following exposure to the virus, high molecular DNA was extracted, digested with Asp718 (a KpnI isoschizomer), subjected to Southern blot analysis and hybridized with a labeled insert of a molecular clone of the* env *region of HIV/BH10 containing a 1.3 kb BglII fragment. Lanes A and F contain DNA from ATH8 cells that were exposed to the virus and not protected by S-dC$_{28}$. Lanes B and G, C and H, and D and I contain DNA from ATH8 cells cultured with 1, 5, and 7 μM S-dC$_{28}$, respectively. Lanes E and J contain DNA from ATH8 cells treated with 50 μM ddA, and lane **K** contains DNA from ATH8 cells that were not exposed to the virus. Sizes are shown in kb. The 2.7 kb* env*-containing internal KpnI fragment of the virus genome was detected only in lanes A and F. [Reproduced from Matsukura et al., 1987, with permission of the publisher.]*

virus binding. It is, however, possible that S-oligo has multifaceted effects on HIV replication, including inhibition of the reverse transcription. Experimental tests of this proposal require further studies.

IV. SEQUENCE-SPECIFIC ANTI-HIV ACTIVITY

A. Methods

1. Viral gene expression inhibition assay (Fig. 10). To generate chronically infected H9 cells (H9/III$_B$), H9 cells were exposed to HIV-1$_B$ (formerly known as HTLV-III$_B$) at an inoculum of 500 virus particles per cell The H9 cells were almost all infected on day 10 after the exposure to the virus. However, since infected cells shortly after exposure to HIV generally tend to be fragile and fluctuate in their viral production, H9/III$_B$ cells were kept in culture for several months or more in complete medium (RPMI 1640 supplemented with 15% fetal calf serum, 4 mM L-glutamine,

50 nM 2-mercaptoethanol, 50 units of penicillin, and 50 μg of streptomycin per ml) prior to the experiment. To detect inhibitory activities (antisense activity) of various oligomers against viral expression and the toxicities of the oligomers in a *quantitative manner*, we first assessed the effect of the number of inoculated cells on cell growth and viral production.

Different numbers of H9/III$_B$ cells (1,250–10,000 cells per well in 96-well culture plate) were inoculated, and both cell growth (by ^{3}H-thymidine uptake) and the viral production in the supernatant (by p24 gag protein *antigen* assay: ELISA or RIA) were quantitated. Figures 11 and 12 show that an inoculum of H9/III$_B$ at 1,250 cells per well in a 96-well culture plate gave a linear increase of p24 gag protein in the supernatant during a 5 day culture and that ^{3}H-thymidine uptake is still increasing on day 5, suggesting that the culture is still in good condition at the end of assay. Increasing the inoculum above 1,250 gave non-linear curves (reaching plateau levels) of cel-

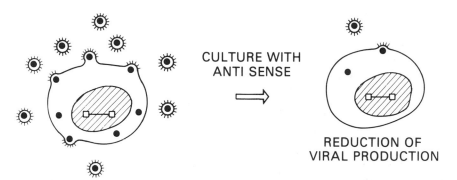

CULTURE WITH
ANTI SENSE

REDUCTION OF
VIRAL PRODUCTION

Fig. 10. *Scheme showing the concept that, if antisense oligomer is active in inhibiting viral gene expression of HIV, reduction of HIV production should be observed with the antisense in this assay system.*

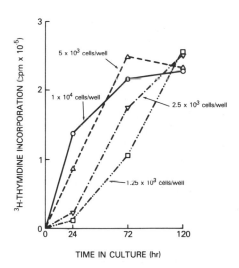

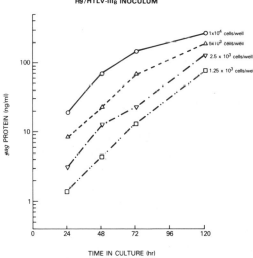

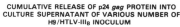

Fig. 11. *About 1,250 cells of H9/III$_B$ per well in a 96-well flat culture plate as an inoculum are still growing on day 5 in the experiment. Culture with inoculums of more than 2,500 cells per well reach plateau in growth on day 5. [Reproduced from Matsukura et al., 1991.]*

Fig. 12. *The culture with 1,250 cells per well as an inoculum shows an exponential increase of p24 gag production in supernatant during the 5 day culture period, but other cultures with more cells do not. [Reproduced from Matsukura et al., 1991.]*

TABLE III. Densitometric Analysis of Nuclease Protection Assay

RNA species	Day 5			Day 28		
	No compound	S-dC$_{28}$	S-α-rev	No compound	S-dC$_{28}$	S-α-rev
Genomic (9.2 kb)	0.89	0.56	0.04	0.53	0.66	0.01
env (4.3 kb)	1.98	0.96	0.77	1.09	1.22	0.81
tat/rev (2.0 kb)	2.71	1.21	0.71	1.52	1.17	0.47

Values are normalized to the density of γ-actin message in the same lane (γ-actin = 1.00; see Fig. 17). On days 5 and 28, the ratio of unspliced mRNA to spliced mRNA in the presence of 25 μM S-α-rev compared with that of the no-drug control culture and control culture with 25 μM S-dC$_{28}$ was examined by scanning the autoradiography of the RNase protection assay (Fig. 17). Note the drastic reduction (>95% from the value of the no-drug control) of unspliced genomic HIV-1 mRNA in samples treated with 25 μM S-α-rev compared with the spliced mRNAs such as env and tat/rev. [Reproduced from Matsukura et al., 1989, with permission of the publisher.]

lular growth and p24 gag protein production. Lower levels of inoculum than the optimum also gave nonlinear curves in p24 gag production (data not shown). As a consequence, 1,250 cells of H9/III$_B$ per well were employed for the assay as an optimal condition.

2. Radioimmunoprecipitation assay (RIPA). Cell cultures of H9/III$_B$ are metabolically labeled with 2.5 mCi of [^{35}S]-methionine and [^{35}S]-cysteine for 4 hours. Equivalent amounts of trichloroacetic acid–precipitable radioactivity were treated with the control antibodies or monoclonal antibodies against HIV viral proteins.

3. Northern blot analysis. Cytoplasmic RNA are extracted by the vanadyl-ribonucleoside complex method from H9/III$_B$ on day 5 or day 28 of culturing in the presence or absence of phosphorothioate oligomers. RNA (10 μg) was subjected to electrophoresis on a formaldehyde/agarose gel, transferred to Zeta probe membrane (Bio-Rad), and hybridized with a nick-translated ^{32}P-labeled DNA probe.

4. RNase protection assay. The method was as reported previously [Melton et al., 1984]. A 0.6 kb EcoRI–KpnI fragment (5,776–6,377) from the cloned HIV BH10 genome was subcloned into pGEM4 (Promega), and a uniformly labeled RNA probe was synthesized with phage T7 RNA polymerase. The human γ-actin probe was synthesized from HinfI-digested pSP6-actin plasmid [Enoch et al.

1986]. Two micrograms of cytoplasmic RNA and ~2 × 10^6 cpm each of HIV and human γ-actin probe were hybridized, and RNase protected fragments were analyzed.

B. Results

1. Inhibitory effect against viral production. Two representative sequences, S-anti-rev$_{28}$ and S-dC$_{28}$, were studied first. S-dC$_{28}$ was one of the most potent sequences tested in the cytopathic effect inhibition assay against de novo infection of HIV (see section III). In chronically infected cells, however, S-dC$_{28}$ did not inhibit p24 gag protein production at the concentrations tested, whereas S-anti-rev$_{28}$ inhibited viral protein production as measured in the supernatant (~90% at 25 μM), in a dose-dependent manner without significant toxicity (Fig. 13). All of the control sequences tested (phosphorothioate senses, random, homocytidine, N-Me-TdR containing anti-rev$_{28}$ S-oligomer, and nuclease-sensitive normal oligomer antisense) failed to inhibit viral expression (Fig. 14), thus confirming both the sequence specificity of the antisense activity of S-anti-rev$_{28}$ and the necessity of nuclease resistance for the antisense activity of S-anti-rev$_{28}$. Another viral protein, envelope glycoprotein, was assayed using radioimmunoprecipitation in the presence of monoclonal antibody against gp160 [Matsushita et al. 1988]. The production of envelope was inhibited to a similar extent to

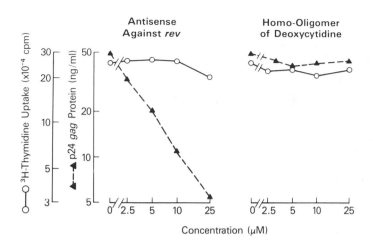

Fig. 13. *Only s-anti-*rev *(28-mer) shows a potent dose-dependent inhibition of the viral p24 gag protein production (*left*). S-dC$_{28}$ shows no inhibition even at 25 μM highest concentration (*right*). [Reproduced from Matsukura et al., 1989, with permission of the publisher.]*

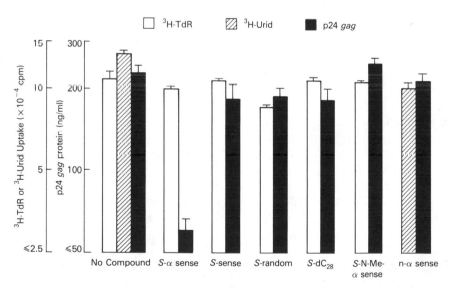

Fig. 14. *Phosphorothioate oligomers (28-mer), including antisense sequence and control sequences, show sequence-specific inhibition of p24 gag production. The concentration of phosphorohioate oligomers used here was 10 μM, which does not seem to cause any toxicity to cells. Error bars represent standard deviations of the data. S-anti-*rev *and n-anti-*rev *are denoted by S-α* rev *and n-α* rev*. [Reproduced from Matsukura et al., 1989, with permission of the publisher.]*

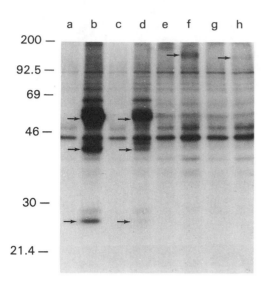

Fig. 15. *S-anti-*rev$_{28}$ *inhibits not only p24 gag protein but also gag protein precursors p55, p38, and envelope glycoprotein gp120 in radioimmunoprecipitation assay. Cell cultures of H9/III$_B$ are metabolically labeled with 2.5 mCi of [^{35}S]methionine and [^{35}S] cysteine for 4 hours. Equivalent trichloroacetic acid–precipitable radioactivity was treated with the following antibodies: 10 μl of control mouse ascites fluid generated by p3 × 63 cells (lanes **a** and **c**), 10 μl of mouse ascites fluid containing the monoclonal antibody to HIV-1 p24 [Robert-Guroff et al., 1987] (lanes **b** and **d**), 5 μg of control mouse IgG (lanes **e** and **g**), and 5 μg of mouse monoclonal IgG antibody to env [Matsushita et al., 1988; Robert-Guroff et al., 1987] (lanes **f** and **h**). Lanes a, b, e, and f, no drug control; c, d, g, and h, samples treated with 25 μM S-anti-*rev$_{28}$*. The gag proteins p55, p38, and p24 indicated by arrows in lane d (s-anti-rev) are greatly reduced in comparison with those in lane b (control), as was gp120 env glycoprotein, indicated by the arrow in lane h (S-anti-rev) in comparison with that in lane f (control). [Reproduced from Matsukura et al., 1989, with permission of the publisher.)*

p24 gag protein and its precursor proteins p55 and p38 (Fig. 15), which indicated that anti-*rev* could inhibit the complete process of viral production rather than the production of only one viral protein. Since this assay system used cellular lysate labeled with [^{35}S]methionine and [^{35}S]cysteine, the results also indicated that S-anti-*rev*$_{28}$ could inhibit not only viral release into the culture supernatant but also the intracellular viral production per se. It is worth noting that the assay systems used for sequence-specific anti-HIV activity sometimes produced

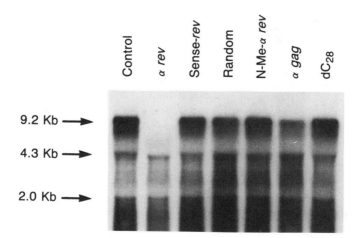

Fig. 16. *Northern blot analysis showing the reduction of genomic mRNA of HIV-1 in the presence of S-anti-*rev$_{28}$*. Cytoplasmic RNAs are extracted from H9/III$_B$ on day 5 in culture in the presence or absence of phosphorothioate oligomers. Note the remarkable change of mRNA profile in S-anti-*rev$_{28}$ (α rev in this figure) but not in other controls including S-anti-gag (α gag). [Reproduced from Matsukura et al., 1989, with permission of the publisher.]*

variable data to a certain extent and therefore repeated experiments were necessary to produce conclusive data such as I describe here. The viral expression of different isolates and clones of HIV-1 other than HIV-1/III$_B$ should theoretically be inhibited by the same S-anti-rev_{28} oligomer for the reasons discussed above. It is still possible, however, that the expression of other isolates or clones of HIV would not be inhibited as well as that of HIV-1/III$_B$, because multiple factors, including possible differences in actual secondary structures of the target mRNA at the *rev* initiation site among the isolates and the clones, could affect the inhibitory activity of S-anti-rev_{28} against the viral expression of HIV-1. These questions should be studied further using multiple isolates.

2. Modification of HIV mRNA profile. In the viral expression of HIV, three sizes of mRNA, namely, the 9.2 kb genomic mRNA (unspliced), 4.3 kb mRNA (partially spliced), and 2.0 kb mRNA (fully spliced), are observed. The 9.2 kb genomic mRNA also serves as template for the synthesis of gag and pol, while partially spliced 4.3 kb mRNA is the template for the synthesis of env. One of the functional viral genes (translated from fully spliced 2.0 kb mRNA), *rev*, is known in HIV to affect the profile of such viral mRNA splicing, perhaps resulting from stabilization of the mRNA in its presence [Felber et al., 1989; Sadaie et al., 1988].

In addition, we hypothesized that S-anti-rev_{28}, which inhibited viral expression in the assay we used, functioned via inhibition of rev protein production. Therefore, a Northern blot analysis and an RNA protection assay were performed to examine the hypothesis. As shown in Figure 16, all of the control sequences tested, but not S-anti-rev_{28}, did not change the mRNA profile. A similar observation was made with the RNase protection assay (Fig. 17). Even after 28 days of continuous exposure to 25 μM oligomers, the modified mRNA profile (almost no genomic mRNA) was still observed with S-anti-rev_{28}, but not with S-dC$_{28}$ (Fig. 17), thus indicating no emergence of virus resistant to the antisense treatment during the time of culture.

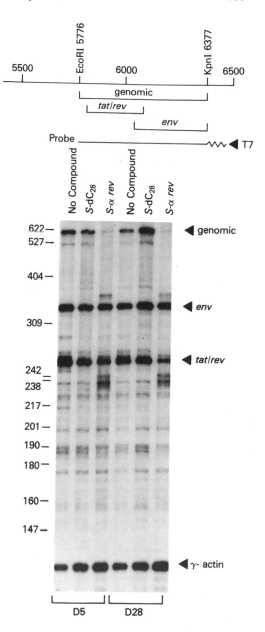

Fig. 17. *RNase protection analysis of HIV-1 cytoplasmic mRNA performed with the samples on days 5 and 28. The expected sizes of fragments protected by the transcripts for gag/pol (genomic), env, and tat/rev are 601, 340, and 268 bp, respectively. Human γ-actin sequence served as an internal control. The numbers on the left indicate the nucleotide lengths of the MspI-digested pBR322 marker. See Table III for the ratio of each HIV-1 mRNA band normalized to the γ-actin mRNA band. [Reproduced from Matsukura et al., 1989, with permission of the publisher.]*

ANTI-SENSE PHOSPHOROTHIOATE OLIGOMER AGAINST
trs/art MODIFIES HIV mRNA PROFILE AND INHIBITS
gag PROTEIN PRODUCTION IN CHRONICALLY HIV-INFECTED
H9 CELLS

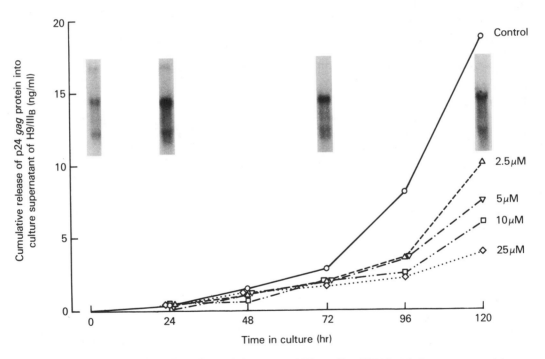

Fig. 18. *After 96 hours, dose-dependent inhibitory effects of S-anti-rev₂₈ to p24 gag protein production are observed. Prior to the change in p24 gag production, there appears to be a significant change in the mRNA profile of HIV-1 with disappearance of the genomic mRNA. [Modified from Matsukura et al., 1989, with permission of the publisher.]*

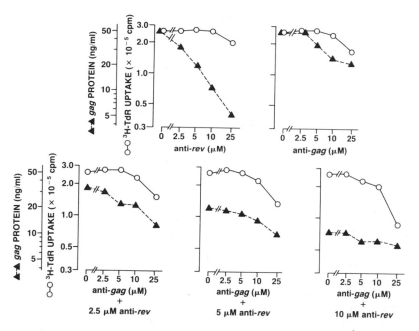

Fig. 19. *S-anti-gag shows a very marginal inhibitory effect against the viral expression. While 25 μM S-anti-rev₂₈ profoundly inhibits the gag protein production (approximately 90%), S-anti-gag shows only 50% inhi-* bition at same concentration. The combination of S-anti-gag with -rev₂₈ does not give synergistic inhibitory effects. [Reproduced from Matsukura et al., 1991.]

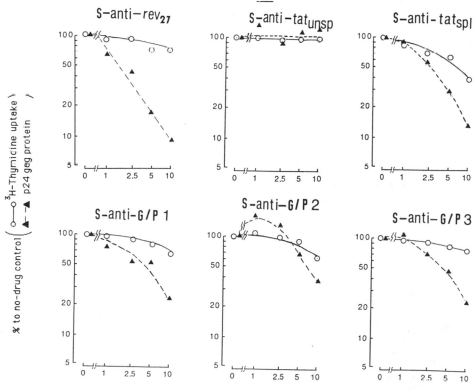

Fig. 20. *Other sequences apart from S-anti-rev₂₇ (see Table II) show less inhibitory effect against viral expression. Note the significant difference in antisense activ-* ities between antisenses against unspliced tat (including splice junction) and spliced tat (mRNA sequence, upstream of initiation site for tat mRNA).

3. Time course. To assess the dynamics of the inhibitory activity of S-anti-rev_{28}, a time course (after 24, 48, 72, 96, and 120 hours of exposure to the oligo) was studied. Significant dose-dependent anti-HIV activity in p24 gag protein production was not observed until 96–120 hours. In the Northern blot assay at as early as 72 hours, there was a drastic change of HIV mRNA profile (disappearance of genomic mRNA, 9.4 kb; Fig. 18). Mechanistically, the time delay between these two time profiles could result from sequential effects, which are translation arrest of *rev* mRNA into rev protein by S-anti-rev_{28} → lower level of intracellular rev protein → change of mRNA splicing pattern → inhibition of production of viral structural proteins. This hypothesis, however, needs to be tested by further experiments using a particular system to detect quantitatively rev protein production in the cell culture-system.

4. Other antisense S-oligos. Other target sequences, including the initiation site of *gag* (S-anti-*gag*), the frameshift region between *gag* and *pol* (S-anti-*G/P*), and the initiation site of *tat* exon 1 (S-anti-tat_{spl}; Fig. 2 and Table II) were tested. S-anti-*gag* shows marginal inhibitory effect in the assay. Even in combination with S-anti-rev_{28}, S-anti-*gag* exhibited comparatively little antiviral activity against HIV (Fig. 19). Figure 20 shows that there was significant inhibition of the viral production by S-anti-tat_{spl}. In contrast, S-anti-tat_{unspl} gave no inhibition of viral production, suggesting that the inhibition of viral protein production by S-anti-tat_{spl} may be from hybridization with the viral mRNA, but probably not with the genomic proviral DNA of HIV-1. S-anti-*G/P*-1, -2, and -3 showed inhibitory activity that was much less than that of S-anti-rev_{27} (Fig. 20).

V. CONCLUSION

It was demonstrated here that phosphorothioate oligomers can exhibit anti-HIV activities by multiple mechanisms, namely, sequence-specific (antisense) and sequence nonspecific (nonantisense) mechanisms. The importance of well-defined experimental conditions (i.e., the distinction between a de novo infection assay using uninfected cells and a chronically infected cell assay) was also emphasized. To establish *real* sequence specificity, it is most important to have control sequences (e.g., sense, random, ^3N-methylthymidine containing antisense, homo-oligomer, and other antisense constructs), as many as is reasonable or necessary. It is also very critical to distinguish the antisense activity from the toxicities of oligomers. The field has shown growth, but much research remains to be done. Since this class of compounds is unprecedented as a therapeutic, the pharmacokinetics and toxicology of phosphorothioate oligomers should be explored extensively. Also, in particular for the development of S-anti-*rev* as an anti-HIV compound, a detailed profile of the possible difference of its anti-HIV activity depending on viral isolates and cell types should be explored.

ACKNOWLEDGMENTS

I thank Drs. Gerald Zon, Samuel Broder, Jack Cohen, Hiroaki Mitsuya, Ichiro Matsuda, Teruhisa Miike, and Shinji Harada for their support during this project and Drs. John Dahlberg, William Egan, and Patrick Iversen for helpful discussion. This was supported in part by a grant from the Yamanouchi Foundation for Research on Metabolic Disorders.

VI. REFERENCES

Agris CH, Blake KR, Miller PS, Reddy MP, Ts'o PO (1986): Inhibition of vesicular stomatitis virus protein synthesis and infection by sequences specific oligodeoxynucleoside methylphosphonates. Biochemistry 25:6268–6270.

Enoch T, Zinn K, Maniatis T (1986): Activation of the human β-interferon gene requires an interferon-inducible factor. Mol Cell Biol 6:801–810.

Feinberg MB, Jarrett RF, Aldovini A, Gallo RC, Wong-Staal F (1986): HTLV-III expression and production involve complex regulation at the levels of splicing and translation of viral RNA. Cell 46:807–817.

Felber BK, Hadzopoulou-Cladaras M, Cladaras C, Copeland T, Pavlakis GN (1989): Rev protein of human immunodeficiency virus type 1 affects the stability and transport of the viral mRNA. Proc Natl Acad Sci USA 86:1495–1499.

Gupta KC (1987): Antisense oligonucleotides provide insight into mechanism of translation initation of two sendai virus mRNAs. J Biol Chem 262:7492–7496.

Izant JT, Weintraub H (1985): Constitutive and conditional suppression of exogegnous and endogenous genes by anti-sense RNA. Science 229:345–352.

Kyogoku Y, Lord RC, Rich A (1967): The effect of substituents on the hydrogen bonding of adenine and uracil derivatives. Proc Natl Acad Sci USA 57:250–257.

Lemaitre M, Bayard B, Lebleu B (1987): Specific antiviral activity of a poly(L-lysine)-conjugated oligodeoxyribonucleotide sequence complementary to vesicular stomatitis virus N protein mRNA initiation site. Proc Natl Acad Sci USA 84:648–652.

Matsukura M, Mitsuya H, Broder S (1991): A new concept in AIDS treatment: An antisense approach and its current status towards clinical application. In Wickstrom E (ed): Prospects for Antisense Nucleic Acid Therapy of Cancer and AIDS. New York: Wiley-Liss, pp 159–178.

Matsukura M, Shinozuka K, Zon G, Mitsuya H, Reitz M, Cohen JS, Broder S (1987): Phosphorothioate analogs of oligodeoxynucleotides: Inhibitors of replication and cytopathic effects of human immunodeficiency virus. Proc Natl Acad Sci USA 84:7706–7710.

Matsukura M, Zon G, Shinozuka K, Robert-Guroff M, Shimada T, Stein CA, Mitsuya H, Wang-Staal F, Cohen JS, Broder S (1989): Regulation of viral expression of human immunodeficiency virus in vitro by an antisense phosphorothioate oligodeoxynucleotide against *rev* (*art/trs*) in chronically infected cells. Proc Natl Acad Sci USA 86:4244–4248.

Matsukura M, Zon G, Shinozuka K, Stein CA, Mitsuya H, Cohen JS, Broder S (1988): Synthesis of phosphorothioate analogues of oligodeoxyribonucleotides and their antiviral activity against human immunodeficiency virus (HIV). Gene 72:343–347.

Matsushita S, Robert-Guroff M, Rusche J, Koit A, Hattori T, Hoshino H, Javaherian K, Takatsuki K, Putney S (1988): Characterization of a human immunodeficiency virus neutralizing monoclonal antibody and mapping of the neutralizing epitope. J Virol 62:2107–2114.

Melton DA, Krieg PA, Rebagliati MR, Maniatis T, Zinn K, Green MR (1984): Efficient in vitro synthesis of biologically active RNA and RNA hybridization probes from plasmids containing a bacteriophage SP6 promoter. Nucleic Acids Res 12:7035–7056.

Mitsuya H, Broder S (1986): Inhibition of the in vitro infectivity and cytopathic effect of human T-lymphotropic virus type III/lymphadenopathy-associated virus (HTLV-III/LAV) by 2′,3′-dideoxynucleosides. Proc Natl Acad Sci USA 83:1911–1915.

Mitsuya H, Matsukura M, Broder S (1987): Rapid in vitro systems for assessing activity of agents against HTLV-III/LAV. In Broder S (ed): AIDS: Modern Concepts and Therapeutic Challenges. New York: Marcel Dekker, pp 303–334.

Mitsuya H, Popovic M, Yarchoan R, Matsushita S, Gallo RC, Broder S (1984): Suramin protection of T cells in vitro against infectivity and cytopathic effect of HTLV-III. Science 226:172–174.

Robert-Guroff M, Giardina PJ, Robey WG, Jennings AM, Naugle CJ, Akbar AN, Grady RW, Hilgartner MW (1987): HTLV-III neutralizing antibody development in transfusion-dependent seropositive patients with β-thalassemia. J Immunol 138:3731–3736.

Sadaie MR, Benter T, Wong-Staal F (1988): Site-directed mutagenesis of two *trans*-regulatory genes (*tat*-III, *trs*) of HIV-1. Science 239:910–913.

Sarin PS, Agrawal S, Civeira MP, Goodchild J, Ikeuchi T, Zamecnik PC (1988): Inhibition of acquired immunodeficiency syndrome virus by oligodeoxynucleoside methylphosphonates. Proc Natl Acad Sci USA 85: 7448–7451.

Shibahara S, Mukai S, Morisawa H, Nakashima H, Kobayashi S, Yamamoto N (1989): Inhibition of human immunodeficiency virus (HIV-1) replication by synthetic oligo-RNA derivatives. Nucleic Acids Res 17: 239–252.

Smith CC, Aurelian L, Reddy MP, Miller PS, Ts'o POP (1986): Antiviral effect of an oligo(nucleotide methylphosphonate) complementary to the splice junction of herpes simplex virus type 1 immediately early premRNAs 4 and 5. Proc Natl Acad Sci USA 83: 2787–2791.

Sodroski J, Goh WC, Rosen C, Dayton A, Terwilliger E, Haseltine W (1986): A second post-transcriptional *trans*-activator gene required for HTLV-III replication. Nature 321:412–417.

Stec WJ, Zon G, Egan W, Stec B (1984): Automated soild-phase synthesis, separation, and stereochemistry of phosphorothioate analogues of oligodeoxyribonucleotides. J Am Chem Soc 106:6077–6079.

Stec WS, Zon G, Uznanski B (1985): Reversed-phase high-performance liquid chromatographic separation of diastereomeric phosphorothioate analogues of oligodeoxyribonucleotides and other back-bone–modified congeners of DNA. J Chromatogr 326:263–280.

Stein CA, Iversen PL, Subasinghe C, Cohen JC, Stec WJ, Zon G (1990): Preparation of ^{35}S-labeled polyphosphorothioate oligodeoxyribonucleotides by use of hydrogen-phosphonate chemistry. Anal Biochem 188:11–16.

Stein CA, Matsukura M, Subasinghe C, Broder S, Cohen JS (1989): Phosphorothioate oligodeoxynucleotides are potent sequence nonspecific inhibitors of de novo infection by HIV. AIDS Res Hum Retro 5:639–646.

Stephenson ML, Zamecnik PC (1978): Inhibition of Rous sarcoma viral RNA translation by a specific oligodeoxyribonucleotide. Proc Natl Acad Sci USA 75: 285–288.

Wickstrom EL, Bacon TA, Gonzalez A, Freeman DL, Lyman GH, Wickstrom E (1988): Human promyelocytic leukemia HL-60 cell proliferation and c-*myc* protein expression are inhibited by an antisense pentadecadeoxynucleotide targeted against c-*myc* mRNA. Proc Natl Acad Sci USA 85:1028–1032.

Wickstrom E, Simonet W, Medlock K, Ruiz-Robles I (1986): Complementary oligonucleotide probe of vesicular stomatitis virus matrix protein mRNA translation. Biophys J 49:15–17.

Zaia JA, Rossi JJ, Murakawa GJ, Spallone PA, Stephens DA, Kaplan BE, Eritja R, Wallace RB, Cantin EM (1988): Inhibition of human immunodeficiency virus by using an oligonucleotide methylphosphonate targeted to the *tat*-3 gene. J Virol 62:3914–3197.

Zamecnik PC, Goodchild J, Taguchi Y, Sarin PS (1986): Inhibition of replication and expression of human T-cell lymphotropic virus type III in cultured cells by exogenous synthetic oligonucleotides complementary to viral RNA. Proc Natl Acad Sci USA 83:4143–4146.

Zamecnik PC, Stephenson ML (1978): Inhibition of Rous sarcoma virus replication and cell transformation by a specific oligodeoxynucleotide. Proc Natl Acad Sci USA 75:280–284.

Zerial A, Thuong NT, Helene C (1987): Selective inhibition of the cytopathic effect of type A influenza viruses by oligodeoxynucleotides covalently linked to an intercalating agent. Nucleic Acids Res 15:9909–9919.

ABOUT THE AUTHOR

MAKOTO MATSUKURA is currently a senior staff member of the Department of Child Development at the Kumamoto Medical School in Japan, where he teaches pediatric neurology, pharmacology, and virology/immunology. After receiving an M.D. from Kumamoto University in 1975, he spent eight years at the University Hospital, training to become a pediatrician. During that period, Dr. Matsukura developed an interest in clinical pharmacology, focusing on antiepileptic drug monitoring and the pharmacokinetics of other clinically important drugs such as salicylate and theophylline. He spent the next six years in the United States, first studying the metabolism of 6-mercaptopurine at Case Western Reserve University Medical School in Cleveland, Ohio, then moving to the National Cancer Institute (NCI) to study anti-HIV compounds with Drs. Samuel Broder and Hiroaki Mitsuya. As part of the Clinical Oncology Program at NCI, he collaborated with Dr. Gerald Zon on an antisense phosphorothioate oligomer against HIV infection. Now back in Japan, his current research focuses on the development of antisense compounds against other viral infections. Dr. Matsukura's research papers have appeared in such journals as the *Proceedings of the National Academy of Sciences USA* and *Gene*, and he serves on the editorial board of *Antisense Research and Development*.

Antisense RNA and DNA: 305–316
© 1992 Wiley-Liss, Inc.

Alternative Antiviral Approaches to Influenza Virus: Antisense RNA and DNA

Sudhir Agrawal and Josef M.E. Leiter

I. INTRODUCTION

The acute respiratory illness known as *influenza* remains an unresolved problem afflicting both humans and animals. A long-term vaccination approach has failed mainly because of the high degree of antigenic variation among influenza viruses. Amantadine and its analog rimantadine are useful drugs for prophylaxis against certain strains of influenza A virus, but acquisition of resistance is seen frequently and is not associated with attenuation. Hence, alternative approaches against influenza virus infection need to be explored.

We first summarize the basics of influenza virus replication, conventional approaches to prevention and treatment of influenza, and antiviral strategies using antisense RNA and DNA. We then describe experiments using phosphorothioate oligodeoxynucleotides to inhibit influenza virus replication. We end with an outlook on possible "oligonucleotide therapy" for influenza and preliminary toxicity results of oligonucleotides in mice and rats.

II. STRUCTURE AND REPLICATION OF INFLUENZA VIRUSES

The biology of influenza viruses has been reviewed extensively elsewhere [Palese and Kingsbury, 1983; Krug, 1989]. The three types of influenza viruses (A, B, and C) were originally distinguished by differences between their internal antigens (NP protein and M1 protein). They possess a single-stranded and segmented RNA genome of negative polarity. Being negative-strand viruses, they require an RNA-dependent RNA polymerase activity, which is encoded by the virus and present in the virus particle (PA, PB1, and PB2 proteins). The viral core is enclosed in a lipid bilayer membrane, which is derived from the infected cell. Inserted into this envelope are, in the case of influenza A and B viruses, two surface proteins, hemagglutinin (HA) and neuraminidase (NA). Influenza C virus possesses only one type of surface protein, HE, which is functionally equivalent to both HA and NA. Another membrane-associated protein, M2, is present in very small amounts in virus particles, whereas the nonstructural proteins NS1 and NS2 are found only in infected cells but not in purified virions.

A simplified version of the viral life cycle is depicted in Figure 1 and briefly described as follows. The virus attaches to the sialic acid–containing receptor of the host cell via the hemagglutinin molecule and subsequently enters the cell. Following uncoating, the viral core is transported into the nucleus where transcription and replication take place, leading to the formation of three kinds of virus-specific RNA: viral RNA (vRNA), complementary RNA (cRNA), and messenger RNA (mRNA). Finally, after translation of virus-specific mRNA into proteins, the virions are assembled and released from the infected cell.

III. CONVENTIONAL APPROACHES TO PREVENTION AND TREATMENT OF INFLUENZA

Since the first human influenza virus was isolated in 1933, its surface proteins underwent three major rearrangements and numerous minor changes. Mainly because of this high degree of antigenic variation, a long-term vaccination approach has remained unsuccessful thus far [Palese and Young, 1982].

Amantadine (1-aminoadamantane hydrochloride) and rimantadine (α-methyl-1-adamantane methylamine hydrochloride) are currently the only drugs used for prophylaxis and therapy of influenza in humans. They have similar antiviral properties in vitro, although rimantadine appears to cause fewer adverse reactions when used in humans. In cell culture, two distinct antiviral effects of amantadine have been described. At high concentrations (>100 μM), uncoating of several viruses is inhibited by elevation of the pH inside the endosome. At lower concentrations (0.1–5 μM), selective interaction of the drug with the M2 protein and subsequent perturbation of virus assembly may be the mechanism of action [Hay et al., 1986]. At such therapeutic doses, amantadine is effective only against certain subtypes of influenza A virus, but not against other viruses. Influenza virus strains resistant to amantadine develop at high frequency and can be isolated in vitro and in vivo [Belsch et al., 1988]. Resistant mutants can spread readily, which may further limit the broad use of this drug. Other antivirals, including ribavirin [Gilbert and Knight, 1986], interferon [Treanor et al., 1987], and isoprinosine [Longley et al., 1973], were tested against influenza infections in humans but not found to be generally useful [Douglas, 1990].

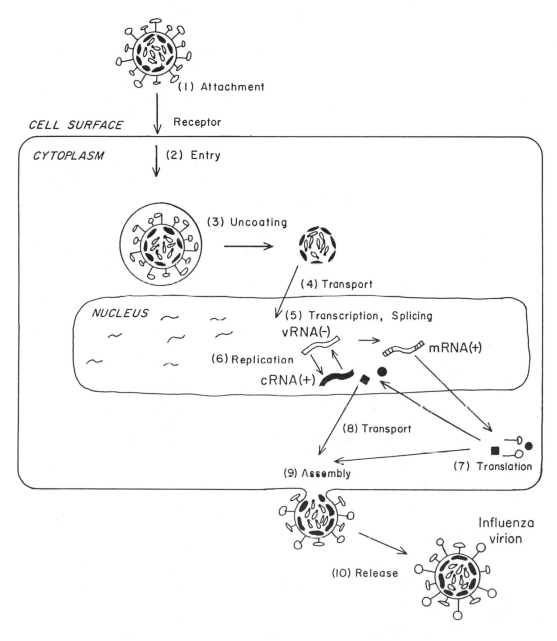

Fig. 1. *Life cycle of influenza virus. Structure and replication of influenza viruses are described in the text.*

IV. ANTISENSE RNA APPROACH TO INHIBIT INFLUENZA VIRUS REPLICATION

Soon after antisense RNA was shown to regulate gene expression in certain prokaryotic systems, it was suggested as a potentially powerful tool to prevent expression of harmful genes, including viral genes [Green et al., 1986]. Since influenza viruses are a major cause of disease not only in humans but also in animals, the development of cell lines and, subsequently, transgenic animals resistant to influenza virus infections would be of great economic importance. We therefore developed a series of cell lines permanently expressing virus specific RNA [Letier et al., 1989]. Two approaches were followed. Murine C127 cells were transformed with bovine papilloma virus vectors containing various influenza virus genes; the resulting cell lines were found to inhibit the replication of influenza virus at low multiplicity of infection. However, it was found that transformed cells contain both plus sense and minus sense RNA transcripts, perhaps because of promoters that are present in both orientations. Furthermore, the inhibitory activity appeared to be associated with a nonspecific, interferon-mediated effect. In the second approach, an expression system was used that involved 293 cells, a human cytomegalovirus-based promoter, and methotrexate-mediated gene amplification. The resulting cell lines were found to express up to 7,500 copies of influenza virus-specific RNA per cell. In this system, no RNA transcripts of the opposite orientation were found. However, all cell lines expressing either plus sense or minus sense viral RNA failed to inhibit viral replication or protein synthesis. Several possible reasons could be given as an explanation, including the dynamics of influenza virus replication, a putative RNA unwinding activity, and/or the protection of virus-specific RNAs by viral proteins.

In summary, the antisense RNA approach was not found to be useful in the influenza virus system. We therefore focused our efforts on using the antisense DNA approach for inhibition of influenza viral replication.

V. ANTISENSE DNA APPROACH TO INHIBIT INFLUENZA VIRUS REPLICATION

The antisense DNA approach is more direct and simpler than the antisense gene approach. Antisense oligonucleotides or their analogs are added to the culture medium in which cells are growing. The oligonucleotides are then taken up into cells by a mechanism that is not yet clearly understood. Once the oligonucleotide is in the cell, it may block translation of the mRNA by "hybridization arrest" in the cytoplasm or may interfere with pre-mRNA processing in the nucleus.

The use of oligonucleotides to inhibit gene expression specifically through intracellular hybridization was first demonstrated by Zamecnik and Stephenson more than a decade ago [Zamecnik and Stephenson, 1978; Stephenson and Zamecnik, 1978]. They showed that a 13-mer oligodeoxynucleotide, complementary to viral RNA of Rous sarcoma virus, inhibited growth and replication of the virus in tissue culture if added to the culture medium. The inhibition of virus replication was sequence specific. They also proposed the possible therapeutic potential of this approach.

Antisense oligonucleotides and various analogs have been used to regulate both cellular as well as viral gene expression as reviewed by Toulmé (this volume). The viral targets include human immunodeficiency virus [Zamecnik et al., 1986; Goodchild et al., 1988; Agrawal et al., 1988, 1989a,b; Matsukura, 1989] (see also Matsukura, this volume), vesicular stomatitis virus [Miller et al., 1985; Agris et al., 1986] (see also Degols et al., this volume), herpes simplex virus [Miller et al., 1985; Smith et al., 1986], and influenza virus [Zerial et al., 1987; Leiter et al., 1990; Kabanov et al., 1990]. Here we further discuss the use of synthetic oligonucleotides to inhibit influenza viral replication. The general use of antisense oligonucleotides has also been reviewed elsewhere [Van der Krol et al., 1988; Uhlmann and Peyman, 1990].

A. Properties of Oligonucleotides

Antisense oligonucleotides can be described in terms of the following properties.

1. Specificity. The oligonucleotide should have an efficient and specific interaction with the target nucleic acid. The specificity depends on the length of the oligonucleotide, which should be sufficient to make it unique for the viral or human genomic segment to be blocked or modulated. Since the genetic alphabet consists of the four letters A, C, G and T, at any one position in a sequence there is one chance in four of occupancy by a particular genetic letter. An 18-mer oligonucleotide of particular sequences will occur by chance one time in 4^{18}, or approximately one chance in 6 billion nucleotides. In the human genome, there are roughly 4 billion genetic letters. Thus we chose to synthesize oligonucleotides 20-mer in length to ensure high specificity at the site of their inhibitory activity. Another aspect of high selectivity of site of action is generally a decrease in toxicity because of reduced interference with some normal metabolic function of the host cell.

2. Stability. The stability of the oligonucleotide in the culture medium and within the cell will influence its final inhibitory effect. It has been demonstrated that unmodified oligonucleotides are of limited use in vitro as well as in vivo because of their sensitivity to cellular nucleases that are present in the culture medium and serum [Wickstrom, 1986]. To overcome this problem, several phosphate backbone-modified analogs have been studied, which include methylphosphonates, phosphorothioates, and various phosphoramidates. These analogs are more resistant to nucleases than unmodified oligonucleotides.

3. Hybridization properties. The antisense oligonucleotide can bind to its target nucleic acid by Watson Crick base pairing. The activity of the binding between the oligonucleotide and its target nucleic acid is characterized by the melting temperature (Tm) of the double-stranded nucleic acid that is formed. In general, the stability of the double strand decreases

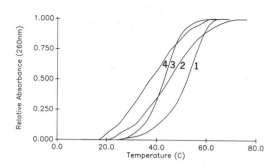

Fig. 2. *Melting curves of duplexes containing unmodified and phosphate backbone-modified oligonucleotides. Oligonucleotide sequence 5'-GTATCAAGGTTACAA (unmodified or phosphate backbone modified) was mixed with equimolar concentration of an unmodified oligodeoxynucleotide sequence, 5'-TTGTAACCTTGATAC, in 0.15 M NaCl–10 µM sodium phosphate buffer, pH 7.4. The samples were heated to 70°C and then cooled to 10°C at 2°C/minute. Melting curves were measured as hyperchromicity at 260 nm from 10° to 80°C (1°C increase/minute) with a Perkin Elmer ×3A spectrophotometer equipped with a thermoelectric cuvette manifold and temperature programmer. The curves shown are 1, unmodified oligonucleotide; 2, methylphosphonate oligonucleotide; 3, phosphorothioate oligonucleotide; and 4, phosphomorpholidate oligonucleotide.*

in the order RNA–RNA > RNA–DNA > DNA–DNA [Inoue et al., 1987]. An oligodeoxynucleotide of 20 bases in length has a melting temperature of about 56°C under physiological conditions, which is well above the 37°C body temperature. However, when the oligonucleotides are modified in the phosphate backbone to stabilize them against nucleases, the duplex stability is also affected. Figure 2 shows the Tm of four different modified oligonucleotides when hybridized with their complementary oligodeoxynucleotides. All three analogs of oligodeoxynucleotide show lower duplex stability than that of the unmodified oligonucleotide, and it is most likely for this reason that the inhibitory effect of various phosphate backbone-modified oligonucleotides has been found to be length dependent [Agrawal et al., 1990b], whereas in the case of unmodified oligonucleotide it is not length dependent [Goodchild et al., 1988].

4. Cellular permeability. The efficiency of cellular uptake and localization of oligonucleotides within the cell will influence its final inhibitory effect. All oligonucleotides and their various analogs studied to date have shown inhibitory effects in tissue culture in different cell lines; however, the mechanism is not yet fully understood. Cellular uptake of oligonucleotides can be enhanced by attaching chemical moieties known for membrane permeability, a question that is covered in greater depth by Degols et al. (this volume).

B. Synthesis and Purification of Oligonucleotides

Both unmodified oligonucleotides and their phosphorothioate analogs were assembled using H-phosphonate chemistry on an automated synthesizer (Milligen/Biosearch, 8700). Oligonucleotide synthesis was carried out on 10 µM scale on controlled pore glass (CPG) support. After the assembly of the required sequence, the intermediate CPG bound oligonucleoside H-phosphonate was treated with either 2% iodine in pyridine/H_2O (98:2, v/v) to generate phosphodiester linkages or 0.2 M sulfur in carbon disulfide/triethylamine/pyridine (9:1:9, v/v) at room temperature for up to 30 minutes to obtain the phosphorothioate internucleotide linkage. Deprotection of the oligonucleotide was carried out with concentrated ammonia at 55°C for 8 to 10 hours. Deprotected oligonucleotides were purified by low pressure reversed phase chromatography (C_{18}; Waters) with a linear gradient of acetonitrile in 0.1 M ammonium acetate (pH 7.0). Reversed phase purified oligonucleotides were further purified by low pressure ion-exchange chromatography on DEAE-cellulose by using a linear gradient of triethylammonium bicarbonate (0.1 to 2 M, pH 8). Finally, oligonucleotides were dialyzed against double-distilled water for 48 hours. After dialysis, oligonucleotides were lyophilized and stored at −20°C. Oligonucleotides and their phosphorothioate analogs were characterized by HPLC [Eadie et al., 1987; Agrawal and Tang, 1990], polyacrylamide gel electrophoresis, and the thermal stability of duplexes.

C. Molecular Targets for Antisense Oligonucleotides for Influenza Therapy

For antisense oligonucleotides, the target should be single-stranded nucleic acids. Influenza virus is a minus strand RNA virus. After infection, the minus sense viral RNA segments are transcribed into mRNAs and replicated into full-length complementary RNAs (Fig. 1). All three forms of virus specific RNA (viral RNA, mRNA, and complementary RNA) are potential targets for antisense oligonucleotides. For the study described here, the target selected was part of the end of the PB1 gene. This target was selected for the following reasons: 1) The ends of the PB1 genes are well conserved among the genomic segments of influenza viruses [Desselberger et al., 1980]; and 2) the influenza virus PB1 protein is part of the polymerase complex that is required for transcription and replication of viral RNA [Palese et al., 1977; Braam et al., 1983]. The PB1 protein is the most highly conserved protein of this complex [Yamashita et al., 1989] and is conserved among influenza A, B, and C viruses.

D. Assays for Inhibition of Influenza Virus by Oligonucleotides

1. Viruses and cells. Influenza A/WSN/33 (ts^+) virus was grown in MDCK cells as described previously. Influenza C/JJ/50 virus was grown in the amniotic sac of 11-day-old embryonated chicken eggs at 35°C.

2. Virus yield assay. About 2×10^5 MDCK cells per well were seeded into 24-well multidishes (Nunc, Thousand Oaks, CA) and allowed to grow overnight. The next day, cells were washed with PBS, and 200 ml MEM-BA containing the oligodeoxynucleotide was added to each well. At various times after addition of compound, 50 µl of virus sample, containing 2×10^5 PFU in MEM-BA, was added per well. In some experiments, only 200 PFU per well were used to allow for multicycle replication. After adsorption for 45 minutes at 37°C, cells were washed twice with PBS, and 1 ml of MEM-BA containing the oligodeoxynucleotide and 3 mg/ml trypsin was added. Cells were incubated at either 37° or 33°C, and samples of medium were taken at various times after infection for hemagglutination assay and titration on MDCK cells.

3. **Plaque formation assay.** MDCK cells were grown in 35 mm dishes. Confluent cells were infected with serial 10-fold dilutions of virus (100 μl/dish). One hour after infection the innoculum was aspirated, and 2.5 ml/dish of agar overlay containing 3 mg/ml trypsin and oligodeoxynucleotide at the indicated concentration was added. Cells were incubated at either 37°C for 2.5 days (influenza A virus) or 33°C for 4 to 5 days (influenza C virus).

E. Inhibition of Influenza Virus Replication by Oligodeoxynucleotides and Their Analogs

Initially, unmodified oligodeoxynucleotides were studied for their antiviral effect against influenza virus. Oligodeoxynucleotides complementary to the PB1 gene of influenza A virus and also to influenza C virus were synthesized (Fig. 3). Also included in the same study were several sequences as "mismatched" control oligonucleotides. Oligonucleotides were added to cells 30 minutes prior to infection and 24 hours after infection. All oligonucleotides tested failed to inhibit influenza virus replication up to a concentration of 80 μM, even at a low multiplicity of infection (multiplicity of infection = 0.001).

Similar results were obtained by another group. An unmodified 11-mer oligonucleotide, complementary to the loop-forming sites of the RNA encoding polymerase 3 of the influenza virus [Kaptain and Nayak, 1982], failed to inhibit influenza virus replication at up to 200 μM concentration [Kabanov et al., 1990]. However, the same oligonucleotide carrying a 5'-undecyl moiety showed some inhibitory activity. Similarly, unmodified oligonucleotides covalently attached to an acridine moiety showed selective inhibition of the cytopathic effect of influenza A virus at a concentration of 100 μM [Zerial et al., 1987]. It has been shown that the unmodified oligonucleotides are ineffective because they are rapidly degraded

by extracellular and intracellular nucleases [Wickstrom, 1986].

Several analogs of oligodeoxynucleotides that have an increased resistance to DNases are now being studied. The analogs include methylphosphonates, phosphorothioates, phosphoramidates, and α-oligonucleotides. In general, modifications can be introduced in the oligonucleotide as long as this does not impair the recognition of the complementary sequence.

Only the phosphorothioate analogs have been studied for their antiviral properties against influenza virus [Leiter et al., 1990]. Phosphorothioate analogs have the advantage of greater nuclease resistance [Eckstein, 1985] and are therefore more stable than unmodified oligonucleotides. Also, the duplex of phosphorothioate DNA and RNA is a substrate for RNase H [Stein et al., 1988; Furdon et al., 1989; Agrawal et al., 1990a], and under appropriate conditions phosphorothioate DNA may act catalytically in cleaving RNA with the help of RNase H.

The phosphorothioate analog of an oligonucleotide complementary to the PB1 gene of influenza A virus (oligonucleotide 5; Fig. 3) showed inhibition of virus replication in tissue culture. The inhibition was dose dependent and was effective at a dose of 1.25 μM (Fig. 4). However, mismatched phosphorothioate oligonucleotides also showed similar inhibition (data not shown). All the phosphorothioate oligonucleotides tested, whether complementary or mismatched, were nontoxic to cells up to a concentration of 80 μM. The nontoxic effect was further confirmed as phosphorothioates delayed influenza A virus replication rather than abolishing it, indicating that cells are still able to support viral replication after as long as 4 days of treatment with phosphorothioates.

There is a precedent for this nonspecific inhibition, since phosphorothioates are known to interact with enzymes, especially with reverse transcriptase [Mazumdar et al., 1989] and have shown sequence non-specific antiviral effects against human immunodeficiency virus [Agrawal et al., 1988]. However, in chronically HIV-infected cells a sequence specific effect could

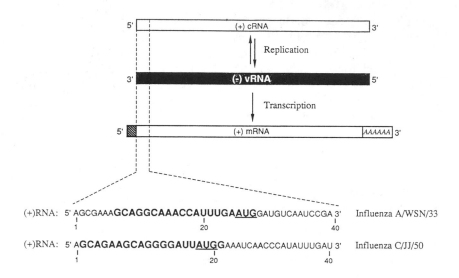

Fig. 3. Top: *Oligonucleotides tested for antiviral activity against influenza virus. Diagram depicts the three different types of RNA present in infected cells (complementary RNA [cRNA], viral RNA [vRNA], and messenger RNA [mRNA]. [Reproduced from Leiter et al., 1990, with permission of the publisher.]* **Middle:** *Part of the RNA sequence of PB1 gene of influenza A/WSN/33 virus [Sivasubramaniam and Nayak, 1982] and influenza C/JJ/50 virus [Yamashita et al., 1989] is shown in the plus sense orientation.* **Bottom:** *The oligonucleotides were synthesized complementary to the AUG site, printed in boldface. Oligonucleotides 1–3 are unmodified, and oligonucleotides 4–12 are phosphorothioate analogues.*

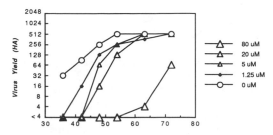

Fig. 4. *Inhibition of influenza A virus replication by phosphorothioate oligonucleotide 5 (see Fig. 3). Cells were infected with influenza A virus (MOI = 0.001) and incubated at 37°C. Oligonucleotide 5 was added to the culture 24 hours prior to infection and was present throughout the experiment. Infected control cells were not treated with oligonucleotide (0 μM). At the times indicated, samples of medium were taken for testing of hemmaglutinin (HA) titres. Concentrations of oligonucleotide 5 tested were 1.25 μM, 5 μM, 20 μM, and 80 μM. (From Leiter et al., 1990.)*

be observed [Matsukura et al., 1989; Agrawal et al., 1989].

While inhibition of influenza A virus replication was sequence independent, a different result was observed in the case of influenza C virus, and a sequence-specific inhibition was found. A phosphorothioate oligonucleotide (oligonucleotide 7), complementary to the vRNA of the PB1 gene of influenza C virus (in the equivalent position to oligonucleotide 5 in the A virus), was found to inhibit virus growth specifically if present at 20 μM concentration. The same sequence had no effect on influenza A virus replication, with which the oligomer has little homology (Fig. 3). Furthermore, the sequence-specific inhibition was confirmed by introducing one mismatch and also three mismatches (Fig. 5). Introduction of one mismatch significantly reduced antiviral activity, while three mismatches rendered the compound inactive. Also, the phosphorothioate oligonucleotide complementary to the plus strand, was not active against influenza C virus. When used at 80 μM concentration, a reduction in the number of virus plaques of up to 10^6 fold could be observed with oligonucleotide 7.

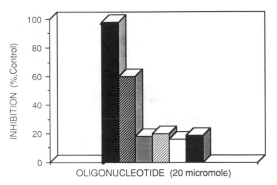

Fig. 5. *Sequence-specific inhibition of influenza C virus replication by oligonucleotide 7. Cells were infected with influenza C virus and oligonucleotides were added to the agar overlay 1 hour postinfection at 20 μM concentration. Control infected cells were not treated with oligonucleotide, cells were incubated at 33°C for 4–5 days, and plaques were then counted. Oligonucleotide 7 inhibited viral replication by up to 97% (solid bar), under the same conditions as oligonucleotide 8 (with one mismatch) inhibited virus replication by approximately 60% (dark crossed bar). Since it has one mismatch in the center, oligonucleotide 8 can still hybridize to the complementary strand by 10 nucleotides. Oligonucleotide 9 (with three mismatches; shaded bar) and oligonucleotides 10 (light crossed bar), 11 (open bar), and 12 (mismatched; far right solid bar) showed very little activity.*

VI. CONCLUSION

The promising results of inhibition of influenza virus replication by oligonucleotide phosphorothioate in cell culture studies suggest that antisense oligonucleotide analogs offer a unique approach for the rational design of drugs with highly specific antiviral activity. To date there have been no published reports on the use of antisense oligonucleotides or their analogs in animals.

Preliminary acute toxicity study of oligonucleotide phosphorothioates in mice and rats have shown that 1) a dose of 100 mg oligonucleotide per kilogram body weight or lower dose, if given to mice daily for 14 days, either intraperitoneally or subcutaneously, is not toxic; and 2) a dose of 150 mg of oligonucleotide per kilogram body weight given intraperitoneally to rat is nontoxic [Agrawal et al., 1988; Agrawal,

1991]. A study is underway to determine the bioavailability, pharmacokinetics, and half-lives of oligonucleotide phosphorothioates in small animals.

NOTE ADDED IN PROOF

The pharmacokinetic study of oligonucleotide phosphorothioate has been carried out [Agrawal S, Temsamani J, Tang J-Y (1991): Pharmacokinetics, biodistribution and stability of oligodeoxynucleotide phosphorothioates in mice. Proc Natl Acad Sci USA 88:7596–7599].

ACKNKOWLEDGMENTS

The authors are indebted to Drs. Paul Zamecnik and to Peter Palese for their encouragement and guidance throughout this work. J.M.E.L. was supported by the Boehringer Ingelheim Foundation (Stuttgart, Germany). Experiments carried out in Paul Zamecnik's lab were supported by NIH grant AI-24846-03 to P.Z. and in Peter Palese's lab were supported by NIH grant AI-24460 and Merit Award AI-18998 to P.P. We thank Ms. Rebecca Ball and Carol Tuttle for excellent secretarial assistance.

VII. REFERENCES

Agrawal S (1991): Antisense oligonucleotides: A possible approach for chemotherapy of AIDS. In Wickstrom E (ed): Prospects for Antisense Nucleic Acid Therapy of Cancer and AIDS. New York: Wiley-Liss, pp 143–158.

Agrawal S, Goodchild J, Civeira MP, Thornton, AH, Sarin PS, Zamecnik PC (1988): Oligodeoxynucleoside phosphoramidates and phosphorothioates as inhibitors of human immunodeficiency virus. Proc Natl Acad Sci USA 85:7079–7083.

Agrawal S, Goodchild J, Civiera MP, Thornton AH, Sarin PS, Zamecnik PC (1989a): Phosphoramidates, phosphorothioates and methylphosphonate analogs of oligodeoxynucleotides: Inhibitors of replication of human immunodeficiency virus. Nucleosides Nucleotides 8:819–823.

Agrawal S, Ikeuchi T, Sun D, Sarin PS, Konopka A, Maizel T, Zamecnik PC (1989b): Inhibition of human immunodeficiency virus in early infected and chronically infected cells by antisense oligodeoxynucleotides

and its phosphorothioate analogue. Proc Natl Acad Sci USA 86:7790–7794.

Agrawal S, Mayrand SM, Zamecnik PC, Pederson T (1990a): Site specific excision from RNA by RNaseH and mixed phosphate backbone oligodeoxynucleotides. Proc Natl Acad Sci USA 87:1401–1405.

Agrawal S, Sun D, Sarin PS, Zamecnik PC (1990b): Inhibition of HIV-1 in early infected and chronically infected cells by antisense oligodeoxynucleotides and their phosphorothioate analogue. J Cell Biochem 14D:145.

Agrawal S, Tang J-Y, Brown DM (1990): Analytical study of phosphorothioate analogues of oligodeoxynucleotides using high performance liquid chromatography. J Chromatography 509:396–399.

Agris CH, Blake KR, Miller PS, Reddy MP, Ts'o POP (1986): Inhibition of vesicular stomatitis virus protein synthesis and infection by sequence-specific oligodeoxynucleoside methylphosphonates. Biochemistry 25:6268–6275.

Belshe RB, Smith MH, Hall CB, Betts R, Hay AJ (1988): Genetic basis of resistance to rimantadine emerging during treatment of influenza virus infection. J Virol 62:1508–1512.

Brahm J, Ulmann I, Krug RM (1983): Molecular model of a eukaryotic transcription complex: Function and movement of influenza P protein during capped RNA primed transcription. Cell 34:609–618.

Desselberger U, Racaniello VR, Zazra J, Palese P (1980): The 3′ and 5′ terminal sequences of influenza A, B, C and virus RNA segments are highly conserved and show partial inverted complementarity. Gene 8: 315–328.

Douglas RG (1990): Prophylaxis and treatment of influenza. N Engl J Med 22:443–450.

Eckstein F (1985): Investigation of enzyme mechanisms with nucleoside phosphorothioates. Annu Rev Biochem 54:367–402.

Froehler B, Ng P, Matteucci M (1988): Phosphoramidate analogues of DNA: Synthesis and thermal stability of heteroduplexes. Nucleic Acids Res 16:4831–4839.

Furdon PJ, Dominski Z, Kole R (1989): RNase H cleavage of RNA hybridized to oligonucleotides containing methylphosphonate, phosphorothioate nad phosphodiester bonds. Nucleic Acids Res 17:9193–9204.

Gilbert BE, Knight V (1986): Biochemistry and clinical application of ribavarin. Antimicrob Agents Chemother 30:201–205.

Goodchild J, Agrawal S, Civiera MP, Sarin PS, Sun D, Zamecnik PC (1988): Inhibition of human immunodeficiency virus replication by antisense oligodeoxynucleotides. Proc Natl Acad Sci USA 85:5507–5511.

Green PJ, Pines O, Inouye (1986): The role of antisense RNA in gene regulation. Annu Rev Biochem 55: 569–597.

Hay AJ, Zambon MC, Wolstenholme AJ, Skehel JJ, Smith MH (1986): Molecular basis of resistance of influenza A virus to amantadine. J Antimicrob Chemother 18(Suppl B):19–29.

Inoue H, Hayase Y, Imura A, Iwai S, Miura K, Ohtsuka E (1987): Synthesis and hybridization studies on two complementary nona (2'-O-methyl)ribonucleotides. Nucleic Acids Res 15:6131–6148.

Kabanov AV, Vinogradov SV, Ovcharenko AV, Krivonos AV, Melik-Nubarov NS, Kiselev VI, Severin ES (1990): A new class of antivirals: Antisense oligonucleotides combined with a hydrophobic substituent effectively inhibit influenza virus reproduction and synthesis of virus-specific proteins in MDCK cells. FEBS Lett 259:327–330.

Kaptain JS, Nayak DP (1982): Complete nucleotide sequence of the polymerase 3 (P3) gene of human influenza virus A/WSN/33. J Virol 42:55–63.

Krug RM (ed) (1989): The Influenza Viruses. New York: Plenum.

Leiter JME, Krystal M, Palese P (1989): Expression of antisense RNA failed to inhibit influenza virus replication. Virus Res 14:141–160.

Leiter JME, Agrawal S, Palese P, Zamecnik PC (1990): Inhibition of influenza virus replication by phosphorothioate oligodeoxynucleotides. Proc Natl Acad Sci USA 87:3430–3434.

Longley S, Dunning RL, Waldman RH (1973): Effects of isoprinosine against challenge with A (H3H-2) Hong Kong influenza virus in volunteers. Antimicrob Agents Chemother 3:506–509.

Matsukura M, Zon G, Shinozuka, Robert-Guroff M, Shimada T, Stein CA, Mitsuya H, Wong-Staal F, Cohen J, Broder S (1989): Regulation of viral expression of human immunodeficiency virus in vitro by an antisense phosphorothioate oligodeoxynucleotides against rev (art/trs) in chronically infected cells. Proc Natl Acad Sci USA 86:4244–4248.

Mazumdar C, Stein CA, Cohen JS, Broder S, Wilson SH (1989): Stepwise mechanism of HIV reverse transcriptase: Primer function of phosphorothioate oligodeoxynucleotide. Biochemistry 28:1340–1346.

Miller PS, Agris CH, Aurelian L, Blake KR, Myrakami A, Reddy MP, Spitz SA, Ts'o POP (1985): Control of ribonucleic acid function by oligonucleoside methylphosphonates. Biochimie 67:769–776.

Mizuno T, Chou M-Y, Inouye M (1983): A unique mechanism of regulating gene expression: Translational inhibition by a complementary RNA transcript (micRNA). Proc Natl Acad Sci USA 81:1966–1970.

Palese P, Kingsbury DW (eds) (1983): Genetics of Influenza Viruses. Vienna: Springer-Verlag.

Palese P, Richtey MB, Schulman JL (1977): Mapping of influenza virus genome II identification of the P1, P2 and P3 genes. Virology 76:114–121.

Palese P, Young JF (1982): Variation of influenza A, B, and C viruses. Science 215:1468–1474.

Sivasubramaniam N, Nayak DP (1982): Sequence analysis of the polymerase 1 gene and the secondary structure prediction of polymerase 1 protein of human influenza virus A/WSN/33. J Virol 44:321–329.

Smith CC, Aurelian L, Reddy MP, Miller PS, Ts'o POP (1986): Antiviral effect of an oligonucleoside methylphosphonate complementary to splice junction of herpes simplex virus type 1 immediately early pre-mRNAs 4 and 5. Biochemistry 83:2787–2791.

Stein CA, Subasinghe C, Shinozuka K, Cohen JS (1988): Physiochemical properties of phosphorothioate oligodeoxynucleotides. Nucleic Acids Res 16:3209–3221.

Stephenson ML, Zamecnik PC (1978): Inhibition of Rous sarcoma viral RNA translation by a specific oligodeoxyribonucleotide. Proc Natl Acad Sci USA 75:285–288.

Treanor JJ, Betts RF, Erb SM, Roth FK, Dolin R (1987): Intranasally administered interferon, as prophylaxis against experimentally induced influenza A virus infection in humans. J Infect Dis 156:379–383.

Uhlmann E, Peyman A (1990): Antisense oligonucleotide: A new therapeutic principle. Chem Rev 90:544–584.

Van der Krol AR, Mol JNM, Stritje AR (1988): Modulation of gene expression by complementary RNA or DNA sequences. BioTechniques 6:958–976.

Wickstrom E (1986): Oligodeoxynucleotide stability in subcellular extracts and culture media. J Biochem Biophys Methods 13:97–102.

Yamashita M, Krystal M, Palese P (1989): Comparison of the three large polymerase proteins of influenza A, B, and C viruses. Virology 171:458–466.

Zamecnik PC, Goodchild J, Taguchi Y, Sarin PS (1986): Inhibition of replication and expression of human T-cell lymphotrophic virus type III in cultured cells by exogeneous synthetic oligonucleotides complementary to viral RNA. Proc Natl Acad Sci USA 83:4143–4146.

Zamecnik PC, Stephenson ML (1978): Inhibition of Rous sarcoma virus replication and transformation by a specific oligodeoxynucleotides. Proc Natl Acad Sci USA 75:280–284.

Zerial A, Thuong NT, Helene C (1987): Selective inhibition of the cytopathic effect of type A influenza viruses by oligodeoxynucleotides covalently linked to an intercalating agent. Nucleic Acids Res 15:9909–9919.

ABOUT THE AUTHORS

SUDHIR AGRAWAL is Foundation Scholar at the Worcester Foundation for Experimental Biology. He earned his degrees from the University of Allahabad, India (M.S. 1975, Ph.D. 1980). He was commonwealth fellow at the MRC Laboratory at Molecular Biology, Cambridge (U.K.) in 1984–1986 in the laboratory of Dr. Michael Gait. Dr. Agrawal has published over 50 scientific papers dealing with various aspects of nucleic acid chemistry, recently in the area of the nucleic acid-based therapeutic approach. His current research is in the area of nucleic acid chemistry, in the nonradioactive nucleic acid-based diagnostic and nucleic acid-based therapeutic approach. He is a member of the American Chemical Society, the American Association for the Advancement of Science, the American Association of Clinical Chemistry, the New York Academy of Sciences and the editorial board of the *Journal of Antisense Research and Development*.

JOSEF M.E. LEITER works for the international management consultancy, McKinsey & Company, Inc., where he takes part in strategy, organization, and operational improvement programs for major private and public sector organizations. He received his M.D. degree from the Karl Franzens University of Innsbruck, Austria, in 1986. A medical student, he performed research in biochemistry leading to a dissertation on histone modifications. Dr. Leiter then pursued postdoctoral research under Dr. Peter Palese at the Mount Sinai School of Medicine, New York, working on various aspects of the molecular biology of influenza viruses. In particular, he was interested in alternative antiviral approaches such as antisense RNA and DNA. His research papers have appeared in such journals as the *Proceedings of the National Academy of Sciences USA*, *Virology*, and *Experimental Cell Research*.

Antisense RNA and DNA: 317–334
© 1992 Wiley-Liss, Inc.

Antisense DNA Control of c-*myc* Gene Expression, Proliferation, and Differentiation in HL-60 Cells

Eric Wickstrom

I. INTRODUCTION

Malignant cells display overexpression or mutant expression of one or more of the genes normally involved in cell proliferation. Such genes are called proto-oncogenes [Bishop, 1991]. Crabtree [1989] has described a cascade of genes, some of which are proto-oncogenes, that are turned on in the course of activation of T lymphocytes. Cell transformation may be caused by inappropriate expres-sion or response of proto-oncogene products during cell proliferation. The implication is that the targets that must be attacked to restore neoplastic cells to normal growth are normal cellular genes that have sustained some activating lesion. Antisense inhibition [van der Krol et al., 1988] may readily be used to study the functions of each proto-oncogene, but may only be applied as a practical model of therapy if the malignant cells are more dependent on the expression of an overexpressed protooncogene than normal cells [Heikkila et al., 1987] or if a specific target sequence exists, in the case of a mutated protooncogene, that differs sufficiently from the normal protooncogene to allow a safe therapeutic ratio [McManaway et al., 1990].

The c-*myc* protooncogene is an evolutionarily conserved gene found in all vertebrates, and Northern blots indicate that c-*myc* mRNA is expressed in most normal, dividing cells [Klein and Klein, 1986] constitutively throughout the cell cycle [Thompson et al., 1985]. The c-*myc* gene codes for a 49 kD polypeptide and expresses a nuclear protein with an electrophoretic apparent molecular mass of 65 kD (p65) [Persson et al., 1984]. Immunostaining studies indicate that p65 is located in the same subnuclear compartments as small nuclear ribonucleoproteins [Spector et al., 1987]. The hydropathy profile of p65 shows 68% homology with adenovirus Ad5 E1A (289 amino acid) protein and significant homology with polyoma large T antigen [Branton et al., 1984]. Either of the latter nuclear proteins may collaborate with c-*ras* to transform primary cell cultures [Land, et al., 1983; Franza, et al., 1986], analogously to c-*myc*. Hence, it is not surprising that *myc* p65 is one of the leucine zipper proteins, which binds with a small partner protein, MAX [Blackwood and Eisenman, 1991], and this heterodimer in turn binds specifically to the sequence GAC CAC GTG GTC which occurs in the regulatory regions upstream of proliferative genes [Halazonetis and Kandil, 1991].

In a wide variety of human leukemias and solid tumors, the proto-oncogene c-*myc* has been found to be amplified, translocated, overexpressed, or abnormally regulated [Klein and Klein, 1986; Bishop, 1991], and overexpression of p65 promotes replication of SV40 DNA [Classon et al., 1987]. Hence, it appears likely that the c-*myc* gene product plays some direct or indirect role in replication. Indeed, inducing quiescent cells to proliferate leads to a rapid accumulation of c-*myc* mRNA upon entering the cell cycle, followed by a decline prior to the onset of DNA synthesis [Kelly et al., 1983]. Furthermore, overexpression of p65 usually correlates with inability of cells to differentiate [Coppola and Cole, 1986].

The HL-60 cell line, which consists predominantly of rapidly dividing cells with promyelocytic characteristics [Collins et al., 1978], contains multiple copies of the c-*myc* gene [Collins and Groudine, 1982] and expresses 35–100 copies of c-*myc* mRNA per cell compared with the normal 5–10 copies per cell (J. Bresser, personal communication). On the other hand, c-*myc* mRNA levels decline in HL-60 cells upon exposure to dimethylsulfoxide (Me$_2$SO) [Westin et al., 1982]. Me$_2$SO inhibits the proliferation of this cell line and induces the cells to differentiate into granulocytic cells that exhibit morphological and chemical properties similar to more mature myelocytes, metamyelocytes, and banded and segmented granulocytes [Collins et al., 1978]. Me$_2$SO also inhibits the ability of HL-60 cells to form colonies in semisolid medium [Filmus and Buick, 1985].

In contrast, phorbol 12-myristyl-13-acetate (PMA) similarly inhibits HL-60 cell proliferation, but induces differentiation along the monocytic line [Rovera et al., 1979]. Furthermore, it is clear that constitutive overexpression of c-*myc* in mouse erythroleukemia cells [Coppola and Cole, 1986], or v-*myc* in human U-937 monoblastic cells [Larsson et al., 1988], blocks induction of differentiation by Me$_2$SO or PMA, respectively. Nevertheless, it is not known whether the reduction in nuclear p65 modulates the differentiation or is simply another manifestation of cellular differentiation. It now appears that c-*myc* p65 derepresses negatively regulated proliferative genes at the transcriptional level [Onclercq et al., 1989]. Hence it is likely that reducing the level of c-*myc* mRNA translation may be sufficient to inhibit cellular proliferation, induce differentiation, and reverse transformation.

Antisense DNA inhibition of translation provides one possible method to downregulate c-*myc* gene expression (Fig. 1). Calculation of a predicted secondary structure for a 400 nucle-

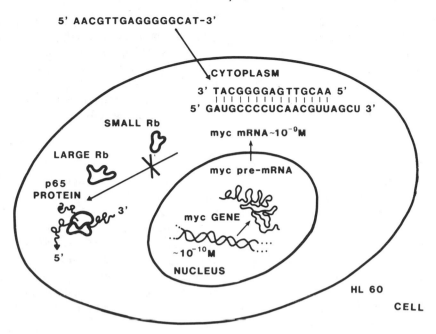

Fig. 1. *Schematic of antisense oligodeoxynucleotide inhibition of* c-myc *gene expression in HL-60 cells showing hybridization between the exogenously added anti-* myc *oligonucleotide and the initiation codon and following four codons of the* c-myc *mRNA preventing translation. Rb = ribosomal subunit.*

otide (nt) portion of human c-*myc* mRNA placed the start codon at the beginning of a large bulge loop in a weakly base-paired region, suggesting that these codons might be readily available for hybridization arrest (Fig. 2) [Wickstrom et al., 1986b]. In preliminary experiments utilizing an antisense oligomer against the predicted initiation codon loop, sequence-specific, dose-dependent inhibition of proliferation, and stimulation of granulocytic differentiation were observed in human HL-60 promyelocytic leukemic cells. Subsequently, it was found that the same anti-c-*myc* oligomer also inhibited expression of the c-*myc* p65 antigen in mitogen-stimulated human peripheral blood lymphocytes, which were then unable to enter S phase, in a sequence-specific, dose-dependent manner [Heikkila et al., 1987]. Harel-Bellan et al. [1988] have extended this observation to interleukin-2–induced T-cell proliferation.

In the studies reported here, treatment of HL-60 cells with the antisense oligodeoxynucleotide directed against the predicted hairpin loop containing the initiation codon of human c-*myc* mRNA elicited sequence-specific, dose-dependent inhibition of c-*myc* gene expression and cell proliferation [Wickstrom et al., 1988]. Furthermore, treatment with the anti-c-*myc* oligomer also elicited a sequence-specific induction of myeloid differentiation with granulocytic characteristics, potentiated inhibition of cell proliferation by Me$_2$SO, and uniformly inhibited colony formation in semisolid medium [Wickstrom et al., 1989].

II. MATERIALS AND METHODS

A. Cell Culture

HL-60 cells [Collins et al., 1978] were grown in RPMI (Roswell Park Memorial Institute) 1640 medium with 10% heat-inactivated fetal bovine serum (FBS), and Q8/MC29 cells [Bister et al., 1977] were grown in RPMI 1640/Dulbecco's modified essential medium (DMEM) (1:1) with 5% heat-inactivated newborn calf serum, 1% chicken serum, and 0.1%

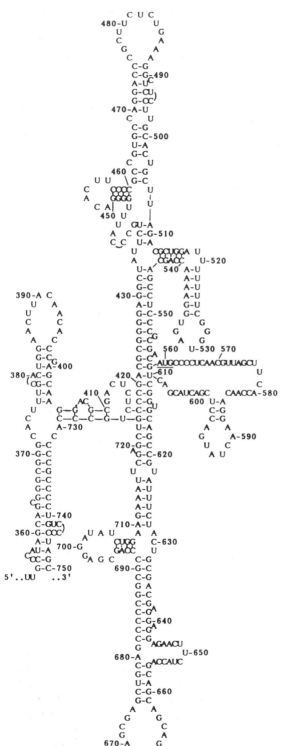

Me₂SO, with 10^5 units/liter penicillin, 0.5 g/liter gentamicin, and 0.1 g/liter streptomycin in both media. Cells were incubated at 37°C in 5% CO_2 saturated with water vapor and maintained in logarithmic growth phase; titers of viable cells were determined by counting Trypan blue excluding cells in a hemocytometer.

B. Secondary Structure and Thermodynamic Predictions

Folding of the 2,121 nt c-*myc* mRNA from K562 cells [Watt et al., 1983] was predicted using the algorithms of Jacobson et al. [1984], with the nearest neighbor dependent free energies of Freier et al. [1986] at 37°C in 1 M NaCl. Thermodynamic calculations of association constants for synthetic antisense oligodeoxynucleotides hybridizing to single-stranded regions of mRNA were based on the same free energies. No correction of the free energies was made for RNA–DNA hybridization, instead of RNA–RNA hybridization, nor was any correction made for topological constraints on RNA loops or bulges.

C. Oligodeoxynucleotide Synthesis

Pentadecamers were prepared for antisense hybridization arrest experiments. This length was selected in order to allow very strong binding to single-stranded regions of mRNA, or adequate binding to regions that already have some secondary structural or tertiary structural limitations. A probe with 15 residues should be just barely sufficient for theoretical uniqueness in the human genome, particularly for those sequences expressed as mRNA. Previous investigations have demonstrated efficacy with antisense oligodeoxynucleotides of 8–26 residues [van der Krol et al., 1988]. Oligode-

---

Fig. 2. *Predicted secondary structure of human c-*myc *mRNA from K562 cells [Watt et al., 1983], from 200 nt upstream of the AUG initiation codon to 200 nt downstream, calculated with the RNAFLD program [Jacobson et al., 1984] using free energies of base pairing at 37°C [Freier et al., 1986]. The principal initiation codon (underlined) begins at nt 559, and the alternative CUG initiation codon for the larger minor product begins at nt 514 [Hann et al., 1988].*

oxynucleotides were synthesized [Sinha et al., 1983] automatically and purified by reversed phase liquid chromatography.

D. Oligodeoxynucleotide Uptake and Stability

For studies of oligodeoxynucleotide uptake and stability in cells, oligodeoxynucleotides were 5' labeled with $5'$-$[\gamma$-^{35}S]thioATP (1276 Ci/mmol, Du Pont/NEN) using T4 polynucleotide kinase, and purified by denaturing gel electrophoresis [Wickstrom, 1986]. For each time point, 5×10^5 cpm of $[5'$-^{35}S]oligodeoxynucleotide were added to 4×10^6 HL-60 cells in 0.5 ml of RPMI 1640 with 10% heat-inactivated fetal calf serum. Each sample was incubated at 37°C for 1, 4, 8, or 24 hours, and then sedimented for 3 minutes at 15,000g. The supernatant was removed and saved, and the cell pellet was washed once in 0.5 ml of 10 mM Na$_2$ HPO$_4$, pH 7.4, 150 mM NaCl (PBS), and sedimented again. The supernatant was removed and saved, and the cell pellet was lysed in 0.1 ml of 10 mM Tris-HCl, pH 7.4, and 150 mM NaCl (TBS) with 1% NaDodSO$_4$ and then extracted with 0.1 ml phenol. The aqueous phase was removed, and the phenol phase was extracted with another 0.1 ml of water. Aliquots of the combined aqueous extracts, cell wash, and culture medium supernatant were analyzed by liquid scintillation counting. The percent of oligodeoxynucleotides taken up by the cells was calculated by dividing counts in the combined aqueous phases of the cell pellet extract by the total of the counts in the cell pellet, cell wash, and culture medium supernatant. To determine oligodeoxynucleotide stability, aliquots of the combined aqueous phases and of the culture medium supernatant fraction were lyophilized, redissolved in 30 μl of 80% deionized formamide including 0.01% xylene cyanole FF and 0.01% bromophenol blue, and then electrophoresed on a denaturing 20% polyacrylamide gel and fluorographed.

E. Inhibition of c-*myc* p65 Expression

Translation of c-*myc* mRNA into p65 was detected by indirect immunofluorescence and radioimmunoprecipitation. For indirect immunofluorescence, samples of 1.0×10^6 HL-60 cells were each resuspended in 1 ml of fresh medium and grown for 6 hours with target or control oligodeoxynucleotides added to the culture medium. Each sample was pelleted, washed, fixed with 1% paraformaldehyde, preadsorbed with 1% normal goat serum, incubated with normal rabbit IgG (Sigma), rabbit anti-β-actin (ICN), or rabbit anti-p65 [Watt et al., 1985], followed by fluorescein-conjugated goat anti-rabbit IgG (Sigma), washed, resuspended in 1% paraformaldehyde as described elsewhere [Lanier et al., 1981], and observed by light and fluorescence microscopy. For radioimmunoprecipitation, samples of 2.0×10^6 HL-60 cells were each resuspended in 2 ml of fresh medium and grown for 6 hours in the presence of target and control oligodeoxynucleotides added to the culture medium. Each sample was sedimented after 6 hours to remove medium containing oligodeoxynucleotide, washed, and then resuspended in 0.5 ml RPMI 1640 lacking cysteine (Gibco) but containing 250 μCi/ml [^{35}S]Cys (1022 Ci/mmol; Du Pont/NEN). Each culture was grown for an additional 1.5 hours and then sedimented, lysed, immunoprecipitated, electrophoresed, and fluorographed as described [Beimling et al., 1985]. Fluorograms were scanned with an LKB Ultroscan laser densitometer.

F. Inhibition of Cell Proliferation

HL-60 cells were diluted in 1 ml of culture medium to a concentration of 10^5 cells/ml, and Me$_2$SO (Sigma), PMA (Sigma), and/or DNA oligomers were then added directly to cell suspensions at concentrations described below. Each sample was dispensed in 0.30 ml aliquots into three wells of a 96-well microtiter plate. After 5 days of growth, the cells in each well were resuspended in 0.3 ml PBS, and 0.1 ml aliquots were removed for staining with 0.1 ml 0.4% Trypan blue and counting. Treated and untreated cells showed 98–100% viability after 5 days growth, and untreated cells typically showed a 10-fold greater titer, about 10^6 cells/ml Q8/MC29 cells were treated similarly, multiplying 10-fold in only 3 days. At this point, the culture medium was aspirated off, and the cells in each well were trypsinized and resuspended in 0.30 ml PBS prior to counting as above. Cell counts were converted to percent inhibition by the calculation $100 \times (N_n - N)/(N_n - N_0)$, where N_0 is the normal titer at the beginning of the experiment, N_n is the titer for untreated cells after n days growth, and N is the titer for treated cells after n days.

G. Inhibition of Colony Formation

For measurements of colony formation, semisolid medium was prepared by adding 10 g of 4,000 mPa methylcellulose (Fluka) to 250 ml of sterile boiling H_2O, adding 250 ml of $2 \times$ RPMI 1640 (Sigma) with 20% BSA, and stirring overnight at $0°-4°C$ [Graf et al., 1981]. Aliquots of HL-60 cells ($<10\,\mu l$) containing 10^4 cells were diluted to 1 ml in semisolid RPMI 1640 with 10% FBS containing no addition, 1% Me_2SO (Sigma), 16 nM PMA (Sigma), anti-c-*myc* oligomer, or anti-VSV oligomer. Cells were visualized microscopically, and colonies of two or more cells were counted and compared with single cell counts. Percentages of cells forming colonies were determined daily by counting 200 colonies/cell in the same quadrant of the well.

H. Morphological Analysis of Differentiation

Treated cells were grown for 5 days, and then 400–600 μl aliquots of cell suspensions were sedimented onto glass slides using a Shandon Cytospin I (Surrey, England) and Wright-Giemsa stained for differential counts performed under light microscopy [Collins et al., 1978]. HL-60 cells consist predominantly of promyelocytes with large and round nuclei, prominent nucleoli, dispersed nuclear chromatin, high nuclear/cytoplasmic ratio, and basophilic cytoplasm with prominent azurophilic granules. Cells in this category and cells actively undergoing mitosis were scored as undifferentiated. A moderate percentage of the uninduced cells and a large percentage of Me_2SO-induced cells differentiate into forms with a more mature appearance, including a smaller size, lower nuclear/cytoplasmic ratio, less prominent cyto-plasmic granules, reduction or disappearance of nucleoli, and marked indentation, convolution, or segmentation of the nuclei. Cells in this category were scored as differentiated [Collins et al., 1978]. Cells with monocytic characteristics are not observed in uninduced HL-60 cells, or those induced with Me_2SO, but only among cells induced with PMA. The percent differentiation in 200 cells counted was calculated from (differentiated cells/200) $\times$ 100, $\pm$ standard error.

I. Cytochemical Analysis of Differentiation

Nitroblue tetrazolium (NBT) is reduced intracellularly to an insoluble formazan dye by superoxide, which is generated by phagocytosis-associated oxidative metabolism in normal granulocytes [DeChatelet et al., 1976] and Me_2SO-differentiated HL-60 cells [Collins et al., 1979] that have been exposed very briefly to PMA. Hence cells treated with oligodeoxynucleotides were grown for 2 days; then 400 μl aliquots of cell suspensions were added to an equal volume of 0.2% NBT (Sigma) in calcium/magnesium-free phosphate-buffered saline and incubated for 20 minutes in the presence of 160 nM PMA, precisely as described by DeChatelet et al., [1976]. The cells were sedimented onto glass slides and Wright-Giemsa counterstained as described above. Cells containing any reduced blue-black formazan dye were scored as NBT positive. The percent NBT-positive cells in 200 counted was calculated from (NBT positive cells/200) $\times$ 100 $\pm$ standard error.

III. RESULTS

A. Prediction of Antisense Targets and Thermodynamics

Calculation of a predicted secondary structure for 400 nt, centered on the initiation codon, of human c-*myc* mRNA from K562 human leukemic cells [Watt et al., 1983] placed the initiation codon and eight following codons in a bulge loop (Fig. 2), suggesting that these codons might be readily available for hybridization arrest by an antisense oligodeoxynucleotide. From the predicted structure, the initiation codon and the next four codons, nt 559–573, were selected as a single-stranded target, and the complementary pentadecadeoxynucleotide, 5'-dAACGTTGAGGGGCAT-3' was synthesized. To control for sequence-specific effects, two other pentadecadeoxynucleotides were prepared; 5'-dTTGGGATAA CACTTA-3', complementary to nt 17–31 of VSV M protein mRNA [Rose and Gallione, 1981], which nonspecifically inhibited translation of VSV mRNAs [Wickstrom et al., 1986a] but differed from the anti-c-*myc* sequence in 13 out of 15 residues; and 5'-dCA-TTTCTTGCTCTCC- 3', complementary to nt 5,399–5,413 of the HIV *tat* gene [Ratner et al., 1985], differing in 12 of 15 residues.

Thermodynamic calculations for the association of 5'-dAACGTTGAGGGGCAT-3' with its complement on c-*myc* mRNA yielded an association constant of 3.2×10^{16}. This approximate result suggests that stoichiometric inhibition should be possible, i.e., at a ratio of one oligodeoxynucleotide per mRNA, for cells in culture in an ideal case assuming no degradation of oligodeoxynucleotide, efficient cellular uptake of oligodeoxynucleotides, uninhibited diffusion of oligodeoxynucleotides through the cytoplasm, and no competition of oligodeoxynucleotides with initiation factor proteins or ribosomes.

B. Oligodeoxynucleotide Uptake and Stability

Exogenously introduced oligodeoxynucleotides would not be very effective agents for hybridization arrest, no matter how stable their complexes with mRNA, if they were rapidly hydrolyzed. In previous work, degradation of oligodeoxynucleotides was not seen in rabbit reticulocyte lysate or Dulbecco's modified essential medium with 5% fetal bovine serum over 2 hours at 37°C, but degradation was complete within that time in a HeLa cell postmitochondrial supernatant, and was complete within 15 minutes in undiluted fetal bovine serum [Wickstrom, 1986]. To examine oligodeoxynucleotide uptake by HL-60 cells, aliquots of labeled anti-c-*myc* oligomer or anti-*tat*

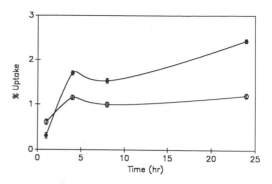

Fig. 3. *Uptake of oligodeoxynucleotides by HL-60 cells. Samples of anti-c-*myc* [5'-^{35}S]oligomer and anti-*tat* [5'-^{35}S]oligomer were incubated with HL-60 cells for 0, 1, 4, 8, and 24 hours as described in the text. Cells were separated from culture medium by sedimentation and extracted with phenol, and the aqueous phases were counted to determine oligodeoxynucleotide uptake as described in the text. ●, Anti-c-*myc* oligomer; ○, anti-*tat* oligomer. [Reproduced from Wickstrom et al., 1988, with permission of the publisher.]*

oligomer were added to HL-60 cells in culture medium and incubated for up to 24 hours.

Radioactivity retained by the washed cell pellets was compared with that left in the culture medium (Fig. 3). In RPMI 1640 with 10% heat-inactivated fetal bovine serum, about 1–2% of the labeled oligomers were found associated with the cell pellet after 4 hours, a proportion that remained about the same up to 24 hours. Denaturing gel electrophoresis of labeled oligomers remaining in the culture medium supernatant revealed significant loss of oligodeoxynucleotide within 1 hour, and disappearance was virtually complete by 8 hours (Fig. 4). In the washed cell pellet, however, labeled oligodeoxynucleotides survived intact for up to 24 hours, decreasing to about one-fourth of the original intensity. These observations agree with an earlier report of oligodeoxynucleotide uptake by HeLa and chick embryo fibroblast cells in culture [Zamecnik et al., 1986].

C. Inhibition of c-*myc* Protein Expression

When HL-60 cells in culture were treated with the anti-*myc* or the anti-VSV oligomer and analyzed by indirect immunofluorescence,

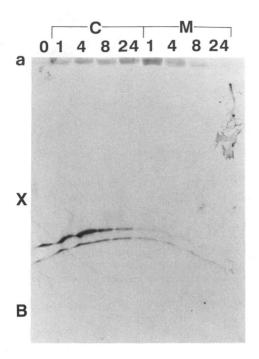

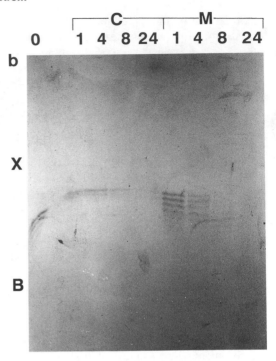

Fig. 4. *Stability of oligodeoxynucleotides in HL-60 cells. Samples of anti-c-myc [5'-^{35}S]oligomer (**a**) and anti-tat [5'-^{35}S]oligomer (**b**) were incubated with HL-60 cells for 0, 1, 4, 8, and 24 hours as described in the text. Cells were separated from culture medium by sedimentation and extracted with phenol, and the aqueous phases were analyzed by denaturating gel electrophoresis, as described in the text. C, washed cell pellet; M, in supernatant medium; X, the mobility of xylene cyanol FF; B, bromophenol blue; numbers indicate hours of incubation. [Reproduced from Wickstrom et al., 1988, with permission of the publisher.]*

neither sequence decreased the level of actin, which was detected only in the cytoplasm (Fig. 5a,d–g). HL-60 cells treated with the anti-VSV oligomer showed the same level of *myc* p65 protein in the nucleus as untreated cells (Fig. 5c, h,i). HL-60 cells treated with 6 μM anti-c-*myc* oligomer showed some reduction in the level of nuclear p65 protein, and significant reduction was observed in cells treated wtih 10 μM anti-c-*myc* oligomer (Fig. 5j, k). Radioimmunoprecipitation, however, showed no significant reduction of p65 antigen with 5 μM anti-*myc* oligomer and about 50% reduction at 10 μM (Fig. 6). The anti-VSV oligomer caused no significant reduction in p65 antigen.

D. Inhibition of Transformed Cell Proliferation

Since c-*myc* p65 antigen appears to be necessary for DNA synthesis, it was of interest to see whether antisense inhibition of p65 expression reduced the rate of cell proliferation. When HL-60 cells in culture were treated with the anti-*myc* oligomer, dose-dependent inhibition of HL-60 cell proliferation was found, with a roughly linear plot indicating 50% inhibition of proliferation at approximately 4 μM (Fig. 7). However, exposure of HL-60 cells either to the anti-VSV sequence or to the anti-HIV sequence had no effect on their proliferation (Fig. 7).

To test further the specificity of inhibition, avian cells transformed by an avian v-*myc* gene were tested for their susceptibility to the human anti-c-*myc* oligomer. The quail embryo fibroblast line Q8 transformed by the replication-defective avian myelocytomatosis virus MC29 [Bister et al., 1977] was treated with both the anti-c-*myc* and the anti-VSV oligodeoxynu-

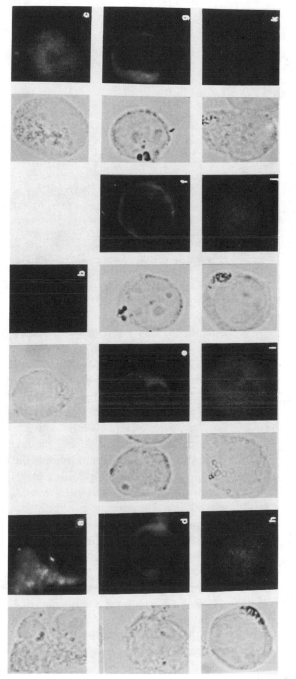

Fig. 5. *Inhibition of human c-myc p65 protein expression by antisense oligodeoxynucleotides, measured by indirect immunofluorescence. HL-60 cells were treated with oligodeoxynucleotides for 6 hours and then immunolabeled with rabbit-produced antibodies, followed by goat-produced antirabbit IgG conjugated with fluorescein, and observed by light and fluorescence microscopy.* **a:** *No oligodeoxynucleotide; anti-actin antibody.* **b:** *No oligodeoxynucleotide; normal rabbit IgG.* **c:** *No oligodeoxynucleotide, anti-c-myc antibody.* **d:** *6 μM anti-VSV oligomer; anti-actin antibody.* **e:** *10 μM anti-VSV oligomer; anti-actin antibody.* **f:** *6 μM anti-c-myc oligomer; anti-actin antibody.* **g:** *10 μM anti-c-myc oligomer; anti-actin antibody.* **h:** *6 μM anti-VSV oligomer; anti-c-myc oligomer; anti-c-myc antibody.* **i:** *10 μM anti-VSV oligomer; anti-c-myc antibody.* **j:** *6 μM anti-c-myc oligomer; anti-c-myc antibody.* **k:** *10 μM anti-c-myc oligomer; anti-c-myc antibody.* ×1,600. [Reproduced from Wickstrom et al., 1988, with permission of the publisher.]*

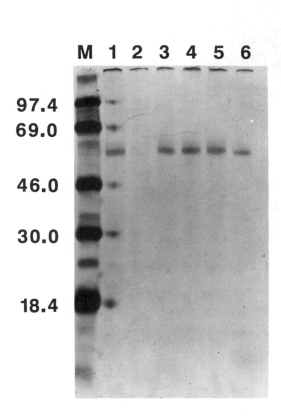

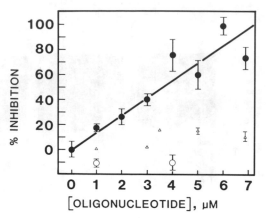

Fig. 7. *Percent inhibition of HL-60 cell proliferation by antisense oligodeoxynucleotides. HL-60 cells were grown for 5 days in medium supplemented with anti-c-myc oligomer, (●), anti-VSV oligomer (○), or anti-tat oligomer (Δ). Cells were counted, and percent inhibition was calculated as described in the text. Error bars on points represent one standard deviation. [Reproduced from Wickstrom et al., 1988, with permission of the publisher.]*

cleotides. The v-*myc* gene in MC29 displays four mismatches in the 15 nucleotides corresponding to the target sequence in human c-*myc*, as does the chicken c-*myc* sequence [Watson et al., 1983]. Neither oligodeoxynucleotide elicited any inhibition of proliferation (Fig. 8).

Treatment of HL-60 cells with Me$_2$SO dramatically reduces the level of c-*myc* mRNA [Westin et al., 1982] and inhibits the proliferation of HL-60 cells [Collins et al., 1978]. Exposure of HL-60 cells to 1% Me$_2$SO resulted in 69% ± 3% inhibition of proliferation (Fig. 9), as seen before [Collins et al., 1978], comparable to the inhibition of proliferation at 6 μM anti-c-*myc* oligomer that is shown in Figure 7. Treatment of HL-60 cells with 4 μM anti-c-*myc* oligomer reduced cell proliferation by 55% ± 5%, while the combination of 4 μM anti-c-*myc* oligomer with 1% Me$_2$SO reduced cell proliferation by 93% ± 10%. Hence the two reagents appear to have complementary inhibitory activity toward proliferation. In a negative control experiment, the anti-VSV oligomer was not inhibitory by itself, nor did it potentate the effect of 1% Me$_2$SO.

Fig. 6. *Inhibition of human c-myc p65 protein expression by antisense oligodeoxynucleotides, measured by immunoprecipitation. HL-60 cells were treated with oligodeoxynucleotides for 6 hours, labeled with [^{35}S]Cys for 1.5 hours, lysed, immunoprecipitated, electrophoresed, and quantitated as described in the text. Lane M, [^{14}C]molecular weight standards; lane 1, no oligodeoxynucleotide, anti-c-myc antibody; lane 2, no oligodeoxynucleotide, normal rabbit serum; lane 3, 5 μM anti-VSV oligomer, anti-c-myc antibody; lane 4, 10 μM anti-VSV oligomer, anti-c-myc antibody; lane 5, 5 μM anti-c-myc oligomer, anti-c-myc antibody; lane 6, 10 μM anti-c-myc oligomer, anti-c-myc antibody. [Reproduced from Wickstrom et al., 1988, with permission of the publisher.]*

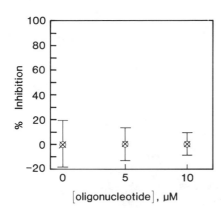

Fig. 8. *Percent inhibition of Q8/MC29 cell proliferation by antisense oligodeoxynucleotides. Q8/MC29 cells were grown for 3 days in medium supplemented with anti-c-*myc* oligomer (○) or anti-VSV oligomer (×). Cells were counted, and percent inhibition was calculated as described in the text. Error bars on points represent one standard deviation.*

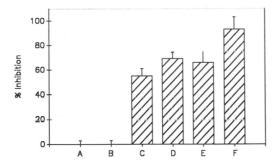

Fig. 9. *Percent inhibition of HL-60 cell proliferation by combinations of antisense oligodeoxynucleotides and Me$_2$SO. HL-60 cells were grown, treated, and analyzed as described in Figure 7, with the addition of 1% Me$_2$SO in some experiments as a control. **A.** Untreated. **B.** 4 μM anti-VSV oligomer. **C.** 4 μM anti-c-*myc* oligomer. **D.** 1% Me$_2$SO. **E.** 1% Me$_2$SO plus 4 μM anti-VSV oligomer. **F.** 1% Me$_2$SO plus 4 μM anti-c-*myc* oligomer. [Reproduced from Wickstrom et al., 1989, with permission of the publisher.]*

E. Inhibition of Colony Formation

A second method for evaluating the growth potential of a cell line is to add the cells to a semisolid medium so that dividing cells will form colonies [Graf et al., 1981]. HL-60 cells treated with a single dose of 5 μM anti-c-*myc* oligomer exhibited a sequence-specific decrease in colony formation after 5 days of growth (Fig. 10). Colony size also decreased significantly, down to two to four cell colonies, and the effect was uniform throughout the culture; no resistant subpopulation was observed. In contrast, no significant decrease in colony number or size was noted in cells treated with 5 μM anti-VSV oligomer, while cells treated with Me$_2$SO or PMA created few colonies (none larger than four cells).

Upon counting colonies and cells in each cell well, it was found that 79% ± 12% of the untreated cells formed colonies, and 74% ± 11% of cells treated with the anti-VSV oligomer continued to form colonies (Fig. 11). However, colony formation by cells treated with anti-c-*myc* oligomer was reduced to 62% + 10%. In contrast, only 23% ± 5% of cells treated with Me$_2$SO formed colonies, and only 4% ± 1% of those were treated with PMA.

F. Morphological Analysis of Differentiation

HL-60 cells induced by Me$_2$SO to differentiate with granulocytic characteristics [Collins et al., 1978] (Fig. 12B) and cells induced by PMA to differentiate with monocytic characteristics [Rovera et al., 1979] (Fig. 12C) were compared with cells that were induced by the anti-c-*myc* and anti-VSV oligomers. Oligomers were added to a concentration of 5 μM because previous work indicated that the antigen inhibition (Figs. 5, 6) and antiproliferative (Fig. 7) responses were about half-maximal in the 4–6 μM range. Cells treated with 5 μM anti-VSV oligomer (Fig. 12D) were as undifferentiated as untreated cells (Fig. 12A), but 5 μM anti-c-*myc* oligomer induced some differentiation along the granulocytic line (Fig. 12E).

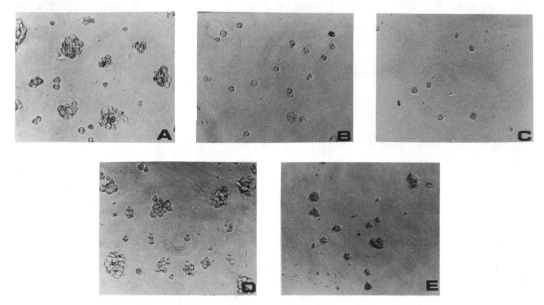

Fig. 10. *Colony formation by HL-60 cells in semisolid medium treated with Me₂SO, PMA, or antisense oligomers.* **A.** *Untreated.* **B.** *1% Me₂SO.* **C.** *16 nM PMA.*

D. *5 μM anti-VSV oligomer.* **E.** *5 μM anti-c-myc oligomer. [Reproduced from Wickstrom et al., 1989, with permission of the publisher.]*

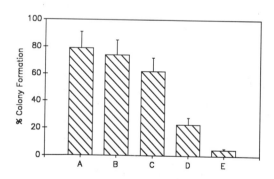

Fig. 11. *Percentages of HL-60 cells forming colonies in semisolid medium treated with Me₂SO, PMA, or antisense oligomers.* **A.** *Untreated.* **B.** *5μM anti-VSV oligomer.* **C.** *5 μM anti-c-myc oligomer.* **D.** *1% Me₂SO.* **E.** *16 nM Pma. [Reproduced from Wickstrom et al., 1988, with permission of the publisher.]*

Differential counts of treated cells demonstrated that 4 μM anti-c-*myc* oligomer, the concentration that inhibited cell proliferation by 50% (Fig. 7), doubled the fraction of differentiated cells resembling granulocytes from 10% ± 2% in untreated cells to 22% ± 3% in treated cells (Fig. 13). No cells with monocytic characteristics were observed. However, 4 μM anti-c-*myc* oligomer was not nearly as effective at inducing differentiation as 1% Me₂SO, which gave 78% ± 6% differentiated cells, as seen before [Collins et al., 1978], and anti-c-*myc* oligomer does not seem to potentiate the induction of differentiation associated with Me₂SO. The negative control sequence against the VSV matrix protein mRNA had no effect on the differentiation.

G. Cytochemical Analysis of Differentiation

A high level of NBT reduction correlates with terminally differentiating granulocytic cells, a low level of NBT reduction correlates with

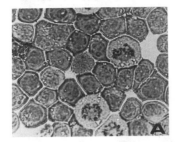
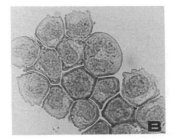
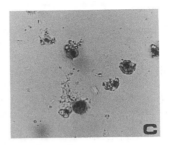
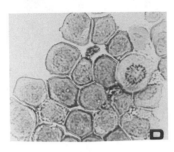
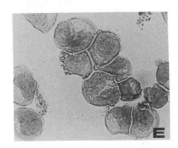

Fig. 12. *Morphological differentiation of HL-60 cell populations treated with Me₂SO, PMA, or antisense oligomers.* **A.** *Untreated.* **B.** *1% Me₂SO.* **C.** *16 nM PMA.* **D.** *5 μM anti-VSV oligomer.* **E.** *5 μM anti-c-myc oligomer. [Reproduced from Wickstrom et al., 1989, with permission of the publisher.]*

immature differentiating cells, while the complete lack of NBT reduction is associated with monocytic differentiation in HL-60 cells [Collins et al., 1979]. Accordingly, untreated HL-60 cells were 33% ± 5% positive for NBT reduction after 2 days, as were cells treated with 5 μM anti-VSV oligomer, whereas cells treated with 5 μM anti-c-*myc* oligomer rose to 48% ± 7% positive, and cells treated with Me₂SO rose to 65% ± 8% (Fig. 14). In contrast, cells treated with PMA decreased to less than 5% ± 1% positive. The uninduced HL-60 cells used in these experiments regularly displayed levels of NBT-positive cells in the range of 30%, by our criteria, while the Me₂SO–induced cells attained high levels similar to those reported by Collins et al. [1979].

IV. DISCUSSION

In the absence of data from nuclease mapping, chemical probing, base pair replacement, or phylogenetic comparisons, secondary structure predictions are often misleading [Auron et al., 1982]. The free energies used for base pairing calculations are still approximate, and we lack algorithms to predict tertiary structure interactions. Nevertheless, calculation of an oncogene mRNA secondary structure provides a starting point for selecting initial targets for hybridization arrest with antisense oligodeoxynucleotides. The effectiveness of the anti-c-*myc* oligomer predicted above to inhibit c-*myc* protein expression and transformed cell proliferation suggests that calculating mRNA secondary structures may be useful in this regard.

One would not have expected that unprotected oligodeoxynucleotides could enter a cell and remain intact long enough to have an impact, as described above. Now that several laboratories have observed the phenomenon [van der Krol et al., 1988], it is necessary to study the mechanisms of uptake, compartmentalization of the oligodeoxynucleotides, and their metabolism. Convincing evidence has appeared for receptor-mediated endocytosis of normal oligodeoxynucleotides [Loke et al., 1989; Yakubov et al., 1989].

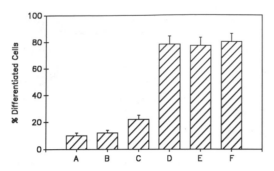

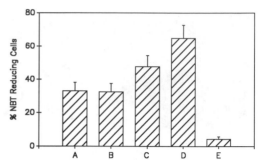

Fig. 13. *Differential counts of HL-60 cell populations treated with antisense oligomers and/or Me₂SO.* **A.** *Untreated.* **B.** *4 μM anti-VSV oligomer.* **C.** *4 μM anti-c-myc oligomer.* **D.** *1% Me₂SO;* **E.** *1% Me₂SO plus 4 μM anti-VSV oligomer.* **F.** *1% Me₂SO plus 4 μM anti-c-myc oligomer. [Reproduced from Wickstrom et al., 1989, with permission of the publisher.]*

Fig. 14. *NBT reduction by HL-60 cells treated with Me₂SO, PMA, or antisense oligomers.* **A.** *Untreated.* **B.** *5 μM anti-VSV oligomer.* **C.** *5 μM anti-c-myc oligomer.* **D.** *1% Me₂SO.* **E.** *16 nM PMA. [Reproduced from Wickstrom et al., 1989, with permission of the publisher.]*

Antisense DNA inhibition resulted in dramatic ablation of c-*myc* p65 antigen in the treated HL-60 cells. Indirect immunofluorescence of cells treated with antisense oligodeoxynucleotides for 6 hours showed greater reduction of p65 protein than did radioimmunoprecipitation of cells treated for 6 hours, and then pulsed with [^{35}S]Cys for 1.5 hours. In view of the rapid degradation of intracellular oligodeoxynucleotides shown in Figure 5, it is likely that the effective concentration of anti-*myc* oligomer in the cytoplasm decreases sufficiently by 6 hours to reduce the efficiency of hybrid arrest. It is also possible that a negative feedback system regulates *myc* transcription, leading to elevated *myc* mRNA levels as a result of hybrid arrest during the first 6 hours.

Although thermodynamic calculations predicted virtually stoichiometric binding of the anti-c-*myc* oligomer to its target on c-*myc* mRNA, the concentration range of anti-*myc* oligodeoxynucleotide required for 50% inhibition of protein expression or cell proliferation, 4–10 μM, is much higher than the concentration of c-*myc* mRNA in HL-60 cells. These cells typically contain 30–50 copies of c-*myc* mRNA during log phase growth (J. Bresser, personal communication) and have a diameter of approximately 20 μm. Cytoplasm

represents only about 20% of HL-60 cell volume, so we calculate a cytoplasmic c-*myc* mRNA concentration of about 0.1 nM, roughly 10^{-5} of the effective anti-c-*myc* oligomer concentration.

In view of the model that c-*myc* protein is specifically required for DNA synthesis [Classon et al., 1987], it was important to examine the impact of c-*myc* inhibition on the cell cycle in normal human cells. In normal human peripheral blood lymphocytes [Heikkila et al., 1987], the anti-c-*myc* oligomer, at 30 μM, inhibited expression of c-*myc* protein and entry into S phase, following stimulation by the mitogen phytohemagglutinin. No effect was seen with the anti-VSV oligomer, the original sense version of the c-*myc* oligomer, or a randomly scrambled version of the anti-c-*myc* oligomer. In contrast to the HL-60 case, c-*myc* inhibition could not be detected below 15 μM anti-c-*myc* oligomer, but was almost complete at 30 μM; that is, normal cells required three times the dose to achieve the effects seen in HL-60 cells. While the anti-c-*myc* oligomer inhibited S phase entry, it did not prevent G_0 to G_1 traversal, consistent with the model and with our results above.

Downregulation of c-*myc* gene expression by antisense DNA not only slowed proliferation of HL-60 cells, but in addition induced granulocytic differentiation of the cells. Many of the HL-60 results have been replicated by Holt et al. [1988], who also reported myeloid differentiation of the treated cells. Similarly, Yokoyama and Imamoto [1987] induced antisense c-*myc* RNA expression in HL-60 cells, but reported cytochemical data implying monocytic differentiation characteristics rather than granulocytic, experiments that are discussed in more detail by Yokoyama (this volume).

Robinson-Benion et al. [1991] recently examined the interaction of oligonucleotide-mediated inhibition of c-*myc* expression in HL-60 cells and the downregulation of c-*myc* mRNA by the growth inhibitors TGFα and TGFβ, but found no additive effect of combining the oligomer with either growth inhibitor.

Both Me_2SO and antisense oligomers inhibit c-*myc* expression, and it appears that either mode of c-*myc* inhibition allows granulocytic differentiation and inhibition of colony formation to occur in HL-60 cells. The lack of synergism between the two agents in stimulating differentiation suggests that they may operate by different mechanisms. Furthermore, the existence of synergism where antiproliferation is concerned suggests that Me_2SO may exert pleiotropic effects on replication, more than just reducing the level of c-*myc* p65. Finally, the uniform inhibition of colony formation in semi-solid medium implies that the effects of the anti-c-*myc* oligomer are felt by the entire HL-60 culture and that there is no resistant subpopulation.

V. FUTURE DIRECTIONS

Given these results, it is now necessary to elucidate the actual mechanism of anti-c-*myc* oligomer inhibition of HL-60 cell proliferation and c-*myc* protein expression, which could be at the level of mRNA transcription, processing, translation, or degradation and the factors that reduce the predicted effectiveness of the antisense oligodeoxynucleotide, including degradation, uptake, intracellular diffusion, compartmentalization, and competition with ribosomes and initiation factors. Further prob-

ing with oligodeoxynucleotides complementary to other sites in the c-*myc* mRNA, and with oligodeoxynucleotides of varying complementarity to the first five codons, may better identify accessible portions of c-*myc* mRNA and the sequence specificity requirements of c-*myc* inhibition [Bacon and Wickstrom, 1991a].

While Me_2SO is too toxic for administration in animals at a dose of 1% by volume, the anti-c-*myc* oligomer at the concentrations used here did not inhibit c-*myc* expression in mitogen-stimulated peripheral blood lymphocytes and was not noticeably toxic to them [Heikkila et al., 1987]. Hence antisense inhibitors may have therapeutic potential as inducers of differentiation. It is clearly necessary to try adding antisense DNA inhibitors to cells every day rather than only once at the beginning of the experiment and to examine the morphological, histochemical, and surface marker effects of the added oligomers every day [Bacon and Wickstrom, 1991b].

Previous studies of oligomer degradation, uptake by cells, and survival within cells showed that oligomer degradation was complete within 15 minutes in undiluted fetal bovine serum at 37°C [Wickstrom, 1986], within 8 hours in the culture medium used here, RPMI 1640 with 10% heat-inactivated fetal bovine serum, and was about three-fourths complete within 24 hours for oligomers taken up by HL-60 cells (Fig. 4). Given these results, it would be worthwhile trying antisense oligomer inhibition in a serum-free medium or with nuclease-resistant oligodeoxynucleotide derivatives. Finally, several other cell lines that overexpress c-*myc* should be studied for their susceptibility to antisense inhibitors of mRNA translation.

ACKNKOWLEDGMENTS

Most importantly, I thank my coworkers on these studies, Erica Wickstrom, Thomas Bacon, and Dr. Gary Lyman, for their dedicated efforts. I also thank Dr. Karin Mölling for a culture of Q8/MC29 cells and for the hospitality of her laboratory during our work with that cell line, Dr. Rosemary Watt for a sample of antiserum against c-*myc* p65 protein, Dr. Grace Ju for a sample of recombinant c-*myc* p65 protein, Dr. Michael Zuker for a copy of RNAFLD, Dr. Julie Djeu for valuable discussions and suggestions, Lois Wickstrom for secondary structure predictions, David Pushkin for oligodeoxynucleotide purification, and Dennis Freeman for technical advice and assistance. This work was supported by U.S. National Institutes of Health grants CA 42960 and RR 07121, by the Leukemia Society of America, by the American Cancer Society Florida Division, by the University of South Florida Research Council, and by the University Medical Services Association.

VI. REFERENCES

Auron PE, Rindone WP, Vary CPH, Celentano JJ, Vournakis JN (1982): Computer-aided prediction of RNA secondary structure. Nucleic Acids Res 10: 403–419.

Bacon TA, Wickstrom E (1991a): Walking along human c-*myc* mRNA with antisense oligodeoxynucleotides: Maximum efficacy at the 5′ cap region. Oncogene Res 6:13–19.

Bacon TA, Wickstrom E (1991b): Daily addition of an anti-c-*myc* DNA oligomer induces granulocytic differentiation of human promyelocytic leukemia HL-60 cells in both serum-containing and serum-free media. Oncogene Res 6:21–32.

Beimling P, Benter T, Sander T, Mölling K (1985): Isolation and characterization of the human cellular *myc* gene product. Biochemistry 24:6349–6355.

Bishop JM (1991): Molecular themes in oncogenesis. Cell 64:235–248.

Bister K, Hayman MJ, Vogt PK (1977): Defectiveness of avian myelocytomatosis virus MC29: Isolation of long-term nonproducer cultures and analysis of virus-specific polypeptide synthesis. Virology 82: 431–448.

Blackwood EM, Eisenman RN (1991): Max: A helix-loop-helix zipper protein that forms a sequence-specific DNA-binding complex with myc. Science 251:1211–1217.

Branton PE, Bayley ST, Graham FL (1984): Transformation by human adenoviruses. Biochim Biophys Acta 780:67–94.

Classon M, Henriksson M, Sümegi J, Klein G, Hammarskjöld ML (1987): Elevated c-*myc* expression facilitates the replication of SV40 DNA in human lymphoma cells. Nature 330:272–274.

Collins SJ, Ruscetti FW, Gallagher RE, Gallo RC (1978): Terminal differentiation of human promyelocytic leukemia cells induced by dimethyl sulfoxide and other polar compounds. Proc Natl Acad Sci USA 75: 2458–2462.

Collins SJ, Ruscetti FW, Gallagher RE, Gallo RC (1979): Normal functional characteristics of cultured human promyelocytic leukemia cells (HL-60) after induction differentiation by dimethylsulfoxide. J Exp Med 149:969–974.

Collins S, Groudine M (1982): Amplification of endogenous *myc*-related DNA sequences in a human myeloid leukemia cell line. Nature 298:679–681.

Coppola JA, Cole MD (1986): Constitutive c-*myc* oncogene expression blocks mouse erythroleukemia cell differentiation but not commitment. Nature 320:760–763.

Crabtree GR (1989): Contingent genetic regulatory events in T lymphocyte activation. Science 243:355–361.

DeChatelet LR, Shirley PS, Johnston RB (1976): Effect of phorbol myristate acetate on the oxidative metabolism of human polymorphonuclear leukocytes. Blood 47:545–554.

Filmus J, Buick RN (1985): Relationship of c-*myc* expression to differentiation and proliferation of HL-60 cells. Cancer Res 45:822–825.

Franza BR Jr, Maruyama K, Garrels JI, Ruley HE (1986): In vitro establishment is not a sufficient prerequisite for transformation by activated *ras* oncogenes. Cell 44:409–418.

Freier SM, Kierzek R, Jaeger JA, Sugimoto N, Caruthers MH, Neilson T, Turner DH (1986): Improved free-energy parameters for prediction of RNA duplex stability. Proc Natl Acad Sci USA 83:9373–9377.

Graf T, von Kirchbach A, Beug H (1981): Characterization of the hematopoietic target cells of AEV, MC29 and AMV avian leukemia viruses. Exp Cell Res 131:331–343.

Halazonetis TD, Kandil AN (1991): Determination of the c-myc DNA-binding site. Proc Natl Acad Sci USA 88:6162–6166.

Hann SR, King MW, Bently DL, Anderson CW, Eisenman RN (1988): A non-AUG translational initiation in c-*myc* exon 1 generates an N-terminally distinct protein whose synthesis is disrupted in Burkitt's lymphomas. Cell 52:185–195.

Harel-Bellan A, Ferris DK, Vinocour M, Holt JT, Farrar WL (1988): Specific inhibition of c-*myc* protein biosynthesis using an antisense synthetic deoxy-oligonucleotide in human T lymphocytes. J Immunol 140:2431–2435.

Heikkila R, Schwab G, Wickstrom E, Loke SL, Pluznik DH, Watt R, Neckers LM (1987): A c-*myc* antisense oligodeoxynucleotide inhibits entry into S phase but not progress from G_0 to G_1. Nature 328:445–449.

Holt JT, Redner RL, Nienhuis AW (1988): An oligomer complementary to c-myc mRNA inhibits proliferation of HL-60 promyelocytic cells and induces differentiation. Mol Cell Biol 8:963–973.

Jacobson AB, Good L, Simonetti J, Zuker M (1984):

Some simple computational methods to improve the folding of large RNAs. Nucleic Acids Res 12:45–52.

Kelly K, Cochran BH, Stiles CD, Leder P (1983): Cell-specific regulation of the c-*myc* gene by lymphocyte mitogens and platelet-derived growth factor. Cell 35:603–610.

Klein G, Klein E (1986): Conditioned tumorigenicity of activated oncogenes. Cancer Res 46:3211–3224.

Land M, Parada LF, Weinberg RA (1983): Tumorigenic conversion of primary embryo fibroblasts requires at least two cooperating oncogenes. Science 222:771–778.

Lanier LL, Warner NL (1981): Paraformaldehyde fixation of hematopoietic cells for quantitative flow cytometry (FACS) analysis. J Immunol Methods 47:25–30.

Larsson L-G, Ivhed I, Gidlund M, Pettersson U, Vennström B, Nilsson K (1988): Phorbol ester-induced terminal differentiation is inhibited in human U-937 monoblastic cells expressing a v-*myc* oncogene. Proc Natl Acad Sci USA 85:2638–2642.

Loke SL, Stein CA, Zhang XH, Mori K, Nakanishi M, Subasinghe C, Cohen JS, Neckers LM (1989): Characterization of oligonucleotide transport into living cells. Proc Natl Acad Sci USA 86:3474–3478.

McManaway ME, Neckers LM, Loke SL, Al-Nasser AA, Redner RL, Shiramizu BT, Goldschmidts WL, Huber B, Magrath IT (1990): Tumour-specific inhibition of lymphoma growth by an antisense oligodeoxynucleotide. Lancet 335:808–811.

Onclercq R, Lavenu A, Cremisi C (1989): Pleiotropic derepression of developmentally regulated cellular and viral genes by c-*myc* proto-oncogene products in undifferentiated embryonal carcinoma cells. Nucleic Acids Res 17:735–753.

Persson H, Hennighausen L, Taub R, DeGrado W, Leder P (1984): Antibodies to human c-*myc* oncogene product: evidence of an evolutionarily conserved protein induced during cell proliferation. Science 225:687–693.

Ratner L, Haseltine W, Patarca R, Livak KJ, Stacich B, Josephs SF, Doran ER, Rafalski JA, Whitehorn EA, Baumeister K, Ivanoff L, Petteway SR, Pearson ML, Lautenberger JA, Papas TS, Ghrayeb J, Chang NT, Gallo RC, Wong-Staal F (1985): Complete nucleotide sequence of the AIDS virus, HTLV-III. Nature 313:277–284.

Robinson-Benion C, Salhany KE, Hann SR, Holt JT (1991): Antisense inhibition of c-*myc* expression reveals common and distinct mechanisms of growth inhibition by TGFβ and TGFα. J Cell Biochem 45: 188–195.

Rose JK, Gallione CJ (1981): Nucleotide sequences of the mRNAs encoding the vesicular stomatitis G and M proteins determined from cDNA clones containing the complete coding regions. J Virol 39:519–528.

Rovera G, Santoli D, Damsky C (1979): Human promyelocytic leukemia cells in culture differentiate into

macrophage-like cells when treated with a phorbol diester. Proc Natl Acad Sci USA 76:2779–2783.

Sinha ND, Biernat J, Köster H (1983): β-Cyanoethyl N,N-dialkylamino/N- morpholinomonochloro phosphoamidites, new phosphitylating agents facilitating ease of deprotection and work-up of synthesized oligonucleotides. Tet Lett 24:5843–5846.

Spector DL, Watt RA, Sullivan NF (1987): The v- and c-*myc* oncogene proteins colocalize in situ with small nuclear ribonucleoprotein particles. Oncogene 1:5–12.

Thompson CB, Challoner PB, Neiman PE, Groudine M (1985): Levels of c-*myc* oncogene mRNA are invariant throughout the cell cycle. Nature 314:363–366.

van der Krol AR, Mol JNM, Stuitje AR (1988): Modulation of eukaryotic gene expression by complementary RNA or DNA sequences. BioTechniques 6:958–976.

Watson DK, Reddy EP, Duesberg PH, Papas TS (1983). Nucleotide sequence analysis of chicken c-*myc* gene reveals homologous and unique coding regions by comparison with the transforming gene of avian myelocytomatosis virus MC29 *gag-myc*. Proc Natl Acad Sci USA 80:2146–2150.

Watt R, Stanton LW, Marcu KB, Gallo RC, Croce CM, Rovera G (1983): Nucleotide sequence of cloned cDNA of human c-*myc* oncogene. Nature 303:725–728.

Watt RA, Statzman AR, Rosenberg M (1985): Expression and characterization of the human c-*myc* DNA-binding protein. Mol Cell Biol 5:448–456.

Westin EH, Wong-Staal F, Gelmann EP, Dalla-Favera R, Papas TS, Lautenberger JA, Eva A, Reddy EP, Tronick SR, Aaronson SA, Gallo RC (1982): Expression of cellular homologues of retroviral *onc* genes in human hematopoietic cells. Proc Natl Acad Sci USA 79: 2490–2494.

Wickstrom E (1986): Oligodeoxynucleotide stability in subcellular extracts and culture media. J Biochem Biophys Methods 13:97–102.

Wickstrom E, Simonet WS, Medlock K, Ruiz-Robles I (1986a): Complementary oligonucleotide probe of vesicular stomatitis virus matrix protein mRNA translation. Biophys J 49:15–17.

Wickstrom EL, Wickstrom E, Lyman GH, Freeman DL (1986b): HL-60 cell proliferation inhibited by an anti-c-*myc* pentadecadeoxynucleotide. Fed Proc Fed Am Soc Exp Biol 45:1708.

Wickstrom EL, Bacon TA, Gonzalez A, Freeman DL, Lyman GH, Wickstrom E (1988): Human promyelocytic leukemia HL-60 cell proliferation and c-*myc* protein expression are inhibited by an antisense pentadecadeoxynucleotide targeted against c-*myc* mRNA. Proc Natl Acad Sci USA 85:1028–1032.

Wickstrom EL, Bacon TA, Gonzalez A, Lyman GH, Wickstrom E (1989): Anti-c-*myc* DNA oligomers increase differentiation and decrease colony formation by HL-60 cells. In Vitro Cell Dev Biol 25:297–302.

Yakubov LA, Deeva EA, Zarytova VF, Ivanova EM, Ryte AS, Yurchenko LV, Vlassov VV (1989): Mechanism of oligonucleotide uptake by cells: Involvement of specific receptors? Proc Natl Acad Sci USA 85:1028–1032.

Yokoyama K, Imamoto F (1987): Transcriptional control of the endogenous *MYC* protooncogene by antisense RNA. Proc Natl Acad Sci USA 84:7363–7367.

Zamecnik PC, Goodchild J, Taguchi Y, Sarin PS (1986): Inhibition of replication and expression of human T-cell lymphotropic virus type III in cultured cells by exogenous synthetic oligonucleotides complementary to viral RNA. Proc Natl Acad Sci USA 83:4143–4146.

ABOUT THE AUTHOR

ERIC WICKSTROM is Professor of Chemistry, Biochemistry and Molecular Biology, and Surgery at the University of South Florida, Tampa, where he teaches protein and nucleic acid structure and function. After receiving his B.S. honors from the California Institute of Technology in 1968, he received his Ph.D. under Ignacio Tinoco, Jr., at the University of California, Berkeley, where he studied biophysical chemistry and the secondary structures of transfer RNA and messenger RNA ribosome binding sites. Dr. Wickstrom then pursued postdoctoral research at the University of Colorado, Boulder, probing the tertiary structure of transfer RNA with rigid, variable-length oligoproline crosslinking reagents, "molecular rulers," in the laboratory of Michael Yarus. His current goal lies in the development of gene-specific antisense and antigene DNA therapeutics for treatment of human solid tumors and leukemias. His research papers have appeared in such journals as *Nature,* the *Proceedings of the National Academy of Sciences USA*, *Biochemistry, Nucleic Acids Research*, and *Antisense Research and Development*. Dr. Wickstrom is a founding member of the editorial board of *Antisense Research and Development* and the editor of a book, *Prospects for Antisense Nucleic Acid Therapy of Cancer and AIDS*. He has been a European Molecular Biology Organization Fellow at the Rijksunivsiteit, Leiden, The Netherlands, a Guest Researcher at the National Cancer Institute, Bethesda, Maryland, and a Guest Researcher at the Max Planck Institut für Molekular Genetik, Berlin, Germany.

Antisense RNA and DNA: 335–352
© 1992 Wiley-Liss, Inc.

Antisense RNA Induces a Nuclear Transcription Factor to Repress the Expression of Endogenous *myc* Gene

Kazushige Yokoyama

I. INTRODUCTION

HL-60 promyelocytic leukemia cells [Collins et al., 1978] represent a model for studying the role of protooncogenes in cellular proliferation and differentiation. Several mutations in specific protooncogenes have been identified in HL-60 cells that may account for the transformed phenotype [Murray et al., 1983]. The nuclear protooncogene c-*myc* is amplified 8- to 30-fold and is highly expressed in these cells [Collins et al., 1982; Dalla-Favera et al., 1982]. HL-60 cells differentiate along the granulocytic pathway when they are treated with dimethyl-sulfoxide (DMSO) [Collins et al., 1978] or retinoic acid [Breitman et al., 1980] and along the monocytic pathway when treated with phorbol esters [Rovera et al., 1979] or a vitamin D analog [Reitsma et al., 1983]. A common result of differentiation induction is a profound decrease in *myc* expression [Cole, 1986; Filmus and Buick, 1985; Grosso and Pitot, 1985; Westin et al., 1982]. It is presently unclear, however, whether decreased *myc* expression is a cause, or a closely linked consequence, of HL-60 differentiation.

The introduction of an antisense RNA complementary to a specific mRNA into cells can effectively create or mimic a null-mutant phenotype and provide a cell line useful for investigating gene products of unknown physiological function [Green et al., 1986; Weintraub et al., 1985]. Yokoyama and Imamoto [1987] recently applied the use of the antisense RNA approach to inhibit *myc* expression and to study its role in HL-60 cell differentiation.

In this review I examine the effects of antisense *myc* transcripts on the constitutive expression of the *myc* gene and show that antisense *myc* gene introduced into the human promyeloleukemia cell line HL-60 can inhibit not only *myc* protein synthesis but also transcription of the endogenous c-*myc* gene. A 74 kilodalton (kD) nuclear transcriptional regulatory protein is identified in the antisense *myc* transformant. This protein can decrease transcription of *myc* and commit HL-60 cells to monocytic differentiation without the help of a differentiation inducer. These results suggest that a 74 kD nuclear protein is one of several negative repressors of *myc* gene transcription and one of the factors that direct cell differentiation.

II. SELECTION OF THE ANTISENSE *myc* TRANSFORMANTS

Initially we constructed the antisense *myc* plasmid using a genomic DNA clone. However, the suppression efficiency of endogenous *myc* expression was quite low, possibly because of the complicated structure formed by the complementary sequence between exon 1 and intron 1 [Cole, 1986] and therefore we decided to use the cDNA sequence for plasmid construction.

Although the structural significance of the noncoding exon 1 of the *myc* gene is not clear, it has possible regulatory roles [Cole, 1986]: 1) A transcription pausing effect may occur at the end of exon 1 [Bentley and Groudine, 1986]; 2) a possible stem–loop structure between exon 1 and exon 2 may suppress the translation frequency of *myc* protein; and 3) the rate-limiting region of *myc* mRNA stability may be located in exon 1. Given these possibili-

ties, we did not use the sequence of exon 1 for the antisense plasmid construction, but took a functional *myc* cDNA lacking exon 1 as a BamHI fragment from the plasmid pC5–8 (S. Tonegawa, unpublished data).

Antisense and sense *myc* plasmids pSVgpt-C5–8 and pMMTVgptC5–8 (Fig. 1) contain the *myc* coding sequences cloned in the antisense and sense orientation relative to a Simian virus 40 (SV40) promoter or mouse mammary tumor virus promoter (MMTV), plus the selective marker gene *Ecogpt (Escherichia coli* xanthine/ guanine phosphoribosyltransferase). The strategy for all constructions was to clone *Ecogpt* into pSVMdhfr [Lee et al., 1981] and to replace the dihydrofolate reductase gene with *myc* [Nishioka and Leder, 1979; Weiss et al., 1983; Yokoyama and Imamoto, 1987]. These antisense and sense plasmid DNAs were introduced into the promyeoleukemia cell line HL-60 by the protoplast fusion method [Sandri-Goldin et al., 1981].

A. Short-Term Transfection Assay

We tested two different types of antisense *myc* HL-60 transformants. One was the inducible expression system. We constructed antisense *myc* plasmids in which the antisense was driven by the MMTV long terminal repeat (LTR) promoter or by the mouse metallothionein promoter (in MTII$_a$) and transfected into HL-60 cells. The second type was the constitutive expression system, in which the antisense *myc* expression vectors were constructed with the promoter of SV40 early region and of Rous sarcoma virus long terminal repeat (RSV LTR). However, none of the antisense *myc* clones established by the short-term transfection assay demonstrated a reduction in *myc* protein production.

B. Long-Term Transfection Assay

Stable antisense *myc* HL-60 transformants were obtained by transfecting with pSVgptC5–8 and selecting for mycophenolic acid resistance. We used two strategies to try to increase the level of antisense *myc* transcripts in transformants [Yokoyama and Imamoto, 1987]. First, the

Inducible Expression

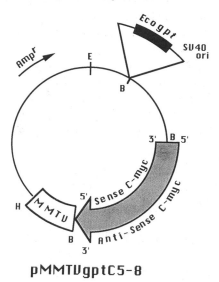

pMMTUgptC5-8

Constitutive Expression

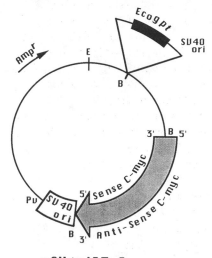

pSUgptC5-8

Fig. 1. *Plasmid construction. Antisense and sense* myc *plasmids pSVgptC5–8 and pMMTVgptC5–8 contain the* c-myc *coding sequences cloned in the anti-sense and sense orientation relative to a simian virus 40 promoter or a mouse mammary tumor virus promoter, plus the selective-marker gene* Ecogpt.

resistance to mycophenolic acid of a primary antisense transformant (AM 93) was gradually increased by successive rounds of selection with increasing drug concentration over a 6 month period. Kim and Wold [1985] reported using the same strategy to increase the antisense transcript to a level sufficient to inhibit tk gene expression. In the second method, secondary (AM 93-4) and tertiary (AM 93-4-12) transformants were established by successively retransfecting AM 93 (after 6 months) and AM 93-4 with antisense *myc* DNA. These procedures resulted in elevated levels of antisense transcripts (see Fig. 2).

RNA extracted from AM93 after 1 month (Fig. 2, lanes 5 and 9), 2 months (lane 8), 3 months (lane 7), and 6 months culture (lanes 2 and 6) showed gradually increasing amounts of a 2.5 kilobase (kb) transcript. The clone cultured for 6 months contained 5–10 times more antisense *myc* RNA than the clone cultured for 1 month. RNA from the secondary (lane 3) and tertiary (lane 4) transformants showed a prominent band at 2.5 kb; these clones contain 10–20

times more antisense *myc* RNA than AM93 cultured for 1 month. A comparison of the amounts of antisense and sense *myc* RNA in various clones is given in Table I. These data clearly show that selection for resistance to a high concentration of mycophenolic acid and repeated transfection with antisense *myc* plasmid DNA both result in increased levels of antisense *myc* RNA. Examination of genomic DNA isolated from the clones demonstrated that amplification of the antisense *myc* gene had occurred, resulting in 100–250 copies in the tertiary transformants (Table I). An RNase protection study confirmed the presence of antisense RNA–sense RNA duplex in the nuclei [Yokoyama and Imamoto, 1987]. The gpt control clone pSV2gpt[+] contained no detectable double-stranded *myc* RNA, whereas a 2.2 kb band was observed in the antisense clones. This protected band was found in the nuclear rather than in the cytoplasmic RNA fraction of the cells [Yokoyama and Imamoto, 1987].

The half-lives of antisense *myc* transcript and the duplex RNA formed between the antisense

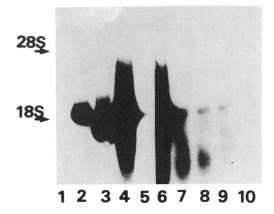

28S →
18S →

1 2 3 4 5 6 7 8 9 10

Fig. 2. *Hybridization of RNA from HL-60 cells and antisense* myc *transformants to a probe specific for the antisense* myc *transcript. Total RNA (10 μg) from the clones indicated was electrophoresed on a 1.2% agarose gel containing 1 M formaldehyde and hybridized to a single-stranded antisense PstI–ClaI fragment of* myc *exons 2 and 3. Lanes 1, untransfected HL-60; 2, 6, primary antisense clone AM93 cultured for 6 months (110 μg/ml of mycophenolic acid); 3, secondary antisense clone AM93-4 cultured for 3 months; 4, tertiary antisense clone AM93-4-12; 5, 9, AM93 cultured for 1 month (35 μg/ml); 7, AM93 cultured for 3 months (95 μg/ml); 8, AM93 cultured for 2 months (65 μg/ml); 10, control clone pSV2gpt⁺. Arrows indicate the 28S and 18S rRNAs. Autoradiography was on Kodak X-Omat S film using Dupont Cronex Quartz intensifying screens at −70°C for 6 hours (lanes 1–5) or 24 hours (lanes 6–10). [Reproduced from Yokoyama and Imamoto, 1987, with permission of the publisher.]*

and sense endogenous *myc* were quite long, about 3–4 hours, in contrast to the short half-life of sense c-*myc* transcript. The relative extent of duplex formation between antisense and sense *myc* RNA is approximately 10% (unpublished data).

III. myc PROTEIN LEVEL IN ANTISENSE *myc* TRANSFORMANTS

Antisense and sense *myc* transformants were incubated with [³⁵S]methionine for 8 hours to analyze p64 *myc* protein synthesis under steady-state conditions. Six representative antisense transformants showed a significant reduction in the amount of *myc* gene product compared with control clones (Table I and Fig. 3). Assuming an equivalent methionine pool size in all clones, about 70% less p64 was observed in lysates from antisense transformants than in controls. Correcting for cell number, it is estimated that the amount of p64 per cell is reduced by >90% in the antisense transformants. This reduction in the steady-state level of p64 (*myc* protein) is probably not due to increased turnover of the protein in the antisense transformants because the half-life of p64 is about the same in the antisense and control clones. The relative amount of p64 in antisense and control clones was not significantly changed by correcting for differences in the half-life of p64 and in the pool sizes of [³⁵S]methionine (data not shown). Control experiments showed that the reduction in the amount of *myc* protein in the antisense transformants apparently depends specifically on the presence of antisense *myc* sequences. Control antisense clones were established by transfecting HL-60 cells with plasmids identical to pSVgptC5−8 except that they contained antisense H-2Kᵇ, antisense α-globin 68-mers, or antisense *Ecogpt⁺* in place of antisense *myc*. They were screened by the identical protocol as the antisense *myc* transformants. These clones exhibited no detectable reduction in p64 synthesis and showed no differentiation phenotypes, indicating that decreased *myc* expression requires antisense *myc* sequences and is not merely an artifact of mycophenolic acid selection or of culture conditions. A second control experiment showed that the level of actin protein synthesis is essentially unchanged in all transformants (Table I). According to the results of cell growth measurement, the inhibition of synthesis of the *myc* gene product might cause a substantial alteration in the growth machinery of the cell, since the cell cycle increases from 30–36 hours in normal HL-60 cells to 95–135 hours in cells expressing antisense *myc* (Table I).

TABLE I. Summary of Sense and Antisense Transformants

Cells	mRNA (cpm × 10⁻³)			DNA (copy number)		p 64 protein (cpm × 10⁻³)		Growth rate (hr per cell cycle)
	S *myc*	Actin	AS *myc*	S *myc*	AS *myc*	*myc*	Actin	
HL-60	20.5	18.2	0.1	25	—	2.7	4.2	33
Control pSV2gpt$^+$	17.5	17.2	0.1	22	—	2.5	4.4	36
Control sense transformants*								
pSVgptC5–8 (sense *myc*)	49.2	16.9	0.4	165	—	3.8	5.3	38
pSVH-2K^b	16.9	16.3	—	25	—	2.5	4.8	36
pSVα-globin 68	18.5	17.0	—	20	—	2.6	4.4	34
Antisense transformants**								
AM93 1 month	17.5	17.2	3.1	20	20	2.2	3.9	—
2 months	14.4	—	9.3	—	25	2.0	—	—
3 months	11.3	—	19.4	—	40	1.8	—	—
6 months	6.2	15.4	31.1	20	55	1.6	4.1	—
AM93-4	5.1	16.0	50.6	22	100	1.0	4.0	—
AM93-4-12	3.3	14.9	77.8	20	120	0.6	4.0	135
AM2-3-91	10.3	17.0	27.2	20	170	1.7	3.9	95
AM46-3-2	4.2	15.2	66.9	20	160	0.8	4.0	120
AM451-6-30	4.5	15.6	62.2	22	250	0.8	4.0	125
AM763-11-4	5.1	16.3	46.7	25	200	1.0	4.0	110
AM966-10-64	6.8	15.8	42.8	20	200	1.4	4.1	105
Control antisense transformants								
pSVH-2K^b (antisense H-2K^b)	17.5	17.1	—	20	—	2.6	4.1	36
pSVα-globin68 (antisense α-globin)	18.1	16.0	—	25	—	2.6	4.5	38
pSV2gpt$^+$ (antisense gpt)	18.5	17.2	—	20	—	2.6	4.1	34

Values are means of four experiments. Protein levels and mRNAs of antisense (AS) and sense (S) *myc* and actin were measured by counting radioactivity in bands cut out of SDS PAGE gels (protein) or out of RNA gel blot filters (mRNA). Copy number of the integrated DNA was determined by comparison with known amount of *myc* plasmid DNA (equivalent to 10⁻⁶ to 10⁻⁹ molecules). Antisense and sense plasmids pSVH-2K^b and pSVα-globin carry the H-2K^b gene on a 5.0 Kb NruI–EcoRI fragment or a 68-mer of the 5′ end of α-globin [Nishioka and Leder, 1980] in place of *myc*.

*Plasmids having same sequences as antisense plasmids, but in sense orientation. Note plasmid name is unchanged for sense and antisense orientations.

**With pSVgptC5–8 (antisense myc).

[Adapted from Yokoyama and Imamoto, 1987, with permission of the publisher.]

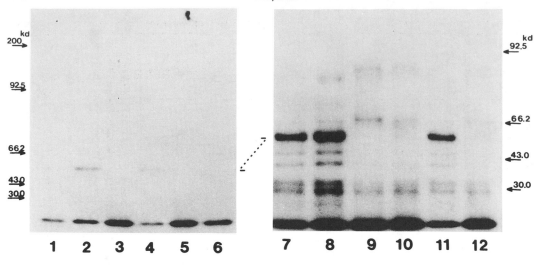

Fig. 3. *Sodium dodecylsulfate polyacrylamide gel electrophoresis (SDS PAGE) of the myc protein (p64). [³⁵S]methionine radiolabeled cells were lysed with 1% Triton X-100 and immunoprecipitated with antihuman myc antibody (Oncor Co.) (lanes 1–6) or rabbit anti-* *human myc peptide antibody (lanes 7–12), and 5 × 10³ cpm of radioactivity was applied to the gel. Lanes 1, 7, control pSV2gtp⁺ clone; 2, 8, untransfected HL-60; 3, 9, AM93-4-12; 4, 11, AM120-1-4; 5, 10, AM451-6-30; 6, 12, AM763-11-4.*

IV. REDUCED *myc* GENE EXPRESSION COMMITS HL-60 CELLS TO DIFFERENTIATE INTO MONOCYTES

The HL-60 cell line, which consists predominantly of rapidly dividing cells with promyelocytic characteristics [Collins et al., 1978], conains multiple copies of the c-*myc* gene [Collins and Groudine, 1982] and expresses a 20–30-fold increased level of *myc* RNA per cell when compared to normal cells. Chemical reagents like TPA and DMSO inhibit HL-60 cell proliferation and commit cells to differentiate along two different pathways. Induction of HL-60 cells to differentiate with reagents leads to a decline in *myc* mRNA [Westin et al., 1982]. To see whether decreased *myc* expression is a cause or a closely linked consequence of HL-60 cell differentiation, we analyzed the phenotype of antisense *myc* transformants.

Immunofluorescence staining and cytochemical studies demonstrated a close correlation between a decrease in the amount of *myc* pro-

tein in cells and an increase in the number of cells with a monocytic phenotype (Table II). This correlation suggests that a reduction in *myc* expression may be the initial event in the commitment of HL-60 cells to differentiate into adherent monocytes. Although reduced *myc* expression may not be sufficient to commit HL-60 cells to monocytic differentiation, the presence of antisense *myc* appears to change the developmental potential of HL-60, biasing the cells toward the monocytic pathway in preference to granulocytic development.

V. TRANSCRIPTIONAL CONTROL OF THE *myc* GENE

During the isolation of antisense *myc* transformants, we found that most showed decreased levels of endogenous *myc* mRNA (approximately three- to sevenfold reduction below normal level; Table I). This finding prompted us to analyze the effect of antisense *myc* on the promoter activity of the endogenous c-*myc* gene.

TABLE II. Comparative Human myc Protein Level and Relative Cell Number of the Differentiated Phenotypes

Cells	S myc expression (p64/mRNA/cell; ($\times 10^3$)*	Percent reactive cells			
		M[†]	G[‡]	OKM-1[§]	OKM-5
HL-60	1.11	0	0	0	0
HL-60 + PMA	—	99	0	95	99
HL-60 + Me$_2$SO	—	0	98	3	4
Control pSV2gpt$^+$	1.18	12	15	5	6
Transformants					
Sense pSVgptC5–8	1.45	15	18	8	10
Antisense pSVgptC5–8					
AM2-3-91	0.22	27	10	32	29
AM46-3-2	0.10	56	13	61	63
AM93-4-12	0.08	72	12	80	88
AM451-6-30	0.08	58	14	67	74
AM763-11-4	0.22	34	15	49	53
AM966-10-64	0.10	41	17	54	69

Values are means of four experiments. Phorbol 12-myristate-13-acetate (PMA) and dimethylsulfoxide (Me$_2$SO) were used as specific inducers of monocyte and granulocyte, respectively. S, sense.

*The ratios of the levels of myc protein per mRNA per cell were calculated from data such as in Table I, correcting for differences in cell number and incorporation rate of methionine.

[†]Percent of positive cells by staining of α-naphthyl acetate esterase.

[‡]Percent of positive cells by staining of naphthol AS-D chloroacetate esterase.

[§]Indirect immunofluorescence staining was performed with fluoroscein isothiocyanate-conjugated goat antihuman IgG (Miles) and monoclonal antibodies OKM-1 and OKM-5 (Ortho Diagnostics).

[Reproduced from Yokoyama and Imamoto, 1987, with permission of the publisher.]

A. Transcriptional Repression of the Endogenous c-*myc* Gene by Antisense RNA

The effect of antisense *myc* RNA on constitutive transcription of the *myc* gene was analyzed by measuring promoter activity with gene fusions to the *E. coli* chloramphenicol acetyl transferase (CAT) gene. Two fragments of the *myc* promoter, the HindIII–PvuII fragment (3 kb) and the PstI–PvuII enhancer-containing fragment (920 bp), were tested in a promoter-probe CAT vector. In parental HL-60 cells and in the pSV2gpt$^+$ control clone, weak CAT activity was observed with the larger fragment of the *myc* promoter (HindIII–PvuII), while the smaller fragment (PstI–PvuII) produced strong CAT activity. Surprisingly, in antisense *myc* transformants, the enhancer carrying promoter fragment (PstI–PvuII) gave less than 10% of the CAT activity obtained with the larger Hind III–PvuII fragment.

Run-on transcription assays using isolated nuclei confirmed that this regulation of promoter activity was at the transcriptional level [Yokoyama and Imamoto, 1987]. To determine the precise nucleotide sequences required for this inhibitory activity we constructed various deletion mutants with ExoIII. These constructs were transfected both into HL-60 cells and into the antisense *myc* transformants (AM93-4-12), and their relative CAT activities were measured.

The results shown in Figure 4 indicate that the target sequence for antisense *myc*-induced suppression is localized within a 56 bp fragment of the *myc* gene promoter region. An internal deletion between − 158 and − 102 also did not cause the negative regulation in the antisense *myc* cell line (data not shown). These data suggest the presence of a target sequence for antisense *myc*-induced gene suppression located between − 158 and − 102.

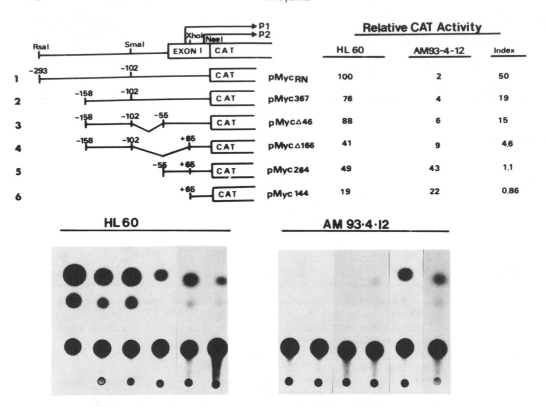

Fig. 4. *Expression of the pMyc-CAT deletion constructs in HL-60 and the antisense* myc *clone AM93-4-12. CAT activity was measured by thin-layer chromatography. Lower signal, chloramphenicol; upper signal, acetylated chloramphenicol; P1 and P2, transcriptional start sites of* myc *gene.*

B. Specific factors in nuclear extracts from antisense *myc* transformants

To investigate whether nuclear factors in extracts derived from AM93-4-12 were capable of binding to CACC–TCC sequences, DNase I footprinting was carried out. Double-stranded oligodeoxynucleotides corresponding to this sequence were prepared and used as a probe for DNase I footprinting experiments. The protected region, using nuclear extract from antisense *myc* transformant AM93-4-12, was identified at position − 151 to − 117 in the promoter region (Fig. 5). As expected, this region was not protected with extracts from HL-60 cells.

We next synthesized a 45-mer oligodeoxynucleotide to be used as a probe in a gel retardation assay (Fig. 6). The shifted band was detected in the extract from AM93-4-12, and the binding was completely inhibited by competitor oligodeoxynucleotides containing the CACC–TCC sequence. The nuclear extracts from HL-60 contained a lesser amount of the binding proteins that bind to this sequence.

Thus, in the nuclear extracts from AM93-4-12, specific DNA-binding proteins were induced and found to bind to the nucleotide sequence CACC–TCC repeat in the promoter region of the c-*myc* gene. Thus the evidence suggests that this binding protein might a negative regulator of c-*myc* gene transcription.

To characterize the nuclear proteins that interact with the CACC–TCC elements, Southwestern blotting [Miskimins et al., 1985; Fainsod et al., 1986] was carried out. Nuclear extracts

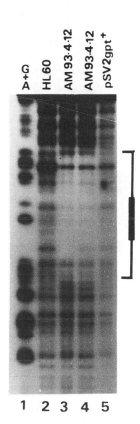

AGTCTCCTCCCCACCTTCCCCACCCTCCCCACCCTCCCCATAAGCGCCCCTCCC
CACC–TCC like Box

Fig. 5. *DNase I footprint analysis of CACC–TCC-binding protein. A radioactively labeled promoter fragment (−158 to −102) was incubated with nuclear extracts prepared from HL-60, antisense* myc *transformant, and pSV2gpt⁺ transformant and analyzed by the DNase I footprint [Katoh et al., 1990]. Lanes 1, G + A; 2, 65 μg of nuclear extract from HL-60; 3, 35 μg of nuclear extract from AM93-4-12; 4, 50 μg of nuclear extract from AM93-4-12; 5, 65 μg of nuclear extract from pSV2gpt⁺ clone. The location of the protected region is indicated to the right, and the protected nuclear sequence is indicated below.*

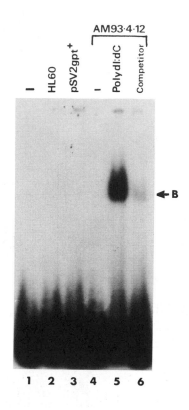

CACC-TCC like Box
Oligodeoxynucleotides 45mers

Fig. 6. *Gel retardation assay of CACC–TCC-binding protein. T4 kinase labeled oligonucleotides (45-mers) containing the binding site of CACC–TCC protein were incubated with the nuclear extracts (5 µg) from HL-60, pSV2gpt⁺ clone, and AM93-4-12 by a method described elsewhere [Katoh et al., 1990]. The shifted band is indicated (B), and the gel retardation is abolished by specific (lane 6) but not the nonspecific (lane 5) competitor poly dI:dC.*

from HL-60 and AM93-4-12 cells were fractionated on a heparin agarose column. The proteins eluted from this column by 0.4 M KCl were applied to a DEAE-Sepharose column and eluted with a stepwise gradient of KCl. These fractionated proteins were then electrophoresed by sodium dodecyl sulfate polyacrylamide gel electrophoresis (SDS PAGE). The nylon filters, after transfer of proteins by Western blotting,

were incubated with ^{32}P-radiolabeled probe containing the CACC–TCC element in order to detect specific binding proteins.

As shown in Figure 7A, the binding of a 74kD protein was seen in nuclear fractions from both HL-60 and from AM93-4-12 cells. However, the DNA-binding activity of 74 kD protein from AM93-4-12 extracts was much higher than that from HL-60 cells. The probes not containing the CACC–TCC sequence did not show specific bands (Fig. 7B,C). The proteins eluted from DEAE-Sepharose were directly applied to an affinity column containing CACC–TCC repeated sequences. The bound fraction was recycled twice. The eluted fractions were analyzed by two-dimensional SDS PAGE. As shown in Figure 7D, E, the distinct spots of 74 kD protein were observed in the eluted fraction of AM93-4-12 nuclear extracts in addition to bands of a 110 kD protein. In the fraction of HL-60 nuclear extracts, a faint band of 74 kD protein was observed. These results indicate that the synthesis of both 74 and 110 kD proteins are specifically induced in the antisense *myc* transformant, and these proteins might be candidates for suppressors of the endogenous c-*myc* gene expression. Given the results of Southwestern blotting, it may be unlikely that the 110 kD molecule is a DNA-binding protein. It is not known how this protein might couple with 74 kD protein and how it might modulate the transcriptional activity of the endogenous c-*myc* gene.

We next attempted to purify this 74 kD negative regulator from AM93-4-12 cells using a combination of column chromatography and DNA-affinity chromatography. The nuclear extracts from cultured AM93-4-12 cells were applied to a heparin–agarose column. The fraction eluted by 0.4 M KCl was then applied to a DEAE-sepharose column and to S-300 column chromatography. Each fraction eluted from the column was assayed by gel retardation using a CACC–TCC DNA fragment as a probe. The final product (1.5 µg) was analyzed for homogeneity on SDS PAGE, as shown in Figure 8.

We also found a 38 kD protein on the gel under reducing conditions. The gel bands rep-

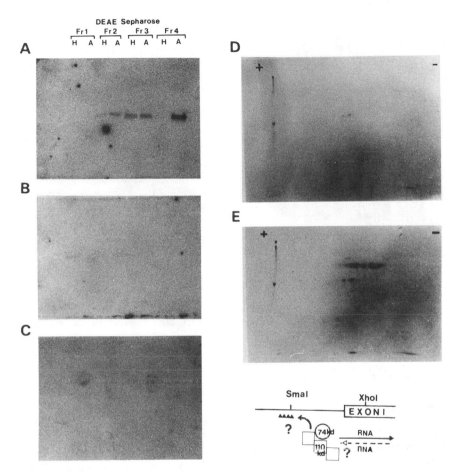

Fig. 7. *Binding of various ³²P-labeled DNA to nuclear proteins from HL-60 and AM93-4-12. A crude nuclear extract from cells of IIL-60 (indicated as H) and AM93-4-12 (indicated as A) was fractionated on a heparin–agarose column and both 0.4 M KCl-eluted fractions were applied to DEAE-Sepharose columns. Both fractions were eluted by stepwise elution of NaCl and applied to a SDS PAGE gel. Fraction 1, 0.1 M NaCl; fraction 2, 0.25 M NaCl; fraction 3, 0.3 M NaCl; fraction 4, 0.4 M NaCl. Binding of ³²P-labeled DNA fragments (− 158 to − 102, **A**; PstI–PvuII frag-* *ment, **B**; and pUC 19 BamH I fragment, **C**) were carried out as described [Miskimins et al., 1985]. HL-60 and AM93-4-12 cells were radiolabeled by [³⁵S]methionine and fractionated as described above. The proteins of fraction 4 on DEAE-Sepharose column were then applied to a DNA-affinity column (CACC–TCC oligonucleotide–agarose) and recycled twice [Kadonaga and Tijian, 1986]. The bound proteins from HL-60 (**D**) and AM93-4-12 (**E**) cells were analyzed by 2D-gel electrophoresis.*

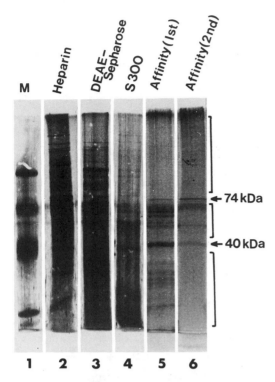

Fig. 8. *SDS PAGE analysis of 74 kD protein. Lanes 1, molecular weight marker; 2, 0.4 M KCl fraction; 3, DEAE-Sepharose fraction; 4, S-300 fraction; 5, DNA-affinity column (CACC–TCC fraction) (1st); 6, DNA-affinity column (CACC–TCC fraction) (2nd).*

resenting the 38 kD protein were cut out, eluted, and renatured by the successive treatment of 6 M guanidine–HCl and dialysis against 10 mM Tris-buffered saline in the presence of protease inhibitors. The binding activity to the CACC–TCC DNA fragment was found only in the 38 kD protein fraction. In the absence of a reducing reagent, the major band observed was the 74 kD protein (data not shown). Thus it is highly possible that the 74 kD protein is composed of a dimer of the 38 kD subunit. Southwestern blotting indicated that the DNA-binding activity was observed only in the fraction of 74 kD protein. Other fractions did not bind to DNA with a CACC–TCC sequence. Thus the specific negative regulator may be a 74 kD protein that is probably a dimer of a 38 kD protein subunit.

C. Function of *myc*-Negative Repressor

In an attempt to see whether the 74 kD protein complex affected the promoter activity of the *myc* gene, we performed a CAT assay of mutant constructs using nuclear extracts in the presence or absence of the affinity-purified 74 kD protein.

As shown in the *in vitro* promoter assay in Figure 9, the CACC–TCC box is revealed to be a negative control element of *myc* promoter (lane 3). Without addition of CACC–TCC binding protein (74 kD protein), we did not detect any decreased activity of the *myc* gene promoter. By contrast, the nuclear extract from AM93-4-12 containing 74 kD protein caused a significant reduction of *myc* promoter activity. This negative activity is evident even when a fraction containing the enhancer-promoter binding factors of the c-*myc* gene like E2F or Sp1 [Nishikura, 1986; Thalmeier et al., 1989] was added to the nuclear extracts, as shown in Figure 9, lane 8. The presence of 74 kD protein in nuclei was shown in other antisense *myc* transformants as well as AM93-4-12 (unpublished results). These results indicate that the negative regulation of the *myc* gene is mediated by the induced 74 kD nuclear protein factor in the antisense *myc* transformant.

To identify the biological activity of the 74 kD protein, liposome-mediated protein transfer was used to introduce 74 kD protein into HL-60 cells. This was done to determine whether transcription of endogenous c-*myc* gene could be decreased by this protein and result in the triggering of cell differentiation. As shown in Table III, the 74 kD protein can commit HL-60 cells to differentiate into macrophages as detected by monoclonal antibody or esterase staining. The level of endogenous *myc* in these cells was also significantly reduced. Although the concentration of 74 kD protein in the cells was estimated to be less than 10^{-6} M, it is clear that the introduction of excess of 74 kD protein results in the decreased expression of the c-*myc* gene and the commitment of cells to differentiate into either monocytes or macrophages.

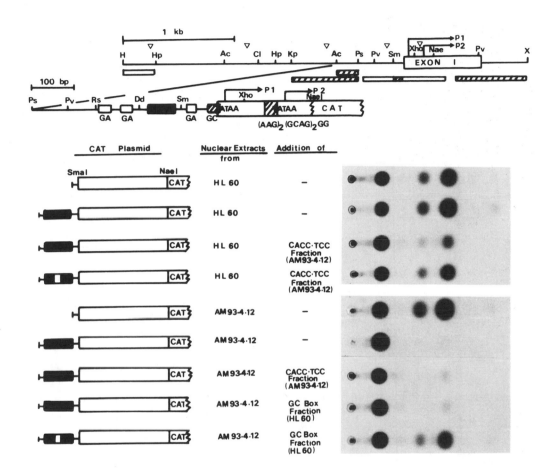

Fig. 9. *Effect of 74 kD protein on the level of CAT expression directed by pMyc-CAT$_{SN}$ (CACC–TCC)$_3$. A CAT plasmid, pMyc-CAT$_{SN}$ (CACC–TCC)$_3$, was constructed by adding three tandem repeats of the 45-mer including CACC–TCC sequence (closed rectangles) or mutated sequences (closed double boxes) to pMyc-CAT$_{SN}$ vector and transfected into either HL-60 or*

AM93-4-12. Nuclear extract (50 µg) was mixed with the partially purified 74 kD protein (1.2 µg) or GC box binding proteins (2.4 µg). The CAT assay was as previously described [Yokoyama and Imamoto, 1987]. The upstream region of the c-myc gene is illustrated, with the transcription starts P1 (minor) and P2 (major) and enlarged below the fusion made to the CAT gene.

TABLE III. Comparative Studies of the Human _myc_ Protein Production and the Differentiated Phenotype of HL-60 Transformants

Cells	_myc_ expression (p64/mRNA/cell; $\times 10^3$)	DNA synthesis in vitro (%)	α-Naphtyl acetate esterase	Monoclonal antibody			Naphthol AS-D chloroacetate esterase
				OKM-1	OKM-5	Anti-neutrophil	
HL-60	1.11	100	0	0	0	0	0
Induced by							
TPA	—	—	99	95	99	8	0
DMSO	—	—	0	3	4	98	98
Control pSV2gpt$^+$	1.18	85	12	5	6	3	15
Sense _myc_ (pSV2gptC5–8)	1.45	187	15	8	10	6	18
Antisense _myc_							
AM93-4-12	0.08	11	72	80	88	9	12
AM451-6-30	0.08	10	58	67	74	8	14
HL-60 incorporated with partial purified 74 kD protein complex introduced by liposome (µg/ml)							
0.3	0.61	60	19	21	27	2	10
2.6	0.57	44	22	26	21	10	9
8.9	0.40	35	34	37	49	9	13
28.0	0.16	18	68	73	82	7	12
HL-60 cultured with the culture medium from AM93-4-12	0.98	95	17	10	14	13	14

Values are means of three experiments. Protein and mRNA levels were measured as described in Table I. Percent of positive cells by staining of esterase was shown as described in Table II. In vitro DNA synthesis was measured as reported elsewhere [Studzinski et al., 1986].

VI. HOW DOES ANTISENSE RNA REPRESS THE TRANSCRIPTION OF THE c-*myc* GENE?

I have shown here that high levels of cellular antisense *myc* RNA can stably reduce the accumulation of sense *myc* RNA and the synthesis of p64 *myc* protein. Plasmids capable of producing *Ecogpt* and antisense *myc* RNA were introduced into HL-60 cells by DNA transfer. Resistance to high concentrations of mycophenolic acid usually involves overexpression of *Ecogpt*, and, in the clones isolated in this study, it also resulted in an increase in the amount of antisense *myc* RNA. Selection for resistance to increasing concentration of mycophenolic acid followed by repeated transfection with plasmid DNA caused an increase in antisense RNA resulting from gene amplification (Fig. 1, Table I).

The gene amplification of the dihydrofolate reductase *(dhfr)* sequence was carried out in *CHOdhfr⁻* cells [Huberman and Riggs, 1968]. Our study was the first to show gene amplification using HL-60 cells. The exact mechanism of amplification of the *Ecogpt* gene is not known. One possible explanation is that the *myc* gene, which is already amplified to 20 to 30 copies in HL-60, is localized to replicating submicroscopic circular DNA molecules in HL-60 cells [Von Hoff et al., 1988]. Plasmids driven by the SV40 promoter can result in an increase in the copy number of the integrated plasmid DNA (K. Yokoyama, unpublished results). A previous study by Ariga et al. [1987] showed that the *myc* protein can substitute for the function of large T antigen in cells to promote DNA replication directly. Thus it might be that *myc* protein can bind to the large T-binding site of the SV40 promoter region and substitute for the function of large T antigen. Given the above findings, DNA molecules carrying the SV40 promoter may replicate autonomously in cells to high copy numbers. The selection by increasing the concentration of mycophenolic acid and the positive regulation of *myc* protein to replicate the SV40 promoter might increase the copy numbers of antisense *myc* DNA and result in the higher expression of antisense *myc* RNA.

The particular promoter and selection system we have used is important. Other promoters apart from the SV40 early promoter (e.g., metallothionein promoter, MMTV promoter) do not function in HL-60 cells. The use of G418 selection for neomycin resistant clones (with the neomycin phosphotransferase gene) can cause reduced survival of cells. Selection by G418 may have a toxic effect on HL-60 cells and may also cause cell differentiation by the action of the selection drug itself (K. Yokoyama, unpublished data).

Growth inhibition was observed in cells that contained 100–250 copies of the antisense *myc* DNA sequence. The amount of sense *myc* DNA in the antisense transformants is 20–25 copies per cell. We found that a minimum ratio of 10:1 antisense to sense DNA is required for inhibition of *myc* expression in transfection studies; at this DNA ratio, there is 15–25-fold more antisense *myc* RNA than sense *myc* RNA. The steady-state level of *myc* protein in the antisense *myc* clones is reduced by >90% per cell (see Section III) compared with that in HL-60 cells and the level of mRNA by up to sevenfold (Section V). These results suggest that antisense RNA exerts an inhibitory effect on p64 protein synthesis by the formation of a stable antisense–sense RNA hybrid in vivo [Yokoyama and Imamoto, 1987] and that this leads to inhibition of endogenous *myc* gene expression, not only as a consequence of "conventional" antisense RNA effects on translation (Tables I–III) as discussed in the first part of this volume, but also by reducing *myc* gene transcription by a novel mechanism (Fig. 4).

Reduced expression of *myc* transcripts correlates with the triggering of HL-60 cell differentiation. It is well known that *myc* mRNA is no longer present in HL-60 cells stimulated into granulocytic or monocytic differentiation by exposure to chemical inducers [Westin et al., 1982]. Still unknown, however, is whether modulation in oncogene expression is required for cell-cycle progression or terminal differentiation of hematopoietic cell types.

The results in Table II show a strong correlation between lowered *myc* transcription and the monocytic phenotype, suggesting that antisense *myc* RNA directs HL-60 cells into the monocytic pathway in preference to the granulocytic pathway.

However, the recent study discussed by Wickstrom (this volume) demonstrates a reduction of *myc* expression by antisense *myc* oligonucleotides that might be sufficient to allow terminal granulocytic differentiation. This discrepancy may be due to our use of constitutive inhibition of *myc* gene expression, while Wickstrom and co-workers used a transient inhibition of *myc* gene expression.

The accumulation of sense *myc* mRNA was reduced in antisense *myc* transformants. In vitro run-on assay and the promoter-specific CAT assay demonstrated that the antisense transformants were clearly defective in promoter activity of the endogenous c-*myc* gene and in the elongation of *myc* RNA in isolated nuclei. These results suggest that *myc* RNA transcription in the antisense clones is regulated by the formation of an RNA–RNA duplex in the nucleus. The regulatory sequence that seems to be recognized by antisense RNA was localized to the CACC–TCC repeated sequence of c-*myc* gene promoter, which was confirmed by DNase I footprinting assay and gel retardation assay (Figs. 5,6).

Although the molecular mechanism of the repression of the *myc* gene is not known, the synthesis of *trans*-acting negative regulatory elements that bind to the CACC–TCC repeats might be induced by RNA–RNA base pairing, by additional regulatory factors induced by the antisense RNA, or as a consequence of the differentiation of the HL-60 cells which occurs as *myc* expression is inhibited.

We have identified 74 and 110 kD proteins in the nuclei of antisense transformants that could be candidates for negative regulation of *myc* gene expression. While the action of the 74 kD protein as a DNA-binding protein has been characterized, the biological function of the 110 kD protein is not yet known.

The results of CAT assays with *myc*-CAT constructs with addition of 74 kD protein purified by affinity column show that this 74 kD protein causes inhibition of *myc* promoter activity. This suppressive activity is evident even when the endogenous E2F, or Sp1-like proteins which function with the enhancer-promoter of the *myc* gene, are present. The introduction of 74 kD protein into HL-60 parental cells using liposome-mediated protein transfer resulted in decreased activity of *myc* and the commitment of cells into macrophage differentiation. This protein therefore seems essential in this regulation of *myc* promoter and the differentiation pathway of HL-60 cells. We do not know how the RNA duplex induces this 74 kD protein. One possible explanation is that nuclear events induced by interferon might be involved in the *myc* gene regulation as a result of RNA–RNA duplex formation [Stewart, 1979].

Ariga et al. [1987] reported that a myc protein is involved in DNA synthesis. Therefore, by interfering with *myc* gene expression, antisense *myc* transformants might inhibit DNA synthesis by regulation of myc protein. The 74 kD protein (and/or 110 kD protein) might be closely linked to this regulatory mechanism of myc protein on *myc* gene transcription. The production of 74 kD protein was paralleled by a decreased level of myc protein (K. Yokoyama, unpublished results). Further study is required to determine which function, the inhibition of DNA synthesis or the commitment to differentiation, is the direct target of 74 kD protein-complex–induced gene regulation. The biological function of 74 kD protein should be characterized to understand the negative regulatory mechanism of *myc* gene expression in the differentiation pathway of HL-60 cells.

VII. CONCLUDING REMARKS

A plasmid carrying antisense human *myc* DNA and the gene encoding *E. coli* xanthine–guanine phosphoribosyltransferase (*Ecogpt*) was introduced into human promyelocytic leukemia cell line HL-60 by protoplast fusion. High-level expression of antisense *myc* RNA was obtained by selecting cells resistant to progressively higher levels of mycophenolic acid over a period of >6 months. The constitutive production of myc protein in clones producing high levels of antisense *myc* RNA was reduced by 90% compared with parental HL-60 cells, and these cells showed increased commitment toward monocytic differentiation. Inhibition of *myc* expression was seen at both the translational and the transcriptional levels, implying that antisense RNA can regulate transcription of the *myc* gene. However, since monocytic differentiation is associated with downregulation of c-*myc* expression, it cannot be ruled out that the induction of the 74 kD protein and reduction of c-*myc* transcription are secondary events, triggered by an initial reduction in c-*myc* expression by "conventional" antisense regulation at the posttranscriptional level.

The CACC–TCC repeats in the c-*myc* leader sequence are the primary transcriptional target of the antisense RNA. The 74 kD nuclear protein, which appears to be induced in the antisense *myc* transformants, was able to bind to this sequence in these transformants. The suppression of endogenous *myc* gene expression by either antisense RNA or this 74 kD protein decreases cell proliferation and triggers monocytic differentiation.

ACKNOWLEDGMENTS

I thank Dr. R. DiNicolantonio for critical reading of the manuscript and Ms. T. Yamauchi for secretarial work. This work was supported by the Life Science Research Project and by the Frontier Research Program of RIKEN.

VIII. REFERENCES

Ariga SMM, Itani T, Kiji Y, Ariga H (1987): Possible function of the c-*myc* product: Promotion of cellular DNA replication. EMBO J 6:2365–2371.

Ariga SMM, Itani T, Yamaguchi M, Ariga H (1987): C-*myc* protein can be substituted for SV40 T antigen in SV40 DNA replication. Nucleic Acids Res 15: 4889–4899.

Bentley DL, Groudine M (1986): A block to elongation is largely responsible for decreased transcription of c-*myc* in differentiated HL-60 cells. Nature 321: 702–706.

Bishop JM (1987): The molecular genetics of cancer. Science 235:305–311.

Breitman TR, Selonick SE, Collins SJ (1980): Induction of differentiation of the human promyelocytic leukemia cell line (HL-60) by retinoic acid. Proc Natl Acad Sci USA 77:2936–2940.

Cole MD (1986): The *myc* oncogene: Its role in transformation and differentiation. Annu Rev Genet 20: 361–384.

Collins SJ, Ruscetti FW, Gallagher RE, Gallo RC (1978): Terminal differentiation of human promyelocytic leukemia cells induced by dimethylsulfoxide and other polar compounds. Proc Natl Acad Sci USA 75: 2548–2462.

Collins S, Groudine M (1982): Amplification of endogenous *myc*-related DNA sequences in a human myeloid leukemia cell line. Nature 298:679–681.

Dalla-Favera R, Wong-Staal F, Gallo RC (1982): Oncogene amplification in promyelocytic leukemia cell line HL-60 and primary leukemic cells of the same patient. Nature 299:61–63.

Fainsod A, Bogarad LD, Ruusala T, Lubin M, Crothers DM, Ruddle FH (1986): The homeo domain of a murine protein binds 5' to its own homeo box. Proc Natl Acad Sci USA 83:9532–9536.

Filmus J, Buick RN (1985): Relationship of c-*myc* expression to differentiation and proliferation of HL-60 cells. Cancer Res 45:822–825.

Green PJ, Pines O, Inouye M (1986): The role of antisense RNA in gene regulation. Annu Rev Biochem 55:569–597.

Grosso LE, Pitot HC (1985): Transcriptional regulation of c-*myc* during chemically induced differentiation of HL-60 cultures. Cancer Res 45:847–850.

Huberman JA, Riggs AD (1968): On the mechanism of DNA replication in mammalian chromosomes. J Mol Biol 32:327–337.

Kadonaga JT, Tijian R (1986): Affinity purification of sequence-specific DNA binding proteins. Proc Natl Acad Sci USA 83:5889–5893.

Katoh S, Ozawa K, Kondoh S, Soeda E, Israel A, Shiroki K, Fujinaga K, Itakura K, Gachelin G, Yokoyama K (1990): Identification of sequences responsible for positive and negative regulation by E1A in the promoter of H-2K^{bm1} class I MHC gene. EMBO J 9:127–135.

Kim SK, Wold BJ (1985): Stable reduction of thymidine kinase activity in cells expressing high levels of antisense RNA. Cell 42:129–138.

Lee F, Mulligan R, Berg P, Ringold G (1981): Glucocorticoids regulate expression of dihydrofolate reductase cDNA in mouse mammary tumour virus chimaeric plasmids. Nature 294:228–232.

Miskims, WK, Roberts MP, McClelland A, Ruddle FH (1985): Use of a protein-blotting procedure and a specific DNA probe to identify nuclear proteins that recognize the promoter region of the transferrin receptor gene. Proc Natl Acad Sci USA 87:6741–6744.

Murray MJ, Cunningham JM, Parada LF, Dautry F, Lebowitz P, Weinberg RA (1983): The HL-60 transforming sequence: A *ras* oncogene coexisting with altered *myc* genes in hematopoietic tumors. Cell 33:749–757.

Nishikura K (1986): Sequences involved in accurate and efficient transcription of human c-*myc* gene microinjected into frog oocytes. Mol Cell Biol 6:4093–4098.

Nishioka Y, Leder P (1979): The complete sequence of a chromosomal mouse α-globin gene reveals elements conserved throughout vertebrate evolution. Cell 18: 875–882.

Reitsma DH, Rothberg PG, Astrin SM, Trial J, Bar-Shavit Z, Hall A, Teitelbaum SL, Kahn AJ (1983): Regulation of *myc* gene expression in HL-60 leukemia cells by a vitamin D metabolite. Nature 306:492–495.

Ringold G, Dieckmarrn B, Lee F (1981): Coexpression and amplification of dihydrofolate reductase cDNA and the *Escherichia coli XGPRT* gene in Chinese hamster ovary cells. J Mol Appl Genet 1:165–175.

Rovera G, O'Brien TG, Diamond L (1979): Induction of differentiation in human promyelocytic leukemia cells by tumor promoters. Science 204:868–870.

Sandri-Goldin RM, Golin AL, Levine M, Glorioso JC (1981): High-frequency transfer of cloned herpes simplex virus type 1 sequences to mammalian cells by protoplast fusion. Mol Cell Biol 1:743–752.

Stewart WE II (1979): The Interferon System. New York: Springer Verlag.

Studzinski GP, Brelvi ZS, Feldman SC, Watt RA (1986): Participation of c-*myc* protein in DNA synthesis of human cells. Science 234:467–470.

Thalmeier K, Synovzik H, Mertz R, Winnacker E-L, Lipp M (1989): Nuclear factor E2F mediates basic transcription and *trans*-activation by Ela of the human *myc* promoter. Genes Dev 3:527–536.

Von Hoff DD, Needham-VanDevanter DR, Yucel J, Windle BE, Wahl GM (1988): Amplified human *myc* oncogenes localized to replicating submicroscopic circular DNA molecules. Proc Natl Acad Sci USA 85:4804–4808.

Weintranb H, Izant JG, Harland RM (1985): Anti-sense RNA as a molecular tool for genetic analysis. Trends Genet 1:22–25.

Weiss E, Golden L, Zakut R, Mellor A, Fahrner K, Kvist S, Flavell RA (1983): The DNA sequence of the H-2K^b gene: Evidence for gene conversion as a mechanism for the generation of polymorphism in histocompatibility antigens. EMBO J 2:453–462.

Westin EH, Wong-Staal F, Gelmann EP, Dalla-Favera R, Pappas TS, Lautenberger J, Eva A, Reddy EP, Tronick SR, Aaronson SA, Gallo RC (1982): Expression of cellular homologues of retroviral oncogenes in human hematopoietic cells. Proc Natl Acad Sci USA 79: 2490–2494.

Yokoyama K, Imamoto F (1987): Transcriptional control of the endogenous *myc* protooncogene by antisense RNA. Proc Natl Acad Sci USA 84:7363–7367.

ABOUT THE AUTHOR

KAZUSHIGE YOKOYAMA is a Research Scientist at RIKEN, where he is involved in the Human Genome Project and the Frontier Research Program of Japan. After receiving his B.A. from Shizuoka University in 1973, he received his Ph.D. in 1979, under Toshiaki Osawa at the Department of Science of the University of Tokyo, where he concentrated in the study of the B-lymphocyte triggering mechanism using pokeweed mitogen. Dr. Yokoyama then pursued postdoctoral research at Albert Einstein College of Medicine, first studying the structure and function relationship of major histocompatibility complex (MHC) Class I antigen in the laboratory of Stanley G. Nathenson and then the molecular biology of thymus leukemia (TL) antigen with the collaborating director, Lloyd J. Old, at Sloan Kettering Institute. After four years, he moved to the laboratory of Keiichi Itakura at Beckman Research Institute of City of Hope to study the molecular genetics of hepatocarcinogenesis, and then was appointed Assistant Professor. In 1985 Dr. Yokoyama returned to Japan for a position at RIKEN. His current research involves the projects of antisense DNA/RNA technology, protooncogene and cell differentiation, transcriptional regulation of MHC class I gene by adenovirus E1A, and human genome mapping. His research papers have appeared in such journals as *EMBO Journal*, *Science*, the *Proceedings of the National Academy of Sciences USA*, *Biochemistry*, the *Journal of Immunology*, the *European Journal of Immunology*, and *Genomics*.

Antisense RNA and DNA: 353–372
© 1992 Wiley-Liss, Inc.

Selective Amplification Techniques for Optimization of Ribozyme Function

Gerald F. Joyce

I. INTRODUCTION

As initially formulated, the antisense method for specific inactivation of a nucleic acid target was very straightforward: One utilizes an antisense oligodeoxynucleotide or RNA that hybridizes to a complementary region within a target RNA, thus preventing the RNA from carrying out its normal function [Zamecnik and Stephenson, 1978; Goodchild et al., 1988]. Recently, however, it has become clear that antisense DNA can also operate by a more complex mechanism involving cleavage of the target RNA by ribonuclease H that is already present in the cell [Minshull and Hunt, 1986; Dash et al., 1987; Walder and Walder, 1988] (see also Minshull and Hunt, this volume). The antisense DNA hybridizes to target RNA and creates a heteroduplex structure that is a substrate for RNase H.

These results have added a new dimension to the antisense method. The problem now becomes to design an oligodeoxynucleotide or oligonucleotide analog that not only binds to the target site but also directs cleavage of the target at a rate that exceeds the rate of degradation of the antisense oligomer. One approach is to rely on the activity of cellular RNase H, using a nuclease-resistant DNA analog as the antisense oligomer [Matsukura et al., 1987, 1989; Agrawal et al., 1988] (see also Miller, this volume; Matsukura, this volume; Agrawal and Leiter, this volume). A second approach is to attach a reagent with endonuclease activity to the antisense oligomer so that binding of the antisense molecule is necessarily accompanied by delivery of the cleavage agent. For example, an Fe^{II}–EDTA complex attached to the 5' end of an antisense DNA has been used to generate hydroxyl radicals that result in cleavage of the complementary target [Dreyer and Dervan, 1985; Chu and Orgel, 1985]. A third approach involves the use of antisense ribozymes that are capable of recognizing the target and carrying out the cleavage reaction themselves. This approach potentially offers a high degree of specificity and has the advantage that the antisense agent is an RNA molecule that can be synthesized within the cell.

Four distinct classes of RNA enyzmes have been recognized, all of which are capable of cleaving an RNA substrate in a sequence-specific manner [Kruger et al., 1982; Guerrier-Takada et al., 1983; Peebles et al., 1986; Uhlenbeck, 1987; McClain et al., 1987; Sharmeen et al., 1988; Zaug et al., 1988; Hampel et al., 1990]. The *Tetrahymena* group I intron and the hammerhead self-cleaving RNA have been studied most extensively for use as a general-purpose endoribonuclease. In the case of the *Tetrahymena* ribozyme, the specificity of cleavage is determined by binding of the substrate to a complementary region (the ''internal guide sequence'') that lies near the 5' end of the enzyme [Been and Cech, 1986; Waring et al., 1986]. Cleavage of the substrate occurs at the phosphodiester bond that lies immediately downstream from the target sequence. Recently it was shown that the relationship between guide sequence and substrate can be generalized to include virtually any complementary pairing [Doudna and Szostak, 1989]. By optimizing the reaction conditions it is possible to achieve a high level of substrate specificity in the

RNA-catalyzed cleavage reaction using a variety of different substrate/guide sequence combinations [Murphy and Cech, 1989].

We have developed a system for optimizing ribozyme function, not by adjusting the reaction conditions, but by mutagenizing the ribozyme and selecting those variants that are most reactive under a particular set of reaction conditions [Joyce, 1989b; Robertson and Joyce, 1990]. The system can be thought of as a selective breeding program for ribozymes. A population of ribozyme variants is put through repeated cycles of selective amplification in order to obtain those individuals with the most desirable catalytic properties. The selected individuals are used to generate a new population of random variants, and the process begins again. I discuss the selective amplification technique in detail, focusing on its use as part of an in vitro system for evolutionary engineering of antisense ribozymes.

II. GROUP I CATALYTIC RNA
A. Specific Endonuclease Activity

The particular RNA enzyme we are working with is the self-splicing group I intron derived from the large ribosomal RNA precursor of *Tetrahymena thermophila*. This molecule is able to catalyze sequence-specific cleavage–ligation reactions involving RNA substrates [Zaug and Cech, 1985; Kay and Inoue, 1987]. Recently we found that the *Tetrahymena* ribozyme is able to cleave single-stranded (ss) DNA as well [Robertson and Joyce, 1990]. A detailed kinetic analysis by Herschlag and Cech [1990] has shown that, for the wild-type ribozyme, DNA cleavage is about 10^5-fold less efficient than RNA cleavage. Nonetheless, this represents a 10^9-fold rate enhancement compared with spontaneous DNA cleavage. Thus group I ribozyme could be used as antisense agents for the specific cleavage of either RNA or DNA targets.

Cleavage of DNA is mechanistically similar to cleavage of RNA. In both cases, the specificity of cleavage is determined by binding of the substrate to the internal guide sequence

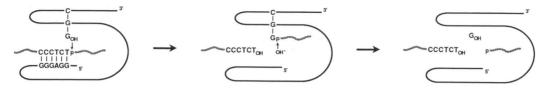

Fig. 1. *Specific cleavage of a DNA substrate catalyzed by the* Tetrahymena *ribozyme.*

of the ribozyme. Cleavage occurs via a phospho-ester transfer mechanism involving nucleophilic attack by guanosine 3′-OH. The products of the reaction are the released 5′ portion of the substrate and guanosine attached to the 5′ end of the 3′ portion of the substrate (Fig. 1). The guanosine nucleophile can be either free guanosine [Bass and Cech, 1984; Herschlag and Cech, 1990] or a guanosine residue that lies at the 3′ end of the ribozyme itself [Joyce, 1989b; Robertson and Joyce, 1990]. The phosphodiester bond between guanosine and the 3′ portion of the substrate is highly labile, undergoing RNA-catalyzed site-specific hydrolysis [Inoue et al., 1986] to release the 3′ portion of the substrate.

It has yet to be shown that the DNA cleavage reaction catalyzed by the *Tetrahymena* ribozyme can be generalized with respect to substrate sequence. However, by analogy to the RNA cleavage reaction, this could be done by changing the internal guide sequence and then optimizing the reaction conditions. Both the RNA–RNA duplex and the RNA–DNA heteroduplex tend to exist in the helical A form [Milman et al., 1967; Wang et al., 1982], so that RNA and DNA substrates are expected to bind to the internal guide sequence in a similar manner. Apparently there are detailed stereochemical differences that account for the lower efficiency of DNA cleavage versus RNA cleavage. It remains to be seen to what extent these differences can be overcome by making compensatory mutations within the ribozyme.

B. Structural Requirements for Catalytic Activity

The *Tetrahymena* ribozyme consists of 413 nucleotides and assumes a well-defined secondary and tertiary structure that is responsible for its catalytic activity. The three-dimensional structure of the molecule is not known. However, a secondary structure model has been developed based on phylogenetic comparison with other group I introns [Burke et al., 1987]. The model suggests that there is a catalytic center consisting of 55 nucleotides supported by a total of 17 stem-loop structures (Fig. 2). We recently completed a comprehensive deletion analysis of the *Tetrahymena* ribozyme showing that 13 of the supporting stem–loop structures can be deleted in a piecewise fashion without loss of catalytic function [Beaudry and Joyce, 1990]. This extends previous work that has been conducted along these lines [Price et al., 1985; Szostak, 1986; Joyce and Inoue, 1987; Barfod and Cech, 1988; Joyce et al., 1989] and defines the minimum secondary structure requirements for catalytic activity of a group I ribozyme. While it has not been possible to combine all of the deletions to produce a naked reaction center, a variety of combined deletions, totalling as many as 236 nucleotides, have been shown to result in a molecule that retains catalytic activity (Fig. 2).

This detailed structural analysis is relevant in two ways to the use of selective amplification techniques for optimization of ribozyme function. First, it provides a number of reactive structural variants of the ribozyme, each of which can serve as a starting point for generating random mutants. Second, it defines those regions of the molecule that are not required for catalytic activity so that random mutations can be directed to the remaining areas that are most likely to influence catalytic function. We begin with the set of 12 reactive structural variants (see Fig. 2) and choose those

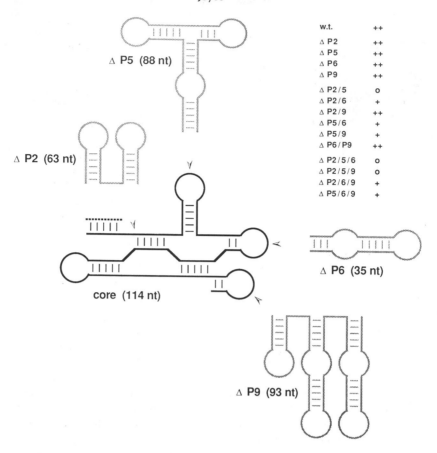

w.t.	++
Δ P2	++
Δ P5	++
Δ P6	++
Δ P9	++
Δ P2/5	o
Δ P2/6	+
Δ P2/9	++
Δ P5/6	+
Δ P5/9	+
Δ P6/P9	++
Δ P2/5/6	o
Δ P2/5/9	o
Δ P2/6/9	+
Δ P5/6/9	+

Fig. 2. *Diagrammatic representation of the secondary structure of the* Tetrahymena *ribozyme showing those regions that can be deleted without a loss of catalytic activity. Arrowheads indicate the sites at which four large internal deletions were made. The substrate (shown as a dashed line) is bound to a complementary sequence located at the 5′ end of the ribozyme.*

Catalytic activity of the various single-, double-, and triple-deletion mutants is tabulated at the upper right. "Activity" refers to the ability of the molecule to catalyze a sequence-specific phosphoester transfer reaction involving the RNA substrate GGCCCUCU–A_{13} under conditions that favor duplex stability. The quadruple-deletion mutant ΔP2/5/6/9 is not reactive.

forms that have the highest level of activity in a particular target reaction. Each of the chosen forms is then used to generate 10^9–10^{12} mutants, including all possible one, two, and three error base substitutions scattered throughout the core portion of the ribozyme. Through repeated rounds of selective amplification, the most reactive individuals are culled from the population. These selected individuals can then be used to generate a new population of ribozyme variants and the entire process begins again.

III. RAPID AMPLIFICATION OF RNA

RNA can be amplified in vitro using a reciprocal primer method [Kwoh et al., 1989; Joyce, 1989a]. The method is similar to the polymerase chain reaction (PCR) [Saiki et al., 1988], except that instead of alternating between plus strand and minus strand DNA, one alternates between plus strand RNA and cDNA (Fig. 3). The first primer hybridizes at the 3′ end of the RNA and is extended by reverse transcriptase to yield cDNA. The second primer hybridizes

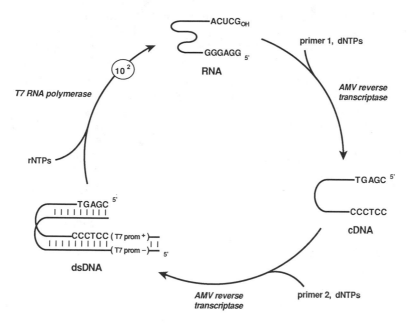

Fig. 3. *Procedure for isothermal amplification of RNA. RNA is reverse transcribed to cDNA, a promoter for T7 RNA polymerase is attached, and the DNA is then transcribed back to RNA. Amplification occurs at the level of transcription because of the high turnover of T7 RNA polymerase.*

at the 3' end of the newly synthesized cDNA and is extended, also by reverse transcriptase, to yield double-stranded DNA (dsDNA). The second primer contains the T7 promoter element within a 5' overhang so that the dsDNA contains a functional promoter that is recognized by T7 RNA polymerase. T7 RNA polymerase has inherent amplification properties, producing 200–1,200 copies of RNA transcript per copy of DNA template [Chamberlin and Ryan, 1982].

A. Isothermal RNA Amplification

An important enhancement of this technique was recently reported [Guatelli et al., 1990]. It turns out that if one mixes all of the components of the RNA amplification system (input RNA, the two primers, the four dNTP, the four rNTP, AMV reverse transcriptase, T7 RNA polymerase, MgCl$_2$, and Tris buffer) amplification proceeds in a continuous manner. RNA is copied to cDNA, cDNA is converted to dsDNA, dsDNA is transcribed to hundreds of copies of RNA, and each copy of RNA can bind a new primer to begin a new round of amplification. Temperature cycling and annealing are not required; the system runs autocatalytically at 37°C.

This "isothermal" RNA amplification technique was developed for the purpose of amplifying very low abundance mRNA. Beginning with 10^{-5}–10^{-1} fmol (1 fmol = 10^{-15} mol) one can achieve roughly 10^6 fold amplification in 1 hour [Guatelli et al., 1990]. We have adapted the technique for use with larger amounts of input RNA, on the order of 10^{-2}–10^1 pmol, so as to accommodate a population of 10^9–10^{12} mutant ribozymes. Our current protocol for RNA amplification is as follows: input RNA is added to a solution containing 1 μmol primers, 2 mM NTP, 0.2 mM dNTP, 10 mM MgCl$_2$, 50 mM Tris (pH 7.5), 5 mM DTT, 0.2 U/μl AMV reverse transcriptase, and 5 U/μl T7 RNA polymerase, and the mixture is incubated at 37°C for 1 hour. This results in 10^3–10^4-fold amplification of the input RNA.

There are several differences between our protocol and that developed by Guatelli et al.

First, we do not add *Escherichia coli* RNase H to the reaction mixture. While RNase H increases the overall level of amplification by about twofold, we prefer to save the time and expense associated with this additional component while avoiding possible contamination by other RNases that might be present within the RNase H preparation. AMV reverse transcriptase has an integral RNase H activity that is sufficient to degrade the RNA template and permit second-strand DNA synthesis to occur. Second, we prefer to use lower concentrations of NTP and dNTP: 2 mM and 0.2 mM, respectively, compared with 4 and 1 mM employed by Guatelli et al. These lower concentrations are optimal for use with amounts of input RNA in the range of 10^{-2}–10^1 pmol. Third, we do not carry out an initial annealing step to promote hybridization of the first primer to the 3' end of the input RNA. Annealing results in about a 1.3-fold increase in the overall level of amplification, but we feel that this is insufficient to justify the added bother. If higher levels of amplification are required, one need only extend the incubation time. Finally, we prefer to work in smaller reaction volumes using lower concentrations of AMV reverse transcriptase than reported by Guatelli et al. We wish to minimize the use of AMV reverse transcriptase, which is the most costly component in the system. Recently we found that Moloney murine leukemia virus reverse transcriptase, a cloned enzyme, can substitute for AMV reverse transcriptase (Cadwell and Joyce, unpublished results). This eases the cost restraint somewhat.

B. Practical Considerations

To simplify the amplification procedure, Eppendorf tubes containing all of the components except the RNA can be premixed, stored at $-20°C$, and thawed as needed. Alternatively, all of the components except the RNA can be mixed with 50–100 mM trehalose, evaporated to dryness, and rehydrated as needed. Trehalose, which consists of two D-glucose residues joined by a $(1\rightarrow1)$ glycosidic linkage, is known to stabilize phospholipids and proteins in the dry state by surrounding them with a protective shell of hydrogen bonds [Crowe et al., 1987]. Trehalose does not interfere with RNA amplification (at concentrations up to 200 mM) and protects the components of the amplification system against dehydration.

As with any reciprocal primer method, the primers used to carry out RNA amplification must be chosen carefully. Primer 1, the primer used to initiate cDNA synthesis, should bind to a unique site at the extreme 3' end of the RNA. The estimated melting temperature for primer binding should be at least 42°C, although for high-specificity selective amplification (see below) we often use primers with an estimated melting temperature of 37°C or less. Primer 2, the primer used to initiate second-strand synthesis, should bind to a unique site at the extreme 3' end of the cDNA. Primer 2 only binds to full-length (or near full-length) cDNA, thus excluding incomplete reverse transcripts from subsequent reaction steps. It is necessary to place several residues usptream from the T7 promoter at the extreme 5' end of primer 2 to ensure synthesis of a viable (fully double-stranded) T7 promoter.

The RNA amplification process is made more powerful by the fact that it can be carried out in an iterative manner. After each round of amplification the amount of RNA produced is quantitated by cold acid precipitation, a small aliquot is removed and diluted by a factor of 10^3–10^4, and a portion of the diluted material is used to seed the next round. In this way one can achieve very high overall amplification rates while maintaining a constant total population size (Fig. 4). Because only a very small fraction of the RNA is required to begin the next round, the bulk of the material remains available for analytical purposes and to create an archive of the system's history.

There are two ways to augment the overall level of amplification. One is to operate the RNA amplification procedure in concert with the PCR [Joyce, 1989a; Tuerk and Gold, 1990; Ellington and Szostak, 1990]. Beginning at the level of cDNA, the same set of reciprocal primers used to carry out RNA amplification can be

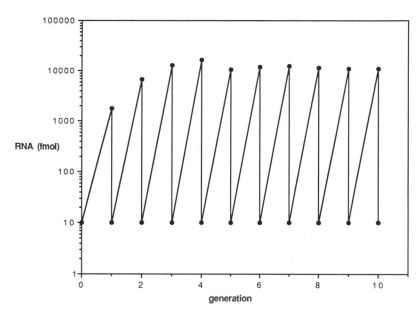

Fig. 4. *Results of a serial transfer experiment involving 10 successive rounds of isothermal amplification. Input RNA was 10 fmol of the ΔP9 single-deletion mutant (see Fig. 2), containing random base substitutions scattered over 38 positions at the 3'-terminal portion of the molecule at a frequency of 5% per position. After each generation a small aliquot was removed and quantitated by acid precipitation, a dilution was made, and 10 fmol of material was used to begin the next round of amplification. Each generation corresponds to 1 hour incubation at 37°C. Reaction conditions: 1 μM (5')-CGAGTACTCCGAC (primer 1), 1 μM (5') - ATCGATAATACGACTCACTATAGGAGGGAAAA GTTATCAGGC (primer 2), 2 mM NTP, 1 μCi/nmol [α³²P]GTP, 200 μM dNTP, 10 mM MgCl₂, 50 mM Tris (pH 7.5), 5 mM DTT, 0.2 U/μl AMV reverse transcriptase, 5 U/μl T7 RNA polymerase; 20 μl volume. Acid precipitation: 2 μl of reaction mixture was added to 98 μl ice-cold salmon sperm DNA (500 μg/ml), a 10 μl aliquot was removed for quantitation, 1 ml ice-cold 100 mM Na₄P₂O₇/1 M HCl was added to the remainder, and the mixture was held at 0°C for 5 minutes; the precipitate was collected on Whatman GF/A paper by vacuum filtration, washed three times with ice-cold 100 mM Na₄P₂O₇/1 M HCl and three times with ice-cold EtOH, and quantitated (in comparison to the initial 10μl aliquot) using a liquid scintillation counter.*

used to perform the PCR, resulting in large amounts of dsDNA that in turn are transcribed to RNA. Another method for obtaining large amounts of material is to subclone the dsDNA into a suitable vector, carry out DNA amplification within a bacterial host, purify the DNA, and transcribe to RNA. We find it advantageous to perform PCR amplification of the dsDNA prior to subcloning rather than subcloning directly from the RNA amplification mixture. A sizeable portion of the dsDNA in the amplification mixture is less than full length because of incomplete second-strand synthesis. While this does not prevent T7 RNA polymerase from producing full-length transcripts [Milligan et al., 1987], it does reduce the fraction of molecules that contain a viable restriction site, especially at the 3' end of the second strand. This problem is remedied by carrying out the PCR.

IV. SELECTIVE AMPLIFICATION OF CATALYTIC RNA

A. General Method

We have developed a method for selective amplification of group I ribozymes based on their catalytic function in vitro [Joyce, 1989b]. The method is compatible with the isothermal amplification procedure described above. Recall

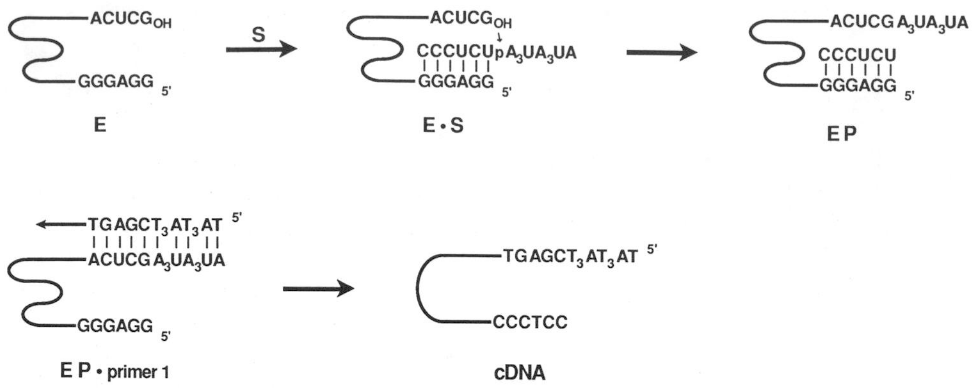

Fig. 5. *Selective amplification of an RNA enzyme (E) based on its ability to react with an oligonucleotide substrate (S). The product of the reaction (EP) offers a unique site for hybridization of the primer used to initiate cDNA synthesis. Amplification occurs when the DNA is transcribed back to RNA.*

that group I introns catalyze phosphoester transfer involving nucleophilic attack by guanosine 3'-OH at a target site within an RNA or DNA substrate. By placing the nucleophilic G_{OH} at the 3' end of the ribozyme, the product of the phosphoester transfer reaction is the ribozyme joined to the 3' portion of the substrate. Selection occurs when an oligodeoxynucleotide primer is hybridized across the ligation junction and used to initiate cDNA synthesis (Fig. 5). The primer is designed such that it does not bind to unreacted starting materials and therefore leads to selective reverse transcription of catalytically active RNAs.

One can amplify a group I ribozyme either nonselectively, using a primer that binds to the 3' end of the ribozyme, or selectively, using a primer that binds across the ligation junction (Fig. 6). Selective amplification proceeds regardless of whether an RNA or DNA substrate is used [Robertson and Joyce, 1990]. In the context of a serial transfer experiment, we alternate between cycles of selective and nonselective amplification. Selective amplification is used to cull the most reactive individuals from a heterogeneous population of mutant ribozymes; nonselective amplification is used to restore the 3'-terminal G_{OH} so that the selected individuals can again be challenged with substrate. These procedures operate on an

hourly time scale, providing a powerful method for surveying a large number of mutant ribozymes in a short period of time.

Design of the primer used to initiate selective cDNA synthesis involves a trade-off between sensitivity and specificity. To achieve high sensitivity, that is, to amplify functional ribozymes that are present in low copy number, we use a primer that is complementary to the last five residues of the ribozyme and to the 3' portion of the substrate. To achieve high specificity, that is, to reject nearly every molecule that has not undergone the target reaction, we use a primer that is complementary to only the 3'-terminal guanosine residue of the ribozyme and to the 3' portion of the substrate. At high sensitivity it is possible to amplify less than 10^{-20} moles of starting material [Guatelli et al., 1990]. At high specificity we can selectively amplify reactive materials that comprise less than 10^{-9} of the total population. The specificity determines our "signal-to-noise ratio" as we search for novel mutants that exhibit some desired catalytic property.

B. Search Strategies

In general there are two search strategies that can be employed during optimization of ribozyme function. The first is *gradualism* [Conrad, 1979; Rada, 1981], relying on gradual and pro-

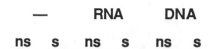

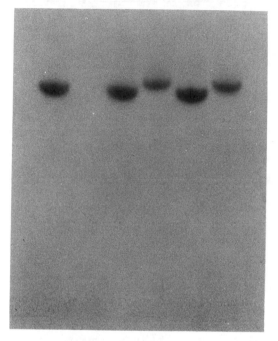

Fig. 6. *Selective isothermal amplification of an RNA enzyme. The* Tetrahymena *ribozyme was incubated with no substrate, with the RNA substrate (5')-GGCCCUCU–A₃UA₃UA, or with the DNA substrate (5')-GGCCCTCT–A₃TA₃TA and then amplified isothermally using either a nonselective or selective primer. The nonselective (ns) primer (5')-CGAGTACTCCAAAAC binds to the 3' end of the ribozyme and amplifies both reactive and unreactive materials. The selective (s) primer (5')-TATTTATTTCGAGT binds across the transesterification junction (see Fig. 5) and amplifies only reaction products. Primer 2 was (5')-ATCGATAAT ACGACTCACTATAGGAGGGAAAAGTTATCAGGC in all reaction mixtures. Reaction conditions: 1 nM ribozyme, 1 μM primers, 2 mM NTP, 1 μCi/nmol [α³²P]GTP, 200 μM dNTP, 10 mM MgCl₂, 50 mM Tris (pH 7.5), 5 mM DTT, 0.5 U/μl AMV reverse transcriptase, 5 U/μl T7 RNA polymerase, 0.05 U/μl E. coli RNase H; 37°C, 1 hour. Products were separated by gel electrophoresis in a 5% polyacrylamide/8 M urea gel, an autoradiogram of which is shown.*

gressive differences in fitness to drive the evolving population toward a novel phenotype. The second is *punctuated equilibrium* [Gould, 1983], seeking rare individuals that move the population toward a novel phenotype in a saltatory manner. For example, beginning with a ribozyme that cleaves a target substrate under condition *A* but not under some desired condition *B*, one could generate mutants and either perform repeated rounds of selective amplification while gradually changing from condition *A* to condition *B* or try to move directly to condition *B* without exploring intermediate conditions. A direct jump to condition *B* would obviously be more expedient. However, given the system's signal-to-noise ratio, if significantly less than 10^{-9} of the starting population is able to perform the target reaction under condition *B*, then selective amplification will give an unacceptable number of false-positive results. In such instances it is better to move to an intermediate condition, optimize catalytic function, and then search for activity under condition *B*.

False-positive selective amplification products can arise in several ways. One possibility is that the primer used to initiate selective cDNA synthesis binds to the 3' end of the ribozyme, even though the ribozyme has not undergone the reaction to become joined to the 3' portion of the substrate. This problem is eliminated by using a primer that binds to only one residue at the 3' end of the ribozyme. Another possibility is that the primer falsely hybridizes at an internal site, producing a truncated cDNA that is amplified more efficiently than full-length materials. These so-called "mini-monsters" [Spiegelman, 1971] are kept at bay by alternating between rounds of selective amplification (using a primer that recognizes the 3' portion of the substrate) and nonselective amplification (using a primer that excludes the 3' portion of the substrate). A third mechanism for the generation of false positives involves two successive RNA-catalyzed cleavage–ligation events that result in joining of the 5' end of one ribozyme molecule to the 3' end of another. The product contains both primer binding sites but little else in between.

This form of mini-monster is amplified very efficiently during selective amplification and nonselective amplification but, lacking catalytic activity, is excluded during the next round of selective amplification.

Finally, we have come across a fourth type of false positive that is the most troublesome of all. It is possible for the second primer, the primer used to initiate second-strand synthesis, to hybridize falsely to the ribozyme so that it initiates both cDNA synthesis and second-strand synthesis. The result is dsDNA with a functional T7 promoter at both ends, capable of directing the synthesis of both plus strand and minus strand RNA. Such a molecule short-circuits the system entirely because it is amplified independently of the primers used to initiate selective or nonselective cDNA synthesis (Fig. 7). Fortunately, we have been able to control this problem by occasionally carrying out RNA amplification according to our original two-step protocol [Robertson and Joyce, 1990]. RNA is reverse transcribed to cDNA in the presence of the first primer alone. The RNA is then digested with alkali, the second primer is added to the mixture, and isothermal amplification is allowed to proceed. In this way any RNA that is not dependent on the first primer for its amplification is eliminated.

V. GENERATION OF MUTANTS

A. Oligonucleotide-Directed Mutagenesis

Our standard method for generating a heterogeneous population of mutant RNAs is to incorporate partially randomized mutagenic oligodeoxynucleotides into a DNA template and transcribe the template directly. The mutagenic oligos are produced on an automated DNA synthesizer using nucleoside $3'$-phosphoramidite solutions that have been doped with a small percentage of each of the three incorrect monomers [Hermes et al., 1990]. The probability P of having k mutations in a doped oligo of length n is given by

$$P(k,n,d) = [n!/(n - k)!\, k!]\, d^k\, (1 - d)^{n-k},$$

where d is the degeneracy at each position. For example, the mutagenic oligo used to produce the starting population for the serial transfer experiment shown in Figure 4 contains 38 doped positions with a degeneracy of 5% per position. Thus the starting population contains 14% zero error, 28% one error, 28% two error, 18% three error, and 12% four-plus error mutants. Given a population consisting of 10 fmol of RNA, each one error mutant is represented by 10^7 copies, each two error mutant by 10^5 copies, and each three error mutant by 10^3 copies.

We have developed an oligonucleotide-directed mutagenesis technique that operates entirely in vitro and is compatible with the selective amplification system described above [Joyce and Inoue, 1989]. Briefly, the technique involves exonucleolytic digestion of the template strand (cDNA portion) of dsDNA, hybridization of one or more mutagenic oligos to the remaining nontemplate strand, and completion of a new template strand using T4 DNA polymerase and T4 DNA ligase. The resulting partial duplex structure, which includes the mutagenic oligo(s) within the newly synthesized template strand, is then transcribed directly using T7 RNA polymerase. T7 RNA polymerase is indifferent to the sequence of the nontemplate strand and produces RNA that are complementary to whatever sequence has been incorporated into the newly synthesized template strand.

To generate a complex population of mutant forms of the *Tetrahymena* ribozyme, we prepared seven partially randomized mutagenic oligos that blanket the catalytic core and adjacent supporting regions of the molecule (Fig. 8). Each mutagenic oligo contains 35 doped positions, with a degeneracy of 5% per position. The first and last position in each mutagenic oligo is nonrandomized to promote its efficient incorporation into the new template strand. The oligos were designed such that they can be used in a variety of combinations, especially those combinations that allow coverage of both halves of a stem–loop structure. We have found certain activities that cannot be obtained using a population based on each of

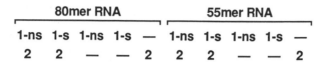

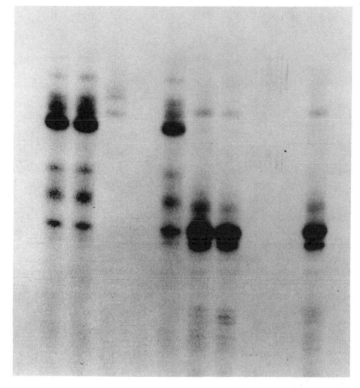

Fig. 7. *Two "mini-monster" RNA species that are dependent on primer 2 alone for their amplification. The RNAs, about 80 nucleotides and 55 nucleotides in length, were isolated by electrophoresis in a 10% poly-acrylamide/8 M urea gel, eluted from the gel, and purified by affinity chromatography on Du Pont Nensorb. Then 10 fmol of the purified RNA was added to a reaction mixture containing either one or both of the primers used to carry out either nonselective or selective isothermal amplification. Nonselective primer 1 (1-ns) was (5')-CGAGTACTCCGAC; selective primer 1 (1-s) was (5')-TTTATTTATTTATTTC; primer 2 was (5') - ATCGATAATACGACTCACTATAGGAGGGGAAA AGTTATCAGGC. Reaction conditions are given in Figure 4. Products were separated by gel electrophoresis in a 10% polyacrylamide/8 M urea gel, an autoradiogram of which is shown.*

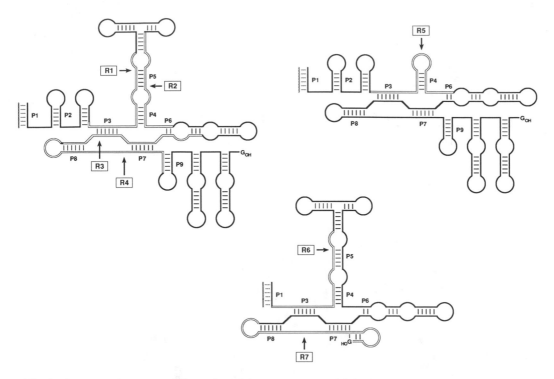

Fig. 8. *Location of seven partially randomized muta-genic oligos within the* Tetrahymena *ribozyme. Each oligo contains 35 doped positions with a mutation frequency (sum of all three possible base substitutions) of 5% per position. R1–R4 blanket the catalytic core of the ribozyme. R5 replaces R1 and produces the ΔP2 deletion (see Fig. 2); R6 replaces both R1 and R2 and produces the ΔP5 deletion; R7 replaces R4 and produces the ΔP9 deletion. There are 16 pairwise combi-* *nations of the seven mutagenic oligos that result in molecules that have catalytic activity in the phospho-ester transfer reaction involving an RNA substrate. There are 15 three-way and 3 four-way combinations that likewise yield catalytically active molecules. The mutagenic oligos were incorporated into dsDNA and transcribed directly using T7 RNA polymerase, as pre-viously described [Joyce and Inoue, 1989; Beaudry and Joyce, 1990].*

the seven mutagenic oligos in isolation, but can be obtained using two or more oligos in combination (Beaudry and Joyce, unpublished results).

B. Mutagenesis Using 5-Br Uracil

The major limitation of the oligonucleotide-directed mutagenesis technique is that after carrying out repeated rounds of selective amplification to reduce the population to a collection of the "fittest" individuals we must determine a consensus sequence and prepare new mutagenic oligos before continuing the search for individuals that have even greater fitness. It would be preferable if there were a way to generate

mutations "on line" during the amplification procedure. We have been developing such a method based on the use of 5-Br uridine and 5-Br deoxyuridine. 5-Br U can form a "wobble" pair with G as well as a Watson-Crick pair with A (Fig. 9). 5-Br U forms a wobble pair charge more readily than U because of the electron-withdrawing effect of Br that reduces the partial negative charge on O^4 and thus reduces the ability of O^4 to serve as a hydrogen-bond acceptor [Iwahashi and Kyogoku, 1977]. Conversely T is *less* likely to form a wobble pair than U because of the 5-Me group of T which has an electron donating effect.

By replacing TTP with 5-Br dUTP at the

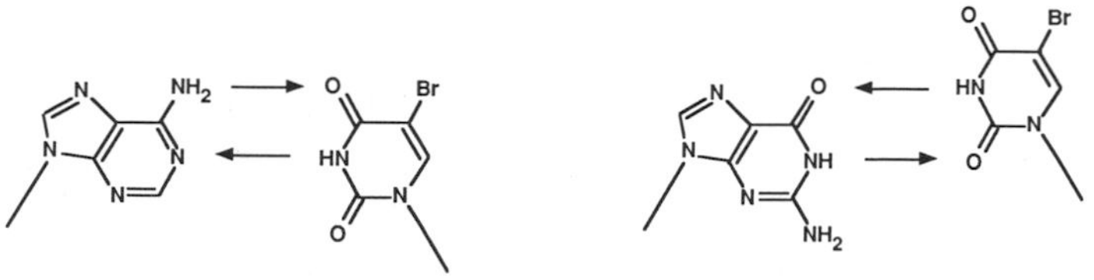

A : 5-BrU G : 5-BrU

(RNA)		(cDNA)		(RNA')		(cDNA')		(RNA")
A	→	5-BrU	→	G				
G	→	5-BrU	→	A				
U	→	A	→	5-BrU	→	G	→	C
C	→	G	→	5-BrU	→	A	→	U

Fig. 9. *Watson Crick and wobble pairing interactions involving 5-Br uracil. rG·5-BrdU pairing results in purine transitions, and dG·5-BrU pairing results in pyrimidine transitions in the context of the isothermal amplification system. RNA → cDNA → RNA' → cDNA' → RNA" refers to successive steps during the amplification cycle.*

level of cDNA synthesis and replacing UTP with 5-Br UTP at the level of RNA synthesis we can promote all four base transitions, as diagrammed in Figure 9. The isothermal amplification procedure continues to operate effectively even when replacement of 5-Br dUTP for TTP and 5-Br UTP for UTP is complete. We find that serial transfer experiments require 2 hours per cycle (vs. 1 hour per cycle with a standard reaction mixture) if we are to achieve 10^3–10^4-fold amplification per cycle. In addition, the ribozyme containing all 5-Br U residues in place of U residues is not catalytically active so that a round of nonmutagenic, nonselective amplification must be performed prior to selective amplification. We are currently quantifying mutation frequencies in order to characterize any sequence bias that might be present in this mutation process.

C. Chemical Mutagenesis

There are a number of established techniques for introducing "random" mutations by chemically modifying ssDNA or dsDNA. Treatment of ssDNA with sodium bisulfite promotes deamination of cytosine to uracil [Shortle and Botstein, 1983], resulting in G→A transitions at the level of RNA. Treatment of dsDNA with hydroxylamine results in GC→AT changes [Chu et al., 1979], which, following PCR amplification and transcription to RNA, leads to G→A and C→U transitions. Treatment of dsDNA with nitrous acid promotes deamination of cytosine, guanine, and adenine, which similarly leads to G→A and C→U transitions as well as a low frequency of other transitions and transversions [Myers et al., 1985]. The main disadvantage of these chemical mutagenesis techniques is that they exert strong AU pressure. Furthermore, they are difficult to apply in a precise, quantitative fashion.

The ideal mutagenesis technique would operate "on-line," perhaps using an engineered form of T7 RNA polymerase that operates with high turnover and low fidelity. A low-resolution crystallographic structure of this enzyme is now available [Chung et al., 1990], so that realization of this goal may not be farfetched. There are several ways to lower the fidelity of an RNA-dependent or DNA-dependent DNA polymerase that involve altering the reaction conditions. For example, substituting Mn^{2+} for Mg^{2+} increases the error rate of AMV reverse transcriptase by about one order of magnitude to 10^{-3} per position [Sirover and Loeb, 1977]. We find that isothermal amplification proceeds at about 20% efficiency in the presence of 1 mM $MnCl_2$ and either 5 or 10 mM $MgCl_2$, but does not occur in the presence of $MnCl_2$ alone. There is a published procedure for carrying out Mn^{2+} mutagenesis in the context of the PCR [Leung et al., 1989]. This method is claimed to yield all possible transitions and transversions at an overall frequency of about 10^{-2} per position. However, the reproducibility of these results is somewhat controversial. In summary, oligonucleotide-directed mutagenesis is at present the most versatile and reliable method for generating a heterogeneous population of mutant RNA.

VI. APPLICATION TO ANTISENSE TECHNOLOGY

The goal of antisense technology is to cleave a target RNA or DNA substrate with high efficiency and high specificity using an agent that does not otherwise interfere with cellular function. An antisense ribozyme, furthermore, should be able to perform the cleavage reaction with high catalytic turnover. The ribozyme must be delivered to the cell and expressed in a stable and controlled manner. The selective amplification techniques described in this chapter address only one aspect of antisense technology, namely, optimization of ribozyme function for the purpose of specifically cleaving an RNA or DNA substrate under conditions that resemble those of the cellular environment.

A. Group I Ribozymes as Antisense Agents

The wild-type *Tetrahymena* ribozyme cleaves RNA substrates at the phosphodiester bond following the sequence CUCU [Zaug et al.,

1986]. By modifying the internal guide sequence (the substrate-binding portion of the ribozyme), a number of ribozyme variants have been constructed, each of which is specific for a different substrate of the form NNNU [Murphy and Cech, 1989]. In principle this work could be extended to include all 64 internal guide sequence/substrate combinations. To achieve altered ribozyme specificity it has been necessary to introduce agents that affect duplex stability, for example, 2.5 M urea [Zaug et al., 1986, 1988], 5 mM spermidine [Doudna and Szostak, 1989], or 0.1–0.6 M NH_4OAc [Murphy and Cech, 1989]. It is not known whether these conditions will prove optimal for substrates other than short synthetic oligonucleotides. In any case, conditions that are required for optimal specificity in vitro are not compatible with the conditions that exist in vivo.

RNA-catalyzed DNA cleavage has been reported only for substrates containing the sequence CTCT or CTCU and only at 50°C [Herschlag and Cech, 1990; Roberston and Joyce, 1990]. By analogy to the work done with RNA substrates, it should be possible to cleave a variety of ssDNA substrates of the form NNNT. However, for the wild-type *Tetrahymena* ribozyme, we find that DNA cleavage does not proceed under physiological conditions (5 mM $MgCl_2$, pH 7.5, 37°C) (Robertson and Joyce, unpublished results). As with RNA, optimal conditions for DNA cleavage in vitro are very different from those that exist in the cell.

B. Evolutionary Engineering of Antisense Ribozymes

Evolutionary engineering, relying on techniques of in vitro mutagenesis and selective amplification, offers the promise of optimizing the structure of the ribozyme rather than the reaction conditions in order to achieve a high level of catalytic activity under physiological conditions. At present this is little more than a promise, buoyed by the fact that all biological catalysts are the product of evolutionary optimization. The program for optimization of ribozyme function is as follows: 1) generate a complex population of ribozyme variants; 2) test the population (as an ensemble) for the ability to cleave a target substrate under a given set of reaction conditions; 3) perform repeated rounds of selective amplification to extract the most reactive individuals from the population; 4) repeat until an optimum is reached. Individuals must then be isolated from the population by shotgun cloning and characterized in detail. Those individuals with the most desirable catalytic properties can then be tested for specific cleavage activity in vivo.

It is important to maintain a *population* of ribozymes throughout the optimization process. Geneticists have long used selection techniques to obtain a mutant organism with desirable phenotypic characteristics. The selected mutant can be isolated and used to generate new mutants that are selected again. Evolution, however, involves more than the repeated selection of individuals. It involves maintenance of a heterogeneous population of individuals, such that the number of copies of each individual is proportional to its fitness under the prevailing selection constraints. It is the tails of the mutant distribution, stretching out for three to four errors in all directions, that tend to give rise to novel mutants with enhanced fitness. Put more simply, today's loser may give rise to tomorrow's winner. For this reason, if one were to select individuals rather than populations of individuals, one would severely restrict the scope of the optimization process.

It is also important to note that selection pressure results not only from the target reaction but also from constraints imposed by the system itself. For example, RNAs that contain strong stop sites for reverse transcriptase will be selected against while RNAs whose secondary and tertiary structure facilitate primer binding will enjoy a competitive advantage. A more serious concern is that by selecting group I ribozymes based on their ability to become joined to the 3' portion of a target substrate, one is selecting *against* their ability to catalyze the subsequent site-specific hydrolysis reaction and therefore selecting against catalytic turnover. Thus one should occasionally select

for site-specific hydrolysis activity as well. This could be done by purifying the ribozyme–RNA or ribozyme–DNA intermediate, allowing it to undergo site-specific hydrolysis, and then isolating the freed ribozyme by gel electrophoresis. Alternatively, one might use a selection procedure that relies on site-specific hydrolysis to create a 3' terminus to which a short oligonucleotide can be ligated in order to create a unique site for hybridization of the primer used to initiate cDNA synthesis.

C. Alternative Selection Regimes

One can imagine a variety of selection procedures, involving either physical separation or differences in chemical properties, that could be used to distinguish between reactive and nonreactive forms of a group I ribozyme. Selective amplification techniques could be applied to the other known classes of ribozymes as well. For example, working with a linear (as opposed to a lariat) form of a group II ribozyme, one could select for the ability of the molecule to react with an RNA substrate consisting of a 5' exon sequence followed by a primer binding site. Analogous to the procedure developed for group I ribozymes, reactive group II ribozymes become joined to the 3' portion of the substrate, creating a binding site for the oligodeoxynucleotide primer used to initiate selective cDNA synthesis.

The RNA component of RNase P catalyzes site-specific hydrolysis of RNA molecules that resemble the CCA stem of tRNA [Guerrier-Takada et al., 1983; McClain et al., 1987]. If one attaches a pre-tRNA to the 5' end of RNase P RNA, one would expect RNA-catalyzed cleavage to yield the pre-tRNA leader sequence plus [5'-P]tRNA–RNase P RNA (Fig. 10a). The latter molecule could then serve as a donor for ligation (by T4 RNA ligase) to an acceptor consisting of the T7 promoter sequence followed by a new leader sequence. The RNA is then reverse transcribed to cDNA, but only those molecules that have undergone RNA-catalyzed cleavage and subsequent protein-catalyzed ligation will yield promoter-containing cDNA. Thus only reactive forms of the RNase

a)

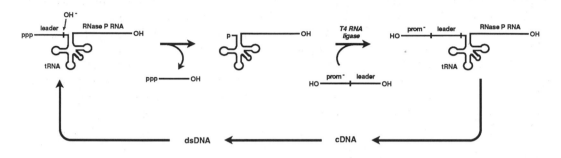

b)

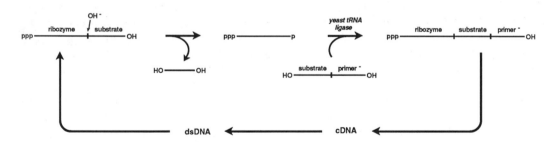

Fig. 10. *Scheme for selective amplification of RNase P RNA and self-cleaving ribozymges.* **a:** *RNase P RNA catalyzes a specific cleavage reaction to produce a donor for ligation to an RNA acceptor consisting of the T7 promoter sequence (5')-NNNUAAUACGACU CACUAUAG followed by the pre-tRNA leader sequence. Primer 1, the primer used to initiate cDNA synthesis, binds to the 3' end of RNase P RNA; primer 2, the primer used to initiate selective second-strand synthesis, binds to the promoter sequence at the 3' end of the cDNA.* **b:** *Self-cleaving RNAs catalyze a spe-cific cleavage reaction to produce an acceptor for liga-tion to an RNA donor consisting of the substrate sequence followed by a site for binding of the primer used to initiate selective cDNA synthesis. Second-strand synthesis is initiated by a primer that binds to the 3' end of the cDNA and contains the T7 promoter sequence within a 5' overhang. As with group I ribozymes, each round of selective amplification must be followed by a round of nonselective amplification to remove the primer binding site from the 3' end of the RNA.*

P ribozyme will be amplified. A simpler but more tedious method for selective amplification of RNase P RNA would involve placing the substrate at either end of the ribozyme, isolating the cleavage products by gel electrophoresis, and then performing isothermal amplification using reciprocal primers that restore the substrate sequence to the appropriate end of the selected RNAs.

The self-cleaving RNAs, which include "hammerheads," "hairpins," and the hepatitis delta agent, catalyze sequence-specific hydrolysis of RNA substrates [Uhlenbeck, 1987; Sharmeen et al., 1988; Hampel et al., 1990]. Selective amplification regimes similar to those described above for RNase P RNA could be employed. Note, however, that, unlike RNase P RNA, self-cleaving RNAs yield products that have a 5′ hydroxyl and a 2′(3′) phosphate. Thus if the substrate is placed at the 5′ end of the ribozyme, the cleavage products are the substrate having a 5′ triphosphate and the ribozyme having a 5′ hydroxyl. The ribozyme must be phosphorylated before it can be used as a donor molecule for ligation (by T4 RNA ligase) to an acceptor consisting of the T7 promoter sequence followed by new substrate sequence. Alternatively, with the substrate located at the 3′ end of the ribozyme, the cleavage products are the ribozyme having a 2′(3′) phosphate and the substrate having a 2′(3′) hydroxyl (Fig. 10b). In this case the freed ribozyme could serve as an acceptor for ligation with a donor molecule containing the primer binding site for selective cDNA synthesis. Now, however, rather than using T4 RNA ligase, one must use the tRNA ligase purified from yeast [Xu et al., 1990] or some related enzyme that joins a donor having a 5′ hydroxyl to an acceptor having a 2′(3′) phosphate.

D. Concluding Remarks

Much has been said about the potential usefulness of ribozymes as antisense agents. Recent work has focused on the "gene shears" ribozyme [Haseloff and Gerlach, 1988], which is a hammerhead that contains three of the four domains required for cleavage, the fourth being provided by the RNA substrate. An engineered ribozyme of this type, targeted against mRNA for chloramphenicol acetyltransferase (CAT), was introduced into a monkey kidney cell line and shown to suppress CAT expression by about 30% [Cameron and Jennings, 1989]. A hammerhead of similar design, targeted against HIV-1 *gag* RNA, was shown to reduce substantially the level of HIV-1 *gag* RNA and *gag*-related p24 antigen in HIV-1-infected HeLa CD4[+] cells [Sarver et al., 1990]. While these results are encouraging, there is at present no evidence to show that antisense ribozymes operate catalytically (rather than by a simple duplex interaction) within the cellular milieu.

If antisense ribozymes are to be useful for specific inactivation of RNA or DNA targets in vivo, then the catalytic efficiency of these agents must be improved. Selective amplification techniques offer one approach to the development of antisense ribozymes that optimally cleave a target substrate under a chosen set of reaction conditions. Existing ribozymes are the product of biological evolution, already optimized to carry out a specific task in their native context. It is not surprising, therefore, that they have suboptimal activity when applied in some other context. It remains to be seen to what extent laboratory evolution can shape the functional properties of an RNA enzyme to fulfill their potential role as antisense agents.

ACKNOWLEDGMENTS

The author is grateful to Amber A. Beaudry for technical assistance. This work was supported by National Aeronautics and Space Administration grant NAGW-1671.

VII. REFERENCES

Agrawal A, Goodchild J, Civeira MP, Thornton AH, Sarin PS, Zamecnik PC (1988): Oligodeoxynucleoside phosphoramidates and phosphorothioates as inhibitors of human immunodeficiency virus. Proc Natl Acad Sci USA 85:7079–7083.

Barfod ET, Cech TR (1988): Deletion of nonconserved helices near the 3′ end of the rRNA intron of *Tetra-*

hymena thermophila alters self-splicing but not catalytic activity. Genes Dev 2:652–663.

Bass BL, Cech TR (1984): Specific interaction between the self-splicing RNA of *Tetrahymena* and its guanosine substrate: Implications for biological catalysis by RNA. Nature 308:820–826.

Beaudry AA, Joyce GF (1990): Minimum secondary structure requirements for catalytic activity of a self-splicing group I intron. Biochemistry 29:6534–6539.

Been MD, Cech TR (1986): One binding bite determines sequence specificity of *Tetrahymena* pre-rRNA self-splicing, *trans*-splicing, and RNA enzyme activity. Cell 47:207–216.

Burke JM, Belfort M, Cech TR, Davies RW, Schweyen RJ, Shub DA, Szostak JW, Tabak HF (1987): Structural conventions for group I introns. Nucleic Acids Res 15:7217–7221.

Cameron FH, Jennings PA (1989): Specific gene suppression by engineered ribozymes in monkey cells. Proc Natl Acad Sci USA 86:9139–9143.

Chamberlin M, Ryan T (1982): Bacteriophage DNA-dependent RNA polymerases. In Boyer P (ed): The Enzymes. New York: Academic Press, pp 87–108.

Chu BCF, Orgel LE (1985): Nonenzymatic sequence-specific cleavage of single-stranded DNA. Proc Natl Acad Sci USA 82:963–967.

Chu C-T, Parris DS, Dixon RAF, Farber FE, Schaffer PA (1979): Hydroxylamine mutagenesis of HSV DNA and DNA fragments: Introduction of mutations into selected regions of the viral genome. Virology 98:168–181.

Chung YJ, Sousa R, Rose JP, Lafer E, Wang B-C (1990): Crystallographic structure of phage T7 RNA polymerase at resolution of 4.0 Å. In Wu FY-H, and Wu C-W (eds): Structure and Function of Nucleic Acids and Proteins. New York: Raven Press, pp 55–59.

Conrad M (1979): Bootstrapping on the adaptive landscape. BioSystems 11:167–182.

Crowe JH, Crowe LM, Carpenter JF, Aurell Wistrom C (1987): Stabilization of dry phospholipid bilayers and proteins by sugars. Biochem J 242:1–10.

Dash P, Lotan I, Knapp M, Kandel ER, Goelet P (1987): Selective elimination of mRNAs in vivo: Complementary oligodeoxynucleotides promote RNA degradation by an RNase H–like activity. Proc Natl Acad Sci USA 84:7896–7900.

Doudna JA, Szostak JW (1989): RNA-catalysed synthesis of complementary-strand RNA. Nature 339:519–522.

Dreyer GB, Dervan PB (1985): Sequence-specific cleavage of single-stranded DNA: Oligodeoxynucleotide-EDTA-Fe(II). Proc Natl Acad Sci USA 82:986–972.

Ellington AD, Szostak JW (1990): In vitro selection of RNA molecules that bind specific ligands. Nature 346:818–822.

Goodchild J, Agrawal S, Civeira MP, Sarin PS, Sun D, Zamecnik PC (1988): Inhibition of human immuno-deficiency virus replication by antisense oligodeoxynucleotides. Proc Natl Acad Sci USA 85:5507–5511.

Gould SJ (1980): Hen's Teeth and Horse's Toes. New York: WW Norton, pp 177–198.

Guatelli JC, Whitfield KM, Kwoh DY, Barringer KJ, Richman DD, Gingeras TR (1990): Isothermal, in vitro amplification of nucleic acids by a multienzyme reaction modeled after retroviral replication. Proc Natl Acad Sci USA 87:1874–1878.

Guerrier-Takada C, Gardiner K, Marsh T, Pace N, Altman S (1983): The RNA moiety of ribonuclease P is the catalytic subunit of the enzyme. Cell 35:849–857.

Hampel A, Tritz R, Hicks M, Cruz P (1990): "Hairpin" catalytic RNA model: Evidence for helices and sequence requirement for substrate RNA. Nucleic Acids Res 18:299–304.

Haseloff J, Gerlach WL (1988): Simple RNA enzymes with new and highly specific endoribonuclease activities. Nature 334:585–591.

Hermes JD, Blacklow SC, Knowles JR (1990): Searching sequence space by definably random mutagenesis: Improving the catalytic potency of an enzyme. Proc Natl Acad Sci USA 87:696–700.

Herschlag D, Cech TR (1990): DNA cleavage catalyzed by the ribozyme from *Tetrahymena*. Nature 344:405–409.

Inoue T, Sullivan FX, Cech TR (1986): New reactions of the ribosomal RNA precursor of *Tetrahymena* and the mechanism of self-splicing. J Mol Biol 189:143–165.

Iwahashi H, Kyogoku Y (1977): Detection of proton acceptor sites of hydrogen bonding between nucleic acid bases by the use of ^{13}C magnetic resonance. J Am Chem Soc 99:7761–7765.

Joyce GF (1989a): Building the RNA world: Evolution of catalytic RNA in the laboratory. In Cech TR (ed): Molecular Biology of RNA. UCLA Symposia on Molecular and Cellular Biology. New York: Alan R. Liss, pp 361–371.

Joyce GF (1989b): Amplification, mutation and selection of catalytic RNA. Gene 82:83–87.

Joyce GF, Inoue T (1987): Structure of the catalytic core of the *Tetrahymena* ribozyme as indicated by reactive abbreviated forms of the molecule. Nucleic Acids Res 15:9825–9840.

Joyce GF, Inoue T (1989): A novel technique for the rapid preparation of mutant RNAs. Nucleic Acids Res 17:711–722.

Joyce GF, van der Horst G, Inoue T (1989): Catalytic activity is retained in the *Tetrahymena* group I intron despite removal of the large extension of element P5. Nucleic Acids Res 17:7879–7889.

Kay PS, Inoue T (1987): Catalysis of splicing-related reactions between dinucleotides by a ribozyme. Nature 327:343–346.

Kruger K, Grabowski PJ, Zaug AJ, Sands J, Gottschling DE, Cech TR (1982): Self-splicing RNA: Autoexcision and autocyclization of the ribosomal RNA intervening sequence. Cell 31:147–157.

Kwoh DY, Davis GR, Whitfield KM, Chappelle HL, DiMichele LJ, Gingeras TR (1989): Transcription-based amplification system and detection of amplified human immuno-deficiency virus type 1 with a bead-based sandwich hybridization format. Proc Natl Acad Sci USA 86:1173–1177.

Leung DW, Chen E, Goeddel DV (1989): A method for random mutagenesis of a defined DNA segment using a modified polymerase chain reaction. Technique 1:11–15.

Matsukura M, Shinozuka K, Zon G, Mitsuya H, Reitz M, Cohen JS, Broder S (1987): Phosphorothioate analogs of oligodeoxynucleotides: Inhibitors of replication and cytopathic effects of immunodeficiency virus. Proc Natl Acad Sci USA 84:7706–7710.

Matsukura M, Zon G, Shinozuka K, Robert-Guroff M, Shimada T, Stein CA, Mitsuya H, Wong-Staal F, Cohen JS, Broder S (1989): Regulation of viral expression of human immunodeficiency virus in vitro by an antisense phosphorothioate oligodeoxynucleotide against *rev (art/trs)* in chronically infected cells. Proc Natl Acad Sci USA 86:4244–4248.

McClain WH, Guerrier-Takada C, Altman S (1987): Model substrates for an RNA enzyme. Science 238:527–530.

Milligan JF, Groebe DR, Witherell GW, Uhlenbeck OC (1987): Oligoribonucleotide synthesis using T7 RNA polymerase and synthetic DNA templates. Nucleic Acids Res 15:8783–8798.

Milman G, Langridge R, Chamberlin MJ (1967): The structure of a DNA–RNA hybrid. Proc Natl Acad Sci USA 57:1804–1810.

Minshull J, Hunt T (1986): The use of single-stranded DNA and RNase H to promote quantitative "hybrid arrest of translation" of mRNA/DNA hybrids in reticulocyte lysate cell-free translations. Nucleic Acids Res 14:6433–6451.

Murphy FL, Cech TR (1989): Alteration of substrate specificity for the endoribonucleolytic cleavage of RNA by the *Tetrahymena* ribozyme. Proc Natl Acad Sci USA 86:9218–9222.

Myers RM, Lerman LS, Maniatis T (1985): A general method for saturation mutagenesis of cloned DNA fragments. Science 229:242–247.

Peebles CL, Perlman PS, Mecklenburg KL, Petrillo ML, Tabor JH, Jarrell KA, Cheng HL (1986): A self-splicing RNA excises as an intron lariat. Cell 44:213–233.

Price JV, Kieft GL, Kent JR, Sievers EL, Cech TR (1985): Sequence requirements for self-splicing of the *Tetrahymena* pre-ribosomal RNA. Nucleic Acids Res 13:1871–1889.

Rada R (1981): Evolution and gradualness. BioSystems 14:211–218.

Robertson DL, Joyce GF (1990): Selection in vitro of an RNA enzyme that specifically cleaves single-stranded DNA. Nature 344:467–468.

Saiki RK, Gelfand DH, Stoffel S, Scharf SJ, Higuchi R, Horn GT, Mullis KB, Erlich HA (1988): Primer-directed enzymatic amplification of DNA with a thermostable DNA polymerase. Science 239:487–491.

Sarver N, Cantin EM, Chang PS, Zaia JA, Ladne PA, Stephens DA, Rossi JJ (1990): Ribozymes as potential anti-HIV-1 therapeutic agents. Science 247:1222–1225.

Sharmeen L, Luo MYP, Dinter-Gottlieb G, Taylor J (1988): Antigenomic RNA of human hepatitis delta virus can undergo self-cleavage. J Virol 62:2674–2679.

Shortle D, Botstein D (1983): Directed mutagenesis with sodium bisulfite. Methods Enzymol 100:457–468.

Sirover MA, Loeb LA (1977): On the fidelity of DNA replication: Effect of metal activators during synthesis with avian myeloblastosis virus DNA polymerase. J Biol Chem 252:3605–3610.

Spiegelman S (1971): An approach to the experimental analysis of precellular evolution. Q Rev Biophys 4:213–253.

Szostak JW (1986): Enzymatic activity of the conserved core of a group I self-splicing intron. Nature 322:83–86.

Tuerk C, Gold L (1990): Systematic evolution of ligands by exponential enrichment: RNA ligand to bacteriophage T4 DNA polymerase. Science 249:505–510.

Uhlenbeck OC (1987): A small catalytic oligoribonucleotide. Nature 328:596–600.

Walder RY, Walder JA (1988): Role of RNase H in hybrid-arrested translation by antisense oligonucleotides. Proc Natl Acad Sci USA 85:5011–5015.

Wang AH-J, Fujii S, van Boom JH, van der Marel GA, van Boeckel SAA, Rich A (1982): Molecular structure of r(GCG) d(TATACGC): A DNA–RNA hybrid helix joined to double helical DNA. Nature 299:601–604.

Waring RB, Towner P, Minter SJ, Davies RW (1986): Splice-site selection by a self-splicing RNA of *Tetrahymena*. Nature 321:133–139.

Xu Q, Phizicky EM, Greer CL, Abelson JN (1990): Purification of yeast transfer RNA ligase. Methods Enzymol 181:463–471.

Zamecnik PC, Stephenson ML (1978): Inhibition of Rous sarcoma virus replication and cell transformation by a specific oligodeoxynucleotide. Proc Natl Acad Sci USA 75:280–284.

Zaug AJ, Been MD, Cech TR (1986): The *Tetrahymena* ribozyme acts like an RNA restriction endonuclease. Nature 324:429–433.

Zaug AJ, Cech TR (1985): Oligomerization of intervening sequence RNA molecules in the absence of proteins. Science 299:1060–1064.

Zaug AJ, Grosshans CA, Cech TR (1988): Sequence-specific endoribonuclease activity of the *Tetrahymena* ribozyme: Enhanced cleavage of certain oligonucleotide substrates that form mismatched ribozyme–substrate complexes. Biochemistry 27:8924–8931.

ABOUT THE AUTHOR

GERALD F. JOYCE is an Assistant Professor in the Departments of Chemistry and Molecular Biology at the Scripps Research Institute. After receiving his B.A. from the University of Chicago in 1978, he entered the M.D./ Ph.D. program at the University of California, San Diego. His thesis research was carried out under Leslie Orgel at the Salk Institute and concerned nucleic acid chemistry and the template-directed synthesis of RNA. After completing his predoctoral studies, Dr. Joyce undertook a medical internship at Mercy Hospital in San Diego and obtained his medical license. He then returned to the laboratory to pursue postdoctoral research with Tan Inoue at the Salk Institute, studying the biochemistry of RNA enzymes. In 1989 Dr. Joyce joined the faculty at the Scripps Research Institute. His current research involves development of directed evolution techniques and application of these techniques to the study and design of RNA enzymes. He also has a longstanding interest in the origins of life and the role of RNA in the early history of life on earth. Dr. Joyce teaches at both the Scripps Research Institute and the University of California, San Diego. He was twice recipient of the Kaiser Award for outstanding teaching. His work has been widely published and has been discussed in both the scientific and lay literature.

Antisense RNA and DNA: 373–381
© 1992 Wiley-Liss, Inc.

Design of Ribozymes Distinguishing a Point Mutation in c-Ha-*ras* mRNA

Makoto Koizumi and Eiko Ohtsuka

I. INTRODUCTION

It is known that RNA can play roles not only as messengers of genetic information but also as catalysts that regulate gene expression, since Cech and coworkers [1981] found that the intervening sequence of *Tetrahymena* rRNA acted as an enzyme. These RNA that catalyze RNA-processing reactions without the involvement of a protein catalyst are called *ribozymes* or *RNA enzymes*.

There are four types of RNA-processing reaction that are catalyzed by ribozymes. One of these reactions is called *self-cleavage in a hammerhead RNA*. This type of ribozyme has been found in plus and minus strands of satellite RNA of tobacco ring spot virus [Buzayan et al., 1986; Prody et al., 1986], plus and minus strands of avocado sunblotch viroid [Hutchins et al., 1986], virusoid of lucerne transient streak virus [Forster and Symons, 1987a], and transcripts of satellite DNA of the newt [Epstein and Gall, 1987]. This specific cleavage is believed to be an essential step of maturation in a rolling circle mechanism, by which these circular RNA replicate.

Cleavage sites of these RNA were identified, and the secondary structure around them was found to form a hammerhead-shaped structure. This domain contains 3 stems and 13 conserved nucleotides, as shown in Figure 1a. Some hammerhead domains were proved to be important for the self-cleavage reaction by investigators using mutant sequences [Forster and Symons, 1987b; Uhlenbeck, 1987; Sheldon and Symons, 1989]. We have also constructed hammerhead RNA duplexes using chemically synthesized 21-mers and changed some conserved sequences into other sequences [Koizumi et al., 1988a]. From the results of these mutagenesis experiments, we were able to design ribozymes for sequence-dependent cleavage of target RNA [Koizumi et al., 1988b, 1989]. We describe construction and design of ribozymes to distinguish a point mutation in c-Ha-*ras* mRNA.

a   b

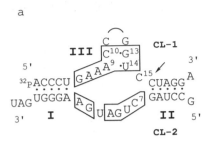
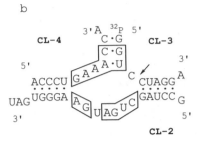
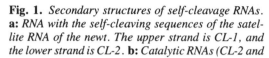

Fig. 1. *Secondary structures of self-cleavage RNAs.* *a: RNA with the self-cleaving sequences of the satellite RNA of the newt. The upper strand is CL-1, and the lower strand is CL-2.* **b:** *Catalytic RNAs (CL-2 and*

CL-4) contain 11 conserved nucleotides. Substrate RNA (CL-3) contains a cleavage site and two conserved nucleotides. [Reproduced from Koizumi et al., 1989, with permission of the publisher.]

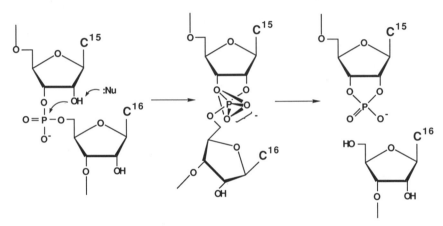

Fig. 2. *Mechanism for self-cleavage reaction. The nucleophile (:Nu) abstracts a proton from the 2'-hydroxyl group of a cleavage site (C^{15}). Nucleophilic*

attack by the 2'-O$^-$ forms a pentacoordinated transition state. 2', 3'-cyclic phosphate and a 5'-hydroxyl group are generated at the 3' and 5' termini, respectively.

II. STRUCTURE AND REACTION MECHANISM OF HAMMERHEAD RNA

The self-cleavage reaction yields a 2',3'-cyclic phosphate and a 5'-hydroxyl group at the termini of the cleaved products. Although the mechanism of this reaction is not clear, it may be similar to that of the Pb(II)-catalyzed hydrolysis found in yeast tRNA^{Phe} [Brown et al., 1985]. X-ray crystallographic investigation of a complex of tRNA^{Phe} and the Pb(II) suggested that Pb(II)-bound hydroxyl group removed a proton from the 2'-hydroxyl group of the cleavage site, and the nucleophilic attack

of the 2'-O$^-$ resulted in transesterification of the cleavage site.

The hammerhead ribozyme also requires metal ions such as Mg^{2+}, Ca^{2+}, or Mn^{2+} [Uhlenbeck, 1987]. In the hammerhead ribozyme, Mg^{2+} may bind to certain unpaired nucleotides, and the Mg^{2+}-bound hydroxyl group may act as the nucleophile that abstracts a proton from the 2'-hydroxyl group of the cleavage site, cytidine 15 (Fig. 2). Mei et al. [1989] have shown, using computational modeling, that Mg^{2+} might bind to certain conserved nucleotides in a hammerhead ribozyme. Using enzymatically synthesized oligonucleotides containing a phosphorothioate diester in the R_p configuration, van Tol et al. [1990] reported that transesterification of a hammerhead RNA occurred by *in-line* attack of the 2′-hydroxyl group. Using chemically synthesized oligonucleotides containing an S_p phosphorothioate diester, we also found that the hammerhead ribozyme catalyzes the cleavage via the same pathway. Heus et al. [1990] showed NMR data for imino protons of hammerhead ribozymes that suggested the presence of a hairpin structure in which stem II was a stable helix.

III. DESIGN OF A HAMMERHEAD RIBOZYME

Base-specific RNases are available for RNA sequencing and for proving the secondary structure of RNA. However, no sequence-specific RNA endonucleases have been found, in contrast to DNA endonucleases such as DNA restriction enzymes. Recently, chimeric oligonucleotides containing 2′-*O*-methylnucleotides and RNase H have been found to be useful for unique cleavage of a complementary RNA [Inoue et al., 1897; Shibahara et al., 1987]. Oligonucleotide-hybrid nucleases also cleaved target RNA in the presence of Ca^{2+} [Zuckerman et al., 1988]. These RNA endonucleases are useful as probes for structural studies of RNA of biological importance. Zaug et al. [1986] showed that a shortened form of the intervening sequence RNA from *Tetrahymena* can act as an RNA endonuclease that recognizes a substrate RNA by binding to the internal guide sequence.

To design ribozymes that cleave target RNA like an endonuclease, we must separate a hammerhead RNA into strands for a substrate and an enzyme, and we must identify essential regions in the conserved sequences of the hammerhead structure. We synthesized a two-strand hammerhead domain containing 3 stems and 13 conserved nucleotides found in a transcript of satellite DNA of newt (Fig. 1a). Specific cleavage occurred at the site consistent with the reaction carried out by the newt satellite RNA.

The upper strand (CL-1) of the hammerhead RNA was separated into two strands between C and G in the loop adjacent to stem III (Fig. 1b). CL-3 (Fig. 1b) contains the cleavage site, and the other (CL-4) is responsible for the catalytic activity. The radiolabeled CL-3 was efficiently cleaved at cytidine 15, as expected [Koizumi et al., 1989].

The substrate strand (CL-3) had two highly conserved sequences (G13 and U14) and the cleavage site (C15). We changed these nucleotides into other sequences in order to find essential base pairs required for efficient ribozymes capable of recognizing various sequences [Koizumi et al., 1988a]. The C:G base pair between C10 and G13 in the duplex (Fig. 1a) could be exchanged with each other to give a reaction comparable to the wild type. However, A10:U13 and U10:A13 yielded less cleavage. Substitution of A9:U14 base pair with G9:C14, C9:G14 or U9:A14 influenced the cleavage reaction (4%, nondetectable, or 11%, respectively). The cleavage site (C15) could be substituted with A15 or U15 but not G15. It is rather strange that the ribozyme cannot recognize a substrate having G15 as a cleavage site. When C7 in the lower strand was substituted with U7, the cleavage at C15 or U15 occurred, but no reaction was obtained if position 15 is occupied with A. The lack of catalytic activity of a mutated duplex containing A15 and U7 may be due to formation of a base pair between position 15 in CL-1 and position 7 in CL-2. The results of these mutagenesis studies are summarized in Figure 3a.

Using information from this study, we constructed ribozymes that are able to cleave RNA by recognizing sequences of 9–15 bases. We consider that four points are important for designing a ribozyme with an RNA endonuclease activity: 1) The cleavage site (X) is A, C, or U but not G. 2) U at the 5′ side of the cleavage site (X) is important for pairing to A in the ribozyme. 3) Stems II and III, which form duplexes between the ribozyme and the substrate RNA, should have sufficient numbers of Watson Crick base pairs (N′:N). 4) The other sequences in the ribozyme comprise the

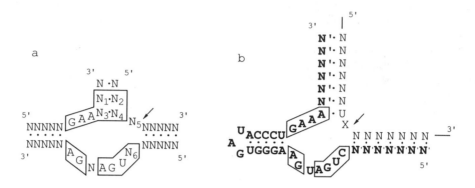

Fig. 3. **a:** *Summary of exchangeable bases in the hammerhead ribozyme. N_1·N_2 can be G:C, C:G, A:U or U:A. N_3·N_4 can be A:U. When N_6 is C, N_5 can be A, C, or U. When N_6 is U, N_5 can be C or U.* **b:** *Secondary structure of a designed ribozyme. Bold letters indicate a designed ribozyme, and the others represent a substrate. N or N' can be A, G, C, or U. X can be A, C, or U. The ribozyme can cleave the substrates containing UX sequences.*

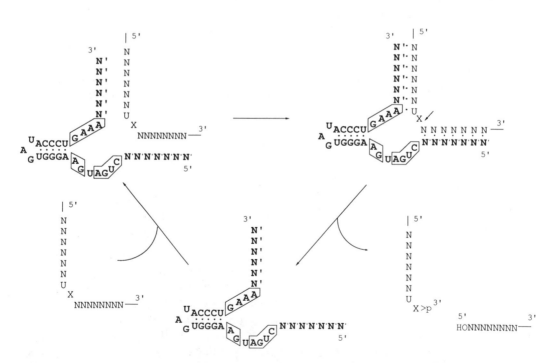

Fig. 4. *Multiple turnover of the cleavage reaction by a designed ribozyme. The ribozyme and the substrate form a hammerhead structure. In the presence of Mg^{2+}, the ribozyme cleaves the substrate, generating a 5' fragment with a hydroxyl group at the terminus and a 3' fragment with 2',3'-cyclic phosphate at the terminus. The ribozyme dissociates from these products in order to associate with the next substrate.*

sequences of the hammerhead RNA containing nine highly conserved sequences. As shown in Figure 3b, information from the results of our mutagenesis provides us with a structure of a hammerhead ribozyme (bold letters in Fig. 3b) with the catalytic activity of cleaving a substrate RNA [Koizumi et al., 1989].

Because ribozymes act as catalysts without being consumed, they should be able to cleave substrate many times. A cycle of cleavage reaction by the ribozyme is shown in Figure 4. The ribozyme binds to the substrate RNA by Watson Crick base pairing, and the paired RNAs form a hammerhead structure. In the presence of Mg^{2+}, the ribozyme cleaves the substrate RNA, generating two fragments ($^5{}'[N]_nUX>p^{3'}$ and $^5{}'HO[N]_n{}^{3'}$). These products dissociate, and the ribozyme associates with the next substrate. This ribozyme can serve as an RNA endonuclease that hydrolyzes at the X position RNA of the sequence $(N)_nUX(N)_n$ (where N = A, G, C, or U; X = A, C, or U).

IV. CLEAVAGE OF AN 11-MER BY A DESIGNED RIBOZYME

An undecamer, pCAGCUAAGUAU, having a sequence similar to the consensus sequence for a 5'-splice site, was selected as the target substrate, and the A in position 6 of this oligomer was chosen as the cleavage site [Koizumi et al., 1988b]. The ribozyme that cleaves this substrate consists of two strands, a 19-mer and a 15-mer (Fig. 5, complex 1). Ribozymes that have a few nucleotides deleted from their 5' or 3' end of one of their component strands are shown in Figure 5 as complexes 2 and 3.

These complexes were hydrolyzed at the expected position in the presence of Mg^{2+}. The half-life of the substrate in complex 1 was found to be about 2 hours at 15°C. At high temperature, the cleavage reaction proceeded slowly, probably because of disruption of the tertiary structure. We obtained similar results previously, in which $[\theta]_{265}$ values in CD spectra decreased with higher temperature [Koizumi et al., 1989]. The cleavage rates of complex 2 and 3, which contain less hydrogen bonds, were

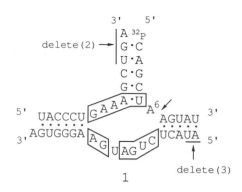

Fig. 5. *Secondary structure of the synthetic ribozyme for cleavage of an 11-mer containing a 5'-splice consensus sequence. The ribozyme consists of a 19-mer (running 3' to 5', left to right) and a 15-mer (5' to 3'). The arrow indicates the site for the cleavage on the 11-mer target. Complex 1 consists of the 11-mer target and the two-strand ribozyme (the 19-mer and the 15-mer). The ribozyme in complex 2 has three nucleotides deleted from the 3' end of the 15-mer (shown as "delete [2]"). The ribozyme in complex 3 has two nucleotides deleted from the 5'-end of the 19-mer ("delete [3]").*

slower than that of complex 1. A Hanes-Woolf plot, which was used to determine the kinetic parameters, showed that the synthetic ribozyme in complex 1 cleaved the substrate catalytically. The K_m and k_{cat} were found to be 0.53 μM and 0.03 min^{-1}, respectively.

V. DESIGN AND SYNTHESIS OF RIBOZYMES FOR CLEAVAGE OF c-Ha-*ras* mRNA WITH A POINT MUTATION

The c-Ha-*ras* gene codes for a protein (p21) whose molecular weight is 21×10^3. The activated c-Ha-*ras* genes, which are found in human cancer cells, result from single mutations at various positions, including the glycine-12, glycine-13, and leucine-61 codons [Tabin et al., 1982]. A point mutation of G to T at codon 12 in the c-Ha-*ras* gene mutates Gly-12 (GGT) to Val-12 (GTT) in p21. We planned to build a hammerhead ribozyme that could recognize a single base mutation in the c-Ha-*ras* mRNA. We designed ribozymes that would cleave only the activated c-Ha-*ras* mRNA

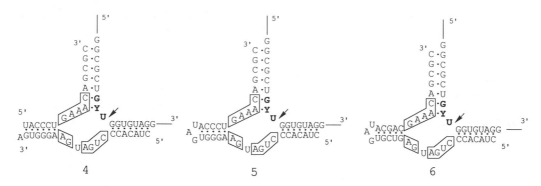

Fig. 6. *Secondary structures of one-strand and two-strand ribozymes interacting with c-Ha-ras mRNA. Y–G refers to the wild type RNA, and Y–U refers to the* *mutant RNA. Refer to text for details. [Reproduced from Koizumi et al., 1989, with permission of the publisher.]*

(Val[GUU]mRNA), but which were inert to the wild type c-Ha-*ras* mRNA (Gly[GGU]mRNA) [Koizumi et al., 1989].

Two target mRNAs (72-mers) were prepared by transcription of two synthetic DNA duplexes coding for position 7-26 of c-Ha-*ras* protein. Figure 6 shows complexes consisting of mRNAs and ribozymes. When complex 4 containing a ribozyme that consists of two oligonucleotides (a 17-mer and a 21-mer) was incubated in the presence of Mg^{2+} at 37°C for 23 hours, the Val(GUU)mRNA was cleaved in a yield of 17% at the 3′ side of Val-12 (GUU). When the Gly(GGU)mRNA was used as a substrate, it was not cleaved, as expected. This can be explained by the fact that the complex between the Gly(GGU)mRNA and the ribozyme contains an A:G mismatch instead of a conserved A:U base pair. The rate of cleavage of complex 4 was very slow. This may suggest that the two-strand ribozyme does not form an active complex very efficiently. In order to improve the catalytic activity and to obtain an effective ribozyme, the two strands were ligated by RNA ligase. The jointed ribozyme (complex 5 in Fig.6) cleaved the Val(GUU)mRNA after 23 hours, as shown in Figure 7.

The rate of cleavage of complex 5 was still rather low. We suspected that this might be due to the formation of a hairpin structure, which came from the 5′-AGGGUG-3′ and 5′-CACCCU-3′ sequences in the ribozyme of complex 5. If this hairpin formed, the ribozyme would not correctly base-pair with the target RNA and could not act as an efficient RNA endonuclease. A new ribozyme having a reformed sequence was prepared as shown in Figure 6 (complex 6) so that the unfavorable hairpin structure could not exist. The half-life of complex 6 was about 30 minutes at 37°C. The rate of cleavage in complex 6 was 20 times faster than that in complex 5. We have therefore shown that the secondary structure of a ribozyme affected the catalytic activity.

We found that the K_m and k_{cat} values for cleavage of the Val(GUU)mRNA in complex 6 were 0.44 μM and 0.011 min^{-1}, respectively. However, the k_{cat}/K_m value of complex 6 was lower than that of complex 1 in spite of the presence of a one-strand ribozyme. Perhaps the catalytic turnover of the ribozyme is inhibited because it is associating with a longer substrate and probably forms a complex with a higher order structure. Fedor and Uhlenbeck [1990] have also shown that secondary structures of a substrate RNA were a determinant of catalytic efficiency. Joyce (this volume) discusses the use of selective amplification and in vitro selection for optimizing ribozyme function, an approach that offers considerable potential for improving the effectiveness of ribozymes designed against a particular sequence.

VI. THE RIBOZYME CAN ACT AS AN RNA ENDONUCLEASE IN VIVO

The hammerhead ribozyme has 30–40 nucleotides, consists of A, G, C, and U, and does not require the presence of modified nucleotides. Therefore such ribozymes can readily be synthesized by the T7 RNA polymerase transcription system, and ribozyme genes that correspond to ribozyme sequences can be expressed in cell lines. It should be possible for therapeutic agents against viruses to be developed, using cleavage of target RNAs with ribozymes.

Haseloff and Gerlach [1988] reported ribozyme-catalyzed cleavage of chloramphenicol acetyltransferase (CAT) mRNA based on the results of mutagenesis of tobacco ring spot virus satellite RNA in vitro. Cameron and Jennings [1989] examined inhibition of CAT expression by a ribozyme cloned into a mammalian expression vector in monkey cells (COS1). They demonstrated that a high molar excess of the ribozyme suppressed CAT expression.

A modified tRNA gene, in which ribozyme coding sequences were placed, was constructed to obtain a ribozyme localized within cells [Cotten and Birnstiel, 1989]. When this gene and its target U7snRNA were coinjected into *Xenopus* oocytes, the gene was transcribed by RNA polymerase III, and the RNA produced cleaved the target RNA in the cytoplasm.

Sarver et al. [1990] designed a ribozyme targeted to HIV-*gag*-1 transcripts as a therapeutic agent. The vector containing the anti HIV-*gag*-1 ribozyme gene was transfected into CD4$^+$ HeLa cells, and these cells expressing the ribozyme were infected with HIV-1. It was shown that the ribozyme of the transformed cells reduced the level of the HIV-*gag*-1 RNA and p24 antigen concentration.

Designed ribozymes may find uses as therapeutic agents for interfering with the expression of genes. Since ribozymes act as endonucleases without being consumed, they may provide a way of regulating RNA levels or function more dramatically than is possible with

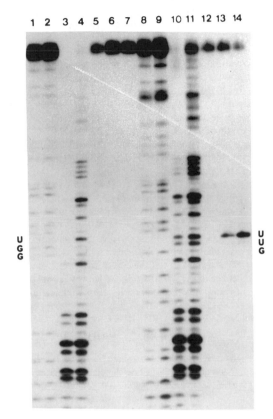

Fig. 7. *Autoradiogram of the cleavage products of complex 5 (Fig. 6). The 5'-labeled substrate (6.25 pmol) was treated with the one-strand ribozyme (7.5 pmol) in 4.6 μl of 40 mM Tris-HCl (pH 7.5), 25 mM MgCl₂, and 20 mM NaCl at 37°C and mixed with loading solution for 10% PAGE in 8 M urea. Lanes 1–7, the substrate was the wild-type mRNA (GGU). Lanes 8–14, the substrate was the mutant mRNA (GUU). 1 and 8, RNase T₁ (0.5 U) digestion; 2 and 9, RNase T₁ (0.1 U) digestion; 3 and 10, RNase PhyM (5 U) digestion; 4 and 11, RNase PhyM (1 U) digestion; 5 and 12, no incubation; 6 and 13, 4 hours incubation; 7 and 14, 23 hours incubation. [Reproduced from Koizumi et al., 1989, with permission of the publisher.]*

antisense RNA. We have shown here that ribozymes can recognize a single base change in RNA and cleaved the substrate catalytically. Specific regulation of the expression of a gene with a single mutation against a background of normal gene expression, such as is the case with the activated c-Ha-*ras* gene, may then be realizable in vivo by the introduction of designed ribozymes. It is unlikely that antisense RNA could show such specificity. Application of these techniques to in vivo experiments is in progress.

VII. REFERENCES

Brown RS, Dewan JC, Klug A (1985): Crystallographic and biochemical investigation of the lead(II)-catalyzed hydrolysis of yeast phenylalanine tRNA. Biochemistry 24:4785–4801.

Buzayan JM, Gerlach WL, Bruening G (1986): Non-enzymatic cleavage and ligation of RNAs complementary to a plant virus satellite RNA. Nature 323:349–353.

Cameron FH, Jennings PA (1989): Specific gene suppression by engineered ribozymes in monkey cells. Proc Natl Acad Sci USA 86:9139–9143.

Cech TR, Zang AJ, Grabowski PJ (1981): In vitro splicing of the ribosomal RNA precursor of *Tetrahymena*: Involvement of a guanosine nucleotide in the excision of the intervening sequence. Cell 27:487–496.

Cotten M, Brirnstiel ML (1989): Ribozyme mediated destruction of RNA in vivo. EMBO J 8:3861–3866.

Epstein LM, Gall JG (1987): Self-cleaving transcripts of satellite DNA from the newt. Cell 48:535–543.

Fedor MJ, Uhlenbeck OC (1990): Substrate sequence effects on "hammerhead" RNA catalytic efficiency. Proc Natl Acad Sci USA 87:1668–1672.

Forster AC, Symons RH (1987a): Self-cleavage of plus and minus RNAs of a virusoid and a structural model for the active sites. Cell 49:211–220.

Forster AC, Symons RH (1987b): Self-cleavage of virusoid RNA is performed by the proposed 55-nucleotide active site. Cell 50:9–16.

Haseloff J, Gerlach WL (1988): Simple RNA enzymes with new and highly specific endonuclease activities. Nature 334:585–591.

Heus HA, Uhlenbeck OC, Pardi A (1990): Sequence-dependent structural variations of hammerhead RNA enzymes. Nucleic Acids Res 18:1103–1108.

Hutchins CJ, Rathjen PD, Forster AC, Symons RH (1986): Self-cleavage of plus and minus RNA transcripts of avocado sunblotch viroid. Nucleic Acids Res 14:3627–3640.

Inoue H, Hayase Y, Iwai S, Ohtsuka E (1987): Sequence-dependent hydrolysis of RNA using modified oligonucleotide splints and RNase H. FEBS Lett 215:327–330.

Koizumi M, Hayase Y, Iwai S, Kamiya H, Inoue H, Ohtsuka E (1989): Design of RNA enzymes distinguishing a single base mutation in RNA. Nucleic Acids Res 17:7059–7071.

Koizumi M, Iwai S, Ohtsuka E (1988a): Construction of a series of several self-cleaving RNA duplexes using synthetic 21-mers. FEBS Lett 228:228–230.

Koizumi M, Iwai S, Ohtsuka E (1988b): Cleavage of specific sites by RNA by designed ribozymes. FEBS Lett 239:285–288.

Mei H-Y, Kaaret TW, Bruice TC (1989): A computational approach to the mechanism of self-cleavage of hammerhead RNA. Proc Natl Acad Sci USA 86:9727–9731.

Prody GA, Bakos JT, Buzayan JM, Schneider IR, Bruening G (1986): Autolytic processing of dimeric plant virus satellite RNA. Science 231:1577–1580.

Sarver N, Cantin EM, Chang PS, Zaia JA, Ladne PA, Stephens DA, Rossi JJ (1990): Ribozymes as potential anti-HIV-1 therapeutic agents. Science 247:1222–1225.

Sheldon CC, Symons RH (1989): Mutagenesis analysis of a self-cleaving RNA. Nucleic Acids Res 17:5679–5685.

Shibahara S, Mukai S, Nishihara T, Inoue H, Ohtsuka E, Morisawa H (1987): Site-directed cleavage of RNA. Nucleic Acids Res 15:4403–4415.

Tabin CJ, Bradley SM, Bargmann CI, Weinberg RA, Papageorge AG, Scolnick EM, Dhar R, Lowy DR, Chang EH (1982): Mechanism of activation of a human oncogene. Nature 300:143–149.

Uhlenbeck OC (1987): A small catalytic oligoribonucleotide. Nature 328:596–600.

van Tol H, Buzayan JM, Feldstein PA, Eckstein F, Bruening G (1990): Two autolytic processing reactions of a satellite RNA proceed with inversion of configuration. Nucleic Acids Res 18:1971–1975.

Zaug AJ, Been MD, Cech TR (1986): The *Tetrahymena* ribozyme acts like an RNA restriction endonuclease. Nature 324:429–433.

Zuckerman RN, Corey DR, Schultz PG (1988): Site-selective cleavage of RNA by a hybrid enzyme. J Am Chem Soc 110:1614–1615.

ABOUT THE AUTHORS

MAKOTO KOIZUMI is researcher at Bioscience Research Laboratories, Sankyo Company Ltd. After receiving his bachelor's degree from Hokkaido University in 1986, he received his Ph.D. in 1991 under Professor Eiko Ohtsuka at Hokkaido University, where he concentrated on studying ribozymes for sequence-dependent cleavage of target RNA. Dr. Koizumi's current research is the studies on hammerhead ribozymes for the cleavage of RNA in vivo and the analysis of structures for hammerhead ribozymes. His research papers have appeared in *Biochemistry* (1991) as "Effects of phosphorothioate and 2-amino groups in hammerhead ribozymes on cleavage rates and Mg^{2+} binding" and in *Nucleic Acids Research* (1989) as "Design of RNA enzymes distinguishing a single base mutation in RNA."

EIKO OHTSUKA is Professor of Bioorganic Chemistry at Hokkaido University, where she teaches nucleic acid chemistry and molecular biology. After receiving her B.A. from Hokkaido University in 1958, she received her Ph.D. under Yoshihisa Mizuno at Hokkaido University, where she collaborated with Yuji Tonomura and Morio Ikehara in studies of ATP analogs in muscle contraction. Dr. Ohtsuka then joined Dr. H.G. Khorana's group as a postdoctoral fellow at the University of Wisconsin, working on synthesis of nucleic acids for the genetic code. Dr. Ohtsuka's current research involves synthetic approaches of the structure–function relationship of catalytic RNA, and protein engineering for molecular recognition in nucleic acids, including synthesis of damaged DNA as substrate of cognate enzymes or as a part of synthetic genes. She has written over 200 research papers, covering such topics as total synthesis of the formylmethionine tRNA, synthesis of a gene for the human growth hormone, and expression of synthetic T4 endonuclease gene. She has received an award from the Pharmaceutical Society of Japan and the twentieth anniversary award of the Princess Takamatsunomiya Cancer Foundation.

Index